# Cracking Phenomena in Welds IV

Thomas Boellinghaus · John C. Lippold
Carl E. Cross
Editors

# Cracking Phenomena in Welds IV

Springer

*Editors*
Thomas Boellinghaus
BAM Bundesanstalt für Materialforschung und -prüfung
Berlin
Germany

Carl E. Cross
Los Alamos National Laboratory
Los Alamos, NM
USA

John C. Lippold
The Ohio State University
Columbus, OH
USA

ISBN 978-3-319-28432-3 ISBN 978-3-319-28434-7 (eBook)
DOI 10.1007/978-3-319-28434-7

Library of Congress Control Number: 2015960236



Cover Illustration: Constitutional liquation associated with Ti-rich, MC carbides in a Waspaloy hot ductility sample heated to a peak temperature of 1300 °C (2370 °F)

Printed on acid-free paper

This Springer imprint is published by SpringerNature
The registered company is Springer International Publishing AG Switzerland

# Preface

This conference, the fourth in a series of workshops starting in 2004, had its format expanded to include both cold and hot cracking in welds. This change is reflected in the revised workshop and book title, *Cracking Phenomena in Welds*. Whereas previous workshops concentrated only on elevated temperature cracking (e.g., solidification cracking, liquation cracking, and ductility dip cracking), the new format includes cracking processes that occur at low temperature (e.g., hydrogen cracking, stress corrosion cracking, and stress relief cracking). Past workshop proceedings include the following: *Hot Cracking Phenomena in Welds* (2005, ISBN 3-540-22332-0), *Hot Cracking Phenomena in Welds II* (2008, ISBN 978-3-540-78627-6), and *Hot Cracking Phenomena in Welds III* (2011, ISBN 978-3-642-16863-5). These popular reference books have served as invaluable guides for understanding how and why weld defects form at temperatures near or above the melting point. In total, the three previous books constitute 64 papers with contributions from over 10 countries. It is our hope that the new expanded format will serve to make this and future compendiums even more useful.

The fourth workshop was held in Berlin, Germany, on April 2–4, 2014, and was hosted by the German Federal Institute for Materials Research and Testing (BAM). There were 77 participants and 29 presentations from 15 countries (Germany, USA, France, Sweden, Italy, UK, South Korea, The Netherlands, Canada, Belgium, Austria, Spain, Japan, Ukraine, and Israel). The conference was organized into seven parts: (i) hot cracking—testing, (ii) hot cracking—design, (iii) hot cracking—nonferrous/ferrous, (iv) hot cracking—nickel base, (v) cold cracking, (vi) stress corrosion cracking, and (vii) elevated temperature solid-state cracking. A variety of different cracking subjects were discussed, including test standards, crack prediction, weldability determination, crack mitigation, stress states, numerical modeling, and cracking mechanisms. Likewise, many different alloys were investigated such as aluminum alloys, copper–aluminum dissimilar metal, austenitic stainless steel, nickel base alloys, duplex stainless steel, creep resistant steel, and high-strength steel.

The papers collected here, together with the 64 weld hot cracking papers from previous workshops, constitute a comprehensive source of information on weld defects. The contributing authors are all highly knowledgeable in their own respective fields and have been given the opportunity (without page limitations) to present their results and freely express their opinions. By viewing the four workshop compendiums in series, one can observe how some ideas and approaches to weldability have evolved over time, providing a perspective and appreciation for current thought.

The editors would like to express their gratitude to all of those who participated. We are particularly grateful to the authors who prepared informative manuscripts, sharing with us their valuable knowledge and insight. In addition, we would like to thank BAM for providing an excellent venue for this workshop and, especially, Frau Jeanette Pannicke for organizing this event and helping with the subsequent assimilation of workshop papers. We are also grateful to other organizations affiliated with this event including The Ohio State University (OSU), International Institute of Welding (IIW), and Los Alamos National Laboratory (LANL).

Berlin, Germany
Columbus, USA
Los Alamos, NM, USA
March 2015

Thomas Boellinghaus
John C. Lippold
Carl E. Cross

# Photograph of Participants

# Contents

# Part I
# Hot Cracking Phenomena I—Testing

# A Historical Perspective on Varestraint Testing and the Importance of Testing Parameters

**Joel Andersson, Jonny Jacobsson and Carl Lundin**

**Abstract** This paper provides a historical perspective on the development of the Varestraint testing method as well as the design of a new Varestraint testing machine, utilized on a daily basis in the aerospace industry to solve hot cracking issues in production. The paper also discloses the importance in choosing the right testing parameters in order to minimize scatter in test results and to be able to make a reliable judgment in susceptibility towards hot cracking. Weld current, weld speed, bending rate and strain level as well as the evaluation procedure with i.e. measurement errors are included in the analyses.

## The Origin of the Varestraint Test

### *The Optimum in a Weld Cracking Test*

The interest in a hot cracking test for welding came about through a study of weld metal solidification in the early 1960's. The Varestraint device was a portion of Professor Lundin's PhD. Thesis [1] with Warren F. "Doc" Savage as his advisor. It was recognized that a hot cracking evaluation device had to incorporate variable strain so as to be applicable to a variety of materials.

J. Andersson (✉)
Department of Engineering Science, University West, 46186 Trollhättan, Sweden
e-mail: joel.andersson@hv.se

J. Andersson · J. Jacobsson
GKN Aerospace Engine Systems, 46181 Trollhättan, Sweden

J. Andersson · J. Jacobsson
Department of Materials and Manufacturing Technology,
Chalmers University of Technology, 41296 Gothenburg, Sweden

C. Lundin
Department of Materials Science and Engineering,
The University of Tennessee Knoxville, Knoxville, TN 37996-2100, USA



Th. Boellinghaus et al. (eds.), *Cracking Phenomena in Welds IV*,
DOI 10.1007/978-3-319-28434-7_1

The 6 points shown below were an integral part of the conceptional design for a new test.

1. Ability to show a direct correlation with actual fabrication and service behavior
2. Reproducibility of results without variation due to the human element
3. Sensitivity to small changes in a test variable
4. Ability to show the effects of several welding variables
5. Economical preparation of the specimens and running of the test
6. Applicability to all welding processes

These criteria are wide reaching and were not met by any known hot cracking test at the time. If all the criteria could be satisfied the test would serve many researchers and industries. Lundin et al. felt that the Varestraint test addressed all of the requirements [2, 3].

## *The "Elegant Version" of the Varestraint Concept*

The initial concept as shown in Fig. 1 was to vary the strain by wrapping a weld plate around a curved die (involute of a circle) with the weld torch remaining fixed. This design was the "Elegant Version" of the Varestraint Concept. The weld pool would be at the point of tangency and the radius of the bend would be constantly changing. Therefore the strain (per Eq. 1) on the plate surface would also change (Variable Strain) as the sample was rotated from a large radius to a smaller radius.

$$\varepsilon = t/2R \qquad (1)$$

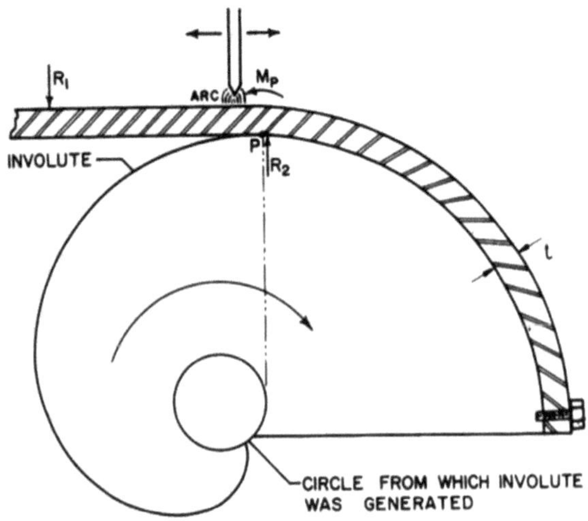


**Fig. 1** Overview of the "Elegant Version" of the Varestraint concept wherein an involute method was used to provide augmented shrinkage strains ($\varepsilon = t/2R$)

where
t the thickness of the material
R the diameter of the die block

A specimen of this type would provide the range of strains necessary to span the crack-no-crack limits of the weld deposit and heat affected zone (HAZ). However, the device proved to be difficult to design and operate because as the die rotated the entire involute die had to move upward. This was not the only problem as the bending of the sample required significant forces, the specimen thickness was chosen to be 1/2 an inch based on a need to accommodate and span the range of strains representing the hot cracking range. Therefore, the design had to be modified by changing the continually variable strain for a series of interchangeable radius dies and that reduced the complexity of both construction as well as the operation.

The modified design that was chosen is shown in Fig. 2 which is the first Varestraint testing machine ever built. The sample is a cantilever beam and is forced to conform to a die of a given radius when the weld pool reaches the point of tangency of the die. The weld continues a small distance past the tangency point before it is terminated.

The loading yoke and air cylinder bend the sample to conform to the die. A range of 4 to 5 die blocks was considered sufficient to provide a appropriate range of strains.

## *Cracking Criteria at Varestraint Testing*

A schematic of a tested weld surface is presented in Fig. 3. All of the relevant features are identified. Significant work went into developing the criteria for evaluation as a function of the strain (different die blocks).

**Fig. 2** Shows the initial Varestraint testing device with a specimen in place

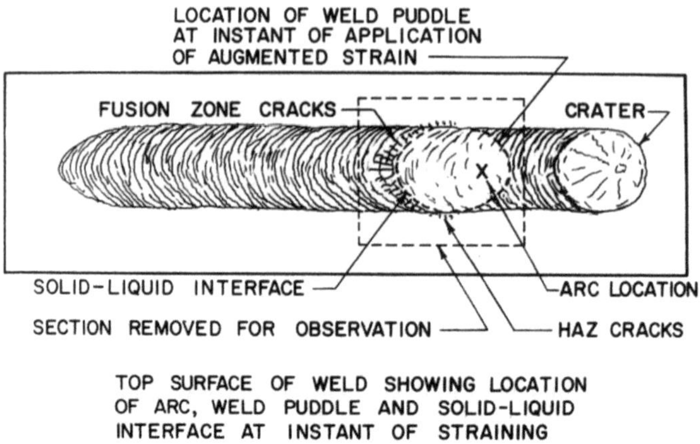


**Fig. 3** A schematic of a tested weld surface with all of the relevant features identified

Three measurements of cracking response resulted in the criteria enumerated below in addition to the threshold strain for cracking limit for the largest radius die block (lowest strain) which was settled on as the definitive measurements.

1. TCL—Total Crack Length
2. MCL—Maximum Crack Length
3. Cracked HAZ Length

These criteria were used in evaluating high alloy material, for example, steel, during production in terms of hot cracking susceptibility of various heats of high grade materials such as the 12Ni-180 alloy and made it possible to terminate the steel manufacturing early on in the process chain if a high cracking susceptibility heat was detected [3]. This saved significant money for the materials production industry. Other studies on AISI 316NG using the Varestraint concept disclosed the effect of solidification mode on hot cracking susceptibility [4].

Another interesting technique that was developed early on was the thermocouple plunge technique, shown in Fig. 4 [1].

Herein, a thermocouple is plunged into the back end of the weld pool, which allowed, when coupled with the maximum crack length, the hot cracking temperature range to be determined.

Another version of the Varestraint testing device developed by Lundin et al. [5, 6], is presented in Fig. 5, which was reduced in specimen and smaller samples size to be able to accommodate necessary strains for sheet materials (reduction of the typical sample thickness to the order of 3 mm). This concept also accommodated testing transverse to the specimen axis.

Many of these type devices of different designs evolved but the one shown in Fig. 5 was the most preferred one and it could also be employed in a weld fusion spot

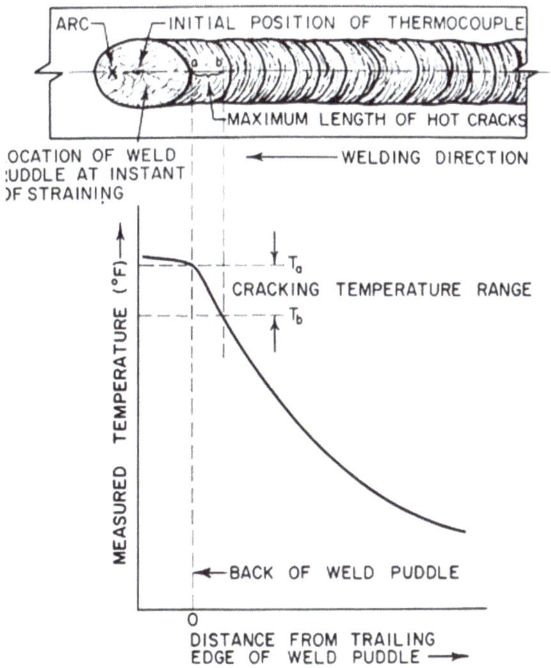


**Fig. 4** Schematic representation of the determination of the hot cracking temperature range

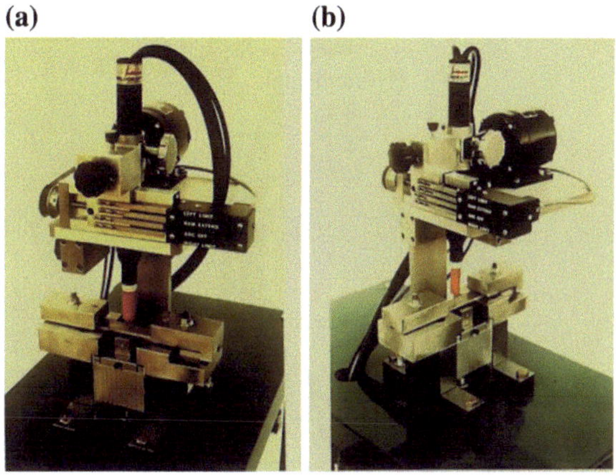

**Fig. 5** Shows a smaller Varestraint device (**a** Longitudinal Varestraint and **b** Transverse Varestraint testing), which accommodates sheet materials and reduces the typical sample thickness to 3 mm

hot cracking mode. At this point it is important to reiterate that as the specimen thickness is reduced the radius of the die block must be reduced to increase the strains for testing ($\varepsilon \sim t/2R$ per Eq. 1). While this may seem trivial it becomes the important feature in a suitable design. A multi-pass Varestraint testing technique has also been developed and used to evaluate the overlapping of weld passes, thus enabling the base metal HAZ and weld bead HAZ's from overlapping to be tested [4].

## A Newly Built Varestraint Testing Device

On the basis of the preceding information it was decided several years ago to design and build a Varestraint testing device at GKN Aerospace Engine Systems in Sweden. The machine, shown in Fig. 6, was designed to meet a range of requirements and consists of three units which are synchronized with each other; a hydraulic part, a 100 ton capability press, and a welding robot.

The testing device in Fig. 6 is apart from Varestraint testing able to operate by means of Transverse and Spot Varestraint testing as well as bending without welding. It may also be used in multi-pass welding studies using filler material. The machine can operate at stroke rates up to 350 mm/s (due to the pressurized accumulator tanks) and still maintaining good accuracy in terms of stop position since the response time of the complete system is only 65 ms.

Since the testing device is new, it was decided to investigate the effect of testing parameters on the variation and reproducibility of the testing device as well as on the evaluation methodology used in evaluating cracking susceptibility, to be confident that the test results are reliable. A similar type of investigation has been performed earlier for Transvarestraint testing [7]. Thus, the aim of this study was specifically to investigate how evaluation methods as well as different testing parameters influence the variation in TCL upon Varestraint testing.

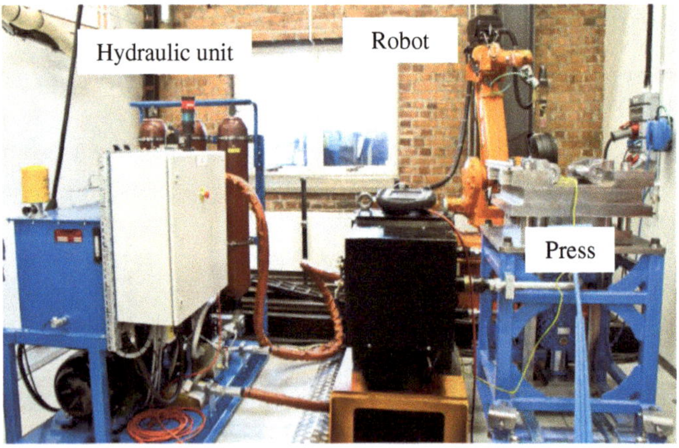


**Fig. 6** A newly built Varestraint weldability testing device

# Methodology

## *Measurement System Analysis*

A measurement system analysis (MSA) was performed using Minitab statistical analysis software. The aim of this MSA was to validate the measurement system in terms of process variation. In this case, four samples of SS304, two operators and five repetitions were used since the precision shall be captured and not the accuracy. Two different means of evaluating the amount of hot cracking was used; stereo-microscope and fluorescence penetrant inspection (FPI). The four SS304 samples comprised the following characteristics:

1. few cracks and a un-machined surface
2. few cracks and a machined surface
3. many cracks and a machined surface
4. many cracks and a un-machined surface

The number of samples used in the MSA determines the process variation. The number of operators will show the measurement variation, which is broken down into reproducibility and repeatability. Reproducibility is the variation that occurs when several operators measure the same part, and repeatability is the variation when the same operator measures one part several times. In other words, to capture the precision, it's important to use several operators while if searching for accuracy, the amount of repetitions is essential.

## *Design of Experiment*

A design of experiment (DoE) was performed to determine what parameters, and on which level, they influenced the variation of the result in terms TCL. In the DoE, ten center points where added, both to discover non-linearity's within the DoE, and to estimate the variation within the process.

In this study, 4 parameters were tested with 3 repetitions each in a 2nd-level full factorial experiment. In a comparison with additional centre point values (the natural variation assessment), conclusions regarding influencing parameters can be drawn.

The tested parameters are: radius die block [40, *50* (*center point*), and 60 mm], welding speed (1, *2*, and 3 mm/s), stroke rate (10, *130*, and 250 mm/s), and weld current (70, *80*, and 90 A). The test sample geometry (150 × 50 × 3.2 mm), material of testing (Haynes 282 sheet, continuous mill annealed condition) and welding process (gas tungsten arc welding) were kept constant during all the tests, as well as the placement of test samples in the equipment. Expendable SS304 bars were used on top of the test samples to make sure the test samples conformed to the specific radius in an ideal way without hinging. Each parameter setting was tested three times (58 tests in total); the results therefore display indications of variation. If the test were to provide the variation with a higher confidence, it would require at least 30 repetitions for each parameter setting.

## Results and Discussion

### *Measurement System Analysis—Stereo Microscopy Versus Fluorescence Penetrant Inspection*

The four SS304 test samples used in the MSA have different characteristics. The test samples were selected with different characteristics in order to analyze the precision of the two measurement systems on differencing aspects. Two test samples; No. 3 and No. 4, with many cracks, and the other two; No. 1 and No. 2, with fewer cracks. One test sample with many cracks; No. 3, and one test sample with few cracks; No. 2 had a machined surface while the other two test pieces had an un-machined surface; No. 1 and No. 4. Each test sample was measured five times by two different operators; this was done in a random order to decrease bias, such as the recognition for the operators of how they have measured the samples previously.

The test procedure was carried out in the same way both for the stereo microscope and the FPI. During both procedures, the operators recorded data of each individual crack on all test samples. This gave them the possibility to analyze the total crack length; number of cracks, maximum crack length and mean crack length, which all are commonly used in evaluation of Varestraint testing. TCL was the main measurement used in this analysis, because this criterion will be independent of how the operator has considered a crack to be divided in many small or one large crack.

Table 1 shows the level of the variation in the measurement system that comes from repeatability and reproducibility. The variation created by the measurement system due to repeatability is 1.5 mm with a confidence interval of 95 % from 1.2 to 2.0 mm and due to reproducibility 1.1 mm with a 95 % confidence interval (CI) from 0.2 to 30.3 mm. The repeatability stands for 11 % of the variation whereas the reproducibility stands for 8 %.

**Table 1** MSA for the microscopy evaluation method

| Source | StdDev (SD) in mm | 95 % CI in mm | 6*SD in mm |
|---|---|---|---|
| Total gage R&R | 1.8 | (1.5; 30.3) | 11.0 |
| Repeatability | 1.5 | (1.2; 2.0) | 8.9 |
| Reproducibility | 1.1 | (0.2; 30.3) | 6.4 |
| Operator | 0.8 | (0.0; 30.3) | 4.9 |
| Operator*parts | 0.7 | (0.0; 3.5) | 4.1 |
| Part-to-part | 13.4 | (7.7; 50.2) | 80.6 |
| Total variation | 13.6 | (7.8; 51.9) | 81.4 |

Table 2 MSA for the penetrant evaluation method

| Source | StdDev (SD) in mm | 95 % CI in mm | 6*SD in mm |
|---|---|---|---|
| Total gage R&R | 1.5 | (1.1;31.1) | 9.1 |
| Repeatability | 0.5 | (0.4;0.6) | 2.8 |
| Reproducibility | 1.5 | (0.9; 31.1) | 8.7 |
| Operator | 0.7 | (0.0; 31.1) | 4.5 |
| Operator*parts | 1.5 | (0.7; 4.7) | 7.5 |
| Part-to-part | 12.8 | (7.4; 48.0) | 77.0 |
| Total variation | 12.9 | (7.4; 50.1) | 77.5 |

Table 2 shows how much of the variation in the measurement system regarding the FPI method that comes from repeatability and reproducibility. The variation due to repeatability is 0.5 mm with a 95 % confidence interval from 0.4 to 0.6 mm and the variation due to reproducibility is 1.5 mm with a 95 % confidence interval from 0.9 to 31.1 mm.

The penetrant fluid has a relatively small variation due to repeatability, which is 0.5 mm. The variation due to reproducibility, 1.5 mm, is substantially higher than that for the repeatability, about three times as high. This indicates that if the testing procedure is performed by one single operator most of the variation in the measurement system would be eliminated. Using only one operator would lead to the condition wherein test samples can't be differentiated because of the measurement system variation would decrease substantially. Subsequently, the target in this case is precision and not accuracy towards a mean value, is best with one operator judging consequentially. For the measurement system, as it is analyzed in this case with two different operators and a larger difference in the extent of cracks on the test samples, it can be seen that the repeatability stands for 3.9 % and the reproducibility stands for 11.6 % of the variation, which mean that even with two operators the testing procedure under these circumstances can be considered reasonably good and with only one operator it can be considered excellent.

It is significantly improved if only one operator performs the measuring procedure for both methods. Considering both measuring methods, it is clear that the variation within the FPI procedure is smaller than the variation within the microscope testing procedure. However, this can result from the high accuracy of the microscope and the subjective judgment of the operator of what is and what is not a crack.

## *Effect of Surface Appearance on the Variation*

The MSA specifies what surface that is beneficial to use when measuring, but not how the hot cracking response is affected by the surface of the test sample. This test is performed on ten stainless steel test samples with machined surfaces and on ten test samples with un-machined (mill) surfaces. The test was performed to determine which surface condition should be employed to acquire a high validity in the final steps.

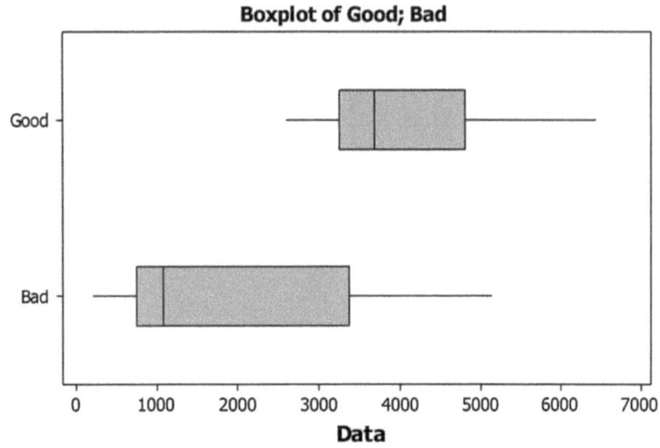


**Fig. 7** Boxplot of crack response on surface variation ("good versus bad" surface appearance)

The standard deviations for the two different sets of samples were 1.2 mm for the test samples with a machined surface and 17 mm for test samples with an un-machined (mill) surface. The range of measured values and mean values for the different surface conditions can be seen in Fig. 7. The data displays that the mean value of TCL is substantially higher for the test pieces with machined surfaces, which has a mean value of 4.0 mm compared to test samples with un-machined surfaces, were the mean value of TCL is 18 mm. The F-test is a hypothesis test with a null hypothesis of that the samples are equal, in this test the confidence is set to 95 %. The *P*-value for the F-test is 0.282, which means that the null hypothesis can't be rejected at a significance level of 95 %.

The difference in standard deviation for the two surfaces is an indication that there might be a difference in variation, even though the null hypothesis can't be rejected at a significance level of 95 %. In order to make sure whether or not there is a difference between the two surfaces, more tests need to be evaluated with the use of the actual material that is to be tested. The difference in mean values between the two surfaces can be an accuracy problem, either occurring because of the measurement system or due to the different surface characteristics. The reason for a lower mean value with the un-machined surface could be due to residuals and oxides that have a higher melting temperature and may be floating on the liquid metal and disturbing the welding process and the cracking response. In short, the test samples have to be prepared identically to be able to distinguish differences in the hot cracking response, and there shall be no sign of residuals on the surface that may interfere with the welding, i.e. the surfaces needs to be treated as equal as possible.

## *Natural Variation in Equipment and Material*

Within all types of processes there is a variation, which can exist due to natural causes or due to assignable causes. The variation that exists due to natural causes can be reduced but not eliminated whereas the variation that occurs due to assignable causes can be eliminated. The natural variation can be reduced through methods such as robust design; a method to reduce a process' sensitivity to interfering factors. The variation that is due to assignable causes can be eliminated, by finding the root causes for the variation and by correcting it.

As the equipment is new, there are no historical data available. To understand if the results are reliable, the variation has to be known in order to see at which level it is actually possible to tell if there is a difference between two tests. This test also ensures that the measurement system is valid by comparing the variations between test samples. The results from this test are the baseline for the analysis of the DoE and will provide the null values from which the parameters are compared. This test is performed inside the DoE with the use of centre points in order to minimize the resource utilization, and avoid bias by utilizing the randomization in DoE.

The natural variation is presumed to follow a normal distribution, and the measurements of the ten center points are presented in Table 3. The data presented in Table 3 is plotted against a normal distribution to define the fit, see Fig. 8.

**Table 3** Results for the center points

| Run order | Results (μm) |
|---|---|
| 8 | 8508 |
| 9 | 8698 |
| 15 | 7724 |
| 24 | 8529 |
| 25 | 10,403 |
| 32 | 7780 |
| 40 | 10,469 |
| 43 | 11,338 |
| 47 | 11,731 |
| 58 | 9198 |

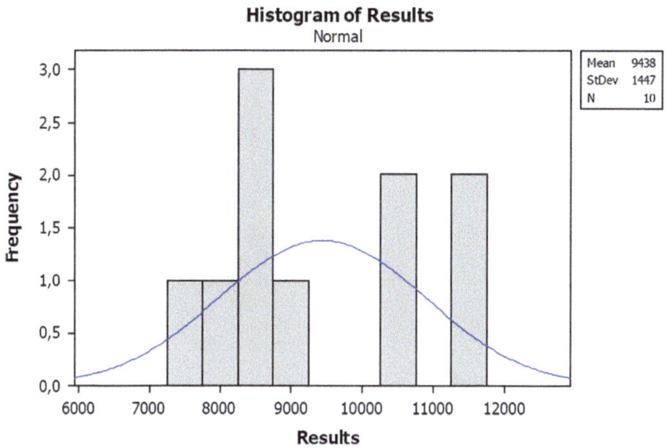


**Fig. 8** Normal probability plot of center points

As disclosed, it is not a perfect fit, but this could be due to a relatively low extent of measurements. The mean value is not interesting since the focus is variation, but the standard deviation is relatively large compared to the mean value (15 %), which indicates that the variation might not only be from natural variation.

An Anderson-Darling test was also performed to see if it can be rejected that the distribution truly follows a normal distribution (*P*-value). Anderson-Darling tests were also performed with other distributions to determine if the variation aligned significantly better. The Anderson-Darling test was performed for a normal-, log-normal-, weibull-, and gamma distribution. The lower the Anderson-Darling value, the better the variation fits the distribution [8]. The main purpose of the distribution determination is to determine that the distribution most likely will reveal a 95 % confidence interval.

The variation in the machine is significant and thus there is a need to reduce variation. However, by relating these results to the box-plot of surface crack response, Fig. 7, the most appropriate distribution is found as the log-normal distribution, Fig. 9, which also aligns with the results from the box-plot.

The MSA has been analyzed previously but in order to confirm that the measurement system contributes to less than 30 % of the natural variation it has to be analyzed with test samples processed equally. Therefore, a MSA was performed on five of the center points from the DoE study. These five points were measured five times and on the basis of previous results only one operator was used, so that the surfaces were equally processed. Just as in previous MSA studies the TCL is the evaluation measure being used.

Chart 1 (top left); Fig. 10 displays that 27.95 % of the natural variation is due to the measurement system, when one operator is used. According to chart 4 (top right); Fig. 10, the measurement system can distinguish some parts as unequal while others are considered equal by the system. The fact that the measurement system

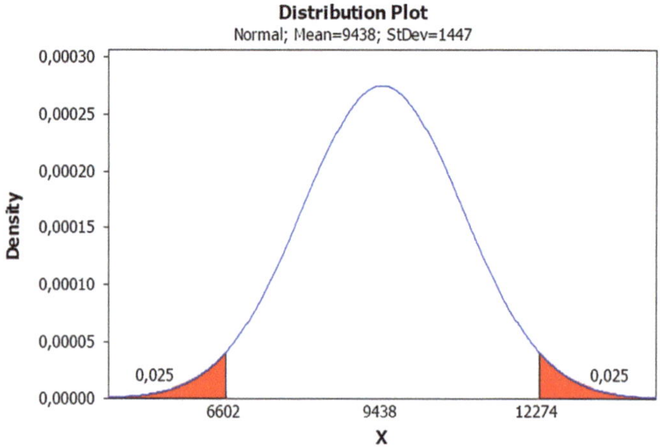


**Fig. 9** Log-normal distribution plot of the normal values

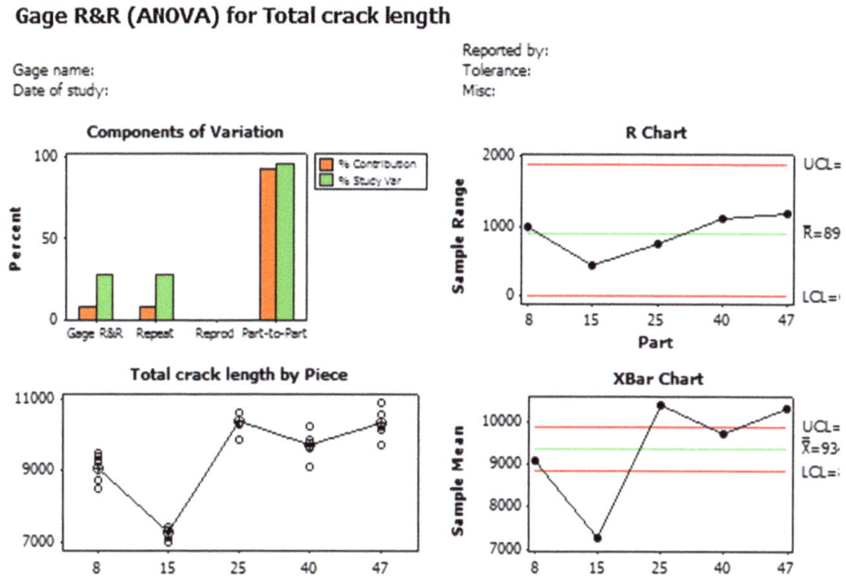


**Fig. 10** MSA on five of the center points from the DoE study

**Table 4** MSA 2 with the center points

| Source | DF | SS | MS | F | P |
|---|---|---|---|---|---|
| Piece | 4 | 33,179,540 | 8,294,885 | 59.9962 | 0.000 |
| Repeatability | 20 | 2,765,138 | 138,257 | | |
| Total | 24 | 35,944,678 | | | |

can differentiate some parts is strengthened by the *P*-value of 0.000 seen in Table 4, which rejects the null hypothesis that all parts are equal.

## *Design of Experiment*

The variability of the parameters in the experiment were analyzed with the assistance of the "least square regression" method to determine which parameters are significantly related to the differences in the variability of the TCL. With this method, main effects and two-way interactions can be evaluated against each other as seen in Fig. 11.

As seen in the normal probability plot of standardized effects in Fig. 11, the main effect A is considered significant, with a standardized effect of −3.1 and an effect of −2.3 with a *P*-value of 0.021. In other words <0.05, which gives an indication that the effect is significant. The other main effects are not considered as significant with

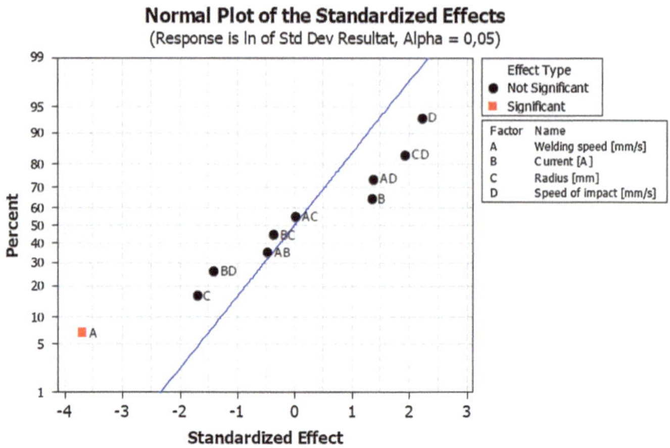


**Fig. 11** Normal probability plot of the standardized effects

a confidence interval of 95 %. If the confidence interval is changed to 80 %, the effects D and CD will also be considered as significant with standardized effects of 1.9 and 1.6 with *P*-values of 0.110 and 0.155, respectively.

If a reduced model is used to create a maximum likelihood estimation (MLE) of the four main effects, all parameters will be considered as significant. This will not provide any further interpretations, why MLE is excluded from the analysis.

To determine how the results vary in relation to each other, a residual plot is made. This is presented by mean values in Table 5 and their residuals in Fig. 12 (number 17 in Fig. 12 is the center point test).

The natural variation (number 17) is relatively large and thereby interferes with the other variations. A small standard deviation in this case, e.g. number 2, might occur due to a small amount of hot cracks in the test sample. Therefore, a calculation of how large the standard deviation is in relation to the mean TCL was accomplished and presented in Table 5.

The data from Table 5 are combined in Fig. 13. The blue line represents the std-dev/mean in %, and the black line represents the std-dev for the test samples. The lines are connected at parameter setting No. 12 as this point is subject to the least std-dev/mean (2 %).

By examining at the combined figure, we can perceive that some test samples, with low scores in both values, will be suitable settings to use to test materials, with a low variation.

Parameter setting No. 2, 6, and, 8 contain only one crack each and are therefore not considered as suitable settings to employ in evaluating hot cracking susceptibility. The parts with the smallest std-dev/mean combined with a variation below the mean value (black horizontal line) are number **5** (3 %), **10** (6 %), **12** (2 %), and **16** (10 %). Two additional points that also can be included are numbers **13** (18 %) and **14** (20 %). These six settings are, together with the centre points, summarized in Table 6.

**Table 5** Mean values and standard deviation over the mean value of the 17 parameter settings used in the DoE

| Part ID | Mean of variation (µm) | Mean of variation (%) |
|---|---|---|
| 1 | 1664 | 17 |
| 2 | 36 | 65 |
| 3 | 3832 | 20 |
| 4 | 1517 | 134 |
| 5 | 144 | 3 |
| 6 | 13 | 36 |
| 7 | 2273 | 15 |
| 8 | 8 | 35 |
| 9 | 2234 | 24 |
| 10 | 437 | 6 |
| 11 | 3588 | 42 |
| 12 | 206 | 2 |
| 13 | 1146 | 18 |
| 14 | 939 | 20 |
| 15 | 1895 | 29 |
| 16 | 637 | 10 |
| 17 | 1446 | 15 |

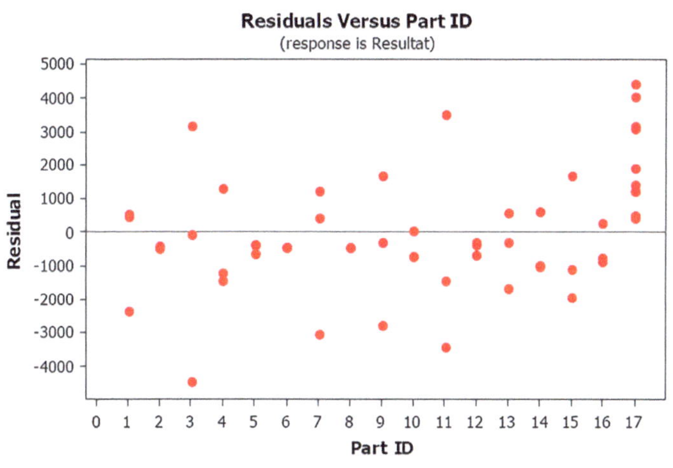


**Fig. 12** Residual plot of the 17 parameter settings used in the DoE

In Table 6, parameter setting number 5 consists of the lowest std-dev of only 144 µm. But, the mean value of number 5 is 4525 µm, which is lower than the mean value for the center points, on which the MSA 2 was performed. The MSA 2 indicates that the measurement system stands for 27.8 % of the variation, and as concluded previously in the MSA 1, a greater amount of hot cracks will provide a smaller variation in the measurement system. This means that number 5 might be subject to a larger measurement error, and is preferably used only if a fifth parameter

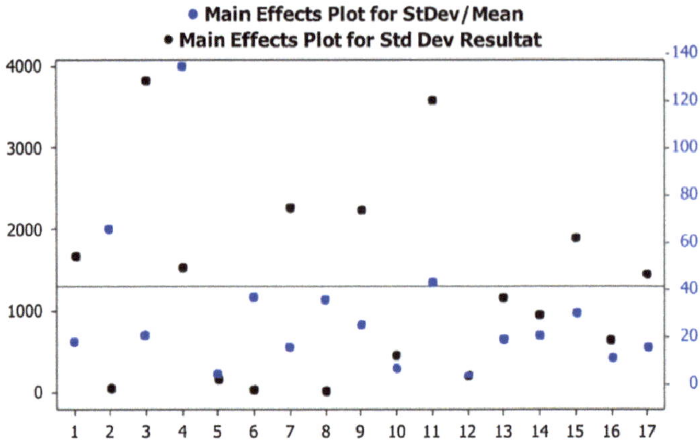


**Fig. 13** Main effects plot of std-dev/mean and std-dev

**Table 6** Chosen testing parameter settings based on the DoE study

| Parameter setting | Welding speed (mm/s) | Weld current (A) | Radius (mm) | Stroke rate (mm/s) | StdDev | StdDev/mean (%) |
|---|---|---|---|---|---|---|
| 5 | 1 | 70 | 60 | 10 | 144.9 | 3 |
| 10 | 3 | 70 | 40 | 250 | 437.2 | 6 |
| 12 | 3 | 90 | 40 | 250 | 206.0 | 2 |
| 13 | 1 | 70 | 60 | 250 | 1146.9 | 18 |
| 14 | 3 | 70 | 60 | 250 | 939.9 | 20 |
| 16 | 3 | 90 | 60 | 250 | 637.9 | 10 |
| 17 | 2 | 80 | 50 | 130 | 1446.9 | 15 |

can be found that forces a greater amount of hot cracks to occur, or if the measurement system error is reduced.

The four main parameters are plotted in Fig. 14 to visualize their respectively contribution to the variation. Plotted values are the standard deviation, and not the mean.

Theoretically; 3 mm/s, 70 A, 60 mm, and 10 mm/s, will provide the least variation, but in practice this means only one hot crack or no hot crack response at all in the material. The stroke rate, as seen in the main effects plot, will not affect the variation in any great extent and is therefore not considered to affect the variation.

By comparing test parameter setting number 12 and 14, it is shown that the natural variation isn't sufficiently accurate, since it interferes with the result. By relating number 14 according to theoretical interpretations of the interactions plot in Fig. 15, it shows that the weld current and the radius die mandrel would provide a smaller std-dev, than the parameter setting for number 12. However, number 14 is subject to a mean value of 4600 μm, which is similar to number 5. This indicates

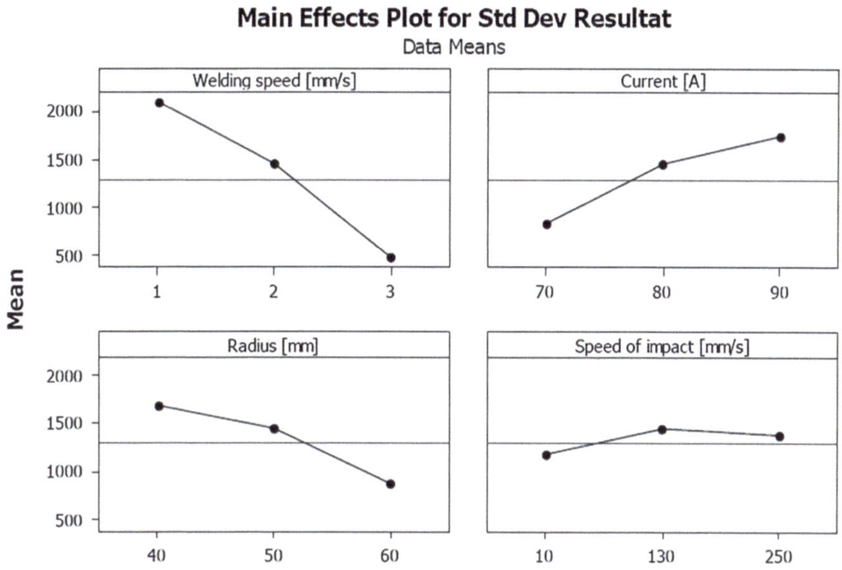


**Fig. 14** Main effects plot for the standard deviation

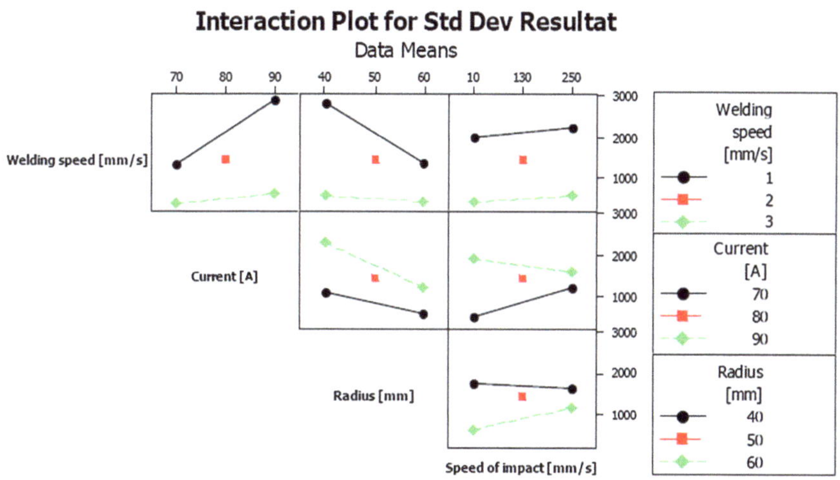


**Fig. 15** Interaction plot of the different parameter setting for the standard deviation

that the variation in the measurement system error has to be reduced in order to draw precise conclusions from the DoE.

Figure 15 displays the interaction plot for the std-dev. In other words how each parameter setting influences the other.

As explained earlier, there are indications that the use of certain parameter settings will reduce the variation in Varestraint testing. The settings that gave the strongest indications of a reduced variation are number 10 and 12. The settings that should give the smallest variation according to the results of the DoE was parameter setting number 14.

Figure 16 shows the indications of how the different normal distributions appear; note that the scale in the figures differs.

From these indications it is confirmed that with the usage of certain parameter settings, a significantly reduced difference in TCL between two different superalloys (e.g. austenitic Nickel-base type of superalloys) can be identified as a viable difference. For example, in Fig. 16d, if one superalloy has a mean value in TCL of 9000 μm and another superalloy has a mean value in TCL of 9450 μm, there is an indication that with 95 % confidence interval, it can be supposed that there is a significant difference in sensitivity to hot cracking between the two superalloys.

It is common practice to perform at least three tests for each superalloy at each parameter setting and it is recommended to continue with this since it increases the validity of the results. The parameter settings number 10 and 12 are recommended to use on the basis on the results from the DoE. However, since these distributions (Fig. 16a–d) are also constructed from three tests, it is recommended that the validation of these parameter settings be accomplished. Parameter setting number 14 is recommended on the basis of the individual parameter results from the DoE, and it is recommended to also validate this parameter setting.

There are also indications that certain parameter settings will generate large normal distributions, one of these settings is number 11. The number 11 normal

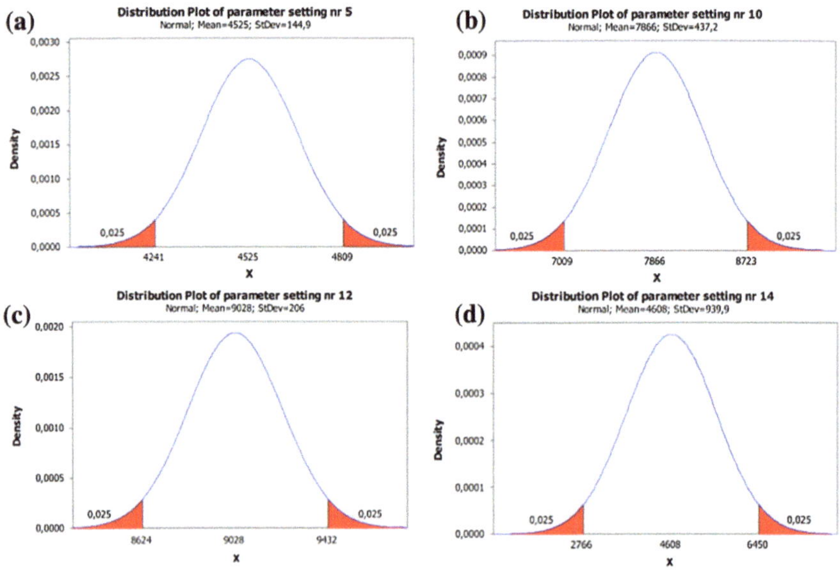


**Fig. 16** Distribution plot of parameter setting No. 5 (**a**), 10 (**b**), 12 (**c**), and 14 (**d**)

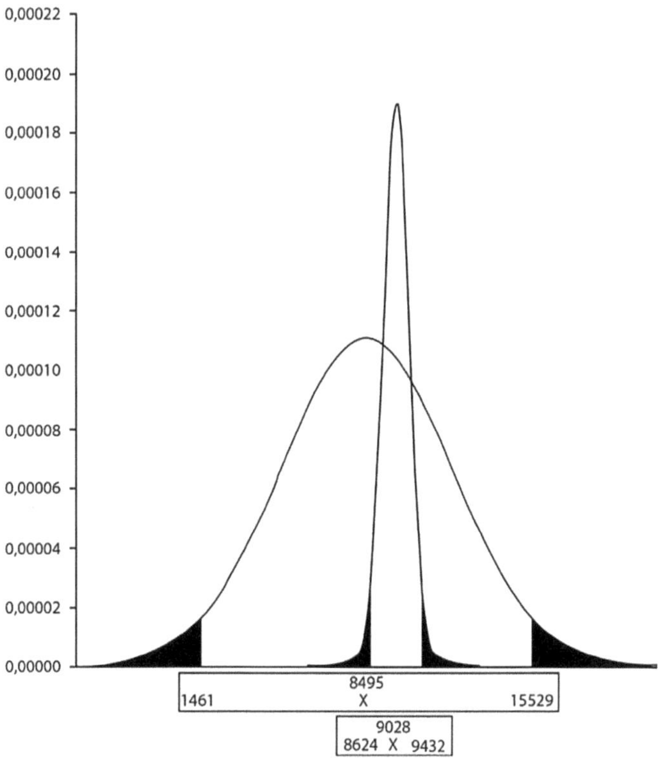


**Fig. 17** Distribution plot of testing parameter setting 11 and 12

distribution is presented in Fig. 17 provides a clear indication that the choice of parameter setting when comparing superalloys are very central in order to be able to distinguish possible differences in sensitivity to hot cracking. To relate parameter setting numbers 11 and 12 to each other, Fig. 17 displays the two normal distributions in the same graph. This graph clearly indicates that there is a possibility to further improve the testing equipment.

## Conclusions

1. The surface of the material affects the hot cracking evaluation of a material.
   (a) A machined surface will generate a smaller variation in measured response with a greater TCL as compared to an un-machined (mill) surface. However, the most important factor is to ensure that the surfaces be as identical as possible when attempting different alloy responses.

2. Measurements should be performed on test samples with as large TCL as possible to minimize variation, preferably with a TCL above ~9 mm.
   (a) This level is obtained from the MSA, which discloses that, a measurement variation of ~28 % with an average TCL of 9.4 mm is advisable.
   (b) A low average of TCL will result in a larger variation.
3. Two or more operators, as compared with one operator, results in a greater variation in reproducibility.
   (a) This is due to the subjectivity of the measurement procedure and also due to the lack of training.
4. The welding speed is according to the DoE the cause for the greatest variation.
5. The parameters that can be used to compare two materials are based on indications from the DoE.
   (a) These results from only three samples gives a variation, and should therefore be considered carefully.
   (b) The DoE parameter settings 5, 10, 12 and 14 can be used to compare different samples on a 95 % confidence are:
      (i) **5** (welding speed of 1 mm/s, weld current of 70 A, radius of 60 mm, bending rate of 10 mm/s)
      (ii) **10** (welding speed of 3 mm/s, weld current of 70 A, radius of 40 mm, bending rate of 250 mm/s)
      (iii) **12** (welding speed of 3 mm/s, weld current of 90 A, radius of 40 mm, bending rate of 250 mm/s)
      (iv) **14** (welding speed of 3 mm/s, weld current of 70 A, radius of 60 mm, bending rate of 250 mm/s)
   (c) The lowest average values of TCL are found for parameters 5 and 14

**Acknowledgements** Mr. Johannes Nilsson and Mr. Oscar Nilsson are highly appreciated for their input and support in the statistical analyses of this work.

## References

1. "A new Approach to the Study of Hot Cracking in Fusion Welds", Carl D. Lundin, PhD. Dissertation, Rensselaer Polytechnic Institute, Materials Science and Engineering, July 1966.
2. "The Varestraint Test," Welding Journal, 44 (10), 1965, pp. 433s-442s, Carl D. Lundin and W. F. Savage
3. "Application of the Varestraint Technique to the Study of Weldability", Welding Journal, 45 (11), 1966, pp. 497s-503s, Carl D. Lundin and W. F. Savage.
4. "Weldability of Nuclear Grade Stainless Steels," A Group Sponsored Program, Welding Research Council Bulletin # 509, February 2006, Carl D. Lundin, C. H. Lee, C.Y.P. Qiao and G. W, Batten

5. "Effect of Iron on the Hot Cracking of Uranium Weld Metal - Part I" Welding Journal, December 1970, V. 49 (12), pp. 29-39, Carl D. Lundin and P. W. Turner.
6. "The Varestraint Test," Welding Research Council Bulletin, 280, August 1982., pp 1-19, Carl D. Lundin, A. C. Lingenfelter, G.E. Grotke, G.G. Lessmann and S. J. Mathews.
7. Standardization of the Transvarestraint test, M.S. Thesis, The Ohio State University, 2003, Finton, T.
8. "Six Sigma – The Pragmatic Approach", Studentlitteratur, Sverige, 2003, Magnusson K., Bergman B. Häyhänen P., Kroslid D., Findlay Mills D.

# Improved Understanding of Varestraint Testing—Nickel-Based Superalloys

**Joel Andersson, Jonny Jacobsson, Anssi Brederholm and Hannu Hänninen**

**Abstract** By Varestraint testing, the susceptibility of an alloy to hot cracking during welding can be evaluated on test plates when they are bent at the same time as welding takes place. The strains imposed by welding can thus be augmented by the strains imposed by the bending action to find the strain limits when hot cracks appear and also the sensitivity to hot cracking by counting the number and measuring the length of the individual cracks as a means to differentiate between the weldability of different alloys. Supports are usually recommended to avoid hinging and to use test plates thicker than 10 mm in order to minimize the influence of the compression strains (lower part of the bent specimen) on the weld cracking at the bending. The cracking response of two precipitation hardening Nickel-based superalloys—ATI 718Plus® and Haynes® 282®—was analysed in the context of the actual tensile/compression ratio imposed and measured by strain gauges attached to the upper and lower surface of the test plates. It was found that no influence of the compressive strains on the cracking response in Varestraint testing takes place. It was also seen that the hot cracking susceptibility of Haynes® 282® is lower compared to that of ATI 718Plus®.

J. Andersson (✉)
Department of Engineering Science, University West,
461 86 Trollhättan, Sweden
e-mail: joel.andersson@hv.se

J. Andersson · J. Jacobsson
GKN Aerospace Engine Systems, 461 81 Trollhättan, Sweden

J. Andersson · J. Jacobsson
Department of Materials and Manufacturing Technology,
Chalmers University of Technology, 412 96 Gothenburg, Sweden

A. Brederholm · H. Hänninen
Aalto University School of Engineering, Helsinki, Finland



Th. Boellinghaus et al. (eds.), *Cracking Phenomena in Welds IV*,
DOI 10.1007/978-3-319-28434-7_2

## Varestraint Testing

Bending tests, such as the Varestraint testing, are among the most frequently used tests for evaluating the ductility of a metal or welded joint by measuring its ability to resist cracking during welding.

The Trans- and Varestraint tests, Fig. 1, are the two weldability testing methods that were developed by Savage and Lundin in the sixties [1] and are commonly employed when determining the weldability of a specific material. In the Trans-Varestraint testing method a weld is made transverse to the loading direction at the same time as a ram is pulling the plate downwards so that it is supposed to closely adhere to the mandrel located underneath the left-hand side of the plate. The die mandrel can be changed depending on what degree of strain one would like to achieve. Smaller radius gives higher strain. The crack length is then plotted as a function of strain. This testing method is primarily used for investigating the susceptibility towards solidification cracking.

The Varestraint testing which is primarily used for investigating heat affected zone (HAZ) cracking also addresses solidification cracking to some extent. The weld pass is performed in the longitudinal direction of the specimen as shown in Fig. 1.

Due to the heating from the welding torch, e.g. by GTAW, the material softens dramatically and a kink easily develops where the momentum is highest at the contact with the mandrel. The material does not fully adhere to the die located underneath the test material; as a consequence the material is not exposed to the intended ideal augmented strain, Fig. 2. When thin sheets are tested it is unfortunately not possible to avoid kinking as shown in the side view of Fig. 2. The material at the kink does not experience the expected amount of strain which therefore makes the results difficult to interpret.

To avoid kinking it is a common practice to use expendable support plates which force the test plate to ideally conform to the die mandrel since the effect of the soft spot due to the heating in the test material is reduced. However, these support plates do not impose tensile stresses in the test plates as schematically shown in Fig. 3, as bending is the process by which a straight length is transformed into a curved

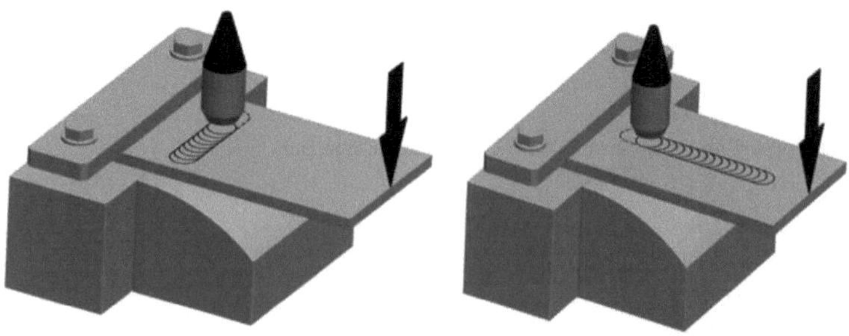

**Fig. 1** Trans- and varestraint test, respectively

**Fig. 2** Sideviews revealing how the testing plate kinks in varestraint testing

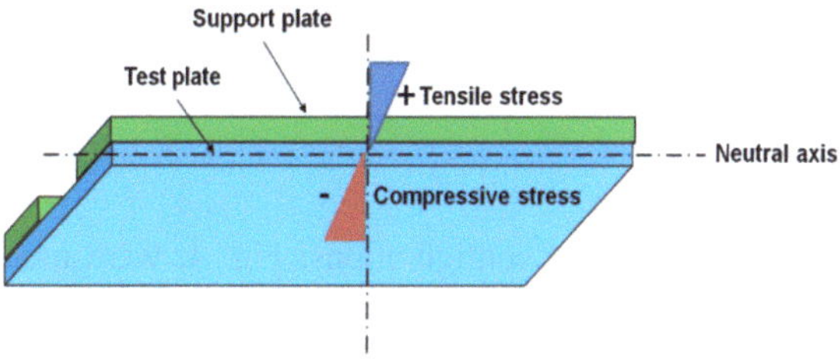


**Fig. 3** Schematic sketch of a test plate with two support plates on top to avoid kinking in varestraint testing

length. The fibers of the metal on the outer (convex) surface of the bend are stretched, thus inducing tensile strains. Simultaneously, the fibers on the inner (concave) surface of the bend are exposed to compressive (negative) strains. Due to the physical nature of the bending tests both tensile (upper half of the plate) and compressive (lower half of the plate) stresses/strains are encountered during bending, Fig. 3.

This means that compressive stresses will develop within the weld in testing of sheet thicknesses below ~7 mm (assuming a weld depth of ~3 mm using GTA welding process). Thickness above ~7 mm is seldom used in aerospace applications and the value of Varestraint testing may be limited in this context and this limitation is addressed in the present study. The aim of the present study is to investigate how the state of stress/strain affects the cracking response in Varestraint testing.

## *Varestraint Testing of Precipitation Hardening Nickel-Based Superalloys*

ATI 718Plus® and Haynes® 282® are two newly developed precipitation hardening Nickel-based superalloys with promising characteristics. Both alloys benefit from the gamma prime phase which provides high temperature strength up to 704 and 800 °C, respectively [2–4]. Welding of the precipitation hardening Ni-based superalloys is of great importance in the manufacturing of structural components used for aero engines [5]. It is therefore necessary to investigate how these alloys behave and respond to different welding operations. One tool is the Varestraint testing to investigate on a more fundamental level how the alloys perform in terms of weldability. There is very limited amount of Varestraint studies for these two alloys. However, previous Varestraint studies performed [5] on ATI 718Plus® indicate that hot cracking resistance seems to be at par with Alloy 718. Hot cracking resistance was associated with Nb-rich MC and Laves phase constituents. No Varestraint studies have been published on Haynes® 282® but Gleeble test results [5] indicate that Haynes® 282® possesses good hot ductility. Liquation was associated with different types of MC.

# Experimental

## *Varestraint Testing*

Two different Varestraint testing configurations, Fig. 4, were used to investigate how the tensile/compressive stress ratio influences the cracking response in testing. The configuration on the left-hand side has the tack-welded support plates on the bottom side of the test plates whereas the configuration on the right-hand side (the normal way) has the support plates lying free on top of the test plate.

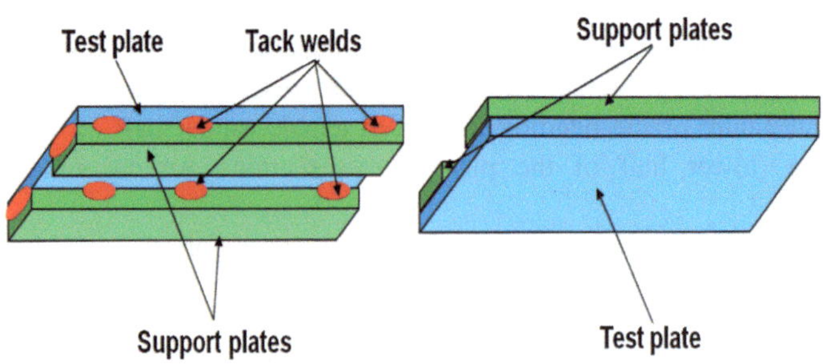


**Fig. 4** Varestraint testing configurations in the present study

**Table 1** Test parameters at bending without welding

| Material | Test plate thickness (mm) | Support plate thickness (mm) |
|---|---|---|
| 304 stainless steel | 2 | 6 |
| 304 stainless steel | 3 | 5 |
| 304 stainless steel | 5 | 3 |

**Table 2** Test parameters in standard varestraint testing

| Material | Test plate thickness (mm) | Support plate thickness (mm) | Welding parameters | Stroke rate (mm/s) |
|---|---|---|---|---|
| ATI 718Plus® and Haynes® 282® | 3.2 | 5 | TIG welding source, 85 A current and 2 mm/s welding speed | 15 |

The configuration on the left-hand side will only impose tensile stresses in the test plate upon bending while the right-hand side configuration will impose both tensile and compressive stresses upon bending.

Strain gauges were used on top and bottom side of the test plates (both configurations) in pure bending (without welding) to verify what was assumed regarding the tensile/compressive stress ratio.

The actual testing parameters in pure bending and Varestraint testing are shown in Tables 1 and 2, respectively.

## *Material*

Expendable AISI 304 stainless steel with different thicknesses (Table 2) was used in the bending experiments to investigate the tensile and compressive stress ratio. The materials used in Varestraint testing were in the mill-annealed condition with an average Vickers hardness (1 kg load) of HV 330 (ATI 718Plus®) and HV 215 (Haynes® 282®). The grain size (GS) of ATI 718Plus® and Haynes® 282® was 10 and 65 μm, respectively (Fig. 5).

The chemical compositions (wt%) are shown in Table 3.

## *Metallographic Procedures*

Scanning electron microscopy (SEM) and energy dispersive X-ray spectroscopy (EDS) was used on samples ground, polished and etched (electrolytically with oxalic acid at 2–5 V for 3–10 s) according to common practice. Fluorescence penetrant inspection together with digital imaging software was used to evaluate the amount of cracking at Varestraint testing.

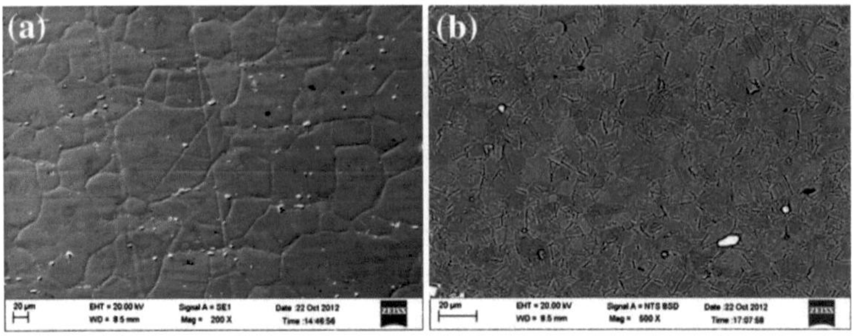

**Fig. 5** The microstructure of as-received material: **a** Haynes® 282® (GS 65 μm, HV 215) and **b** ATI 718Plus® (GS 10 μm, HV 330)

**Table 3** Chemical compositions in wt% of ATI 718Plus® and Haynes® 282®

| Element | ATI 718Plus® | Haynes® 282® |
|---|---|---|
| Ni | Bal. | Bal. |
| Cr | 18.0 | 19.63 |
| Fe | 9.6 | 0.35 |
| Co | 8.9 | 10.35 |
| Mo | 2.6 | 8.56 |
| Al | 1.52 | 1.41 |
| Ti | 0.74 | 2.21 |
| Nb | 5.47 | – |
| C | 0.02 | 0.068 |
| P | 0.010 | 0.002 |
| B | 0.005 | 0.004 |
| Mn | 0.12 | 0.08 |
| S | 0.0003 | 0.002 |
| Si | 0.08 | – |
| W | 1.0 | – |

# Results and Discussion

## *Strain Measurement*

The strain measurements in pure bending of the test plate without tack-welded support plates are shown in Fig. 6. Here, the target strain was calculated according to Eq. 1 below and estimated to ~2 %.

$$\varepsilon = t/2R \tag{1}$$

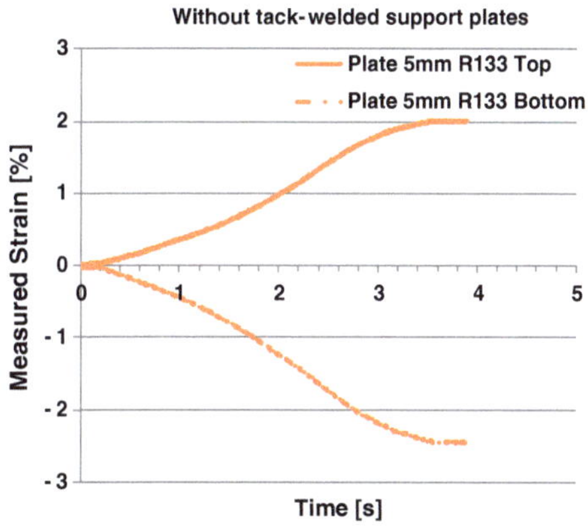


**Fig. 6** Example of strain measurement in pure bending with a target strain of ~2 % using a test plate thickness of 5 mm and radius of 133 mm (R133) with support plates lying free on top of the test plate. The *solid line* reveals measurements from the *top side* of the test plate, whereas the *dotted line* represents measurements from the *bottom side* of the same test plate

where
t = thickness of the plate (mm)
R = the size of the radius die mandrel (mm)

It is clearly revealed that the upper half of the plate thickness is in tension (about 2 %) whereas the lower half is exposed to compressive strains (about −2 %) in bending.

The bending tests with tack-welded support plates are shown in Fig. 7. A constant total thickness of 8 mm (support plate plus test plate) was used throughout the testing. A smaller test plate thickness was thus compensated by a larger tack-welded support plate thickness. So, while decreasing the test plate thickness and increasing the thickness of the tack-welded support plates, the compressive strains imposed upon bending will decrease. In fact, support plates of 6 mm and a testing plate of 2 mm is the only case which reveals any compressive strains in bending. This means that the following target strains were aimed for:

- Sample 1–2: (top surface), 2 %—(bottom surface), −0.25 %
- Sample 3–4: (top surface), 2 %—(bottom surface), 0.25 %
- Sample 5–6: (top surface), 2 %—(bottom surface), 0.5 %

All the target strains coincide fairly well with the results presented in Fig. 7 except the bottom side of the sample 5–6 (~1 % on the bottom side) which seems to exceed the estimated target strain of 0.5 %. The reason for this is not clear but could be associated with lack of fixturing and/or error in the mounting of strain gauge.

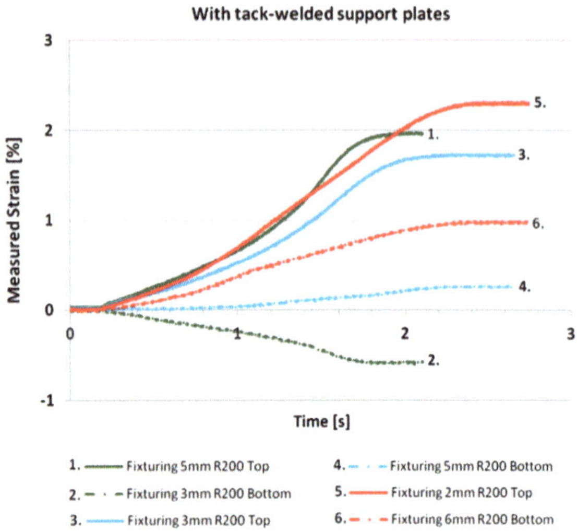


**Fig. 7** Strain measurement in pure bending using a radius of 200 mm (R200) with tack-welded support plates on the *bottom side* of the test plate. The *solid lines* are measurements from the *top side* of the test plates, whereas the *dotted lines* with the same colour are measurements from the bottom side of the same test plate (color figure online)

## *Varestraint Testing of ATI 718Plus® and Haynes® 282®*

The total crack length (TCL) response versus ideal augmented strain of ATI 718Plus® and Haynes® 282® in Varestraint testing using both test configurations (that is with and without tack-welded support plates) is shown in Figs. 8 and 9, respectively. Surprisingly, no distinction between the different plate configurations could be made in Varestraint testing of ATI 718Plus® and Haynes® 282® despite the presumably different strain situations within the materials. There is certainly a difference in state of strain in Varestraint testing compared to pure bending where no weld fixturing interferes. The exact strain situation when welding influences to

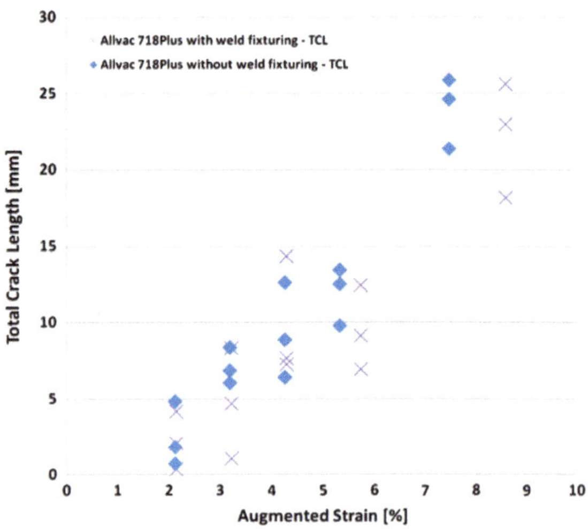


**Fig. 8** The crack response in terms of TCL of ATI 718Plus® in varestraint testing using both test configurations

**Fig. 9** The crack response of Haynes® 282® in varestraint testing using both test configurations

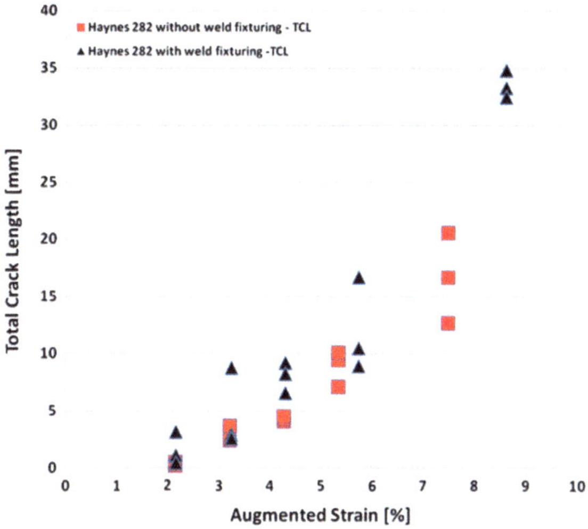


the bending process is complex and difficult to visualize. Another fact is that in the evaluation of the Varestraint test materials, only cracks visible on the surface are considered and not the cracks underneath the surface. Since only surface cracks are measured and both test configurations are exposed to the same magnitude of strain at the top surface, it seems that whatever state of strain under the surface it does not influence the cracking response at the surface. Thus, it does not matter if the lower half of the test plate is exposed to compressive strains and very thin plates can consequently be tested as long as support plates are used to avoid kinking.

Hot cracking (FZ and HAZ) results for both Haynes® 282® and ATI 718Plus®, as shown in Fig. 10, indicate comparable susceptibility to hot cracking. Remembering the positive effect of small grain size on weldability and the large difference in grain size, ASTM 5 for Haynes® 282® and the very fine ASTM 10 grain size for ATI 718Plus®, it can definitely be expected that Haynes® 282® possesses a better hot cracking resistance compared to that of ATI 718Plus®.

Looking closer into the weld microstructure and especially the liquated grain boundaries within the HAZ of ATI 718Plus®, Fig. 11, it appears as if NbC has experienced constitutional liquation.

The constitutional liquation of NbC is supported and validated by the elemental phase map presented in Fig. 12 which indicates the presence of NbC and Laves phase. It should be noted that it is not possible to quantify C using SEM-EDS analysis, but the C-levels between the matrix phase and the NbC/Laves constituents are different indicating the presence of C in the first one. Both of these two phases (NbC and Laves) contain large amount of Nb [5]. As temperature increases rapidly and as the MC is in contact with the γ-matrix a local equilibrium promotes the transformation to γ-Laves mixture due to the constitutional liquation.

Haynes® 282® on the other hand did not reveal significant liquation in the HAZ as ATI 718Plus®, Fig. 13. However, in SEM examination of the HAZ it is seen that

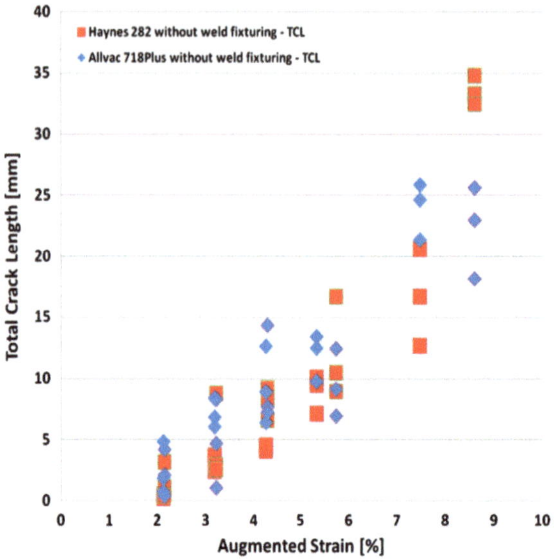


**Fig. 10** Comparison in cracking response between Haynes® 282® and ATI 718Plus® in varestraint testing using both test configurations

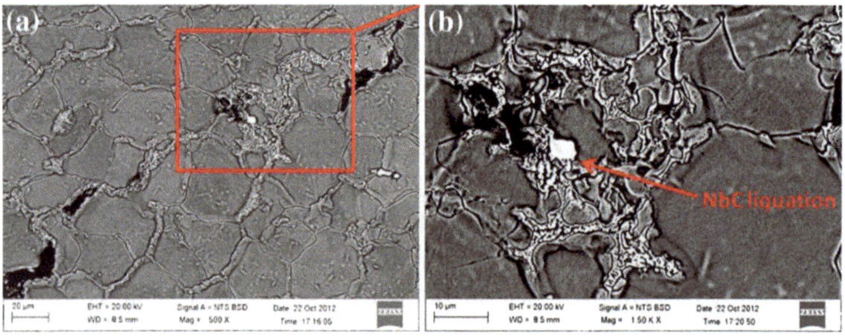


**Fig. 11** Constitutional liquation of NbC in ATI 718Plus®

also in this alloy constitutional liquation of MC takes place locally due to the rapid heating, see Fig. 13.

The MC liquation in Fig. 13 does reveal enrichment of Mo, C and S, as seen in Fig. 14, indicating the presence of MoC [6]. However, care should be taken with the indication of S since the Mo and S energy peaks overlap and may result in an erroneous interpretation of the results.

The situation is complex, since many elements are involved and a non-equilibrium situation is encountered due to the fast heating and cooling. However, it is possible that a ternary interaction between Mo(C) and Ni-S occurs. It is well known that Ni-S and Ni-Mo have eutectic transformations at ~600 and 1315 °C, respectively [7–9]. It is not unreasonable that a ternary eutectic may be

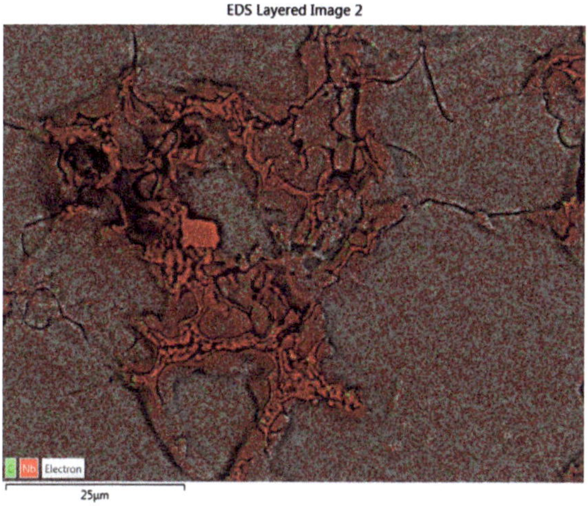


**Fig. 12** SEM-EDS elemental map of ATI 718Plus® revealing enrichment of Nb and C

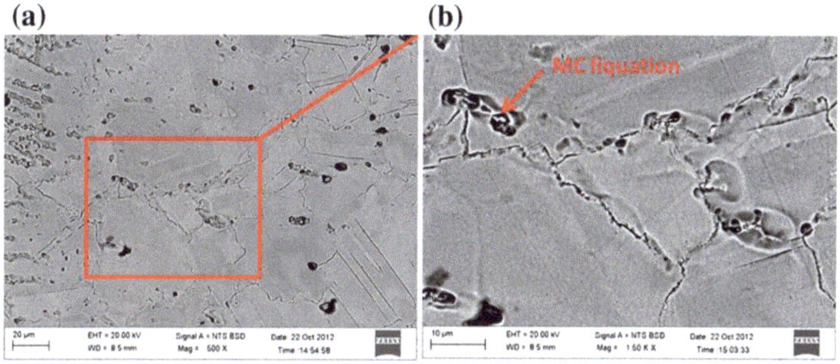


**Fig. 13** Constitutional liquation of MC in Haynes® 282®

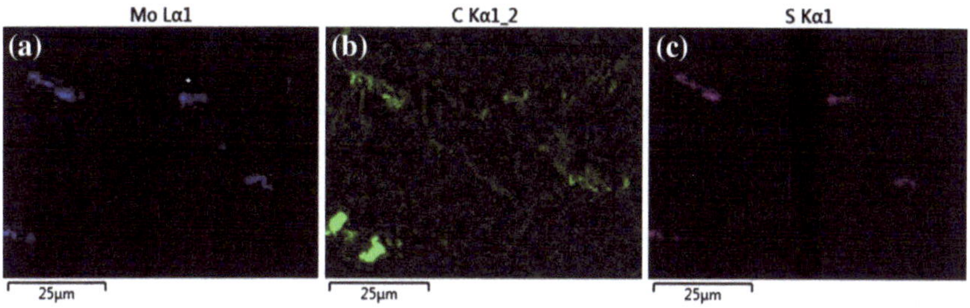


**Fig. 14** SEM-EDS elemental maps of a liquated grain boundary in Haynes® 282® revealing enrichment of **a** Mo, **b** C and **c** S

present due to the non- equilibrium diffusion of S and causing the liquation in Haynes® 282®. Since only traces of S are present in the two alloys of our study the non- equilibrium diffusion of S offers an explanation, but this is still to be verified by the use of more sophisticated analysis tools.

## Conclusions

1. No influence of the compressive strains on the cracking response in Varestraint testing was observed.
2. Thin test plates can be used as long as kinking is eliminated when support plates are used.
3. Hot cracking susceptibility of Haynes® 282® is lower compared to that of ATI 718Plus® especially when the grain size effects are considered.
4. ATI 718Plus® liquates through constitutional liquation of NbC.
5. A hypothesis regarding liquation of MoC through eutectic transformation of Ni-Mo or Ni-Mo-S in Haynes® 282® is possible.

**Acknowledgements** We thank Mr. Heikki Vestman at Aalto University School of Engineering for his help with Varestraint testing. We also thank Professor Lars-Erik Svensson and Leif Karlsson at University West as well as Dr. Göran Sjöberg at GKN Aerospace Engine Systems Sweden for fruitful input and guidance.

## References

1. Savage W. F., and Lundin C.D.; “The Varestraint Test”, The Welding Journal, October, 1965, pp. 433- 442.
2. Cao W.D.; “Nickel-Base Alloy”, U.S. Patent 6,730,264 B2, May 04, 2004.
3. Kennedy R. L.; ”Allvac 718Plus, Superalloy for The Next Forty Years”, Superalloys 718, 625, 706 and Various Derivatives, TMS, Warrendale, PA, 2005, pp. 1-14.
4. Pike L. M.; ”Development of a Fabricable Gamma – Prime (γ’) Strengthened Superalloy”, Superalloys 2008, eds. R.C. Reed, K.A. Green, P. Caron, T. P. Gabb, M. G. Fahrmann, E. S. Huron, S. A. Woodard, The Minerals, Metals & Materials Society, 2008, pp. 191-200.
5. Andersson J.; “Weldability of Precipitation Hardening Superalloys – Influence of Microstructure”, PhD Thesis, Chalmers University of Technology, Gothenburg, Sweden, 2011.
6. Osoba, L.O., Ding, R.G., and Ojo, O.A.; “Microstructural Analysis of Laser Weld Fusion Zone in Haynes 282 Superalloy”, Materials Characterization, 65 (7), 2012, pp. 93.
7. Hansen M., and Anderko K.; “Constitution of Binary Alloys”, McGraw-Hill, 1958.
8. R.P. Elliot, “Constitution of Binary Alloys”, First Supplement, McGraw-Hill, 1965, pp. 968.
9. Okamoto H.; “Ni-S (Nickel-Sulfur)”, Journal of Phase Equilibria andDiffusion, 30 (1), 2009, pp. 123.

# Towards Establishment of Weldability Testing Standards for Solidification Cracking

**N. Coniglio and C.E. Cross**

**Abstract** The establishment of standards for weldability testing would be desirable to both industrial and academic research laboratories. This would have the obvious advantage of allowing data to be reliably compared between different research labs. But making decisions regarding standards requires some careful thought and agreement on (i) how test parameters affect test results, (ii) what exactly needs to be measured, and (iii) how test results should be interpreted and reported. Our depth of understanding on these points has matured significantly over time and while there is not always universal agreement, it is at least possible to identify factors important to standards. This paper examines these factors, including:

1. *Welding Parameters*. When comparing different alloys having different thermal characteristics, the use of constant welding parameters (common practice) will result in variable weld penetration and weld pool shape. This can influence grain shape and mushy zone size, which can result in inequitable weldability comparisons.
2. *Restraint*. Welding on test coupons having different dimensions can affect restraint, which will influence the strain fields around a moving weld pool. Variation in test fixtures may also affect restraint. High restraint does not always result in higher crack susceptibility.
3. *Travel Speed*. Use of high travel speed gives tear-drop shaped weld pools that are more susceptible to cracking, and it reduces the size of the mushy zone. However, high speed can also result in grain refinement in certain alloys, giving improved weldability. Speed can also shift the location of compression/tension regions behind the weld pool.
4. *Rate of Loading*. Extrinsic tests, such as the Varestraint Test, involve the controlled application of a strain during welding. The rate of loading, relative to the

N. Coniglio
MSMP Laboratory, Arts et Metiers Paris Tech, Aix en Provence, France

C.E. Cross (✉)
Los Alamos National Laboratory, Los Alamos, NM, USA
e-mail: cecross@lanl.gov

 
Th. Boellinghaus et al. (eds.), *Cracking Phenomena in Welds IV*,
DOI 10.1007/978-3-319-28434-7_3

weld travel speed, can influence crack length due to crack extension during loading.

5. *Cracking Index.* Selection of an appropriate cracking index is required for data analysis. Quantification of crack length and brittle temperature range are common indexes used for comparison. Threshold strain and critical strain rate are additional indexes. How well these indexes actually represent weldability remains unclear.

This paper will examine and quantify these issues in detail, thus providing the reader with a comprehensive appreciation of all things that must be considered when preparing a standardized procedure for weldability testing.

**Nomenclature**

| | |
|---|---|
| BTR | Brittle temperature range |
| $Cr_{eq}/Ni_{eq}$ | Chromium–nickel equivalent ratio |
| CRW | Controlled restraint weldability |
| CST | Critical strain for temperature drop |
| CTR | Coherent temperature range |
| CTW | Controlled tensile weldability |
| GTAW | Gas tungsten arc welding |
| HAZ | Heat-affected zone |
| MCD | Maximum crack distance |
| MCL | Maximum crack length |
| MISO | Mean of in situ observation |
| MVT | Modified varestraint test |
| PLTS | Pre-loading tensile strain |
| PVR | Programmable deformation cracking |
| SB-TVT | Slow bending transvarestraint test |
| SCTR | Solidification cracking temperature range |
| TCL | Total crack length |
| THCT | Tensile hot cracking test |
| TVT | Transvarestraint test |
| VDR | Variable deformation rate |
| VTST | Variable tensile strain test |

## Introduction

The establishment of standards for weldability testing is an on-going topic of interest to industry. However, in comparison to cold cracking, very little has been accomplished regarding the establishment of test standards for hot cracking and, in particular, tests for solidification cracking. Such standards would presumably enable a more reliable comparison of cracking susceptibility between different alloys and welding conditions. Standardization would also allow a meaningful comparison among different labs to help confirm results. The establishment of such test procedures would need to represent, to the best extent possible, the weldability of alloys experienced in practice.

The absence of such standards is not due to any lack of testing or published information, where significant work has already been devoted to characterize weldability [1–3]. In addition, there is no huge disparity found in alloy rankings among different tests and laboratories as evidenced by round-robin studies and data comparisons [4]. Also, there does not appear to be a large disparity among test rankings and perceived real world behavior [3]. In short, numerous existing tests appear to do a reasonable job of providing a rough comparison (i.e. ranking) of alloy weldability in spite of the absence of standards.

The problem arises in not knowing how variations in testing procedure may affect test rankings and how test results relate to real welding applications. Difficulty is sometimes encountered in deciphering small differences in weldability among similar alloys. There is also the inability to predict, with any certainty, whether or not cracking will occur in a specific application. In addition, exceptions have been found to general trends relating crack measurement to weldability (e.g. crack susceptibility normally, but not always, increases with solidification range). This has led one to question the validity of these trends, if tests are being performed correctly, or if the correct things are being measured. In essence, this begs a bigger question of how test measurements actually relate to weldability.

A case in point is the analysis of crack length, a common metric used in weldability tests. Although considered intuitive, it is not clear exactly how total crack length relates to weldability. It is generally assumed that maximum crack length observed in testing indicates reduced weldability. Understanding why this is the case is worthy of consideration. Recent developments defining cracking mechanisms provide insight to this question and provide a strategy for assessing weldability.

The goal of the present paper is to provide a perspective on establishing future standards for the weldability testing of different alloys and welding conditions. A status review of current weldability testing in the literature is compiled to identify important test parameters and test limitations. The meaning behind weldability data is examined. A check-list summary is provided that facilitates the comparison of weldability data.

## Alloy Composition

When considering weldability, minute details regarding alloy composition and impurities can be of utmost importance. For this reason, it is necessary to accurately measure and document base metal, filler metal, and weld metal compositions. Certain alloys systems are notorious for poor weldability, and so appropriate filler alloys must be selected and dilution controlled to minimize cracking susceptibility. For this reason, dilution should always be determined and reported. For flux-based welding processes, flux composition can likewise have a major effect on weld metal composition. Possibilities for inadvertent weld contamination must also be considered, including shielding gas, joint preparation, weld fixtures, and weld spatter. Outlined below are examples for two alloy systems, although by no means comprehensive, that demonstrate the importance of alloy composition on weldability.

## Aluminum Alloys

When welded autogenously, certain aluminum alloys are known to be highly susceptible to weld solidification cracking [5, 6]. In such instances, filler metals having compositions different from the base metal are selected to shift the weld pool composition to a less crack-sensitive alloy regime. It is well established that use of an appropriate filler metal can improve weldability as demonstrated in Fig. 1. In particular, the Al-Si 4043 filler is popular in this regard.

Shifting the weld metal alloy composition is not the only approach available. The addition of grain refiners to the weld pool, added by way of the filler metal in small quantities (less than 0.1 wt%), can significantly improve the weldability of aluminum alloys. An example of this can be seen in Fig. 1, where filler metal Alloy 2319 is identical to base metal Alloy 2219 except for the addition of a grain refiner. Grain refinement increases the number of grain boundaries, thereby reducing the strain experienced at each grain in the mushy zone. Improved weldability of the following alloys has been attributed to grain refinement: 6060 [7], 7108 [5], Al-6 wt% Cu [8], and Al-2.2Li-2.7Cu [9].

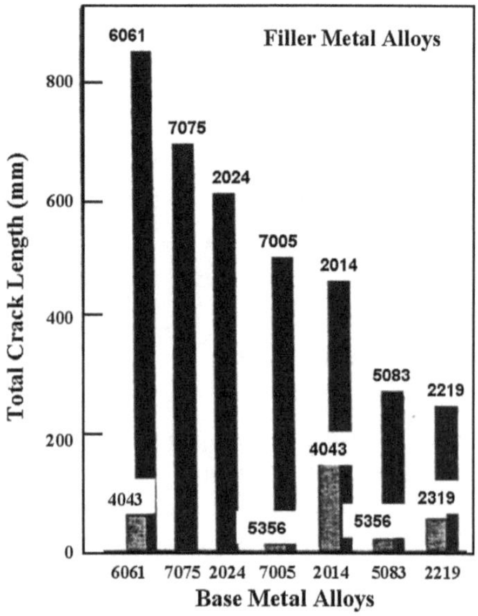


**Fig. 1** Total crack length for different GMA aluminum base metals showing reduced cracking in T-joint weldability test when using a filler metal [6]

## *Stainless Steel Alloys*

The weldability of stainless steel alloys has been shown to be predominantly controlled by solidification mode (determined by alloy content) and the presence of impurities sulfur or phosphorous [10]. These impurities are known to be associated with solidification cracking because they extend the solidification range by forming low melting eutectic liquid [11]. Boron and copper can play a similar role. Impurity-rich liquid segregates to grain boundaries where liquid films are exposed to local tensile strains resulting from thermal contraction and solidification shrinkage.

Alloys that solidify as primary austenite ($\gamma$) and remain fully austenitic (A-solidification) are most susceptible to cracking. The well-defined, straight and continuous nature of the grain boundaries associated with this mode of solidification is believed to be at least partially responsible for its high cracking susceptibility [12]. This corresponds to chromium-nickel equivalent ratios ($Cr_{eq}/Ni_{eq}$) of less than 1.48 for arc welds as shown in Fig. 2, with chromium and nickel equivalents defined by Hammar and Svensson [13, 14].

For alloys with $Cr_{eq}/Ni_{eq}$ ratios higher than 1.48, where primary solidification occurs as ferrite ($\delta$) and then austenite forms subsequently as a peritectic-eutectic (FA-solidification), the resistance to cracking is exceptional. For these alloys, weld metal grain boundaries are irregular and ill-defined making crack propagation more difficult [12]. Also, it has been argued that impurity liquid films do not wet austenite/ferrite boundaries as well as austenite/austenite boundaries [15]. It should be noted that the quantity of impurity (P + S) needed for cracking increases as the $Cr_{eq}/Ni_{eq}$ ratio approaches 1.48; whereas above this ratio, no amount of impurity will cause cracking.

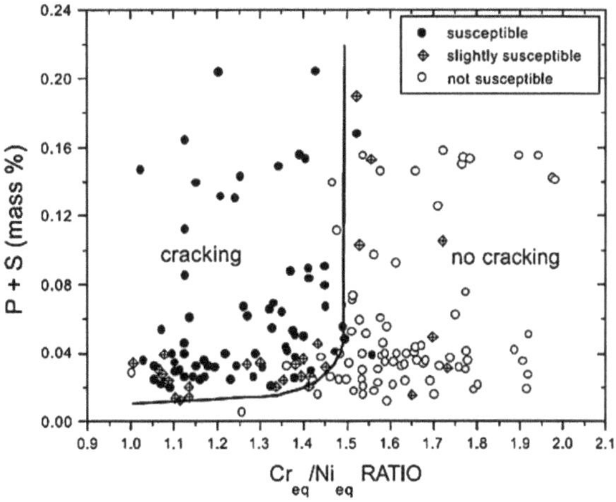


**Fig. 2** Effect of $Cr_{eq}/Ni_{eq}$ ratios on stainless steel weldability for different phosphorous and sulfur contents [13]

# Test Selection

Selecting the most appropriate weldability test requires careful consideration of many different aspects including cost, availability, and type of cracking expected. Because tests are typically not available commercially, this requires in-house design and construction, leading to unique test variations from one laboratory to another. Test designs can vary in complexity from a simple turn-table to make a Circular Patch Test (cost: ≈ $1K), to a hydraulic Varestraint Test (cost: ≈ $50K), to a variable speed Controlled Tensile Weldability Test (≈ $100K).

## *Test Type*

A broad variety of weldability tests have been developed, all with the specific intension of promoting weld cracking by exposing the weld mushy zone to severe conditions not normally encountered during welding. This is accomplished by imposing controlled restraint and/or stress-strain during welding in the vicinity of the weld pool [1–3]. Table 1 lists different weldability tests according to classifications discussed below.

## *Intrinsic Versus Extrinsic*

Intrinsic (or self-restrained) tests rely upon internally generated stresses and strains to cause cracking. While simple and less expensive to perform, intrinsic tests typically involve very complex and non-uniform loading that cannot be easily quantified. This limits the collection of information that would help define critical conditions needed for cracking. Even so, these tests provide an easy and inexpensive way to rank alloys and are useful for a "go-no go" analysis, if that is all that is desired. The severity of loading can often be systematically varied, for example, by changing coupon dimensions, slot depths (Houldcroft and Lehigh Tests), or circle diameter (Circular Patch Test). A detracting aspect of these tests is the difficulty in separating out effects of welding parameters and material properties, since the stress/strain experience is strongly linked to these values.

Extrinsic (or externally loaded) tests, on the other hand, involve the controlled application of an external stress or strain during welding that enables loading to be more independent of material properties or welding parameters. These tests are more expensive to build, but allow for the quantification of loading and/or straining. The simplest of these tests are the cross-weld, pre-load tests such as the Sigmajig Test [35]. Here the stresses of welding add to the pre-applied load in order to promote cracking. A critical applied stress needed for cracking can be found in this manner and can be considered a measure of weldability.

**Table 1** Weldability tests developed to investigate solidification cracking

| Test | Cast versus weld | Intrinsic versus extrinsic | Bending versus tensile | Solidification cracking index | References |
|---|---|---|---|---|---|
| Cast ring test | Cast | Intrinsic | – | Crack length | [16–18] |
| U-Shape casting test | Cast | Intrinsic | – | Crack length | [19] |
| Harp (backbone) mold test | Cast | Intrinsic | Tensile | Crack length<br>Specimen length<br>Specimen diameter | [20, 21] |
| Warrington test | Cast | Intrinsic | – | Crack length | [8] |
| Cast pin tear test | Cast | Intrinsic | Tensile | Crack length<br>Specimen length<br>Specimen diameter | [22–24] |
| Circular patch test | Weld | Intrinsic | Tensile | Crack length | [25, 26] |
| T-Test | Weld | Intrinsic | – | Crack length | [27] |
| Lehigh test | Weld | Intrinsic | Tensile | Crack length<br>Slot depth | [28] |
| Houldcroft test | Weld | Intrinsic | Tensile | Crack length<br>Slot depth | [29–31] |
| Fan-shaped test | Weld | Intrinsic | Tensile | Crack length<br>Plate width | [32] |
| Free-edge test | Weld | Intrinsic | Tensile | Crack length<br>Distance from free edge | [33] |
| Variable restraint box test | Weld | Intrinsic | Tensile | Restraint | [34] |
| Sigmajig test | Weld | Intrinsic | Tensile | Preload | [35] |
| Pre-loading tensile strain (PLTS) test | Weld | Intrinsic | Tensile | Preload | [36] |
| Varestraint test | Weld | Extrinsic | Bending | Strain<br>Ductility curves-BTR | [37] |
| Transvarestraint test | Weld | Extrinsic | Bending | Strain<br>Ductility curves-BTR | [38] |
| Spot varestraint test | Weld | Extrinsic | Bending | Strain | [39] |
| Modified varestraint test (MVT) | Weld | Extrinsic | Bending | Strain<br>Strain rate | [40] |
| Variable tensile strain hot cracking test | Weld | Extrinsic | Tensile | Strain rate | [41, 42] |

(continued)

**Table 1** (continued)

| Test | Cast versus weld | Intrinsic versus extrinsic | Bending versus tensile | Solidification cracking index | References |
|---|---|---|---|---|---|
| Slow bending transvarestraint test | Weld | Extrinsic | Bending | Strain<br>Strain rate<br>Ductility curves-BTR | [43, 44] |
| Murex test | Weld | Extrinsic | Bending | Angular speed | [45] |
| Variable deformation rate (VDR) test | Weld | Extrinsic | Tensile | Angular speed | [46] |
| Programmable deformation cracking (PVR) test | Weld | Extrinsic | Tensile | Crack length | [47] |
| Wrap bend test | Weld | Extrinsic | Bending | Crack length | [48] |
| Controlled tensile weldability (CTW) test | Weld | Extrinsic | Tensile | Strain rate | [7, 49, 50] |
| Controlled restraint weldability (CRW) test | Weld | Extrinsic | Tensile | Preload | [51] |

Many variations to the Sigmajig Test can be found. The Controlled Restraint Weldability (CRW) Test [51] has the pre-load confined to the central region of the coupon, which avoids problems encountered with crack initiation at coupon edges (see Fig. 3). Some cross-weld loading tests allow for precise control of strain application similar to what is done in a tensile test. In the Controlled Tensile Weldability (CTW) Test [7, 49, 50], strain is applied during welding at a fixed displacement rate. Here the critical strain rate required for cracking can be found.

Bend-type extrinsic tests can also be used to apply controlled strain to a weld, e.g. by wrapping a coupon over a curved mandrel during welding as is done in the

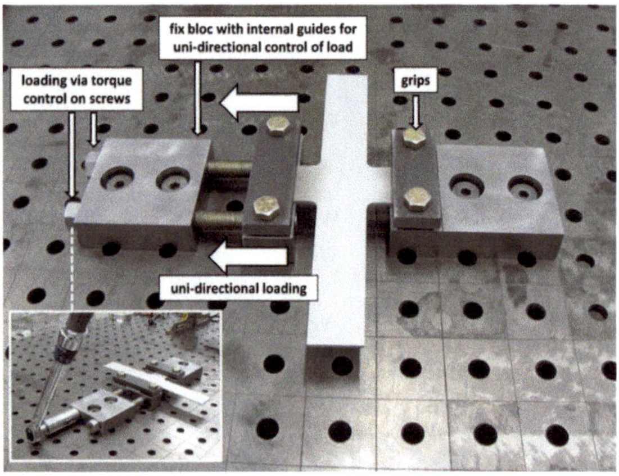


**Fig. 3** Overview of controlled restraint weldability (*CRW*) test with superimposed overview of preloading procedure by torque control [51]

Varestraint Test [35]. Here the maximum strain experienced in the vicinity of the weld pool ($\varepsilon$) is determined by the radius of the mandrel ($R$) and coupon thickness ($t$):

$$\varepsilon = \frac{t}{2R} \times 100\ \% \qquad (1)$$

Mandrels having different radii can be exchanged in a series of tests to vary the maximum applied strain, normally between 0.25 and 4 %. A minimum amount of strain is needed to get cracking depending upon the material. Higher amounts of applied strain, above the minimum, result in higher amounts of total accumulated crack length (TCL). When TCL is plotted against augmented strain, the resulting curve provides a useful means of comparing alloy behavior. Although, TCL is probably not as good an indicator for weldability as maximum crack length (MCL), to be discussed later.

## *Bend Versus Tensile*

While bending tests are easy to perform, they do not lend themselves to the precise determination of critical conditions needed for cracking. The amount of strain applied, and the rate at which it is applied, is generally an order of magnitude greater than most real welds will ever experience. Added to this is the non-uniform distribution of strain and strain rate within the coupon during strain application, which varies over the duration of the test and throughout the coupon thickness. "Hinging" is another problem that can occur in bend tests, arising from the non-conformity of the specimen with the die block as depicted in Fig. 4, where plastic deformation is concentrated in hot material along the weld seam. MISO measurements have revealed that the weld metal is subjected to much higher strains and strain rates than expected using Eq. 1 [42].

The Varestraint Test is commonly operated in a rapid mode of strain application. This near-instantaneous strain generates cracks whose maximum length represents the size of the crack-sensitive region within the mushy zone. Even at high rates of straining, however, there is finite crack growth as the weld advances during

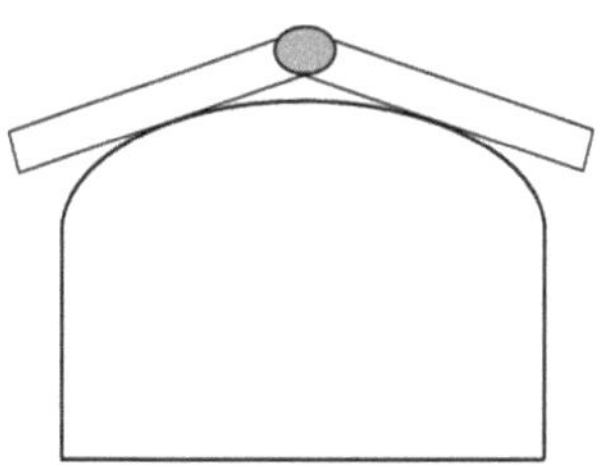

**Fig. 4** Schematic illustration showing hinge effect in Transvarestraint Test

bending, potentially affecting crack length results [52, 53]. Accordingly, ram speed is something that should always be measured and documented for these types of tests. If operated at a slow speed, long continuous cracks can sometimes be grown, particularly when using the Transvarestraint Test. When operated in the slow mode, this test has been used to determine critical strain rates needed for cracking based upon surface strain measurements [43, 44, 54].

Cross-weld tensile tests in comparison are much better suited for measuring critical values of strain and strain rate needed for cracking. It is advantageous to have controlled strain applied normal to the centerline grain boundaries, where cracking is most likely to occur. Also, at any point in time, the load, strain and strain rate are uniform through the coupon thickness. However, these tests also have their unique drawbacks. High welding speeds, such as those encountered in laser welds, require large displacement rates that result in rapid strain accumulation and premature failure in the base metal [50]. Also, there are numerous questions regarding test procedure that still need to be examined including effects of coupon size, weld length, and when strain should first applied (relative to weld start).

## *Castability Versus Weldability*

Solidification defects in castings (i.e. hot tears) are similar in nature to solidification cracking observed in welds. It follows that castability tests developed to produce hot tears should likewise indicate susceptibility to solidification cracking in welds. Castability evaluations have the advantage of allowing the testing of small heats of experiment alloys without having to produce rolled base plate or draw experimental filler wire. Thus, this approach is particularly attractive in early stages of alloy development and has been successfully applied to aluminum and ferrous alloys [8, 16–22]. It has the obvious disadvantage, however, of not exactly replicating the same thermo-mechanical or steady-state solidification conditions experienced in a weld.

The solidification conditions of temperature gradient, cooling rate and growth rate for castings compared to welds can differ by an order of magnitude, depending upon mold material. It follows that for castability tests to most accurately reflect upon weldability behavior, the solidification microstructure (grain size and dendrite spacing) should be similar. This requires, roughly, 100 μm grains and 10 μm dendrite spacing. These solidification conditions can be replicated using metal chill molds (e.g. copper) to rapidly extract heat, plus mold preheat to fine-tune the microstructure.

In castability tests, the mold cavity is designed in such a way as to accumulate solidification shrinkage and thermal contraction within the last region to solidify. In the Cast Ring Test [16–18], hoop stresses are generated as an annular shaped casting shrinks around a central core, which results in circumferential cracking. In the Harp Test [20, 21], uniaxial strain is accumulated within the gage section of a cast tensile bar, restrained by end pieces supported by the mold (Fig. 5). When the

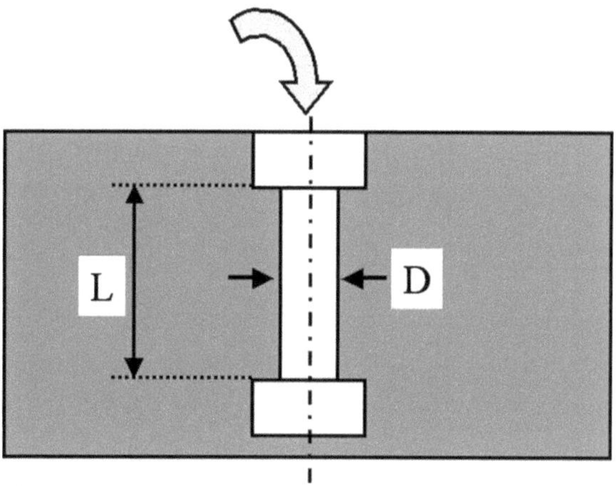


**Fig. 5** Schematic cross-section of harp test [20]

gage length is increased, the amount of strain is increased and eventually circumferential cracking is observed. A similar approach has been refined using a Cast Pin Tear Test [22–24], also discussed in a separate paper in this workshop (see Przybylowicz, et al.).

## How Test Parameters Affect Results

### *Effect of Loading Direction*

The direction in which load is applied to the weld, relative to the welding direction, will tend to favor one type of hot crack over another. Cracks will preferentially grow along grain boundaries oriented perpendicular to the load. Longitudinal loading, as experienced in PVR and Longitudinal Varestraint Tests, promotes transverse cracking in the HAZ (i.e. liquation cracking) and also near the fusion line in the weld metal. In these locations, grain boundaries are found to be normal to the welding direction. It is sometimes the case in welding that centerline solidification cracks initiate near the fusion line and then grow to the weld center. In such cases, the use of this type of test may best represent weldability. Transverse loading, as experienced in Transvarestraint, CTW, and Circular Patch Tests, promotes centerline cracking in the weld metal, but may not be ideally suited to study HAZ liquation cracking.

### *Effect of Stress*

When welding complex, multiple joint structures, residual stresses can build that influence weldability. The presence of tensile stress formed during welding is

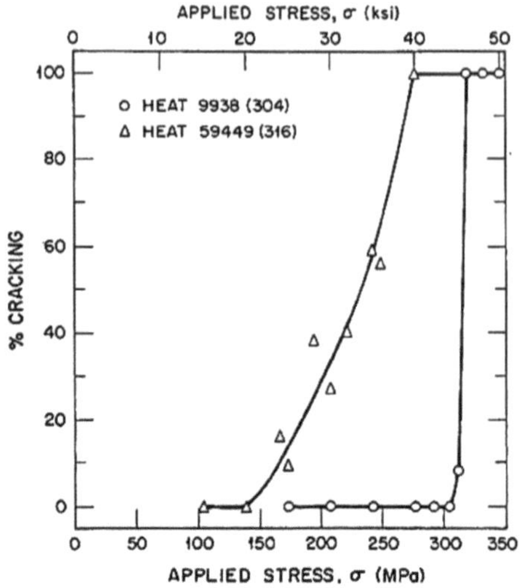


**Fig. 6** Cracking susceptibility of austenitic stainless steels for different pre-loads as measured by the sigmajig test [33]

known to aggravate cracking. Along these lines, several weldability tests consist of applying a transverse preload in the plane of a test coupon prior to welding (e.g. Sigmajig, CRW, PLTS). This allows crack-no crack comparisons and permits a critical preload to be measured, where greater critical preloads indicate better weldability. In some alloys there is gradual increase in amount of cracking with applied stress, as opposed to a sharp crack-no crack transition (see Fig. 6). Although cracking models exist based upon mushy zone strength [55, 56], this property cannot be used to uniquely characterize liquid-solid behavior. High preloads add to the stresses normally associated with welding and result in greater local tensile strains and strain rates. The upper limit of preload is bounded by the yield stress of the metal to be welded. The relation between preload and local strain and strain rate is generally not known, but can be modeled.

## *Effect of Strain and Strain Rate*

Strain has generally been taken to have the primary influence on solidification crack formation, assuming the mushy zone to have limited ductility [57–59]. Extensive weldability test data supports this assumption, where higher applied strain eventually leads to cracking, and even higher applied strain generates more extensive cracking [59–61]. Likewise it has been argued that strain rate is important, since it allows more strain to be accumulated over a fixed time of solidification [62, 63]. It has been demonstrated experimentally that a critical strain rate must be exceeded in order to cause cracking [53].

More recent models [64] propose that cracks originate as a result of interdendritic pressure drop (i.e. cavitation), which is directly influenced by transverse strain rate [65]. Another model suggests that strain rate may play a more direct role in solidification cracking, enabling continuous crack growth [66]. If the transverse strain rate is insufficient to compensate for the rate of liquid back-filling, then cracks cannot grow. What this means is that strain may play only a secondary (non-critical) role in weldability testing, serving only to provide the displacement needed to open up cracks present at grain boundaries, formed due to strain rates that far exceed critical values (e.g. Varestraint Test).

An important point to be made here is that while strain is applied globally in weldability tests (e.g. maximum strain in the Varestraint Test), it is actually the local strains in the vicinity of the weld pool that are of primary importance to cracking models, and these strains cannot be easily measured. Techniques developed to estimate these local strains have included examination of surface oxide displacement (so called "MISO" technique) [42, 60, 67], use of mechanical extensometers [7, 49, 68, 69], optical grid pattern analysis [70–72], and digital image correlation with speckle pattern [54, 73, 74]. The latter method is discussed in a separate paper in this workshop proceeding (see Gollnow, et al.).

## *Effect of Restraint*

Restraint intensity is a measure of the resistance to strain during welding and can be thought of as a spring constant [34]. It is influenced by the rigidity of the welding fixture as well as the weld coupon size and shape. The thermal and shrinkage stresses generated during welding interact with restraining forces to influence strain around the moving weld pool. It is generally the case that high restraint gives high residual stress and low distortion after welding, but how it affects local strains during welding is not as well understood. Some weldability tests that involve the controlled variation of restraint intensity, e.g. Houldcroft (Fig. 7) and Lehigh Tests, incorporate slots milled along the sides of the weld coupon.

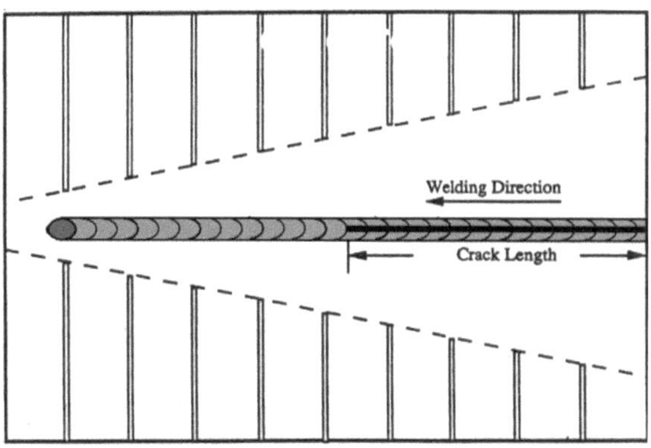


**Fig. 7** Schematic of houldcroft test (*top view*)

In the Houldcroft Test, increasingly longer slots towards the end of the coupon relieve restraint to the point where crack growth can no longer continue. Crack length is taken as a measure of weldability, where long cracks represent crack growth resilience (poor weldability). A distinction should be made here between crack initiation and crack growth when considering weldability (to be discussed later). Crack initiation occurs when the weld is initiated at the coupon edge, a problem commonly encountered whenever welding thin plate or sheet metal.

The Free-Edge and Fan-Shape Tests vary restraint by varying the distance to an un-clamped free edge (see Fig. 8). The closer the weld is to the free edge $a$, there is less restraint and greater possibility for crack initiation. There exists a critical distance $a_{cr}$ below which cracking occurs, taken as a measure of weldability. The Box Restraint Test (Fig. 9) uses a box frame welding fixture where side panels can be exchanged to vary box height ($H$) and degree of restraint. It was found that higher restraint resulted in shorter crack length [34], with cracks initiating at the weld start (edge of coupon).

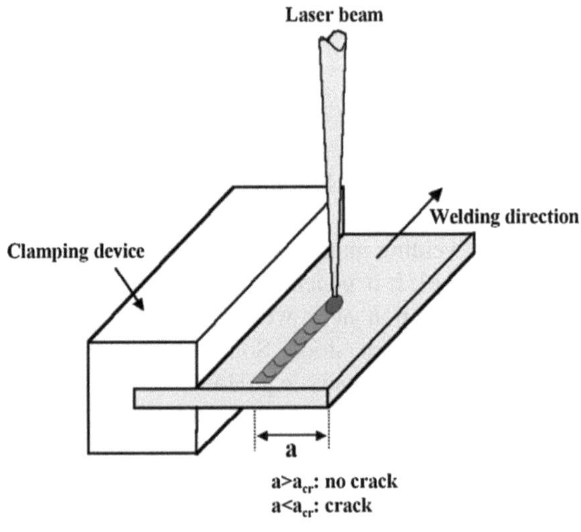


**Fig. 8** Schematic of free edge test

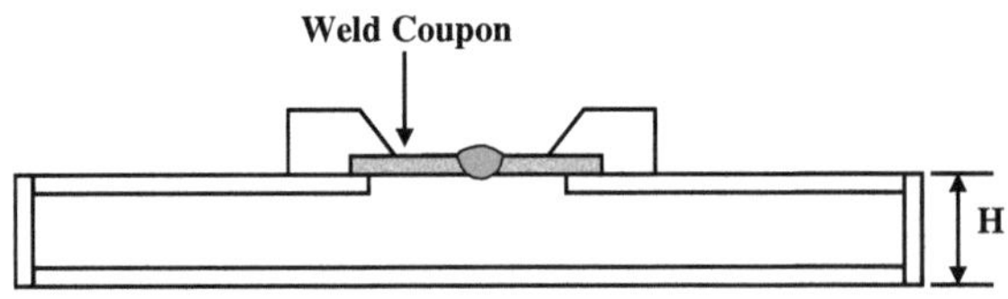


**Fig. 9** Schematic of box restraint test

## *Effect of Welding Parameters*

Welding parameters can have a significant impact on the cracking susceptibility of an alloy, and hence must be considered when performing weldability tests. If ranking alloys for a given application, it makes sense to use a similar welding process and parameters for both test and application if possible. However, some alloys, because of their different thermal properties, may require higher current to achieve the same penetration. For this reason, it may be argued that alloys should be weldability tested each with unique welding parameters to produce a constant penetration. Whichever approach is chosen, it should be appreciated that any change in weld heat input can modify the size and shape of the weld pool, HAZ, and mushy zone, as well as change stress and strain behavior. It can also change temperature gradients and cooling and solidification rates, which will result in different weld metal grain structure and sub-grain microstructure.

*Heat Input.* Welding heat input, defined as energy per unit length of weld (calculation: current × voltage/travel speed), allows a comparative indication of the heat used to make a weld. Making welds with similar heat input for a given material provides roughly similar weld pool volumes and cooling rates, but not necessarily similar penetration. Increasing heat input will increase the weld cross-section and lower the cooling rate, producing a larger mushy zone. It is generally observed that increased heat input gives greater susceptibility to cracking (see Fig. 10).

*Travel Speed.* The influence of torch travel speed on weld solidification cracking has often been studied, although there exists no general consensus on what to expect. Some studies have reported an improvement in weldability when increasing the torch travel speed [46, 70, 75, 76]. These measurements were made for the GTAW process at travel speeds between 2.5 and 13 mm $s^{-1}$. When Chihoski [70, 75]

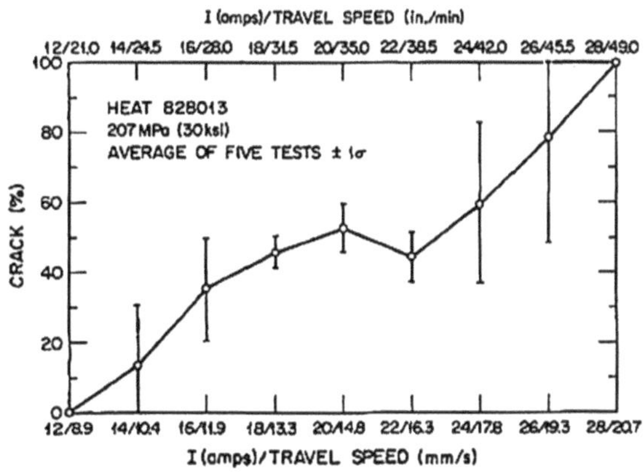


**Fig. 10** Relationship between welding heat input and solidification crack extent [35]

experimentally monitored strain cells during welding, high travel speeds were observed to favor a compressive cell behind the weld pool, thereby avoiding cracking. With decreased travel speed, this compressive cell diminished and was replaced with a tensile cell that favored crack formation. However, some simulations [77] and experimental observations [78] have shown an increase in cracking susceptibility with increased travel speed. But in these situations, the welding speeds are high; between 50 and 100 mm $s^{-1}$ for laser welding [78] and between 16 and 25 mm $s^{-1}$ for GTAW [77].

Besides local strain cells, weld travel speed can also influence weld pool shape, demonstrated to have a strong effect on weldability [79]. Increasing travel speed tends to elongate the weld pool, favoring a shape transition from round, to elliptical, to teardrop. The teardrop shape in particular generates long continuous grain boundaries along the weld centerline, orientated normal to transverse thermal and shrinkage strains. Also, the macro-segregation of impurities (e.g. S and P in steel) is promoted with a teardrop shape, where columnar grains push rejected solute directly to the weld center. Thus the possibility for cracking can occur at below normal impurity levels [49].

Travel speed can also affect weld metal grain structure, where solidification undercooling increases with solidification rate (related to travel speed) and thus favors equiaxed grain formation. This has been observed for aluminum Alloy 6082 containing $TiAl_2$ grain refiner [80] as shown in Fig. 11. At higher speeds, the weld center becomes dominated by equiaxed grains, which significantly improves weldability.

*Weld Dilution.* It has already been pointed out that filler alloy selection and control of dilution is important for crack resistance, something that should be controlled and quantified. An example of this is given in Fig. 12, where weldability of aluminum Alloy 6082 (expressed in terms of a critical strain rate) is shown to vary with the amount of Alloy 4043 filler metal added to the weld pool. Greater amounts of Alloy 4043 and weld metal silicon content gives improved weldability and requires higher strain rates to generate cracking [48].

Curvature of the weld crown is another dilution controlled factor to consider for crack avoidance [81]. While a concave bead surface is stressed in tension during

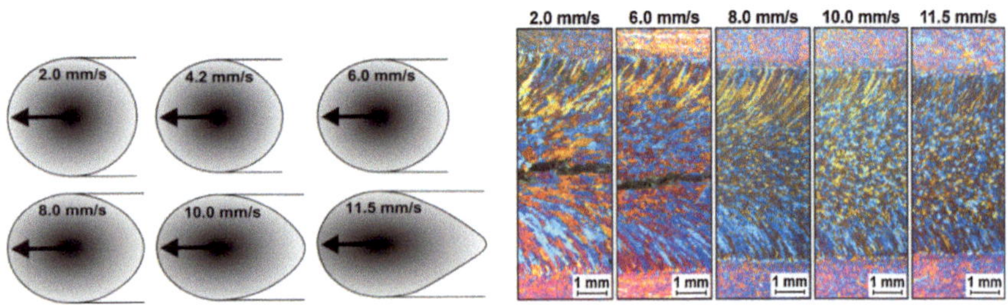


**Fig. 11** Relationship between weld pool shape and weld metal grain structure for aluminum alloy 6082 [80]

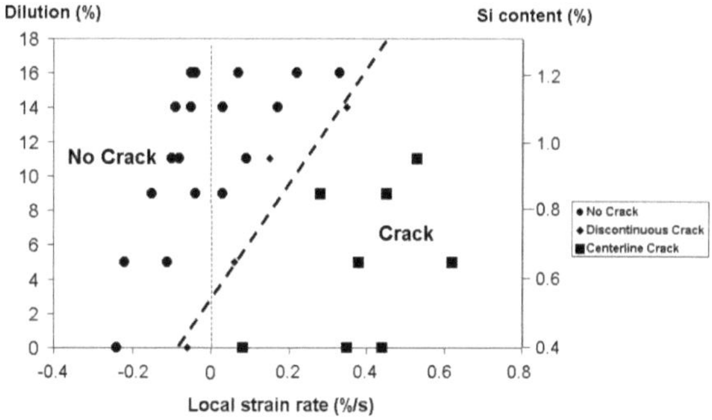


**Fig. 12** Cracking susceptibility of alloy 6060 for variable 4043 dilution shown as a function of local strain rate [49]

solidification, a convex surface can shrink without going into tension and is believed to provide better cracking resistance. A minimum amount of filler addition is required to maintain a convex curvature. Crater cracks are often encountered at the weld termination, where filler is stopped before the arc is extinguished.

## Measurements that Uniquely Define Weldability

The desired outcome of weldability tests is cracking, which is most often quantified in some manner of crack length: maximum crack length-MCL, total (accumulated) crack length-TCL, or percentage of weld length. It is generally assumed that for fixed testing conditions, an alloy having higher cracking susceptibility should result in more extensive cracking in weldability tests [16, 17, 27, 60, 62]. However, there are inherent problems when considering only crack length, due to alloy-dependent variations in thermal gradients and weld pool shape. In essence, it becomes difficult to compare the extent of cracking for different alloys when all test conditions are not exactly the same. Also, there is a confusing issue of crack back-filling, i.e. cracks which form and then are partially back-filled with liquid.

A more appropriate measure of cracking susceptibility that is more characteristic of a particular alloy is its solidification range, i.e. the range between liquidus and non-equilibrium solidus temperatures. Alloys having a large solidification range tend to be more susceptible to cracking, although there are a few exceptions. A larger temperature range relates to higher cracking susceptibility since it provides the possibility for more strain build-up and a more difficult condition for liquid feeding. Even more representative of cracking susceptibility is the temperature range over which cracking is actually found to occur, often referred to as the Brittle Temperature Range (BTR) or Coherent Temperature Range (CTR).

Critical strain rate can also be viewed as something that uniquely defines weldability for a particular alloy. As has been pointed out in the above discussions, strain rate plays a direct role: (i) influencing crack initiation due to interdendritic pressure drop and (ii) enabling crack growth to overcome liquid back-filling.

## Brittle Temperature Range

The BTR is less than the solidification range, bounded by the coherent and non-equilibrium solidus temperatures. The coherent temperature represents the beginning of dendrite interlocking, providing resistance to mechanical shear and liquid back-feeding. It follows that cracking is not found at temperatures above coherency, because of the ease of liquid back-filling and absence of any stress. An alloy's coherent temperature is not something that can typically be found documented in handbooks, but it can be measured experimentally using either thermal recalescence or casting stir-torque techniques [82]. Another approach to measuring BTR is to relate maximum crack length (MCL) in a Transvarestraint (or similar) Test to its associated weld temperature gradient, determined using an implanted thermocouple [30, 41–44, 46, 60]. The temperatures at the two ends of a crack thus define the BTR.

The BTR is something that has been widely reported in the literature and is generally accepted to represent weldability. A collection of BTR values are compiled in Table 2 from numerous different sources in the literature generated using two different cracking tests. Large variations in BTR have been reported for the same material, suggesting problems with inconsistencies in test technique. Lippold [1] has attempted to refine the Transvarestraint technique, taking care to more precisely define a maximum crack distance (MCD) in calculating the temperature range, a distance measured perpendicular to the thermal gradient. He also refers to the resulting calculation as the Solidification Cracking Temperature Range (SCTR) to distinguish it from other means of measuring BTR [83, 84].

To relate crack distance in a Transvarestraint Test to a corresponding temperature range requires the assumption of instantaneous bending in order to neglect any crack growth during bending. As has been pointed out earlier, there can be inaccuracies associated with this assumption. Also, there is a question of what augmented strain in the Transvarestraint Test should be applied to sufficiently open a crack to its extreme, but not extend the crack into solid material. Problems may also be associated with liquid back-filling of cracks. Hence, there are numerous factors that can limit the accuracy and repeatability of BTR measurement.

## Critical Strain Rate

Ductility-based models view strain rate as important for determining how much strain can be accumulated during solidification. This concept was first suggested by Prokhorov [58] and later referred to as the critical strain for temperature drop (CST) [59]. Using this strain build-up concept, the critical strain accumulation, that is the boundary between crack and no-crack conditions, is defined by where the deformation curve becomes tangent to the ductility curve (dashed line-(ii) in Fig. 13a) and is illustrated for three aluminum alloys in Fig. 13b. It should be noted

**Table 2** Brittle temperature range (BTR) for aluminum, steel, and inconel weld metal alloys

| Material | Welding conditions | | | Weldability test[a] | BTR (°C) | SCTR (°C) | Reference |
|---|---|---|---|---|---|---|---|
| | Voltage (V) | Current (A) | Welding speed (mm/s) | | | | |
| Aluminum alloy 1050 | 25 | 200 | 2.5 | TVT | 16 | – | [59] |
| Aluminum alloy 2024 | 25 | 200 | 2.5 | TVT | 190 | – | [59] |
| Aluminum alloy 5083 | 25 | 200 | 2.5 | TVT | 110 | – | [59] |
| Austenitic stainless steel SUS 304L | 18.5 | 65 | 0.17 | THCT | 40 | – | [61] |
| Austenitic stainless steel AISI 310 | 18 | 250 | 2.5 | TVT | 110 | – | [59] |
| Austenitic stainless steel SUS 310S | 18.5 | 65 | 0.17 | THCT | 32 | – | [60] |
| Austenitic stainless steel SUS 310S | 15 | 75 | 0.67 | THCT | 150 | – | [61] |
| Austenitic stainless steel SUS 316 | 18.5 | 65 | 0.17 | THCT | 32 | – | [60] |
| Austenitic stainless steel AISI 316 | 18 | 250 | 2.5 | TVT | 120 | – | [59] |
| Austenitic stainless steel AISI 321 | 18 | 250 | 2.5 | TVT | 60 | – | [59] |
| Austenitic stainless steel SUS 321 | 18.5 | 65 | 0.17 | THCT | 110 | – | [60] |
| Plain carbon steel S15C | 15 | 75 | 0.67 | THCT | 65 | – | [61] |
| Plain carbon steel S35C | 15 | 75 | 0.67 | THCT | 100 | – | [61] |
| Plain carbon steel SK6 | 15 | 75 | 0.67 | THCT | 130 | – | [61] |
| Plain carbon steel SS41 | 18 | 250 | 2.5 | TVT | 100 | – | [59] |
| High carbon steel S55C | 19 | 250 | 2.5 | TVT | 110 | – | [59] |
| High strength stainless steel HT-80 | 19 | 250 | 2.5 | TVT | 90 | – | [59] |
| Inconel alloy 600 | 18.5 | 65 | 0.17 | THCT | 96 | – | [60] |

(continued)

**Table 2** (continued)

| Material | Welding conditions | | | Weldability test[a] | BTR (°C) | SCTR (°C) | Reference |
|---|---|---|---|---|---|---|---|
| | Voltage (V) | Current (A) | Welding speed (mm/s) | | | | |
| Inconel alloy 600 | 15 | 75 | 0.67 | THCT | 120 | – | [61] |
| Alloy C-22 | – | – | – | TVT | – | 50 | [83] |
| 304L SS FN6 | 10 | 180 | 2.1 | TVT | – | 31 | [84] |
| 316L SS FN4 | 10 | 180 | 2.1 | TVT | – | 49 | [84] |
| Al6XN superaustenitic SS | 10 | 180 | 2.1 | TVT | – | 115 | [84] |
| 310 SS | 10 | 180 | 2.1 | TVT | – | 139 | [84] |

[a]*TVT* Transvarestraint test, *THCT* Tensile hot cracking test

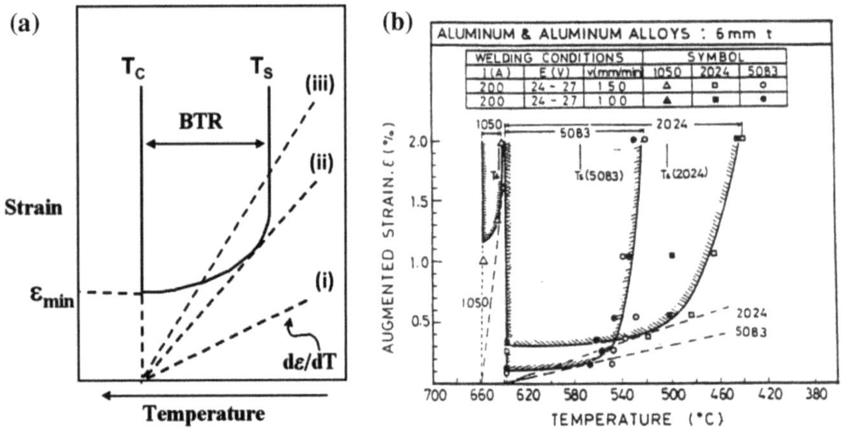


**Fig. 13** Strain rate dependence of solidification crack ductility **a** showed in general schematic and **b** illustrated for three aluminum alloys [59]. *Dashed line* (*ii*) represents critical rate of strain accumulation for cracking

that the ductility curve represents the lower limit of the BTR, and is thus subject to all of the measurement limitations discussed in the above section.

The strain accumulation with temperature drop (*dε/dT*) is related to the cooling rate (*dT/dt*) and strain rate (*dε/dt*), accordingly:

$$\frac{d\varepsilon}{dT} = \frac{d\varepsilon/dt}{dT/dt} \tag{2}$$

The CST criterion has been applied to evaluate the weldability of various alloy systems including aluminum, steel, and nickel (Table 3) and is found to rank the alloys reasonably well according to welding experience. That CST is directly

**Table 3** Critical strain for temperature drop (CST) required for weld solidification crack formation

| Material | Welding conditions | | | Weldability test[a] | CST (1/°C) | Reference |
|---|---|---|---|---|---|---|
| | Voltage (V) | Current (A) | Welding speed (mm/s) | | | |
| Aluminum alloy 1050 | 25 | 200 | 2.5 | TVT | $91 \times 10^{-5}$ | [59] |
| Aluminum alloy 2024 | 25 | 200 | 2.5 | TVT | $3.3 \times 10^{-5}$ | [59] |
| Aluminum alloy 5083 | 25 | 200 | 2.5 | TVT | $1.7 \times 10^{-5}$ | [59] |
| Austenitic stainless steel AISI 304 | 18 | 250 | 2.5 | TVT | $38 \times 10^{-5}$ | [59] |
| Austenitic stainless steel SUS 304L | 18.5 | 65 | 0.17 | THCT | $320 \times 10^{-5}$ | [60] |
| Austenitic stainless steel AISI 310 | 18 | 250 | 2.5 | TVT | $1.6 \times 10^{-5}$ | [59] |
| Austenitic stainless steel SUS 310S | 18.5 | 65 | 0.17 | THCT | $400 \times 10^{-5}$ | [60] |
| Austenitic stainless steel SUS 316 | 18.5 | 65 | 0.17 | THCT | $320 \times 10^{-5}$ | [60] |
| Austenitic stainless steel AISI 316 | 18 | 250 | 2.5 | TVT | $2.5 \times 10^{-5}$ | [59] |
| Austenitic stainless steel SUS 321 | 18.5 | 65 | 0.17 | THCT | $40 \times 10^{-5}$ | [60] |
| Austenitic stainless steel AISI 321 | 18 | 250 | 2.5 | TVT | $7.3 \times 10^{-5}$ | [59] |
| Plain carbon steel 0.08 wt% C | – | – | – | – | $230 \times 10^{-5}$ | [85] |
| Plain carbon steel 0.16 wt% C | – | – | – | – | $170 \times 10^{-5}$ | [85] |
| Plain carbon steel 0.31 wt% C | – | – | – | – | $120 \times 10^{-5}$ | [85] |
| Plain carbon steel 0.50 wt% C | – | – | – | – | $68 \times 10^{-5}$ | [85] |
| Plain carbon steel SS41 | 18 | 250 | 2.5 | TVT | $3.7 \times 10^{-5}$ | [59] |

(continued)

**Table 3** (continued)

| Material | Welding conditions | | | Weldability test[a] | CST (1/°C) | Reference |
|---|---|---|---|---|---|---|
| | Voltage (V) | Current (A) | Welding speed (mm/s) | | | |
| High carbon steel S55C | 19 | 250 | 2.5 | TVT | $3.5 \times 10^{-5}$ | [59] |
| High tensile strain HT-80 | 19 | 250 | 2.5 | TVT | $10 \times 10^{-5}$ | [59] |
| Inconel alloy inconel 600 | 18.5 | 65 | 0.17 | THCT | $80 \times 10^{-5}$ | [60] |

[a]*TVT* transvarestraint test, *THCT* tensile hot cracking test

related to strain rate gives support to critical strain rate as a unique weldability criterion, particularly when considering the role of strain rate in cracking mechanisms.

Measured critical strain rates and deformation rates required for weld metal solidification crack formation are listed in Table 4 for specific alloys and welding conditions. These values were obtained by making local strain measurements at the crack initiation site. The critical strain rates for cracking are higher for commercially pure aluminum and smaller for steels than for aluminum alloys, in agreement with practical experience. Another interesting observation is the occurrence of cracking at negative strain rates for alloy 6060, an alloy known to be very sensitive to solidification cracking when welded autogenously. In this case, the strain was measured with an extensometer spanned across the weld, and negative values indicate inward movement of material to feed shrinkage. Critical strain rate for cracking can be taken as a fundamental quantitative measurement of weldability, with higher rates corresponding to improved weldability.

## Summary

Many weldability tests have been developed to rank cracking susceptibility of different alloys based solely upon relative comparisons of crack length. Although useful in making rough-cut rankings, a better approach is to consider weldability testing as an experimental means to identify critical conditions for solidification crack initiation and growth. Mechanisms for both crack initiation and growth show strain rate and brittle (coherent) temperature range to be controlling factors, backed up by considerable experimental evidence. With advancing ideas regarding cracking mechanisms and improved testing methods, welding science is moving towards a point where the likelihood of cracking may someday be predicted for a given alloy and welding application.

**Table 4** Critical strain and deformation rates required for weld solidification crack formation

| Material (base metal + filler metal) | Welding conditions | | | Weldability test[a] | Critical strain rate (% $s^{-1}$) | Critical deformation rate (mm $s^{-1}$) | Reference |
|---|---|---|---|---|---|---|---|
| | Voltage (V) | Current (A) | Welding speed (mm $s^{-1}$) | | | | |
| Aluminum alloy 1050 | 25 | 200 | 2.5 | TVT | 7.5 | – | [59] |
| Aluminum alloy 1070 | – | – | 1.7 | SB-TVT | 5.00 | – | [44] |
| Aluminum alloy 1100 + 1070 | – | – | 4 | VDRT | – | 0.15 | [46] |
| Aluminum alloy 1100 + 1070 | – | – | 6.7 | VDRT | – | 0.25 | [46] |
| Aluminum alloy 1100 + 1070 | – | – | 10 | VDRT | – | 0.40 | [46] |
| Aluminum alloy 1100 + 1070 | – | – | 13.3 | VDRT | – | 0.50 | [46] |
| Aluminum alloy 2017 | 18 | 230 | 1.7 | SB-TVT | 0.20 | – | [53] |
| Aluminum alloy 2017 | – | – | 1.7 | SB-TVT | 0.15 | – | [44] |
| Aluminum alloy 2024 | 25 | 200 | 2.5 | TVT | 0.72 | – | [59] |
| Aluminum alloy 2219 | – | – | 1.7 | SB-TVT | 0.50 | – | [44] |
| Aluminum alloy 5052 | – | – | 5 | VTST | 0.15 | – | [41] |
| Aluminum alloy 5052 | – | – | 1.7 | SB-TVT | 0.64 | – | [44] |
| Aluminum alloy 5052 + 1070 | – | – | 6.7 | VDRT | – | 0.05 | [46] |
| Aluminum alloy 5083 | 25 | 200 | 2.5 | TVT | 0.15 | – | [59] |
| Aluminum alloy 5083 | 18 | 230 | 1.7 | SB-TVT | 0.35 | – | [86] |
| Aluminum alloy 5083 | – | – | 5 | VTST | 0.20 | | [41] |
| Aluminum alloy 5083 | – | – | 1.7 | SB-TVT | 0.47 | – | [44] |
| | – | – | 6.7 | VDRT | – | 0.18 | [87] |

(continued)

**Table 4** (continued)

| Material (base metal + filler metal) | Welding conditions | | | Weldability test[a] | Critical strain rate (% $s^{-1}$) | Critical deformation rate (mm $s^{-1}$) | Reference |
|---|---|---|---|---|---|---|---|
| | Voltage (V) | Current (A) | Welding speed (mm $s^{-1}$) | | | | |
| Aluminum alloy 5083A + 5083B | | | | | | | |
| Aluminum alloy 5154 | – | – | 1.7 | SB-TVT | 0.70 | – | [44] |
| Aluminum alloy 6060 | 17.8 | 110 | 4 | CTW | −0.06 | −0.006 | [49] |
| Aluminum alloy 6060 + 5 % 4043 | 17.8 | 110 | 4 | CTW | 0.06 | 0.006 | [49] |
| Aluminum alloy 6060 + 9 % 4043 | 17.8 | 110 | 4 | CTW | 0.17 | 0.017 | [49] |
| Aluminum alloy 6060 + 11 % 4043 | 17.8 | 110 | 4 | CTW | 0.22 | 0.023 | [49] |
| Aluminum alloy 6060 + 14 % 4043 | 17.8 | 110 | 4 | CTW | 0.30 | 0.032 | [49] |
| Aluminum alloy 6060 + 16 % 4043 | 17.8 | 110 | 4 | CTW | 0.35 | 0.037 | [49] |
| Austenitic stainless steel AISI 304 | 18 | 250 | 2.5 | TVT | 6.5 | – | [59] |
| Stainless steel SUS 304L | 12 | 70 | 0.83 | THCT | 5.4 | – | [63] |
| Austenitic stainless steel SUS 304L | 18.5 | 65 | 0.17 | THCT | – | – | [60] |
| Austenitic stainless steel AISI 310 | 18 | 250 | 2.5 | TVT | 0.21 | – | [59] |
| Stainless steel SUS 310S | 12 | 70 | 0.83 | THCT | 1.2 | – | [63] |
| Austenitic stainless steel SUS 310S | 18.5 | 65 | 0.17 | THCT | – | – | [60] |

(continued)

**Table 4** (continued)

| Material (base metal + filler metal) | Welding conditions | | | Weldability test[a] | Critical strain rate (% $s^{-1}$) | Critical deformation rate (mm $s^{-1}$) | Reference |
|---|---|---|---|---|---|---|---|
| | Voltage (V) | Current (A) | Welding speed (mm $s^{-1}$) | | | | |
| Austenitic stainless steel SUS 310S | 12 | 100 | 2.5 | SB-TVT | 0.90 | – | [86] |
| Austenitic stainless steel SUS 316 | 18.5 | 65 | 0.17 | THCT | – | – | [60] |
| Austenitic stainless steel AISI 316 | 18 | 250 | 2.5 | TVT | 0.40 | – | [59] |
| Austenitic stainless steel SUS 321 | 18.5 | 65 | 0.17 | THCT | – | – | [60] |
| Austenitic stainless steel AISI 321 | 18 | 250 | 2.5 | TVT | 1.2 | – | [59] |
| Stainless steel SUS 430 | 12 | 70 | 0.83 | THCT | 3.2 | – | [63] |
| Plain carbon steel SS41 | 18 | 250 | 2.5 | TVT | 0.81 | – | [59] |
| High carbon steel S55C | 19 | 250 | 2.5 | TVT | 0.72 | – | [59] |
| High tensile strain HT-80 | 19 | 250 | 2.5 | TVT | 3.4 | – | [59] |
| Inconel alloy inconel 600 | 12 | 70 | 0.83 | THCT | 1.2 | – | [63] |
| Inconel alloy inconel 600 | 18.5 | 65 | 0.17 | THCT | – | – | [60] |

[a] *TVT* transvarestraint test, *SB-TVT* slow bending trans-varestraint test, *THCT* tensile hot cracking test, *CTW* controlled tensile weldability test, *VTST* variable tensile strain test, *VDRT* variable deformation rate test

**Table 5** Checklist to be considered for each weldability test procedure

| Category | Items |
|---|---|
| Alloy | Base metal—filler metal—dilution |
| Temperature | Liquidus–solidus—coherent |
| Crack type | Solidification—liquation—ductility dip |
| Test type | Intrinsic—extrinsic—bending—tensile |
| Load orientation | Longitudinal—transverse |
| Restraint | Coupon dimensions |
| Welding process | Pool shape—cooling rate |
| Welding parameters | Current—voltage—travel speed |
| Linear heat input | Constant penetration—constant heat input |
| Applied strain | Maximum strain—ram speed—strain rate |
| Local strain | Maximum strain—strain rate |
| Solidification temperature range | BTR—SCTR |
| Crack measurement | MCL—TCL—MCD |

This paper has examined numerous weldability test approaches in order to identify important distinctions. Standardization of not only weldability test procedures, but also defining what needs to be measured will help in the long term understanding of alloy weldability. Table 5 is a proposed checklist of all the things one should consider and quantify, if possible, when performing weldability tests.

## References

1. Lippold JC (2005) Recent developments in weldability testing. In: Hot Cracking Phenomena in Welds, Springer: 271-290.
2. Heuser H (2005) Value of different hot cracking tests for the manufacturer of filler metals. In: Hot Cracking Phenomena in Welds, Springer: 305-327.
3. Farrar JCM (2005) Hot cracking tests – The route to international standardization. In: Hot Cracking Phenomena in Welds, Springer: 291-304.
4. Finton TL, Lippold JC, Bowers RJ (1999) Comparison of weld hot cracking tests – summary of an IIW round Robin study. IIW Doc IX-H-459-99 (II-C-175-99).
5. Mousavi MG, Cross CE, Grong Ø (1999) Effect of scandium and titanium- boron on grain refinement and hot cracking of aluminum alloy 7108. Sci. Tech. Weld. Joining 4 (6): 381-388.
6. Dudas JH, Collins FR (1966) Preventing Weld Cracks in High-Strength Aluminum Alloys. Welding Journal 45 (6): 241s-249s.
7. Coniglio N, Cross CE (2008) Weld parameter and minor element effects on solidification crack initiation in aluminum. In: Hot Cracking Phenomena in Welds II, Springer: 277-310.
8. Warrington D, McCartney DG (1989) Development of a new hot-cracking test for aluminum alloys. Cast Metals 2 (3): 134-143.
9. Dvornak MJ, Frost RH, Olson DL (1989) The weldability and grain refinement of Al-2.2Li-2.7Cu. Welding Journal 68 (8): 327s-337s.
10. Lippold JC and Kotecki DJ (2005) Welding metallurgy and weldability of stainless steels, Wiley (ISBN-13: 978-0471473794).

11. Li L, Messler RW (1999) The effects of phosphorus and sulfur on susceptibility to weld hot cracking in austenitic stainless steels. Welding Journal 78: 387s–396s.
12. Brooks JA, Thompson AW and Williams JC (1984) A fundamental study of the beneficial effects of delta ferrite in reducing weld cracking. Welding Journal 63: 71s–83s.
13. Kujanpää V, Suutala N, Takalo T and Moiso T (1979) Correlation between solidification cracking and mirostructure in austenitic and austenitic-ferritic stainless steel welds. Welding Research International 9: 55–76.
14. Suutala N, Takalo T and Moisio T (1979) Relationship between solidification and microstructure in austenitic and austenitic-ferritic stainless steel welds. Metallurgical Transactions 10A: 512–514.
15. Hull FC (1967) Effect of delta ferrite on the hot cracking of stainless steel. Welding Journal 46: 399s–409s.
16. Jennings PH, Singer ARE, Pumphrey WI (1948) Hot-shortness of some high-purity alloys in the systems aluminum-copper-silicon and aluminum-magnesium-silicon. J. Inst. Metals 74: 227-248.
17. Singer ARE, Jennings PH (1947) Hot-shortness of some aluminum-iron-silicon alloys of high purity. J. Inst. of Metals 73: 273–284.
18. Singer ARE, Jennings PH (1947) Hot shortness of the aluminum-silicon alloys of commercial purity. J. Inst. of Metals 73: 197-212.
19. Verö J (1936) The hot-shortness of aluminum alloys. The Metal Industry 48: 431-494.
20. Novikov II, Grushko OE (1995) Hot cracking susceptibility of Al-Cu-Li and Al-Cu-Li-Mn alloys. Mat. Sci. Techn. 11: 926-932.
21. Decultieux F, Vicente-Hernandez P, Levaillant C (1993) Hot tearing test: experiments and FEM modelling. In: Modeling of Casting, Welding, and Advanced Solidification Processes VI, Warrendale TMS: 617-624.
22. Hull FC (1959) Cast-pin tear test for susceptibility to hot cracking. Welding Journal 38 (4): 176s-181s.
23. Alexandrov BT, Lippold JC (2010) Further development of the cast pin tear test for evaluating solidification cracking in welds. In: Hot Cracking Phenomena in Welds III, Springer 317-331.
24. Alexandrov BT, Lippold JC, Nissley NE (2007) Evaluation of weld solidification cracking in Ni-base superalloys using the cast pin tear test. In: Hot Cracking Phenomena in Welds II, Springer 193-213.
25. Borland JC, Rogerson JH (1962) Examination of the patch test for assessing hot cracking tendencies of weld metal. British Welding Journal 9: 464-498.
26. Nelson TW, Lippold JC, Lin W, Baeslack III WA (1997) Evaluation of the circular patch test for assessing weld solidification cracking, part I – Development of a test method. Welding Journal 76 (3): 110s-119s.
27. Dowd JD (1952) Weld cracking of aluminum alloys. Welding Journal 31 (10): 448s-456s.
28. Stout RD, Tor SS, McGeady LJ, Doan GE (1946) Quantitative measurement of cracking tendency in welds. Welding Journal 25 (9): 522s-532s.
29. Houldcroft PT (1955) A simple cracking test for use with argon-arc welding. British Welding Journal 2: 471-475.
30. Matsuda F, Nakata K, Harada S (1980) Moving characteristics of weld edges during solidification in relation to solidification cracking in GTA weld of aluminum alloy thin sheet. Trans. of JWRI 9 (2): 83-93.
31. Matsuda F, Nakata K, Tsukamoto K, Kohzoh A (1983) Effect of additional alloying element on weld solidification crack susceptibility of Al-Zn-Mg alloy. Trans. of JWRI 12 (2): 93-102.
32. Matsuda F, Nakata K (1982) A new test specimen for self-restraint solidification crack susceptibility test of electron-beam welding bead - fan shaped cracking test. Trans. of JWRI 11 (2): 87-94.
33. Ploshikhin V, Prikhodovsky A, Makhutin M, Ilin A, Zoch HW (2005) Integrated mechanical-metallurgical approach to modeling of solidification cracking in welds. In: Hot Cracking Phenomena in Welds, Springer 223-244.

34. Cross CE, Böllinghaus Th (2006) The effect of restraint on weld solidification cracking in aluminum. Welding in the World 50: 51-54.
35. Goodwin GM (1987) Development of a new hot-cracking test: the Sigmajig. Welding Journal 66 (2): 33s-38s.
36. Nishimoto K, Mori H (2004) Hot cracking susceptibility in laser weld metal of high nitrogen stainless steels. Sci. Tech. Adv. Mat. 5: 231-240.
37. Savage WF, Lundin CD (1965) The varestraint test. Welding Journal 44 (10): 433 s-442 s.
38. Bailey N, Regelous GR (1976) Recent developments in transvarestraint testing. Weld. Inst. Res. Bul. 201-204.
39. Turner PW, Lundin CD (1970) Effect of iron on the hot cracking of uranium weld metal – part 1. Welding Journal 49 (12): 579s-587s.
40. Wilken K, Kleistner H (1982) The MVT-test - A new universal procedure for testing the hot cracking sensitivity during welding (in German). Material und Technik 1: 3-10.
41. Tamura H, Kato N, Ochiai S, Katagiri Y (1977) Cracking study of aluminum alloys by the variable tensile strain hot cracking test. Trans. JWS 8 (2): 16-22.
42. Matsuda F, Nakagawa H, Nakata K, Kohmoto H, Honda Y (1983) Quantitative evaluation of solidification brittleness of weld metal during solidification by means of in-situ observation and measurement (Report I) - Development of the miso technique. Trans. of JWRI 12 (1): 65-72.
43. Arata Y, Matsuda F, Nakata K, Katayama S (1977) Solidification crack susceptibility in weld metals of fully austenitic steels (Report II) – Effect of ferrite, P, S, C, Si, and Mn on ductility properties of solidification brittleness. Trans. of JWRI 6 (1): 105-117.
44. Arata Y, Matsuda F, Nakata K, Shinozaki K (1977) Solidification crack susceptibility of aluminum alloy weld metals (Report II) - Effect of straining rate on cracking threshold in weld metal during solidification. Trans. of JWRI 6 (1): 91-104.
45. Jones PW (1957) A summary of recent work on the murex hot cracking test. British Welding Journal 4: 189-197.
46. Matsuda F, Nakagawa H, Nakata K, Okada H (1979) The VDR cracking test for solidification crack susceptibility on weld metals and its application to aluminum alloys. Trans. of JWRI 8: 85-95.
47. Herold H, Streitenberger M, Pchennikov A (2000) Prevention of centerline solidification cracking during one side welding. IIW-Doc. IX-2000-01, II-C-220-01.
48. Schobbert H, Böllinghaus Th, Wolf M (2003) Hot cracking resistance of laser and hybrid welded austenitic stainless steels. In: Trends in Welding Research, ASM: 76-81.
49. Coniglio N, Cross CE, Michael Th, Lammers M (2008) Defining a critical weld dilution to avoid solidification cracking in aluminum. Welding Journal 87 (8): 237s-247s.
50. Cross CE, Coniglio N, Westing EM, Gumenyuk A (2011) Laser weldability of stainless steel. In: Hot Cracking Phenomena in Welds III, Springer 131-144.
51. Coniglio N, Patry M (2013) Measuring laser weldability of aluminum alloys using controlled restraint weldability test. Sci. Technol. Weld. Join. 18 (7): 573-580. (STW2545)
52. Robino CV, Reece M, Knorovsky GA, DuPont JN, Feng Z (2005) Prediction of maximum crack length in longitudinal varestraint testing. In: Proc. 7th Int. Conf. Trends in Welding Research, ASM Int: 313-318.
53. Arata Y, Matsuda F, Nakata K, Shinozaki K (1977) Solidification crack susceptibility of aluminum alloy weld metals (Report III) - Effect of straining rate on crack length in weld metal. Trans. of JWRI 6 (2): 47-52.
54. Coniglio N (2008) Aluminum alloy weldability: identification of weld solidification cracking mechanisms through novel experimental technique and model development. Doctoral Thesis, Otto-von-Guericke University, Magdeburg, Germany. ISBN 978-3-9812354-3-2.
55. Saveiko VN (1961) Theory for hot cracking. Russ. Cast. Prod.: 453-456.
56. Lahaie DJ, Bouchard M (2001) Physical modeling of the deformation mechanisms of semisolid bodies and a mechanical criterion for hot tearing. Met. Mat. Trans. 32B (8): 697-705.
57. Pellini WS (1952) Strain theory for hot tearing. Foundry 80: 125-199.

58. Prokhorov NN (1956) The problem of the strength of metals while solidifying during welding. Svar. Proiz. 6: 5-11.
59. Senda T, Matsuda F, Takano G, Watanabe K, Kobayeshi T, Matsuzaka T (1971) Experimental investigations on solidification crack susceptibility for weld metals with transvarestraint tests. Trans. JWS 2 (2): 141-162.
60. Matsuda F, Nakagawa K, Kohmoto H, Honda Y, Matsubara Y (1983) Quantitative evaluation of solidification brittleness of weld metal during solidification by in-situ observation and measurement (Report II) – Solidification ductility curves for steels with the miso technique. Trans. of JWRI 12 (1): 73-80.
61. Matsuda F, Nakagawa H, Sorada K (1982) Dynamic observation of solidification and solidification cracking during welding with optical microscope (I) – solidification front and behavior of cracking. Trans. JWRI 11 (2): 67–77.
62. Pumphrey WI, Lyons JV (1948) Cracking during the casting and welding of the more common binary aluminum alloys. J. Inst. Metals 74: 439-454.
63. Borland JC (1960) Generalized theory of super solidus cracking in welds and castings – an initial development. British Welding Journal 7 (8): 508-512.
64. Coniglio N, Cross CE (2013) Initiation and growth mechanisms for weld solidification cracking. Int. Mat. Rev. 58 (7): 375-397.
65. Rappaz M, Drezet JM, Gremaud M (1999) A new hot tearing criterion. Met. Mat. Trans. 30A: 449-455.
66. Coniglio N, Cross CE (2009) Mechanisms for solidification crack initiation and growth in aluminum welding. Met. Mat. Trans. 40A: 2718-2728.
67. Matsuda F, Nakagawa H, Tomita S (1986) Quantitative evaluation of solidification brittleness of weld metal during solidification by in-situ observation and measurement (Report III) – Effect of strain rate on minimum ductility required for solidification crack initiation in carbon steels, stainless steels, and inconel alloy. Trans. of JWRI 15 (2): 297-305.
68. Kannengiesser Th, McInerney T, Florian W, Böllinghaus Th, Cross CE (2002) The influence of local weld deformation on hot cracking susceptibility. In: Mathematical Modeling of Weld Phenomena 6, Maney 803-817.
69. Herold H, Streitenberger M, Pchennikov A (2001) Modelling of the PVR Test to examine the origin of different hot cracking types. In: Mathematical Modelling of Weld Phenomena 5, IOM, London 783-792.
70. Chihoski RA (1972) The character of stress field around a weld arc moving on aluminum sheet. Welding Journal 51 (1): 9s-18s.
71. Johnson L (1973) Formation of plastic strains during welding of aluminum alloys. Welding Journal 52 (7): 298s-305s.
72. Johnson L (1974) Moire techniques for measuring strains during welding. Exp. Mech. 14: 145-150.
73. Peters WH, Ranson WF (1982) Digital imaging techniques in experimental stress analysis. Optical Engineering 21 (3): 427-432.
74. Sutton MA, Wolters WJ, Peters WH, Ranson WF, McNeil SR (1983) Determination of displacements using an improved digital correlation method. Image and Vision Computing 1 (3): 133-139.
75. Chihoski RA (1979) Expansion and stress around aluminum weld puddles. Welding Journal 58 (9): 263s-276s.
76. Hunziker O, Dye D, Roberts SM, Reed RC (2005) A coupled approach for the prediction of solidification cracking during the welding of superalloys. In: Hot Cracking Phenomena in Welds, Springer 299-319.
77. Shibahara M, Serizawa H, Murakawa M (2005) Finite element method for hot cracking analysis using temperature dependent interface element. In: Hot Cracking Phenomena in Welds, Springer 253-267.
78. Cicala E, Duffet G, Andrzejewski H, Grevey D, Ignat S (2005) Hot cracking in Al-Mg-Si alloy laser welding - Operating parameters and their effects. Mat. Sci. Eng. 395A: 1-9.

79. Savage WF, Aronson AH (1966) Preferred orientation in the weld fusion zone. Welding Journal 45: 85s-89s.
80. Schempp P, Cross CE, Pittner P, Oder G, Neumann RS, Rooch H, Dörfel I, Österle W, Rethmeier M (2014) Solidification of GTA aluminum weld metal: part I- grain morphology dependent upon alloy compostion and grain refiner content. Welding Journal 93: 53s-59s.
81. Blodgett WO (1966) Design of welded structures. The James F. Lincoln Arc Welding Foundation.
82. Dahle AK, Arnberg L (1997) Development of strength in solidifying aluminium alloys. Acta Mater. 45 (2): 547-559.
83. Gallagher ML, Lippold J (2011) Weld cracking susceptibility of alloy C22 weld metal. In: Hot Cracking Phenomena in Welds III, Springer 367-391.
84. Sowards JW, Alexandrov BT, Lippold J, Frankel JS (2011) Weldability of a new Ni-Cu welding consumable for joining austenitic stainless steels. In: Hot Cracking Phenomena in Welds III, Springer 393-413.
85. Matsuda (1990) Hot crack susceptibility of weld metal. In: Advances in welding metallurgy, American Welding Society/Japan Welding Society/Japan Welding Engineering Society: 19–35.
86. Arata Y, Matsuda F, Katayama S, and Ogata S (1977) Solidification crack susceptibility in welds metals of fully austenitic stainless steels (Report III) – Effect of strain rate on cracking threshold in weld metal during solidification. Trans. of JWRI 6 (2): 197-207.
87. Hatami N, Babaei R, DadashzadehM, Davami P (2008) Modeling of hot tearing formation during welding. J. Mater. Process. Technol. 205: 506–513.

# Use of Computational and Experimental Techniques to Predict Susceptibility to Weld Cracking

A.T. Hope and J.C. Lippold

**Abstract** A range of computational and experimental techniques has been employed to determine relative susceptibility to various types of weld metal cracking. These techniques can be valuable tools that facilitate both filler metal selection and filler metal development. A combination of thermodynamic and kinetic modeling is employed alongside experimental validation. In some cases, a design of experiment can be used to reduce the number of experiments and optimize a filler metal composition. Two examples will be given; one regarding solidification cracking in Ni-base alloys, and the other on local brittle zone formation in dissimilar metal overlay of a Ni-base alloy on carbon steel. High-chromium, Ni-base filler metals are used for construction and repair applications of nuclear power plants based on their exceptional corrosion resistance. Niobium is added to these alloys to improve resistance to ductility dip cracking by formation of NbC. Additions of Nb also cause low melting point eutectics to form at the end of solidification, which increases susceptibility to solidification cracking. Alternative carbide formers have been investigated using a design of experiment methodology along with ThermoCalc-Scheil simulations to determine the potential for solidification cracking based on the magnitude of the solidification temperature range. Compositions were optimized and then verified using a combination of button melting and small scale weldability testing. Dissimilar metal welds are widely used in the petrochemical industry to improve corrosion resistance and facilitate field fabrication of welded structures. For some metal combinations, such as Ni-base alloys to steels, service failure can occur at the fusion boundary. This failure is related to a brittle zone that forms due to carbon diffusion from the steel towards the interface during postweld heat treatment (PWHT). Carbon diffusion during PWHT has been modeled using DICTRA®. Hypothetical alloy combinations have been

---

A.T. Hope (✉) · J.C. Lippold
The Ohio State University, Ohio, USA
e-mail: hope.30@buckeyemail.osu.edu



Th. Boellinghaus et al. (eds.), *Cracking Phenomena in Welds IV*,
DOI 10.1007/978-3-319-28434-7_4

simulated with this model in order to demonstrate the influence of carbon content and PWHT conditions. Examples will be given of dissimilar combinations that reduce the potential for carbon migration during PWHT that avoid brittle zone formation at the interface.

## Introduction

Computational techniques are frequently used to explore effects of alloy content on various thermodynamic responses. These techniques, combined with experimental validation and proper design of experiment, can enable one to choose or optimize alloys to resist various weld related cracking mechanisms. In this paper, two such examples will be detailed. The first example deals with optimizing a high-chromium Ni-base alloy for resistance to solidification and ductility-dip cracking. To accomplish this, a combination of techniques was employed. The Scheil module within Thermo-Calc was used to predict solidification temperature ranges and relative amounts of eutectic phase formation. These data were verified using button melting techniques, Single Sensor Differential Thermal Analysis (SS-DTA), and metallographic image analysis. To optimize a composition, a design of experiment was employed to determine empirical relationships between alloy content and the responses. The second example involves failures at dissimilar metal weld (DMW) fusion boundaries. These failures are related to carbon diffusion during PWHT. A carbon diffusion model was set up using DICTRA® and different base metal and weld metal compositions were simulated. Carbon concentration profiles were verified using electron probe microanalysis.

## High-Cr, Ni-Base Filler Metal Development

### *Background*

The nuclear power industry uses high-Cr, Ni-base alloys for construction and repair in applications where resistance to primary water stress corrosion cracking (PWSCC) is needed [1]. Alloys such as FM52 and FM52M (ERNiCrFe-7/7A) are being used as replacements for FM82 (ERNiCr-3), due to their superior PWSCC resistance [2]. However, FM52 and FM52M have been found to be susceptible to ductility-dip cracking (DDC) in some multipass weld and overlay applications [3].

Multiple theories exist for the mechanism of DDC [4–9]. DDC is known to be an intermediate temperature embrittlement phenomenon. Cracking occurs due to a drop in ductility between 0.5 $T_{solidus}$ and the solidus temperature. Yamaguchi proposed that sulfur segregates to grain boundaries and embrittles the boundaries at these temperatures. Nishimoto determined that rare earth elements can scavenge sulfur and reduce grain boundary embrittlement. Ramirez and Lippold proposed that DDC is a grain boundary creep phenomena. They found that in alloys which form more eutectic at the end of solidification, grain boundaries will be tortuous (non-straight), and susceptibility can be reduced. A schematic of this mechanism is shown in Fig. 1. The grain boundary tortuosity mechanism to reduce DDC in susceptibility will be used in the work presented here.

More recently, alloys such as FM52MSS [10] (ERNiCrFe-13) have been developed which have higher Nb content. This higher Nb content causes more Nb-rich eutectic to form at the end of solidification. This alloy has shown excellent resistance to DDC [9].

While increases in Nb help with DDC resistance, they can be detrimental to solidification cracking resistance [11]. A generally accepted theory for solidification cracking involves the presence of both residual stress and liquid films at the end of solidification [12, 13]. When residual stresses, which build up from thermal shrinkage, overcome the strength of liquid films, cracks will form. There are many metallurgical factors that influence crack susceptibility. Solidification temperature range (STR) is important, because it is related to the width of the mushy zone,

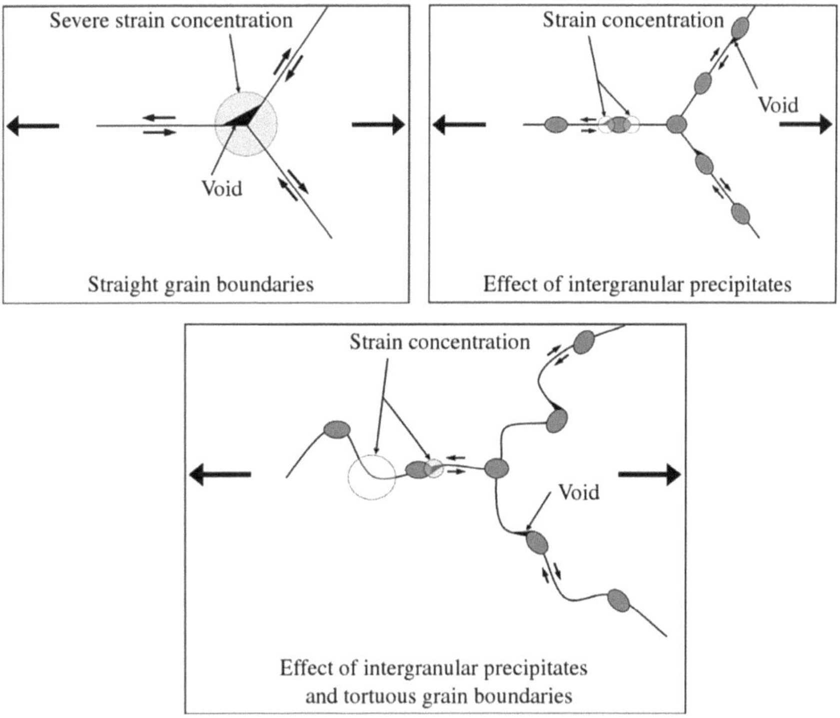


**Fig. 1** Proposed mechanism by Ramirez and Lippold showing the effect of precipitates and grain boundary tortuosity on DDC resistance [9]

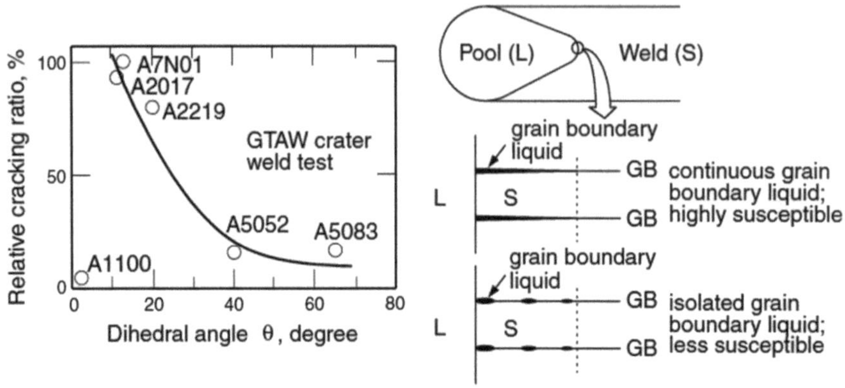


**Fig. 2** *Left* effect of grain boundary liquid wetting on solidification crack susceptibility [14]. *Right* schematic showing the mechanism [15]

which is the region where solidification cracks will form in a weld. The morphology of the liquid present at the end of solidification is also important [14]. Nakata and Matsuda investigated the effect of wetting angle of the final liquid to solidify on cracking resistance. They found when liquid wets out along grain boundaries, cracking was more severe, as shown in Fig. 2, where low dihedral angles represent spreading of the liquid film along the boundary. Metallurgical factors such as impurity concentration or eutectic formation are important, as they can affect both the STR and the wetting characteristics of the final liquid to solidify.

Specifically for Ni-base alloys, there are a few key factors in determining solidification crack susceptibility. DuPont et al. [16] have correlated the solidification temperature range (STR) to crack susceptibility using the Transvarestraint test. The Transvarestraint test is a simple test that bends a sample around a die block while a weld is being made transverse to the loading direction. The amount of strain can be varied and solidification cracks may form in the mushy zone if the strain is high enough. They found alloys with higher STRs had higher crack susceptibility, as shown in Fig. 3. Dupont has also investigated metallurgical factors that influenced the solidification microstructure in Nb bearing Ni-base superalloys [17, 18]. Increasing Nb content caused more eutectic constituent to form. This eutectic can either be γ/NbC or γ/Laves, which starts to form around 1350, and 1250 °C, respectively. Increasing C caused the eutectic to be mostly γ/NbC. Increasing Fe and Si promoted formation of γ/Laves at the expense of γ/NbC. This difference in eutectic formation can also be seen in Fig. 3, where two distinct groups of alloys can be observed. The alloys with STR around 75 °C only formed γ/NbC eutectic, whereas the alloys with STR around 175 °C formed some combination of both eutectics. Thus, it can be concluded that the formation of γ/Laves eutectic is detrimental to solidification cracking susceptibility because it expands the STR by about 100 °C when it forms.

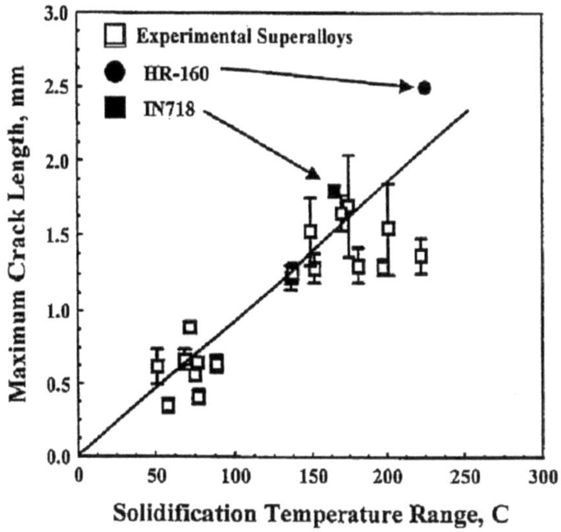


**Fig. 3** Relationship between STR determined by DTA and maximum crack length (*MCL*) in the transvarestraint test [16]

Since γ/Laves eutectic needs to be avoided, it is useful to know what alloying elements play a role in promoting its formation. DuPont established that Fe and Si promote the formation of Laves phase [18]. Work done by Chang found that similar to Fe, Cr promotes Laves phase on solidification [19]. These factors may explain the relatively good DDC resistance of FM82 versus FM52MSS. Both of these alloys have around 2.5 wt% Nb. FM82 is fairly resistant to DDC most likely due to the formation of a fair amount of γ/NbC eutectic [20]. FM52MSS is very resistant to DDC because of formation of large amounts of γ/NbC and γ/Laves, but is quite susceptible to solidification cracking [11]. The extra (30 vs. 20 wt%) Cr in FM52MSS promotes γ/Laves which expands the STR and increases the cracking susceptibility.

Since increases in Fe content will promote γ/Laves, this can also present a problem with dilution. DuPont also found Fe decreased the partition coefficient (k) of Nb, a measure of the segregation that occurs during solidification [17]. This decrease causes Nb to segregate more strongly to the solidification grain boundaries, increasing the amount of eutectic constituent that forms. Studies have been performed by Hope that directly measured the crack susceptibility as a function of dilution [21]. A direct correlation between dilution and cracking susceptibility was observed.

Baeslack et al. [22] investigated substituting tantalum for niobium in Alloy 718 weld metal. They performed spot-Varestraint and mini-Varestraint tests on heats of modified and unmodified 718. Gleeble heat affected zone (HAZ) tests were also performed to determine the liquation temperature. The authors found improvement in liquation and solidification cracking for the alloys with Ta instead of Nb, shown in Fig. 4. They attributed this to the higher liquation temperature of the Ta rich γ/

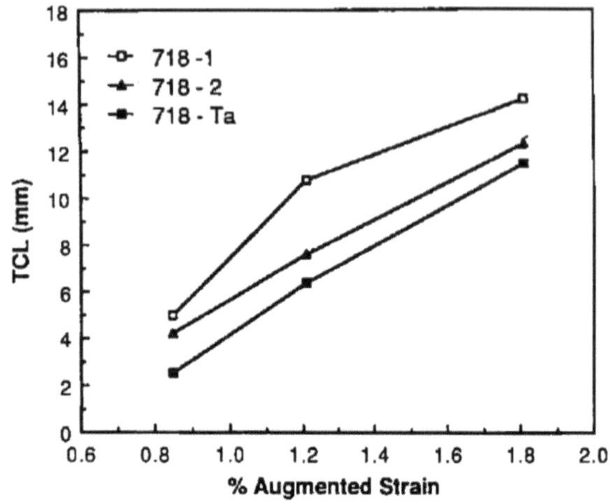


**Fig. 4** Total crack length versus augmented strain in the mini-varestraint test for standard 718 and Ta-modified 718 alloys [22]

Laves eutectic relative to the Nb rich γ/Laves eutectic. The liquation temperatures of the Laves phases were 1225 and 1175 °C, respectively.

Hafnium is another possible replacement for Nb in high-Cr, Ni- base alloys. Historically, Hf has been added to cast superalloys to modify the grain boundary carbide morphology [23]. MC carbides form upon solidification and are generally Hf rich [24]. A study by Dmitrieva et al. found the γ/HfC eutectic to solidify at 1330 °C [25]. Furthermore, a study by Nash and West found no evidence of Laves phase in a ternary Ni–Cr–Hf system [26]. Nash and West melted samples of varying composition and determined the solidus temperature, as well as phases present by XRD. They found two eutectic phases that could solidify, shown in Fig. 5. The first

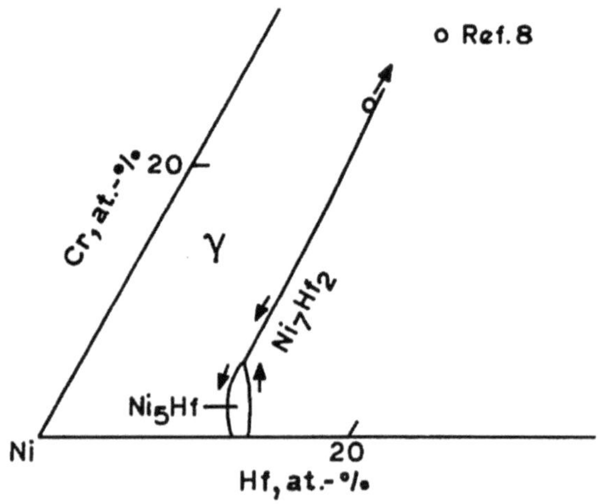


**Fig. 5** Schematic representation of primary solidification regions in Ni-rich portion of Ni–Cr–Hf ternary system [26]

was the binary Ni–Hf eutectic, $Ni_5Hf$, which solidified at 1227 °C. The second, $Ni_7Hf_2$, formed when 30 at.% Cr was present, and also formed at a higher temperature of 1288 °C. This is promising because when carbon is exhausted on solidification, other eutectic phases may form, such as Laves in systems containing Nb or Ta. The $Ni_7Hf_2$ phase which could form when Hf is present and C has been exhausted has a much higher eutectic temperature than both the Nb or Ta rich Laves eutectics. This is expected to improve solidification cracking resistance over Nb or Ta eutectics by narrowing the solidification temperature range.

## *Experimental*

To optimize an alloy composition for further weldability testing, a combination of thermodynamic modeling and experimental testing was used. A design of experiment approach reduced the number of experiments and provided a way to optimize a composition based on chosen outputs. A flow chart is shown in Fig. 6 that outlines the steps taken; these will be described herein. Details regarding the entire procedure has been published elsewhere [27].

To develop a DOE to use for computational Scheil analysis, composition ranges were chosen to cover a range of interest for each alloying element. For example, work done by DuPont showed Fe increased segregation of Nb during solidification [17]. Therefore, Fe was varied from 0–20 to gain some insight into how Fe may affect solidification behavior in these systems. The ranges chosen are shown in Table 1.

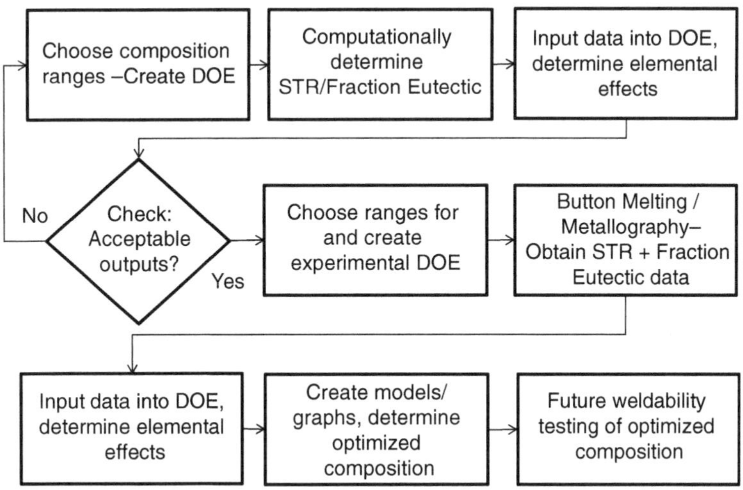


**Fig. 6** Flow chart describing steps taken for alloy composition optimization

**Table 1** Composition ranges used in computational design of experiment

| DOE alloy group | Ni | Cr | Fe | Mo | Hf | Nb | C |
|---|---|---|---|---|---|---|---|
| 1 – Ni–Cr–Fe–Mo–Hf–C | Bal. | 30 | 0–20 | 0–4 | 0–2 | – | 0–0.04 |
| 3 – Ni–Cr–Fe–Mo–Nb–C | Bal. | 30 | 0–20 | 0–4 | – | 0–3 | 0–0.04 |

Values in wt%

These ranges are then input into Design-Expert, a DOE software package, and an appropriate model is chosen. In this case the response surface method was chosen as it allows exploration of response curvature and interactions between elements. The DOE software then generates a number of actual compositions to test.

These compositions are then entered into the Scheil module within Thermo-Calc. The TCNi5 database was used for simulations, as it had phase information for the relevant Hf-rich phases. The Scheil simulation predicts the maximum segregation possible upon solidification, and the corresponding solidification temperature range (STR). The simulation also outputs the mole fraction of eutectic phases that form. These outputs, STR and mole fraction eutectic, were then entered into the DOE software for analysis.

Based on the results from the computational DOE, a refined set of compositions was chosen for further study. A new DOE was constructed according to Table 2. Small charges or buttons of material were cast using high purity metals for each of these compositions using a GTA button melter. These buttons were then subjected to a thermocouple plunging experiment, shown in Fig. 7. In this test, a GTAW spot

**Table 2** Nominal compositions (wt%) of actual buttons cast for analysis

| Alloy | Ni | Cr | Fe | Mo | C | X[a] |
|---|---|---|---|---|---|---|
| X[a]-1,2 | Bal. | 30 | 8 | 0 | 0.02 | 2.0 |
| X-3,4 | Bal. | 30 | 8 | 0 | 0.06 | 0.5 |
| X-5,6 | Bal. | 30 | 8 | 4 | 0.02 | 1.25 |
| X-7,8 | Bal. | 30 | 8 | 4 | 0.04 | 0.5 |
| X-9,10 | Bal. | 30 | 8 | 4 | 0.06 | 2.0 |
| X-11,12 | Bal. | 30 | 13 | 0 | 0.02 | 0.5 |
| X-13,14 | Bal. | 30 | 13 | 0 | 0.06 | 2.0 |
| X-15,16 | Bal. | 30 | 13 | 2 | 0.04 | 1.25 |
| X-17,18 | Bal. | 30 | 13 | 4 | 0.02 | 2.0 |
| X-19,20 | Bal. | 30 | 13 | 4 | 0.06 | 0.5 |
| X-21,22 | Bal. | 30 | 18 | 0 | 0.02 | 2.0 |
| X-23,24 | Bal. | 30 | 18 | 0 | 0.06 | 0.5 |
| X-25,26 | Bal. | 30 | 18 | 2 | 0.06 | 2.0 |
| X-27,28 | Bal. | 30 | 18 | 4 | 0.02 | 0.5 |
| X-29,30 | Bal. | 30 | 18 | 4 | 0.04 | 2.0 |
| X-31,32 | Bal. | 30 | 18 | 4 | 0.06 | 1.25 |

[a]X = Hf or Nb-bearing samples, respectively

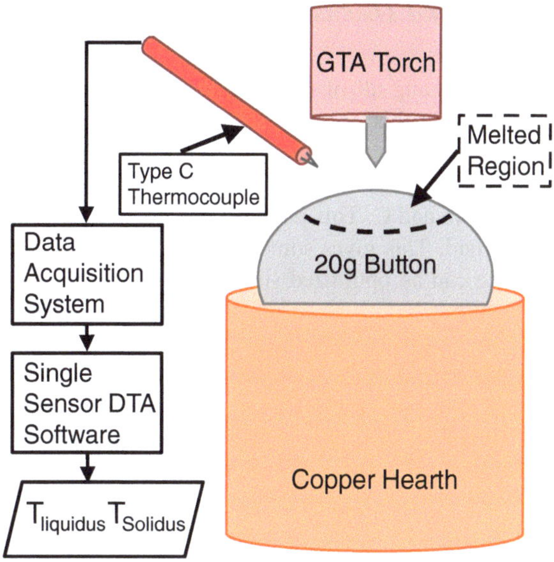


**Fig. 7** Schematic of thermocouple plunging technique used to acquire solidification temperature range

weld is made on the top of the button under argon at 250 A for 4 s. Upon extinguishing the arc, a type C thermocouple is plunged into the center of the molten puddle. The thermal history is acquired at 1000 Hz. The resulting cooling curve is then processed using Single Sensor Differential Thermal Analysis (SS-DTA) software. This yields the STR for each alloy composition. Each test was duplicated for repeatability. The resulting button is then cross sectioned and examined metallographically. Images are taken using light optical and scanning electron microscopy. The resulting images are then processed using Image J, where they undergo a thresholding procedure, shown in Fig. 8. This allows the area fraction of eutectic to be quantified. The actual compositions for each button were measured using optical emission spectroscopy. In this way, any loss of material during melting or due to human error can be accounted for in the final analysis.

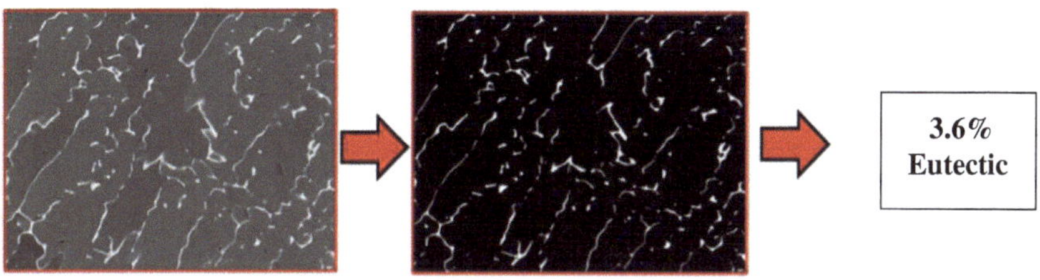


**Fig. 8** Typical SEM image of microstructure converted to binary image for fraction eutectic analysis

## *Results and Discussion*

After collecting all of the data, the results are entered into the DOE software as responses. In this way, the software can calculate relationships between the composition and the responses. Many graphs and visual representations of the relationships can be constructed. Figure 9 shows a contour plot of the STR as a function of both Nb and C. This plot also has fraction eutectic lines as a function of Nb and C overlaid. This gives some insight on how the two responses, STR and fraction eutectic can be optimized simultaneously.

For an alloy to be resistant to both solidification cracking and ductility dip cracking, some criteria were chosen based on previous research. For sufficient resistance to solidification cracking, the STR should be lower than 150 °C. Alloys known to be resistant to DDC had at least 2 % eutectic, based on image analysis. In the case of Nb as a eutectic former, it can be seen in Fig. 9 that sufficient eutectic to resist DDC is not attainable without raising the STR to unacceptable levels.

When Hf was used as the primary eutectic former, the results are much different. In Fig. 10 the hatched region shows how the desired range is attainable for this

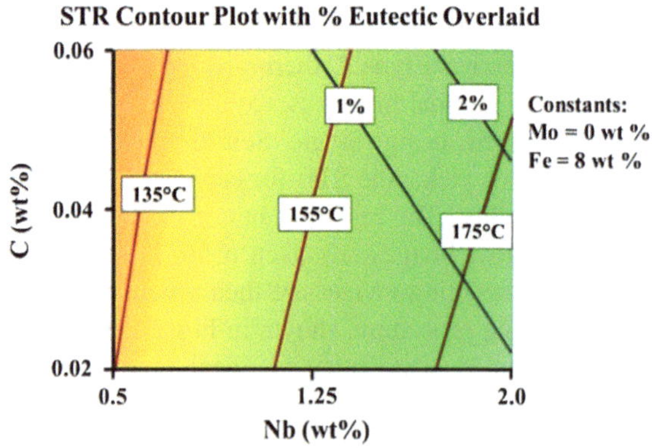


**Fig. 9** Contour plot of STR results as a function of C and Nb, with fraction eutectic contour lines overlaid

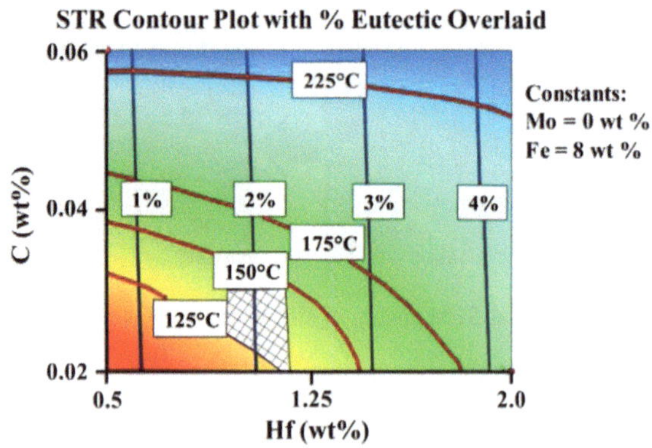


**Fig. 10** Contour plot of STR results as a function of C and Hf, with fraction eutectic contour lines overlaid. The *hatched region* shows an area in which 2 % eutectic and less than 150 °C STR is possible

particular system. This can be attributed to the different eutectics that can form in Hf bearing Ni-base alloys, as discussed previously.

To optimize a composition based on all the factors simultaneously, empirical relationships between the responses and the compositions must be established. The DOE software does this and also reports the respective coefficient of determination ($r^2$) values for each equation. Such relationships are shown in Eqs. 1–4. These relationships are based on data from the experimental thermocouple plunges and image analysis. Only factors which have been determined to be significant ($p < 0.05$) are included in the equations.

**Hf-bearing Alloys**

$$\begin{aligned} \mathrm{STR}(^\circ\mathrm{C}) = {} & 232 * \mathrm{Hf} - 20 * \mathrm{Fe} + \mathrm{Fe}^2 - 94.3 * \mathrm{Hf}^2 \\ & - \mathrm{Fe} * \mathrm{Mo} - 223 * \mathrm{Fe} * \mathrm{C} + 4 * \mathrm{Fe} * \mathrm{Hf} \quad (R^2 = 0.875) \\ & - 483 * \mathrm{Mo} * \mathrm{C} - 593 * \mathrm{C} * \mathrm{Hf} \end{aligned} \tag{1}$$

$$\mathbf{f}_{\mathbf{eut}}(\%) = 1.84 * \mathrm{Hf} \quad (R^2 = 0.974) \tag{2}$$

**Nb-bearing Alloys**

$$\mathbf{STR}(^\circ\mathbf{C}) = 198 + 17 * \mathrm{Fe} * \mathrm{Mo} \quad \left(R^2 = 0.350\right) \tag{3}$$

$$\mathbf{f}_{\mathbf{eut}}(\%) = 0.87 * \mathrm{C} + 1.4 * \mathrm{Nb} + 0.72 * \mathrm{C} * \mathrm{Nb} \quad (R^2 = 0.996) \tag{4}$$

These relationships are then optimized simultaneously with the DOE software. Limits or desired responses are defined, in this case STR < 150 °C and fraction eutectic = 2 %. The optimized composition for the Hf bearing alloy is shown in Table 3. However, for Nb-bearing alloys, a composition could not be optimized due to the low $r^2$ value for the STR equation. The noise in this data was too great to determine an empirical relationship. More STR tests need to be performed to improve the statistical analysis.

The eutectic relationships in Eqs. 2 and 4 give insight into what eutectic constituents may be forming in these systems. Equation 4 for the Nb-bearing alloys shows that carbon and Nb interact to form more eutectic. This is indicative of NbC being the primary eutectic in this composition space. In Eq. 2, Hf-bearing alloys, only Hf was significant in determining the fraction eutectic. This indicates that HfC isn't the primary eutectic. The eutectic that forms is likely $Ni_7Hf_2$, as discussed previously. This has been verified and published in concurrent research at The Ohio State University [28].

**Table 3** Optimized Hf-bearing composition based on DOE regression analysis

| | Fe | Ni | Cr | Hf | Mo | C |
|---|---|---|---|---|---|---|
| Optimized Hf bearing composition | 13 | Bal. | 30 | 0.75 | 0 | 0.03 |

### *Summary and Conclusions*

1. A DOE methodology was employed using Thermo-Calc to determine the effects of Fe, Mo, C, and Hf (or Nb) on the solidification behavior of 30 wt% Cr Ni-base alloys, the results of which helped define a composition space for experimental validation.
2. Thermocouple plunging and image analysis combined with a DOE methodology allowed for experimental determination of elemental effects of Fe, Mo, C and Hf (or Nb) on the STR and fraction eutectic.
3. Regression analysis of STR and fraction eutectic data yielded empirical relationships that were then optimized using DOE software to determine an optimum composition for resistance to both solidification and ductility dip cracking.
4. The target STR and fraction eutectic of 150 °C and 2 %, respectively, was able to be achieved a 30Cr, Hf-bearing alloy, but not for a 30Cr alloy containing Nb.

## Carbon Diffusion Model for DMW During PWHT

### *Background*

Weld overlays of Ni-base alloys on to carbon or low alloy steels are commonly used to eliminate the need for a PWHT in the field. Smaller sections of pipe or parts of a structure can be 'buttered' with a Ni-base alloy and given a proper PWHT in the shop. This PWHT is necessary to reduce the hardness in the HAZ of the steel and to reduce weld residual stresses. The final weld can be made in the field on the overlay without affecting the base metal. A typical overlay of Alloy 625 on AISI 8630 steel is shown in Fig. 11. The compositions of these materials are shown in Table 4.

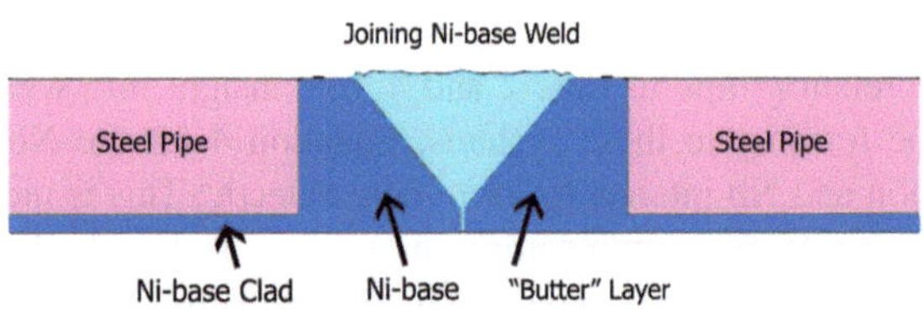


**Fig. 11** Typical joint configuration for 'buttered' steel pipes

**Table 4** Nominal compositions of materials used in carbon diffusion models

| Material | C | Mn | Si | Cr | Mo | Ni | Nb | Fe |
|---|---|---|---|---|---|---|---|---|
| 8630 | 0.3 | 0.75 | 0.25 | 0.5 | 0.2 | 0.5 | – | Bal. |
| F22 | 0.1 | 0.45 | 0.25 | 2.25 | 1 | – | – | Bal. |
| 625 | 0.1 | 0.25 | 0.25 | 21.5 | 9 | Bal. | 3.0–4.0 | <5.0 |

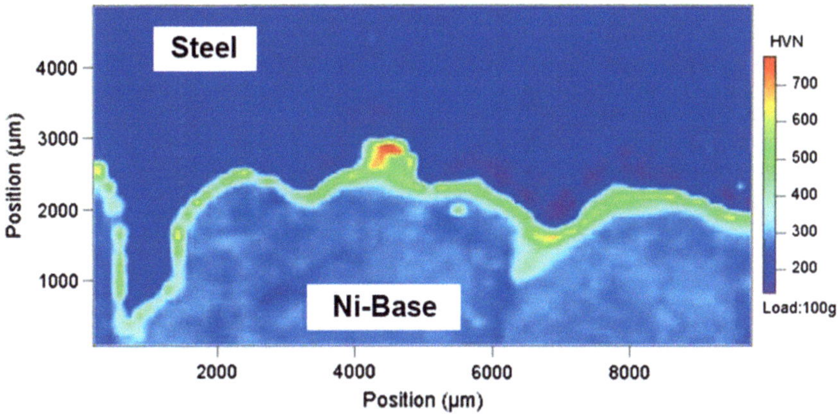


**Fig. 12** Hardness map near the fusion boundary of an 8630/625 overlay that has undergone PWHT

The oil and gas industry occasionally experiences brittle failures of subsea pipes under cathodic protection conditions. Failures have been observed at the fusion boundary of Alloy 625 overlays on 8630 steel. The mechanism has previously been determined to be related to hydrogen embrittlement due to a local brittle zone that forms at the fusion boundary during PWHT [29–31].

The PWHT is required to temper the martensite that forms in the HAZ during overlaying. A typical PWHT schedule for this type of weld is 677 °C (1250 °F) for 8 h. During this treatment, carbon can migrate from the steel into the weld metal [29, 31, 32]. Carbon migrates due to lower chemical potential for carbon in the weld metal compared to the base metal. However, the diffusivity for carbon in the weld metal is much lower than in the steel, and carbon can "pile up" at the fusion boundary. This leads to a softening in the steel HAZ, and a locally hard region in the composition transition zone at the fusion boundary. Figure 12 shows a hardness map at the fusion boundary for an 8630/625 overlay that has undergone PWHT.

Thermodynamic and kinetic models can be used to simulate this carbon diffusion, and to find an alloy combination that will be more resistant to carbon migration during PWHT.

## *Experimental*

DICTRA® was used to simulate the diffusion of carbon during PWHT. The Solid Solutions Database v.4 was used for thermodynamic data and the Mobility Database v.2 was used for kinetic data. A 2000 μm 1-D block of material was set up with an initial composition gradient that closely matched that of an actual as-welded microstructure. Figure 13 shows the initial gradient for Fe for a 625/8630

**Fig. 13** Initial Fe gradient used in the model. *Left side* is 8630 steel, *right* is Ni-base alloy 625. Fusion boundary is at 1000 μm. Other elements use this same gradient for their respective compositions

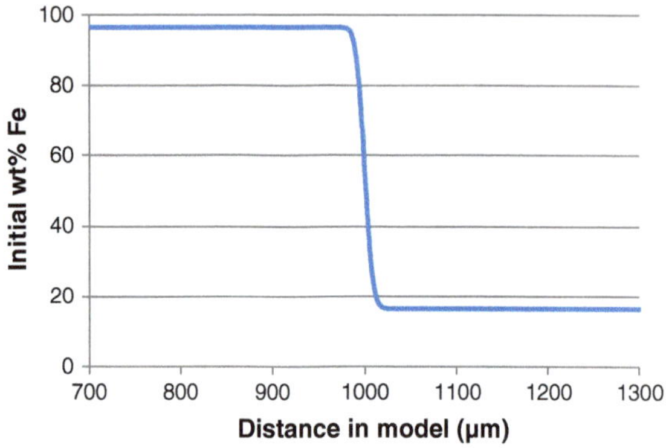


combination. The steel is on the left, and the Ni-base overlay is on the right. The fusion boundary is half way into the 2000 μm block. The DICTRA® Homogenization Model was chosen for simulation, as it can simulate long range diffusion through a multiphase mixture. Since one side of the system is BCC and the other FCC at the PWHT temperature, this type of model was chosen. The model also assumes a local equilibrium at each point. The simulation temperature and time were set to 677 °C and 8 h, respectively.

To verify the accuracy of the model, electron probe microanalysis (EPMA) was performed on as-welded and PWHTed samples of 625/8630. Analysis was performed using a JEOL JXA-8200 microprobe with light element detection capabilities. Area and line scans were acquired across the weld interface with a 2 μm step size.

## *Results and Discussion*

The DICTRA® diffusion simulation was performed for an 8630/625 weld fusion boundary. The results, shown in Fig. 15, show significant carbon migration for a 677 °C/8 h heat treatment. On the steel side, a relatively large carbon depleted zone exists. This region is about 200 μm wide, and is at almost half of the nominal base metal carbon content. This carbon depleted region corresponds well with the drop in hardness seen previously in Fig. 12. At the fusion boundary, the simulation predicts a large spike in carbon content, over 1 wt% at its peak. This peak also corresponds with the high hardness values seen at the fusion boundary.

EPMA area scans are shown in Fig. 14 for as-welded and PWHTed conditions of 8630/625 overlays. The scans clearly show that the carbon buildup at the fusion boundary is not present before heat treatment. Line scans in the PWHT condition across the fusion boundary are shown together with simulation results in Fig. 15. There is good correlation between the scan and the simulation; therefore further

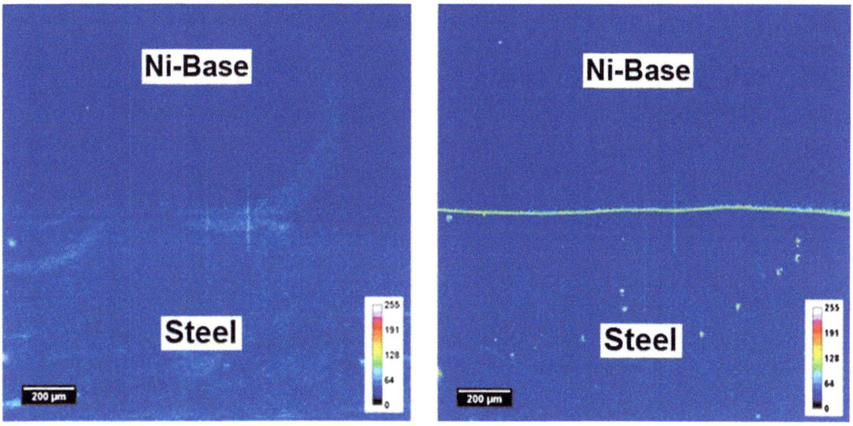


**Fig. 14** EPMA area scans for carbon in the as welded (*left*) and PWHT (*right*) conditions

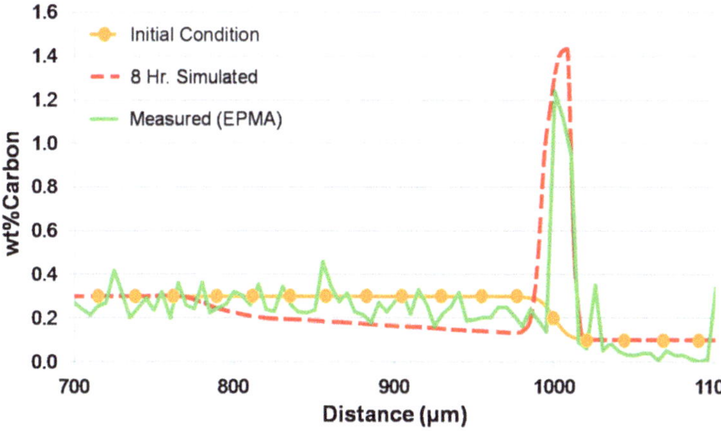


**Fig. 15** Simulated and measured carbon content across fusion boundary of 8630/625 overlay after PWHT

simulations were performed on a hypothetical material combination in an attempt to mitigate the carbon diffusion problem.

The carbon diffusion and pile up during PWHT takes place due to the differences in chemical potential and diffusivity of carbon between the steel and the Ni-base alloy. DICTRA® was used to determine these values for each material, and the results are shown in Table 5. The lower chemical potential for carbon in the 625 is what drives the diffusion of carbon towards the weld metal. However, the much lower diffusivity of C in Alloy 625 means that once the carbon reaches the fusion boundary it cannot migrate far into the 625, resulting in a high carbon concentration just across the fusion boundary. It is known that the chemical potential of carbon is affected by not only the carbon content, but by other elements such as Cr. Cr will

**Table 5** Diffusion coefficient and chemical potential data calculated by DICTRA® for each material

| Material | PWHT temperature and duration (°C/h) | Diffusion coefficient ($m^2/s$) | Chemical potential of carbon (J/mol) |
|---|---|---|---|
| 8630 | 677/8 | $1.59 \times 10^{-10}$ | –5500 |
| 625 | | $7.46 \times 10^{-11}$ | –15400 |
| F22 | | $1.08 \times 10^{-10}$ | –10500 |

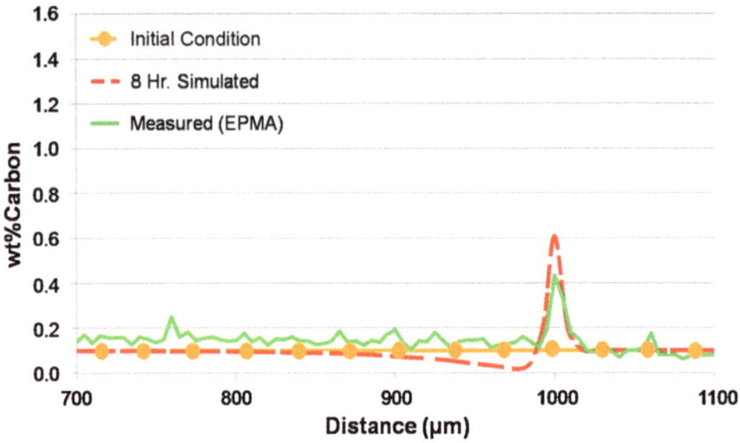


**Fig. 16** Simulated and measured carbon content across fusion boundary of F22/625 overlay after PWHT

decrease the chemical potential of carbon in the matrix. Therefore by choosing a steel with slightly higher Cr content, the chemical potential gradient can be reduced across the fusion boundary. This should reduce the extent of carbon diffusion. For these reasons a new steel, F22 (Fe–2.25Cr–1Mo–0.1C) was chosen and simulated using DICTRA®. Compositions of all the materials used are shown in Table 4.

The carbon diffusion simulation for F22/625, along with EPMA line scan results, is shown in Fig. 16. This weld was given the same heat treat as the 8630/625 combination. The results again show good correlation between the simulation and the EPMA results. A much lower carbon peak was observed, as well as a more narrow carbon depleted zone. This has resulted in reduced susceptibly to hydrogen embrittlement at the fusion boundary during cathodic protection [33].

## *Summary and Conclusions*

1. A DICTRA® model has been developed that can simulate carbon diffusion across a dissimilar weld fusion boundary during PWHT; The accuracy of this model has been verified by EPMA.

2. The carbon diffuses from the steel towards the Ni-base alloy due to differences in chemical potential.
3. A carbon pile up at the fusion boundary forms due to differences in diffusivities. The diffusivity of carbon in the Ni-base alloy is much lower than the steel.
4. The carbon pile up can be reduced by using a steel that has a lower chemical potential for carbon. In this case, a 2.25Cr–1Mo–0.1 steel (F22) was found to have over a 50 % reduction in the peak carbon content at the interface after PWHT.

## References


1. Hickling, J., *Materials Reliability Program Crack Growth Rates for Evaluating Primary Water Stress Corrosion Cracking (PWSCC) of Alloy 82, 182, and 132 Welds*, 2004, EPRI: Palo Alto, California.
2. Jacko, R.J., R.E. Gold, and A. Kroes, *Accelerated Corrosion Testing of Alloy 52 M and Alloy 182 Weldments*, in *11th Int. Conf. Environmental Degradation of Materials in Nuclear Systems*2003.
3. Smith, R., *Experiences with hot cracking in Alloy 52 M overlays*, in *8th International Conference on Welding and Repair Technology for Power Plants*2008: Sanibel Island, FL.
4. Matsuda, F., et al., *Weldability of Fe-36 % Ni alloy report II: Effect of chemical composition on reheated hot cracking in weld metal.* Trans. JWRI, 1984. **13**(2): p. 69–75.
5. Nishimoto, K., K. Saida, and H. Okauchi, *Microcracking in multipass weld metal of alloy 690 Part 1–Microcracking susceptibility in reheated weld metal.* Science and Technology of Welding & Joining, 2006. **11**(4): p. 455-461.
6. Nishimoto, K., et al., *Microcracking in multipass weld metal of alloy 690 Part 2–Microcracking mechanism in reheated weld metal.* Science and Technology of Welding & Joining, 2006. **11**(4): p. 462-470.
7. Ramirez, A.J. and J.C. Lippold, *High temperature behavior of Ni- base weld metal - Part II - Insight into the mechanism for ductility dip cracking.* Materials Science and Engineering a-Structural Materials Properties Microstructure and Processing, 2004. **380**(1-2): p. 245-258.
8. Yamaguchi, S., et al., *Effect of minor elements on hot workability of nickel-base superalloys.* Metals Technology, 1979. **6**(1): p. 170-175.
9. Lippold, J.C. and N.E. Nissley, *Ductility-Dip Cracking in High Chromium, Ni-Base Filler Metals.* Hot Cracking Phenomena in Welds Ii, 2008: p. 409-425.
10. Kiser, S.D. and E.B. Hinshaw, *Nickel alloy welding requirements for nuclear service.* Focus on Nuclear Power Generation, 2005: p. 21-26.
11. Alexandrov, B.T., et al., *Weldability Studies of High-Cr, Ni-Base Filler Metals for Power Generation Applications.* Welding in the World, 2011. **55**(3-4): p. 65.
12. Borland, J.C., *Generalized theory of super-solidus cracking in welds (and Castings).* British Welding Journal, 1960. **7**(8): p. 508-512.
13. Matsuda, F., *Hot crack susceptibilitly of weld metal*, in *Proceedings of the 1st US-Japan Symposium on Advances in Welding Metallurgy*1990.
14. Nakata, K. and F. Matsuda, *Evaluations of ductility characteristics and cracking susceptibility of Al alloys during welding.* TRANSACTIONS-JWRI, 1995. **24**: p. 83-83.
15. Kou, S., *Welding metallurgy*2003, New York: John Wiley & Sons.
16. Dupont, J.N., C.V. Robino, and A.R. Marder, *Solidification and weldability of Nb-bearing superalloys.* Welding Journal, 1998. **77**(10): p. 417s-431s.

17. DuPont, J., et al., *Solidification of Nb-bearing superalloys: Part II. Pseudoternary solidification surfaces.* Metallurgical and Materials Transactions A, 1998. **29**(11): p. 2797-2806.
18. DuPont, J., et al., *Solidification of Nb-bearing superalloys: Part I. Reaction sequences.* Metallurgical and Materials Transactions A, 1998. **29**(11): p. 2785-2796.
19. Chang, K.-M., H.-J. Lai, and J.-Y. Hwang, *Existence of Laves Phase in Nb-Hardened Superalloys*, in *Superalloys 718, 625, 706 and Various Derivatives*, E.A. Loria, Editor 1994.
20. Nissley, N., *Development of the strain-to-fracture test for evaluating ductility-dip cracking in austenitic stainless steels and Ni-base alloys.* Welding in the World, 2002. **46**(7-8): p. 32.
21. Hope, A.T., et al. *The Effect of Dilution on the Cracking Susceptibility of FM52/308L Overlays on Stainless Steel.* in *Welding & Fabrication Technology for New Power Plants and Components* 2011. Orlando, FL.
22. Baeslack Iii, W.A., *Weld cracking in Ta-modified cast Inconel 718.* Scripta Metallurgica, 1988. **22**(5): p. 729.
23. Kotval, P.S., J.D. Venables, and R.W. Calder, *The role of hafnium in modifying the microstructure of cast nickel-base superalloys.* Metallurgical Transactions, 1972. **3**(2): p. 457.
24. Dahl, J.M., W.F. Danesi, and R.G. Dunn, *Partitioning of Refractory Metal Elements in Hafnium-Modified Cast Nickel-Base Superalloys.* Metallurgical Transactions, 1973. **4**(4): p. 1087-1096.
25. Dmitrieva, G.P. and A.K. Shurin, *Investigation of phase equilibria and structure in Ni-ZrC and Ni-HfC alloys.* Metallofizika, 1974. **53**: p. 97-100.
26. Nash, P. and D.R.F. West, *Phase-Equilibria in Ni-Rich Region of Ni-Cr-Hf System.* Metal Science, 1981. **15**(8): p. 353-356.
27. FUSNER, E., A. HOPE, and J. LIPPOLD, *Development of High- Cr, Ni-Based Filler Metals Using Combined Computational and Experimental Techniques.* Welding Journal, 2014. **93**(5).
28. Feng, X., A. Hope, and J.C. Lippold, *Effect of Cr on eutectic phase formation and solidification temperature range in Ni–Cr–Hf system.* Materials Letters, 2014. **116**: p. 79-81.
29. Beaugrand, V., L.S. Smith, and M.F. Gittos, *Subsea dissimilar joints: failure mechanisms and opportunities for mitigation.* CORROSION 2009, 2009.
30. Beaugrand, V.C., L.S. Smith, and M.F. Gittos. *Hydrogen embrittlement of 8630 M/625 subsea dissimilar joints: factors that influence the performance.* in *ASME 2009 28th International Conference on Ocean, Offshore and Arctic Engineering*. 2009. American Society of Mechanical Engineers.
31. Olden, V., et al. *The Effect of PWHT on the Material Properties and Micro Structure in Inconel 625 and Inconel 725 Buttered Joints.* in *ASME 2003 22nd International Conference on Offshore Mechanics and Arctic Engineering*. 2003. American Society of Mechanical Engineers.
32. Gittos, M. and T. Gooch, *The interface below stainless steel and nickel-alloy claddings.* Carbon, 1992. **2**: p. 461-472.
33. Alexandrov, B.T., et al., *A New Test For Evaluation of Susceptibility to Hydrogen Assisted Cracking In Dissimilar Metal Welds.* CORROSION 2012, 2012.

# Part II
# Hot Cracking Phenomena II—Design Considerations

# Considerations for Sound Parameter Studies in Electron Beam Welding of Thick Walled Components

C. Wiednig, N. Enzinger and C. Beal

**Abstract** Joining new materials, material combinations or geometries by means of electron beam welding requires some preliminary parameter studies to assess the proper welding range. Determining parameters for thick walled applications is usually done by performing bead on plate welds. Due to resource efficiency, several welds are placed in a single block. After welding, the block is cut perpendicularly to the welding direction to judge the geometry of the molten zone and the influence of the used parameter combination on the joint quality. In different bead on plate welding experiments for several steels, the cross-sections revealed cracks, located in the centerline of the weld seam. In further experiments it turned out that these failures were not fully repeatable and therefore could not be attributed to the welding parameters or on the material solely. The experimental setup as a whole is crucial. Therefore, experiments were planned and performed to investigate the issues of these centerline cracks in EBW bead on plate welding studies. This contribution documents the performed failure case analysis. It shows that these cracks where not caused by miscalculated welding parameters or due to the metallurgic sensitivity of the material. Cracks appearance occurs however by reasons of a combination of shrinkage strain and dendritic solidification. Furthermore this document provides some recommendations to avoid this defect and beware of misleading interpretation of welding trials.

## Introduction

Electron beam welding (EBW) is a high quality fusion welding process and offers beneficial characteristics especially for thick walled welding. Furthermore the process stability of EBW is remarkable [1].

C. Wiednig (✉) · N. Enzinger · C. Beal
Institute of Materials Science and Welding, Graz University of Technology, Graz, Austria
e-mail: christopher.wiednig@tugraz.at

 
Th. Boellinghaus et al. (eds.), *Cracking Phenomena in Welds IV*,
DOI 10.1007/978-3-319-28434-7_5

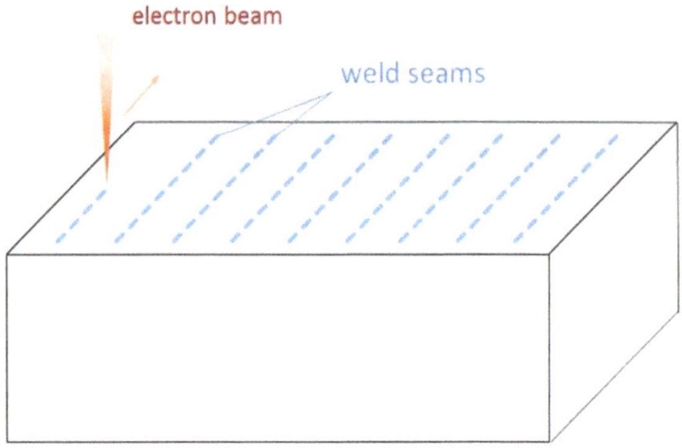


**Fig. 1** Bead on plate welding setup

An EBW machine offers a lot of settable parameters to tailor the welding process for a huge range of applications. A large number of adjustable welding parameters (voltage, welding speed, beam current, focal point beam-figure with oscillation, frequency and amplitude …) can lead to an enormous testing effort. For a new application preparative tests to find proper welding parameters have to be done. Mostly these preparative tests are performed as bead on plate welds (welding trials in full material) [1–4]. To reduce the workload, a proper experiment planning with the Design of Experiment method (see [5]) is strongly recommended.

To save resources but still keep application like heat conduction, several welds are carried out in one bead on plate welding block (Fig. 1). For each welding trial the block has to cool down to have the same initial temperature conditions. After welding, the block is cut at several positions, perpendicular to the welding seams, to investigate the geometry and the quality of the welds.

Nearly all seams in this welding block show center line cracks. These center line cracks are well known in the field of high powered beam welding. Cracks can be caused due to multiple causes.

## Cold Cracks

A high hardness in the heat affected zone and the fusion zone of EB welds is often addressed to cause this welding defects. This so called cold cracks will form when the residual stress in the material will exceed the fracture stress. Residual stresses are based on the thermal cycle of the weld; high temperatures and fast cooling cause high shrinkage stress. In the work of Bibby et al. [6] this cracking phenomenon was investigated. Cold cracks appear as centerline crack and as horizontal crack. During this experiments no horizontal crack should be found.

## Hot Cracks

Another causation of these defects is hot cracking. In general, hot cracking is a result of the rupture of liquid films, present in the mushy zone, in the centre of weld pools under tensile stresses [7]. Hot cracks are formed during and after the weld process. Hot cracks which occur in the fusion zone are called solidification cracks whereas cracks in the fusion line or close to the base material are called liquidation crack [8].

The mechanism of these, in the seam center located solidification cracks shown in Fig. 2. Solidification of a weld pool takes place from the cold base material at the fusion line. Dendrites grow crystallographic towards the seam center till they reach the dendrites from the opposite side of the weld and then grow into each other [8].

During thermal expansion of the fusion zone compressive stresses are introduced, which lead to plastic deformation. When the material starts to solidify, shrinkage takes place in the weld zone and tensile stresses are generated [10]. Especially, in welding procedures with high energy density and therefore high temperature gradients, elevated stress levels follow [11]. If remaining melt is present in the seam center the shrinkage stress cannot be transferred and a solidification crack along the seam center occurs.

A lot of publications are dealing with the different reasons of solidification cracking in electron beam and laser beam welding [8, 12–15]. The basic causes can be summarized as followed:

- Material composition: alloys with a wide solidification temperature range [13] or macro segregations in the last stage of solidification [14]
- Weld geometry: a high depth to wide ratio (>5) [14, 16]

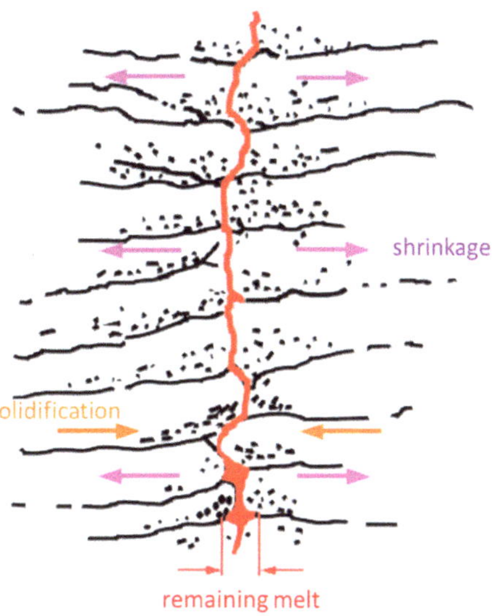


**Fig. 2** Solidification situation in the weld centre [9]

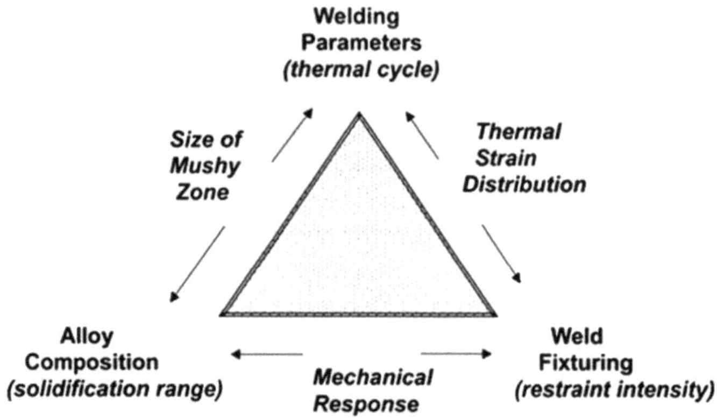


**Fig. 3** Interaction between process parameter affecting weld solidification cracking [13]

- Welding parameters: a tear-drop shaped welding pool, formed at high welding speeds, will form a distinct centerline where defects are frequent [8, 14, 17]
- Residual stress and shrinkage: Dendrites cannot grow into each other, stresses cannot be absorbed [7, 8, 10].

However, a complex interaction of these different metallurgical- and technological reasons with a special emphasis on residual stresses (Fig. 3) influence the cause of hot cracking [7].

## Inconsistent Welding Defects

During EBW bead on plate welding studies, with different materials and different welding depths, in the range between 20 and 110 mm, similar defects occurred. The welds exhibit cracks along the seam centre Fig. 4a, b.

This cracking phenomenon was observed in a martensitic 9 % chromium steel (cost CB2 steel [18]) Fig. 4a, when suitable parameters for a 50 mm deep weld were developed. Welds were placed parallel in one block with 15 mm spacing in-between (seam center to seam center). Weld width is approximately 5 mm. A similar situation could be observed during welding experiments with soft martensitic steel (X4CrNiMo13-4; 1.4313) Fig. 4b. For this material bead on plate welds between 50 mm and 100 m in depth where carried out. The welds were placed with a spacing of 25 mm. Weld width is approximately 8 mm Fig. 4b.

In ongoing welding trials it was observed that defects do not occur in single placed welds. Parameter sets from the former study where tested again but with more spacing between the welds. Spacing was set to 60 % of the intended welding depth. None of these welds showed any centerline cracks (Fig. 5a, b).

To determine which cracking mechanism causes these defects, a fractographic investigation by means of scanning electron microscope (SEM) was carried out.

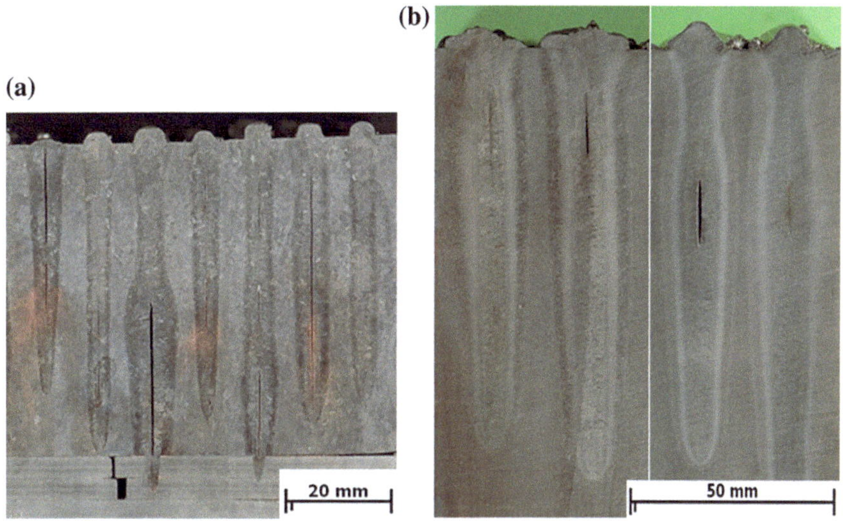


**Fig. 4** **a** Welding block CB2, **b** welding block 1.4313

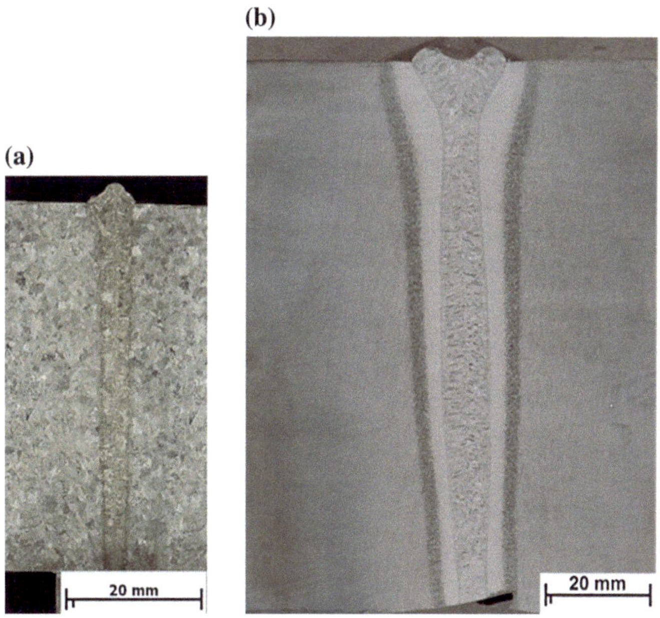


**Fig. 5** **a** Increased spacing CB2, **b** increased spacing 1.4313

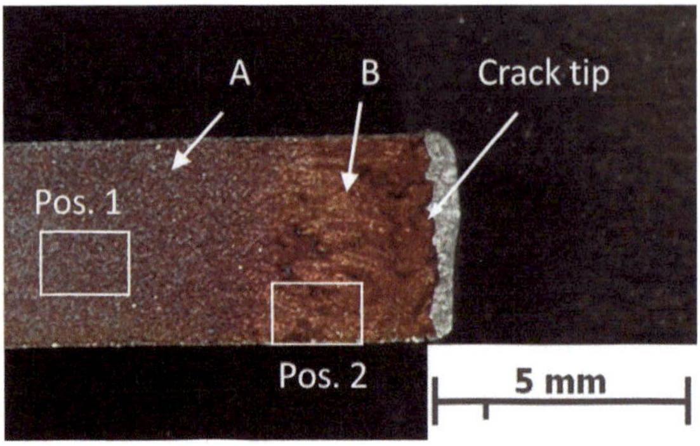


**Fig. 6** Crack surface CB2

The cracks were extracted and opened after cooling in liquid nitrogen. After a subsequent cleaning in an ultrasonic alcohol bath the opened specimens were ready for investigation. For investigation, the non-etched side was used.

Several positions of every crack surface were examined. A first visual investigation of several opened cracks from the CB2 material and the chromium nickel steel exhibit uniform characteristics for all crack surfaces:

- Crack surfaces appear fine grained (area A in Fig. 6)
- Close to the crack edges the surface looks more uneven (area B in Fig. 6)
- Cracks close to the top of the weld show annealing colors.

In Fig. 6 one crack surface is shown. The nitrogen fracture is visible on the left side. Afterwards the more uneven surface close to the crack tip is visible (A). The fine surface (B) is visible 3 mm away from the crack tip. The two rectangles (Pos.1 and Pos.2) mark the positions where a SEM investigation was carried out.

In Fig. 7a the Pos.1 and in Fig. 7b Pos.2 is displayed in high magnification. Two different appearing surface conditions on the fracture areas can be distinguished:

- Dendrites transverse to the growing direction (Pos.1)
- Longitudinal dendritic structure (Pos.2)

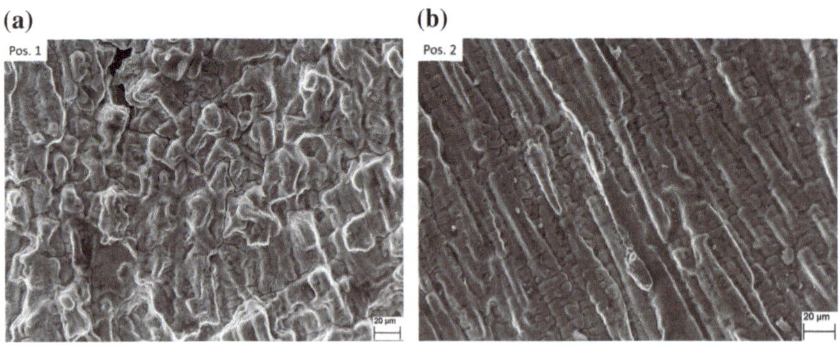


**Fig. 7** **a** *Pos.1* dendrites transverse, **b** *Pos.2* dendrites longitudinal

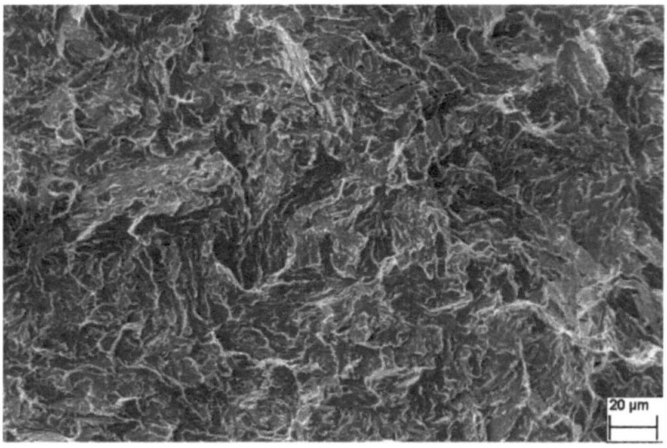


**Fig. 8** Residual nitrogen fracture opened under liquid nitrogen; CB2

Near the crack tips a longitudinal dendritic structure is visible. The longitudinal dendrites are orientated to the edge of the crack and can be found all around the borders of the crack. The width of this area is, depending on the size of the crack, between 100 µm and 3 mm. Beside this crack front, the surface changes to transverse arranged dendrites. The tips of the dendrites point out of the surface and are arranged uniformly over the whole surface. All fracture surface structures appear with smooth contours and no sharp edges.

The areas of the nitrogen fracture were investigated additionally. The surface, illustrated in Fig. 8, is clearly different to the previous pictures. The topography appears brittle.

## Experiment to Induce Cracks

To confirm observations that these cracks are a matter of the experimental setup a further experiment was carried out to try to force this kind of cracks. Four welds with *equal* parameters were performed in one block of CB2 steel (Fig. 9). The numbers next to the welds represent the chronological order. Weld #1 was placed first and fare on the left side of the block to simulate a single weld. This weld was used to prove that proper welding parameters were selected. Weld #2, #3, and #4 are placed close to each other (10 mm spacing between the seams) to simulate the former bead on plate welding experiments. All welds were performed in the cold block.

After welding the experimental block was cut, grinded and etched to make the fusion zones visible. Figure 10 shows the cross-section of the block.

The reference weld #1 is free of defects, indicating that the parameter combination is suitable for this material and welding depth. Weld #2 is also free of defects whereas in welds #3 and #4, centerline cracks could be found (Fig. 10).

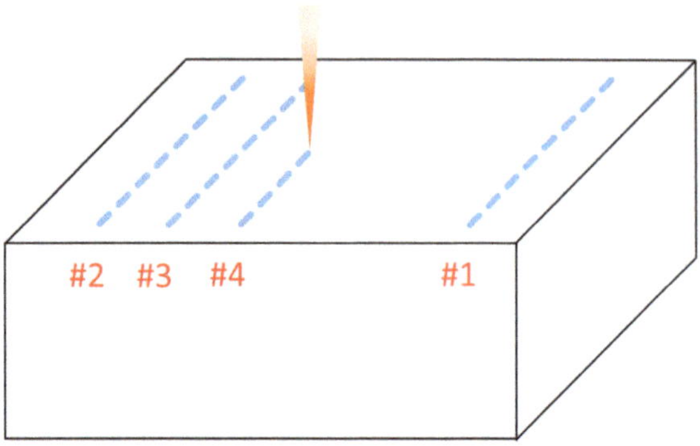


**Fig. 9** Crack-experiment welding block; CB2

**Fig. 10** Crack-experiment welding block; cross section

A in the former experiments theses cracks were opened to investigate the crack surface. Figure 11 displays one opened crack.

Two clear distinguishable surfaces can be recognized in Fig. 11: the dull surface of the crack and the shiny surface of the brittle liquid nitrogen fracture. Due to the quite small size of the crack surface no details of the crack surface can be identified with the naked eye. The rectangle in Fig. 12 marks the position for a further investigation by the means of SEM. Figure 12 provides this area in higher magnification.

Under the SEM the crack surface look equivalent to the before discussed pictures (Fig. 7a, b). Longitudinal arranged dendrites close to the edge of the crack can be seen and about 50–200 µm away from the crack boarders the surface shows transverse dendritic structure.

This experiment could exactly simulate the defects found in the EBW bead on plate parameter studies.

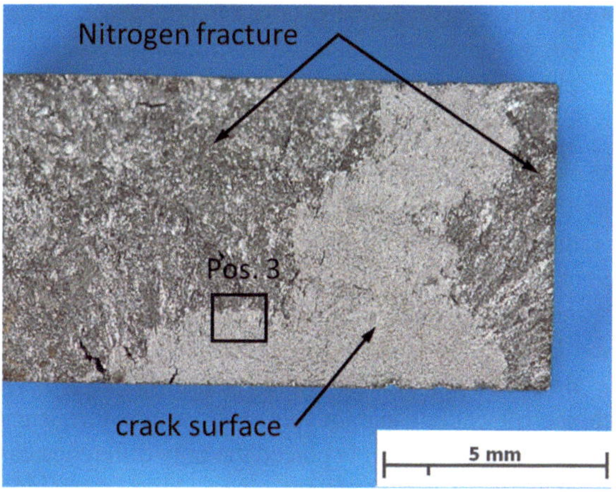


**Fig. 11** Opened crack from crack experiment block

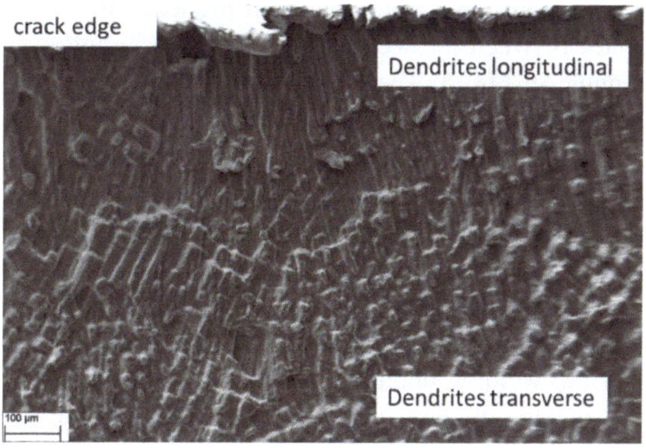


**Fig. 12** Opened crack #4

## Results and Discussion

- In EBW parameter studies with different materials and for different welding depths similar welding defects could be observed.
- When welds are placed next to each other centerline cracks occurred in most of the welds.
- When repeating the trials with similar parameters but with more spacing between the welds, no defects occur.
- A tailored experiment where several welds with equal parameters were performed confirmed these circumstances.
- Due to the chronological order of the welds it can be derived that in a row of several welds the former weld indicate cracks in the next weld. In this experiment weld #2 is inducing cracks in #3 and perhaps also #4. Furthermore #3 is indicating cracks in #4.

- All investigated cracks show a similar crack surface. Based on a SEM analysis it can be determined that theses cracks were developed in the presence of a liquid phase.

The surface structures of the examined welding cracks are characteristic for hot cracks [8, 12]. Another indication that these cracks occurred directly after welding, are the annealing colors. These findings combined with the seam-centered location of the found defects, lead to the conclusion that these cracks are solidification cracks (also called centerline hot cracks) [19–21]. Similar phenomena in the area of beam welding were reported in the work of Schuster et al. [8], Lippold [14], Wang and Lu [22], and Schaumann and Collmann [23].

Based on the present results, these defects cannot be attributed to an incorrect choice of welding parameters or as a metallurgical problem. Considering that no cracks occurred in single placed the only difference can be found in the initial state of the material. The main factor for cracking are residual stresses caused from nearby welds. Residual stresses, whether tension or compression, initiated from welds placed from the former weld will influence the following, even if the heat affected zones between the welds do not. This assumption is affirmed by the hot cracking triangle in Fig. 3. The welding parameter corner as well as the composition alloy corner are fixed, so the defects are caused by reasons of thermal strain and mechanical response. Some papers which investigated residual stresses in EB and laser welds confirm this. For this contribution especially the work of Ueda et al. [11] and Liu et al. [24], who carried out stress measurements in thick walled welding, are relevant and used for interpretation. Their work show that in the fusion zone of thick walled beam welds the residual stress in transverse direction varies. On the top and bottom (root) side of the weld, compressive residual stresses occur whereas around half the welding depth tensile residual stresses dominates. The authors agree also that outside the heat affected zone always tensile stresses are present. Considering the work of Lacki et al. [25], Liu et al. [24] and Raman et al. [26] it can be stated that the stress distribution in the cross section of an electron beam welding is much wider than the visible heat affected zone (Fig. 13). Depending on the welding situation, an up to ten times wider area can be affected [25].

In consideration of the carried out experiment and the reviewed literature two possible cracking modes can estimated. Residual tensile stress from the nearby welds:

- Causes strain which tear apart the liquid or mushy zone in the seam center.
- Causes a shrinkage speed that exceed the growing speed of the dendrites. Dendrites never meet in the seam center and cannot grow into each other.

From the experiments performed in this work it clearly can be deduced that the positioning and the order of the bead on plate welds has a significant influence on the appearance of these solidification cracks. To avoid misleading results in EBW parameter studies the cause of these cracks must be understood to be able to formulate proper countermeasures in blind welding parameter studies.

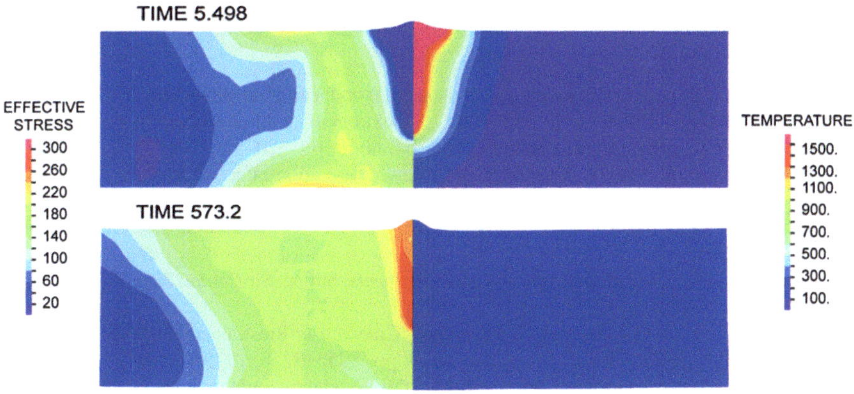


**Fig. 13** Comparison of effective stresses (*left*) and temperature (*right*) for an EBW weld in a 14 mm thick X5CrNi1810 steel plate, during welding (*top*) and after cooling (*bottom*) [25]

## Conclusion

In parameter optimisation for EBW of thick walled components often multiple welds are performed in a single block. It was observed that cracks occurred in a stochastic and not repeatable way. A fractographic analysis revealed that these cracks are solidification cracks. Cracks appear despite well-known welding parameter were used and in materials which are not prone to hot cracking.

Well-designed experiments showed that the crack appearance depends mainly on the residual stresses introduced by previous welds (sequence) and their constraint (location). Proper welding parameters were indicated wrong due to an incorrect experimental setup.

As a result of this investigations, concerning EBW bead on plate parameter studies, the following findings can be drawn:

- Between two EB welds a minimum spacing of 60 % of the intended welding depth should be kept
- Chronological order of seam placing is crucial; former welds influence following welds
- For applications where several EB welds are placed next to each other, a residual stress evaluation and/or a stress releaving heat treatment is recommended
- Further studies addressing the centerline cracking in EBW welding are necessary to fully understand this problem.

## References


1. H. Schultz, *Elektron beam welding*. Cambridge: Woodhead Publishing Ltd, 2004.
2. K. R. Schulze, "Durch dick und dünn - Elektronen strahlschweißen - Verfahren und Technologien," *Schweiss. und Schneid.*, vol. 63, no. 6, pp. 320–325, 2011.
3. Y. Luo, J. Liu, C. Du, H. Xu, and C. Li, "The evaporation effect of front keyhole wall in penetration welding with an electron beam," *Mater. Sci. Forum*, vol. 686, pp. 355–360, 2011.
4. K. Woeste, *Elektronenstrahlschweißen metallischer Werkstoffkombinationen*. Shaker Verlag Aachen, 2005.
5. D. Montgomery, *Design and Analysis of Experiments*, 8th ed. Singapore: John Wiley & Sons, 2013.
6. M. J. Bibby, J. A. Goldak, and G. Burbidge, "Cracking in Rrestrained EB Welds in Carbon and low-alloy Steels," *Weld. J.*, vol. 54, no. 8, pp. S253–S258, 1975.
7. C. E. Cross, "On the Origin of Weld Solidification Cracking," in *Hot Cracking Phenomena in Welds SE - 1*, T. Böllinghaus and H. Herold, Eds. Springer Berlin Heidelberg, 2005, pp. 3–18.
8. J. Schuster, S. Keitel, E. Schulze, and H. Maly, "Entstehung erstarrungsbedingter Risse in Laserstrahlschweißverbindungen an unlegierten und niedriglegierten Baustählen," *Schweißen und Schneid.*, vol. 51, no. 5, pp. 252–257, 1999.
9. Gesellschaft für Schweißtechnik, "Theoretische Ausbildung, Lehrplan und Lernziele SFI / IWE," in *Internationaler Schweißfachingenieurlehrgang*, GSI, Ed. Düsseldorf: DVS-Verlag GmbH, 2009.
10. M. Zain-ul-Abdein, D. Nelias, J.-F. Jullien, and D. Deloison, "Prediction of laser beam welding-induced distortions and residual stresses by numerical simulation for aeronautic application," *J. Mater. Process. Technol.*, vol. 209, no. 6, pp. 2907–2917, Mar. 2009.
11. Y. Ueda, Y. C. Kim, and A. Umekuni, "Measurement of three-dimensional welding residual stresses due to electron beam welding.," *Q. J. Japan Weld. Soc.*, vol. 4, no. 1, pp. 138–142, 1986.
12. R. Phaoniam, K. Shinozaki, and M. Yamamoto, "Solidification cracking susceptibility of modified 9Cr1Mo steel weld metal during hot-wire laser welding with a narrow gap groove," *Weld. ...*, vol. 58, no. 4, pp. 469–476, Mar. 2014.
13. Z. Tang and F. Vollertsen, "Influence of grain refinement on hot cracking in laser welding of aluminum," in *IIW C-IV Annual Assembly*, 2014, pp. 1–14.
14. J. C. Lippold, "Centerline cracking in deep penetration electron beam welds in Type 304L stainless steel," *Weld. J.*, vol. 64, no. 5, p. 127 s–136 s, 1985.
15. S. Sarafan, P. Wanjara, H. Champliaud, L. Mathieu, and J. Lanteigne, "Characteristics of electron beam welded CA6NM," *Mater. Sci. Technol.*, pp. 720–732, 2013.
16. U. Dilthey, *Schweißtechnische Fertigungsverfahren 1. Schweiß-und Schneidtechnologien*, 3rd ed. New York: Springer-Verlag Berlin HEidelberg, 2006.
17. K. R. Schulze, "Durch dick und dünn - Elektronen- strahlschweißen - Verfahren und Technologien," *Schweißen und Schneid.*, vol. 63, no. 6, pp. 320–325, 2011.
18. M. Schuler, S. Baumgartner, R. Schnitzer, and N. Enzinger, "Creep investigation of CB2 joints using similar rutile CB2-flux cored wire," *Weld. World*, vol. 58, no. 6, pp. 903–913, 2014.
19. TWI, "Defects - solidification cracking," 1999. .
20. J. F. Wallace, "Effects of sulphur and phosphorus on weld metal solidification cracking," *Met. Constr. Br. Weld. J.*, vol. 2, no. 8, pp. 333–338, 1970.
21. D. J. Grieve, "Welding Defects," 2003. .
22. J. Wang and F. Lu, "INVESTIGATION ON HOT CRACKING PROPAGATION IN FIBER LASER WELDING ON 6013 ALUMINUM ALLOY," in *IIW C-IV Annual Assembly*, 2013, pp. 1–15.
23. P. Schaumann and M. Collmann, "Influence of Weld Defects on the Fatigue Resistance of Thick Steel Plates," *Procedia Eng.*, vol. 66, pp. 62–72, 2013.

24. C. Liu, J. Zhang, B. Wu, and S. Gong, "Numerical investigation on the variation of welding stresses after material removal from a thick titanium alloy plate joined by electron beam welding," *Mater. Des.*, vol. 34, pp. 609–617, Feb. 2012.
25. P. Lacki, K. Adamus, and P. Wieczorek, "Theoretical and experimental analysis of thermo-mechanical phenomena during electron beam welding process," *Comput. Mater. Sci.*, pp. 1–10, Feb. 2014.
26. P. Venkata Ramana, G. Madhusudhan Reddy, T. Mohandas, and a. V. S. S. K. S. Gupta, "Microstructure and residual stress distribution of similar and dissimilar electron beam welds – Maraging steel to medium alloy medium carbon steel," *Mater. Des.*, vol. 31, no. 2, pp. 749–760, Feb. 2010.

# Consideration of Welding-Specific Component Design on Solidification Crack Initiation

**Christian Gollnow and Thomas Kannengießer**

A commonly used way of minimizing the occurrence of hot cracks, especially solidification cracks during component welding, is mainly to analyse and vary process parameters such as welding speed and consequently the heat input. Metallurgy and component design are however hardly ever considered due to special production requirements and therefore, restricted flexibility in material selection and design. Such conditions, especially crack-critical welding positions are given by slot-welds or welds near pre-deformed areas, for instance bending edges. Hence, it follows that increased local and global residual component stress caused cracking on reaching a solidification crack critical level, which is characterised by solidification crack initiation.

Therefore, this work was focused on design-specific solidification cracking. In order to reproduce real component welding situations, an externally loaded hot cracking test, i.e. the Controlled Tensile Weldability (CTW)-test, was used. By applying an external pre-load to the specimens, realistic component residual stresses (loads) were transferred to the test plates. Furthermore, the local material movement acting in the solidifying weld due to the local material fusion was simulated via the transient global external load in scope of extension rates during welding. The material movement (displacements) close to the weld was identified in situ by Digital Image Correlation (DIC)-technique while performing the CTW hot cracking test.

It is evident that a distinction between applied pre-load and residual stress prior to welding and temporally transient external load (extension rate) during welding should be made. Furthermore, the results indicate a direct consideration of solidification cracks, external specimen loading during welding and hence the measured surface deformation. The investigated cases have shown that much higher applied pre-load than the temporally transient external solidification crack-critical load (crack occurring) could not initiate solidification cracks. That influence of design

C. Gollnow (✉) · T. Kannengießer
BAM Federal Institute for Materials Research and Testing,
Unter Den Eichen 87, 12205 Berlin, Germany
e-mail: christian.gollnow@bam.de

T. Kannengießer
e-mail: thomas.kannengiesser@bam.de



Th. Boellinghaus et al. (eds.), *Cracking Phenomena in Welds IV*,
DOI 10.1007/978-3-319-28434-7_6

aspect has been valued by different material movement close to the weld. Following investigations have to find out whether or not the weld undergoes temporally solidification crack critical stress during welding. Such real case investigations should be based on distinguishing marks of design.

## Introduction

Increasing demands on stress resistance and reliability of components require a higher complexity of design. Neglecting these aspects can yield to solidification cracking (Fig. 1) even for materials that are from metallurgical point of view hot cracking resistant. Hence, the design impact has to be strongly considered as early as at the beginning of every technological development process.

The risk of hot cracking, especially solidification cracking, can be described as a complex interaction between the given main influences of welding parameters, alloy composition and restraint intensity $R_y$ [1]. Design includes e.g. component and weld geometry (local). The component geometry (global) can be described by the restraint intensity. The design and the design-specific boundary conditions of the component being welded may considerably influence the local stress within the weld and hence the solidification cracking behaviour. Existing theories consider different causes and criteria for hot crack initiation [2], for example critical stress [3, 4], strains [5, 6] and strain rates [7]. As Fig. 2 illustrates, the design-specific solidification cracking may be described by near- and far-fields. There have been different measurement techniques, like the measurement by means of in situ observation (MISO)-technique and the use of strain gauges.

Zacharia et al. [3, 4] presented a local stress-based approach to determine the thermo-mechanical behaviour close to the weld pool and the associated hot crack initiation. The model describes the opportunity for interaction in the welding-specific stress field by imposing a pre-load, illustrated in Fig. 3. The size of the compressive and tensile stress fields and hence the crack susceptibility during solidification can be influenced by external pre-load.

Further, design-specific investigations concerning tacking were done by Kannengießer et al. [8], see Fig. 4. It could be concluded that the type of tacking has

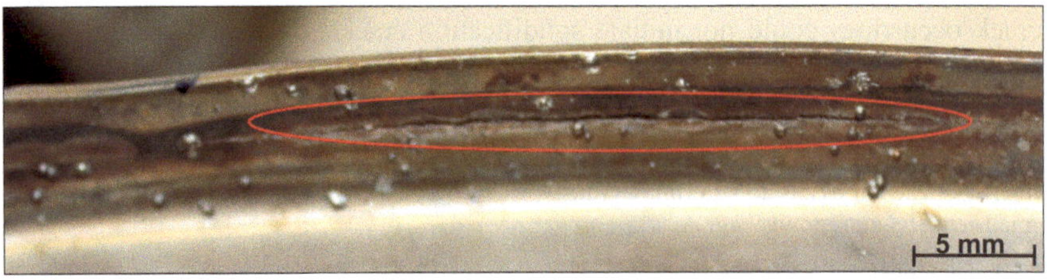


**Fig. 1** Solidification crack occurring in a component resulting from crack critical design, material 1.4828

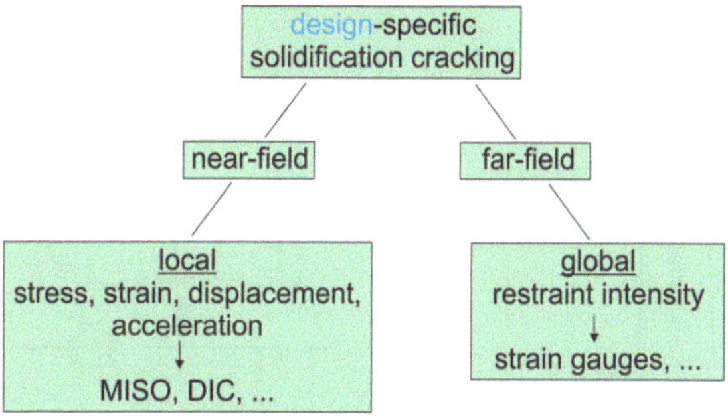


**Fig. 2** Describing design-specific solidification cracking by near- and far-fields and the corresponding measurement techniques

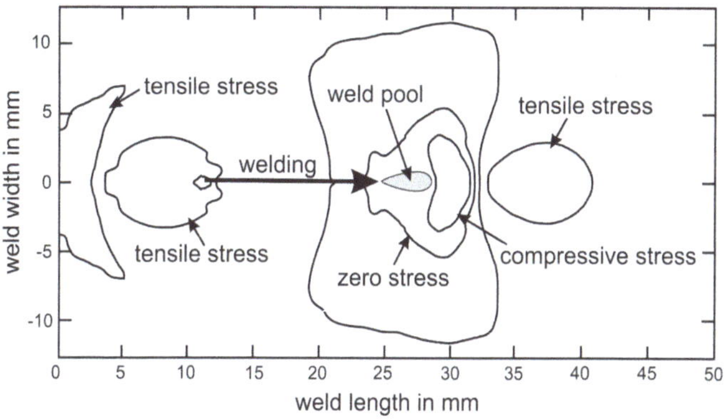


**Fig. 3** Local stress field calculated from the Sigmajig-test by Zacharia et al. [4]

an influence of the restraint intensity and hence on the crack critical transverse displacements. Unfortunately, tack configurations could promote a tack opening representing crack critical displacements during welding and therefore the risk of solidification cracking.

In order to analyse the global influence of external structural restraint $R_{Fy}$, Kannengießer et al. [9] did some investigations using the IRC-test. Increasing restraint intensity resulted in a reduction of transverse displacements, and therefore in the prevention of solidification cracking. Increasing restraint intensity decreases the occurrence of solidification cracks, as shown in Fig. 5. It has to be emphasised that the structural restraint was affected by pressure springs (force-controlled). Force-controlled means in that context the assumption of constant pre-load conditions during the tests (especially during welding). Anyway, the restraint intensity resulted from compressive stresses, which is less critical for solidification cracking [3, 4]. In any case, the restraint intensity does not describe the condition of stress fields, tensile or compressive. In summary, the reduction of solidification cracking

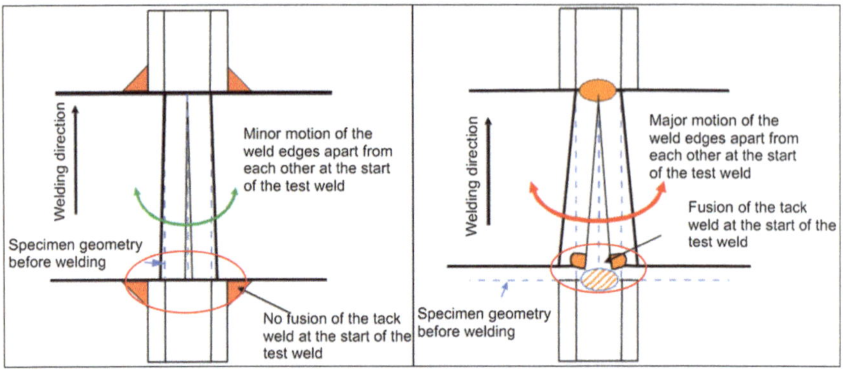


**Fig. 4** Global transverse displacement of the weld edges for corner tack (*left*) and end tack (*right*) [8]

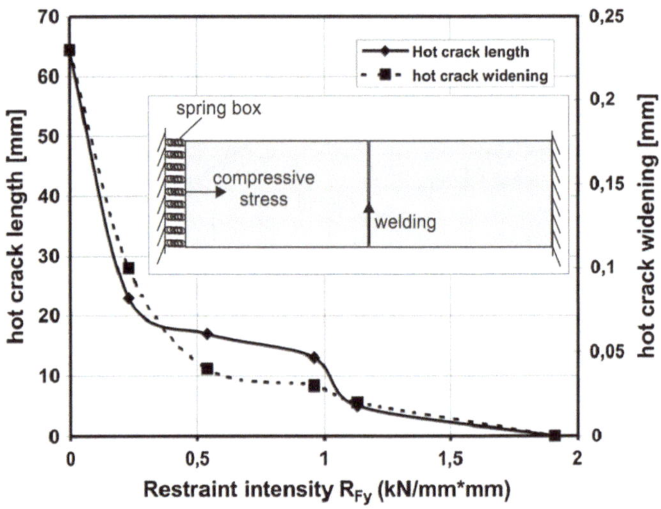


**Fig. 5** Sketch of the clamping arrangement in the IRC-test and the resulting solidification crack length and widening as a function of the global restraint intensity [9]

susceptibility due to additional support by springs and hence the variation of restraint intensity by compressive stress fields was determined.

Other investigations were done by Cross and Böllinghaus [10]. Figure 6 illustrates the restraint box used to simulate different restraint intensities by changing the length H of the side panel, and hence the stiffness. Additionally, the global effect of a transverse compressive stress pre-load (force-controlled) was analysed. It was discovered that at higher restraint intensity crack initiation could be avoided. Assuming that the stiffness of the box is well known, with that test a good possibility to simulate the complex restraint intensity of construction components was shown.

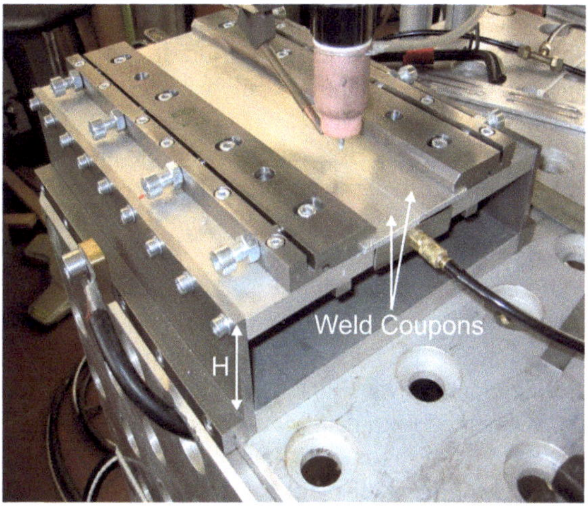


**Fig. 6** Photograph of a welding test with a "variable restraint box fixture" [10]

The purpose of such tests is to get a better comprehension about the solidification crack critical conditions, especially in components. The challenge is to measure the practical design effect and not only the complex interaction between welding parameters, metallurgy and design. Within the scope of such measurements, an external stress test, i.e. the Controlled Tensile Weldability test [11] should be used. That test enables analyses of different design-specific conditions like different pre-loads or loads during welding.

Figure 7 basically depicts the local approach of dealing with these external load tests according to the hot cracking model of Prokhorov [5]. Particularly, the global load during welding may be increased until solidification cracking. Hence, with

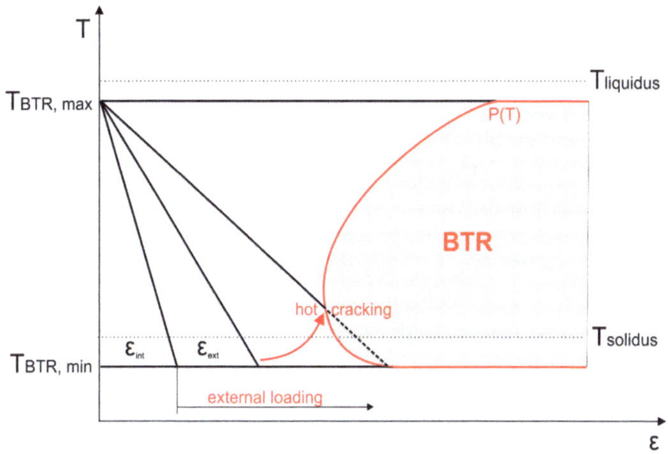


**Fig. 7** Schematic hot crack occurrence, if the external load test exceeds a threshold, according to the model of Prokhorov [5]

increasing external load (F or $v_l$) the design effect (external strains $\varepsilon_{ext}$) dominates the primary metallurgical behavior. It can be mentioned that material with low internal strains $\varepsilon_{int}$ shows a high hot cracking resistance from the metallurgical point of view. The BTR is represented by the deformation curve P(T). The P(T)-curve characterises on one hand the maximum tolerable strain/strain rate and on other hand the temperature range where the material is susceptible to solidification cracking. Prokhorov [5] claimed solidification cracks to occur in case of internal and external strains intersecting the P(T)-curve.

Experimental detection, especially with common procedures, of local deformation during welding is quite difficult due to the high temperature. One of the first optical techniques for determining critical deformations and deformation rates during cracking via visual identification of the displacements $d = d_2 - d_1$ was the MISO method following Matsuda et al. [12, 13]. A major advancement was realised by the DIC-technique [14], which was used in this research work for in situ analyses of local displacement fields in the weld vicinity under different load conditions during welding.

## Test Equipment and Procedure

Based on the work performed by [6, 15–20] the aim of the investigations reported here was to compare different design-specific cases. This was accomplished by detecting the specimen deformation under three different load conditions, meaning free-shrinkage, pre-loads F (tensile stress) and extension rates $v_l$ (tensile stress) during welding. The case of free-shrinkage is presented by unclamped plates. The displacements near the weld pool were detected in situ during welding by the DIC-technique.

Figure 8 illustrates the specimen geometry as well as the TIG weld position. The process sequence was automated by implementing a time trigger in the controller of the TIG welding power source. This was accomplished by keeping the welding speed constant at $v_w = 5$ mm/s and the linear energy at E = 0.22 kJ/mm with a current of 115 A and a voltage of 9.5 V. The welding electrode was thoriated and had a diameter of 4 mm. The shielding gas which was used consists of 97.5 % argon and 2.5 % carbon dioxide.

A TIG run was automatically executed in the middle of the specimen (x = 250 mm, $y_{start} = 25$ mm, $y_{end} = 125$ mm). The length between weld start and end was b = 100 mm. The stochastic pattern and, respectively, the position of the measuring field can be seen as well. For that a white primer (aluminium oxide) and dashes of black iron oxide were applied. Previous investigations revealed no significant differences in the resulting deformation between the measured sides of the weld for these cases. Because of the measuring arrangement the right side of weld was chosen for detecting the deformations.

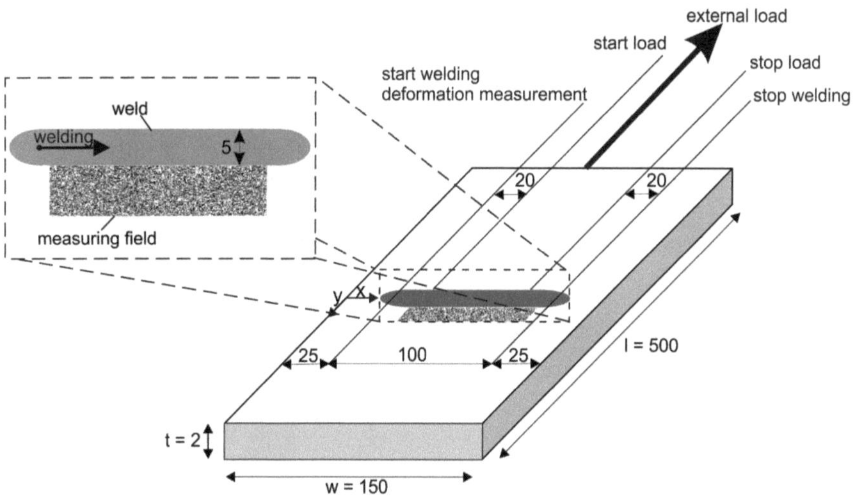


**Fig. 8** Specimen geometry with weld position (units in mm) and the time-triggered process steps

This test series was carried out with austenitic steel (1.4828). Higher-alloyed material shows primarily ferritic solidification under conditions of welding and is therefore relatively insensitive to hot cracking with regard to the metallurgy.

As in this series of experiments the focus was on the influence of design-specific boundary conditions on hot crack initiation, it was necessary to produce geometrically uniform full penetration welds.

The experiments were conducted using the CTW-test, see Fig. 9. In the reported test series the pre-load and load direction was transverse to the welding direction. The specimens were pre-loaded using a defined tensile force of F = 100 kN i.e. ca. 145 % of the yield strength $R_{p0.2}$. In the other case the specimen was subjected during welding to a solidification crack critical extension rate of $v_l = 5$ mm/min. For comparison, the specimen deformation for different load conditions was measured under the same boundary conditions (particular specimen geometry and welding parameter), on the same conditions as free-shrinkage.

Figure 9 represents an overview image of the test station including the CTW-test frame, the stereo camera system and strobe lights for in situ detection of the displacements by means of the depicted DIC-system, the TIG torch connected with a linear drive as well as the specimen provided with the applied stochastic pattern. In order to realise uniform lighting and measuring conditions for the DIC-technique during a welding process, two stroboscopic light sources were used producing flashes at defined time intervals. Additionally, specific light filters were applied to the DIC-camera system to detect the deformation directly at the weld without shielding plate. For proper use of the optical 3D-measuring system, heat resistant stochastic patterns were applied to the specimen surface. With the help of this measuring field the local transverse displacements $d_y$ and the total displacements $d_t$

**Fig. 9** Test equipment consisting of the CTW-test, TIG torch, DIC-system, strobe lights and the specimen with stochastic pattern for deformation measurement

were photogrammetrically determined from the raw data captured by the stereo camera system.

## Results and Discussion

The in situ measured data were all captured from the DIC-measuring field at the same positions for a better comparability at y = 4 mm from the weld centre line, x = 35 mm and x = 50 mm (square brackets in Fig. 10) for the coordinate system, see Fig. 8. It should be noted that the negative transverse displacement correlates with a material movement towards the weld pool. For better apprehension and traceability, only displacements were examined in these investigations. Particularly, as opposed to strain analyses for displacement studies, this measuring system does not require gaged facets for reference lengths. Especially in this case of small scales, the influence of error associated with these reference lengths becomes a large percentage of the measured data.

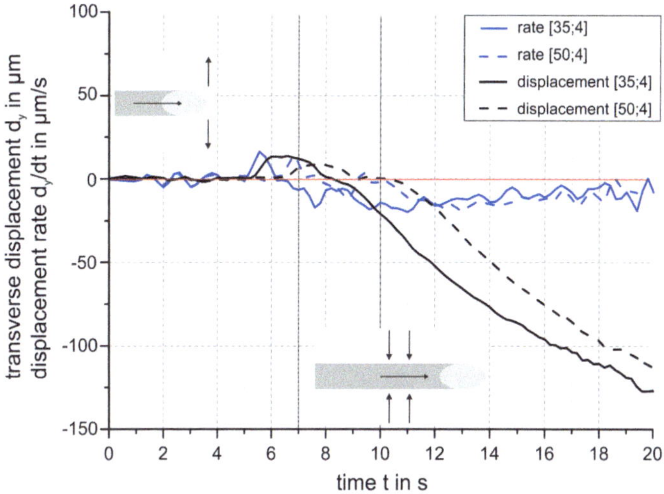


**Fig. 10** Local transverse displacement $d_y$ and displacement rate $d_y/dt$ versus time t, $v_w = 5$ mm/s, free-shrinkage

Figure 10 is a plot of the local transverse displacement $d_y$ and displacement rate $d_y/dt$ at two close different points versus the welding time under free-shrinkage. The values in the square brackets, for instance [35; 4], correspond to the coordinates [x;y] in Fig. 8. The displacement rate is listed in the following diagrams for sake of completeness. The transverse displacement profiles are found to correlate qualitatively with the already mentioned model of Zacharia [3, 4] as well as with the results of previous work [16]. For example, the heat input during welding leads first to positive transverse displacement values due to thermal expansion, suggesting a material movement away from the weld pool (especially at t = 7 s). Maximum positive values means the movement away from the weld (sketch in the lower area of the diagram) are reached after the weld pool has passed the measuring point. After the weld pool has passed these measuring points the transverse displacement decreased with time, due to shrinkage and ultimately changes its sign. Upon completion of welding, shrinkage specific negative transverse displacements (movement towards the weld) between $d_y = -110$ and $d_y = -125$ µm were measured. During welding and under free-shrinkage condition a maximal transverse displacement rate of approx. $d_y/dt = 20$ and $d_y/dt = -20$ µm/s was found.

Figure 11 shows the characterisation of the free-shrinkage stress condition, according to Zacharia [4] and Prokhorov [5] corresponding to Fig. 10. The comparison between Fig. 10 and the model of Zacharia [4] in the left picture, below shows a qualitatively similar curve shape. Therefore, the solidification crack will not initiate in a compressive stress field. Corresponding to the material specific influence $\varepsilon_{int}$ it can be assumed the design specific influence $\varepsilon_{ext}$ is comparatively less, because $\varepsilon_{int} \gg \varepsilon_{ext}$. Hence, the safety area $\varepsilon_{safety}$ up to intersection with the P (T)-curve is sufficient to avoid solidification cracking.

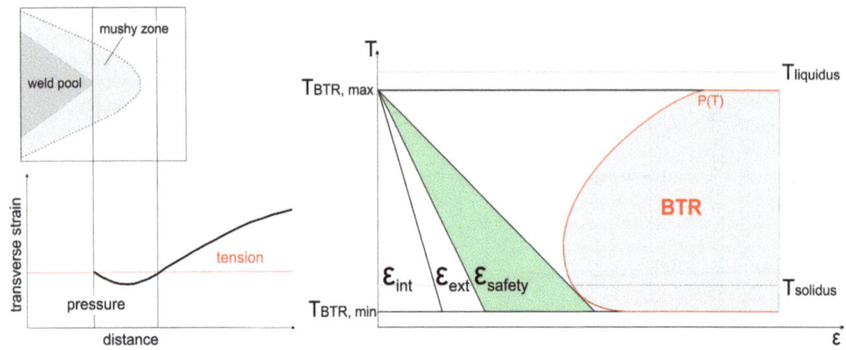


**Fig. 11** Local characterisation of the free-shrinkage stress condition, according to Zacharia [4] and Prokhorov [5] (from *left* to *right*) corresponding to Fig. 10

Figure 12 shows the in situ measured local transverse displacement $d_y$ and displacement rate $d_y/dt$ versus time for a CTW-test performed at an external path-controlled pre-load with the tensile force of F = 100 kN. Path-controlled suggests, the pre-load is fixed only at beginning without readjustment during the welding process. Hence the load condition is variable during testing. This diagram clearly indicates increasing positive displacement values, up to approx. $d_y = 85$ µm, at the time before the weld pool reaches the measuring points. Due to material heating, the pre-load reacts to result in a material movement away from the weld. Additionally, the measured force descends down to ca. F = 22 kN (32 % of $R_{p0.2}$) at this time. After the weld pool has passed the measuring points (t = 7 s and t = 10 s)

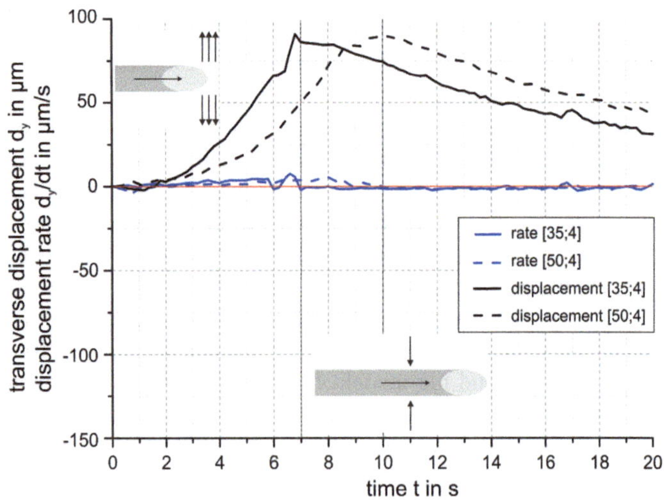


**Fig. 12** Local transverse displacement $d_y$ and displacement rate $d_y/dt$ versus time t, $v_w = 5$ mm/s, pre-load with the force F = 100 kN (path-controlled)

the displacement decreased due to shrinkage, as under free-shrinkage conditions. Because of the reduction of the pre-load from F = 100 kN to F = 22 kN the pre-load is relieved due to the passage of the weld pool. Therefore, it can be assumed that this load condition is not solidification crack critical. During shrinkage the force F starts increasing again. The displacement rate $d_y/dt$ is comparatively uniform and will therefore not be discussed further in this case.

Figure 13 shows the characterisation of the pre-load stress condition, according to Zacharia [4] and Prokhorov [5] corresponding to Fig. 12. In the following illustration the weld pool is in front of the measuring points. The case of time t after the weld pool has passed the measuring points is represented in Fig. 11. The comparison between the various points of time shows that the pre-load-caused high design specific influence $\varepsilon_{ext}$ was reduced during welding. Therefore, $\varepsilon_{safety}$ increases respectively when the solidifying weld is surrounded by a compressive stress field. Hence, no solidification cracks occur. So it can be assumed that the case of path-controlled pre-load condition indicates less effect on the solidification cracking.

In Fig. 14, the in situ measured local transverse displacement $d_y$ and displacement rate $d_y/dt$ versus time are plotted for a CTW-test performed at a solidification crack critical extension rate of $v_a = 5$ mm/min. This crack critical extension rate was identified by determining solidification cracking during welding for these load conditions. Start (t = 4 s) and end (t = 16 s) of the extension rate was measured according to the time scale shown in Fig. 8. Start and end of this loading cause the curves to descend and ascend. At this extension rate, crack initiation was detected. The red marking lines indicate the beginning and end of the solidification crack.

Because of this influence of the extension rate during welding the displacement and displacement rate behaviour is seen to be significantly different from that under free-shrinkage and pre-load conditions (Figs. 10 and 12). The initial positive displacements are now almost zero (compared to Fig. 10), because of superposition

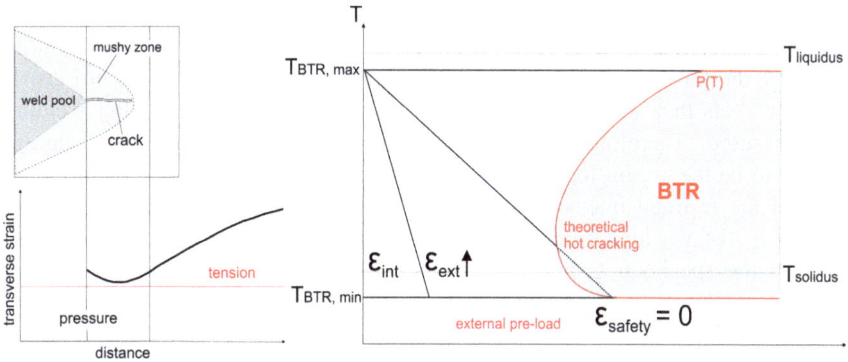


**Fig. 13** Local time and spatial resolution of the given pre-load condition (tensile) in Fig. 12, according to Zacharia [4] and Prokhorov [5] (from *left* to *right*) for the time t before the weld pool has passed the measuring points (path-controlled)

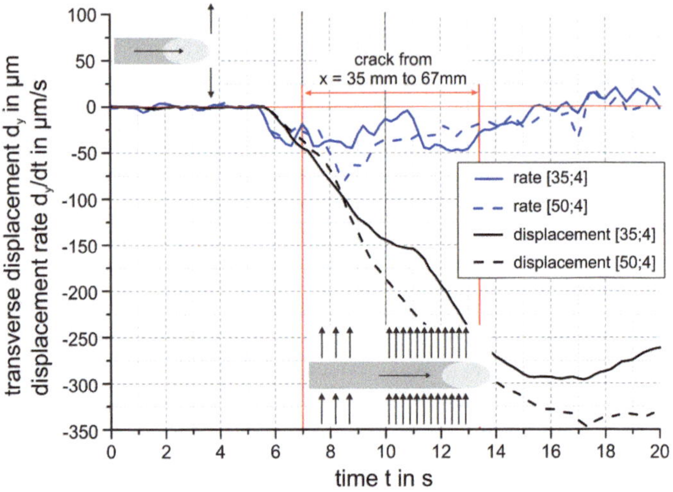


**Fig. 14** Local transverse displacement $d_y$ and displacement rate $d_y/dt$ versus time t, $v_w = 5$ mm/s, $v_l = 5$ mm/min (solidification crack critical)

with the external load (extension rate). Also, it can be seen at first sight that the extension rate leads to more or less linear curves of local transverse displacement. But close to the measuring points (between t = 7 s and t = 12 s) a displacement in the opposite direction was measured. But the main orientation of the displacements is still towards the direction of the extension rate, indicated by the arrows in the sketch in the lower centre of the diagram. Especially, the displacement rate shows rising gradients. In that time period the solidification crack was initiated. Obviously, the opposite material movement affected the crack initiation. The displacement and displacement rate decrease more up to extension rate stop. At the end of extension rate (after t = 16 s) a maximum displacement value of $d_y = -290$ µm respectively $d_y = -340$ µm is reached. As a result of the more cooled material the applied load has a lower effect at the position [35; 4] of the measured displacements. After welding an external load of ca. F = 23 kN (33 % of $R_{p0.2}$) was detected. During shrinkage the force increased again. Comparison of the measured force in Figs. 12 and 14 reveals that almost the same force, especially load directly after welding, was measured. Therefore, not only the load respectively the restraint intensity appears to be the reason for solidification cracking. Particularly, with regard to the value of the restraint intensity of approx. $R_y = 1$ kN/mm$^2$, the applied load was more than 20-times higher than the restraint intensity $R_y$. The restraint intensity is a result of the specimen geometry and can be calculated by Young's-modulus E, specimen thickness t and free specimen length l (without length of clamping):

$$R_y = \frac{E * t}{l} = \frac{\frac{210000\,\mathrm{N}}{\mathrm{mm}^2} * 2\,\mathrm{mm}}{400\,\mathrm{mm}} \approx 1\,\mathrm{kN/mm}^2.$$

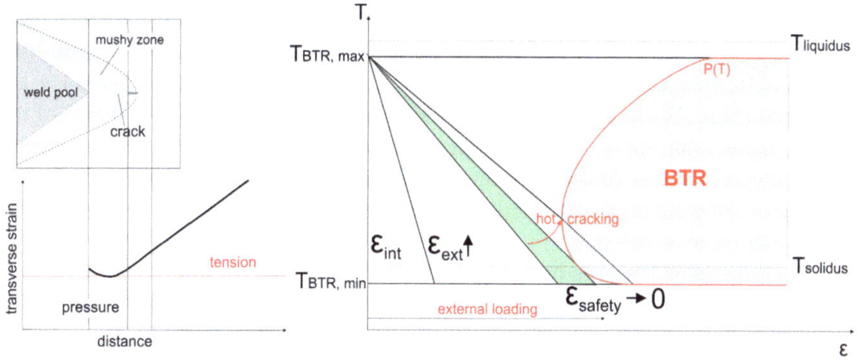


**Fig. 15** Local time and spatial resolution of the solidification crack critical load (extension rate) during welding in Fig. 14, according to Zacharia [4] and Prokhorov [5] (from *left* to *right*) for the time t after the weld pool has passed the measuring points

Figure 15 shows the characterisation of the condition of continuous extension rate during welding, according to Zacharia [4] and Prokhorov [5] corresponding to Fig. 14. In the following illustration the weld pool has already passed the measuring points. The case of time t before the weld pool has passed the measuring points (beginning of welding) is represented in Fig. 14. Corresponding to the material specific influence $\varepsilon_{int}$ it can be seen the design specific influence $\varepsilon_{ext}$ at the beginning (weld pool in front of the measuring points) is comparatively less. Hence, at that time the safety area $\varepsilon_{safety}$ up to intersecting the P(T)-curve is theoretically sufficient to avoid solidification cracking. The comparison between the various points of time shows that continuously loading caused high design specific influence $\varepsilon_{ext}$ was not reduced. Therefore, $\varepsilon_{safety}$ decreases until solidification cracking occurs.

In Fig. 16 the rounded dendrites prove that a solidification crack was detected (Fig. 14). It was formed in the time interval of coexisting liquid and solid phase, without any mutual contact of the dendrite tips.

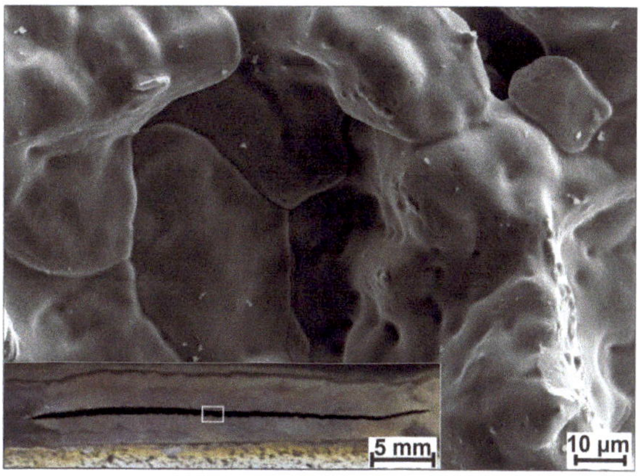


**Fig. 16** SEM image of solidification crack with melted dendrites, $v_w$ = 5 mm/s, $v_l$ = 5 mm/min (solidification crack critical)

Analogous to the above discussed graphs, the two-dimensional total displacement fields $d_t$ as a result of the different load conditions (free-shrinkage, pre-load and solidification crack critical extension rate) at the point in time t = 14 s are compared (Figs. 17, 18 and 19). The total displacement results in the addition of the displacement values in x-, y- and z-direction. The trend of displacement orientations is presented by arrows (not describing tensile or compressive fields in general). The position of the weld pool is marked as well. The local data measured are close to the fusion line up to y = 1 mm.

The comparison between the different load conditions shows clearly that the total displacements $d_t$ increase from Fig. 17 (free-shrinkage) up to Fig. 19 (solidification crack critical extension rate). It has to be noticed that the highly pre-loaded case indicates the lowest gradients respectively in the homogenous total displacement field. Furthermore, the vector orientation under pre-load illustrates a smaller influential area than in Fig. 17, which results only because of the material cooling/shrinkage. The greatest new orientation of the vectors can be seen in Fig. 19 with the applied solidification crack critical extension rate during welding. Also, a new orientation between the front and back of the weld pool position is observed.

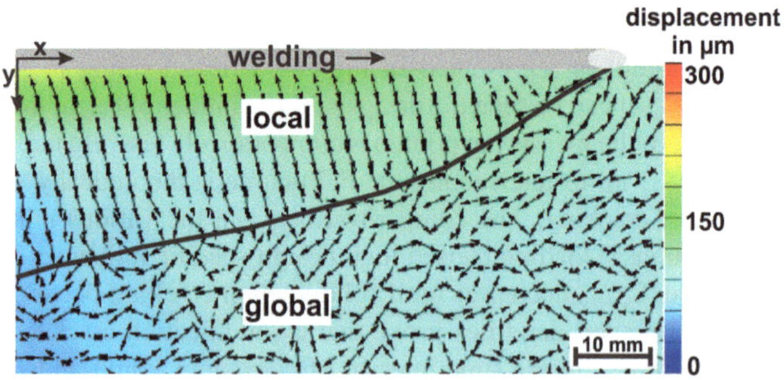


**Fig. 17** Local and global total displacement field $d_t$ at t = 14 s, $v_w$ = 5 mm/s, free-shrinkage

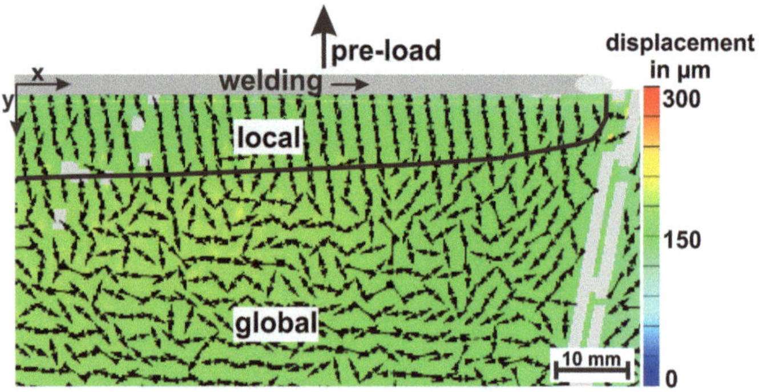


**Fig. 18** Local and global total displacement field $d_t$ at t = 14 s, $v_w$ = 5 mm/s, pre-load F = 100 kN

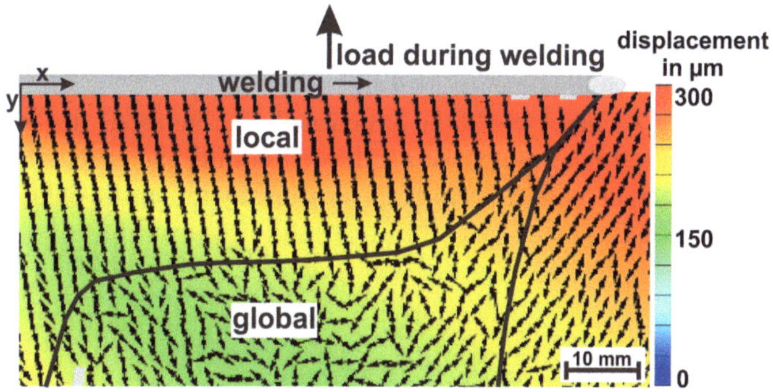


**Fig. 19** Local and global total displacement field $d_t$ at t = 14 s, $v_w$ = 5 mm/s, extension rate $v_l$ = 5 mm/min (solidification crack critical)

The vectors appear uniaxial orientated at a higher distance from the weld after the weld pool has passed the measuring point. From this fact it becomes evident that most of the load (extension rate) seems to act in the weld. Furthermore, the superposition of total displacements $d_t$ can be regarded as a result of load and expansion/shrinkage. In general it can be concluded the local deformation around the weld pool cannot be compared with the global one.

## Conclusions

This research work was focused on different design-specific influences on local displacements and displacement rates measured in situ using the DIC-technique.

- The use of specific light filters in extended investigations following from [15–17] enabled deformation to be detected without any shielding plate at reduced interfering radiation close to the weld pool in direct correlation to the weld pool position.
- In combination with the externally loaded CTW-test it was possible to evaluate the structural hot cracking behaviour. The results show how design-specific boundary conditions influence the forming of solidification cracking. It can be assumed that the acting material movement (displacement) exerts a major influence on solidification cracking due to continuous loading during welding.
- It can be concluded that an applied pre-load by comparison with an applied extension rate during welding shows a different influence on solidification cracking. It is evident that local pre-load is less solidification crack critical than applied load during welding in the case of extension rate.
- The difference stated between this research work and the earlier investigations of Kannengießer [8, 9], Cross and Böllinghaus [10] is the reduced effect of pre-load in the present study. Consequently, the case of pre-load condition has to be decided between a path- and force-controlled application.

- The results indicate that solidification cracking is not only depending on the restraint intensity $R_y$ but also on critical material movement (displacements) which has a detectable influence as well.
- The theory of Zacharia could be confirmed. Furthermore, the theories of Zacharia [3] and Prokhorov [5] were extended in time and space (see Figs. 11, 13 and 15). External loads interact in a different way for tensile fields in specific welding processes.

## References

1. Matsuda, F.; Nakata, K.; Arai, K.; Tsukamoto, K.: Assessment of solidification cracking test for aluminium alloy welds. Transactions of JWRI, vol. 11, no. 1, pp. 67-77, 1982.
2. Borland, J. C.: Generalized Theory of Super-Solidus Cracking in Welds (and Castings). British Welding Journal, pp. 508-512, 1960.
3. Zacharia, T; Aramayo G. A.: Modelling of Thermal Stresses in Welds.International Conference Proceedings on Modelling and Control of Joining Processes, USA, pp. 533-540, 1993.
4. Zacharia, T.: Dynamic Stresses in Weld Metal Hot Cracking, Welding Journal, vol. 73, no. 7, pp. 164-172, 1994.
5. Prokhorov, N. N.: The technological strength of metals while crystallising during welding. Doc IX-479-65, 1965.
6. Wolf, M.; Kannengiesser, T.; Boellinghaus, T.: Determination of critical strain rate for solidification cracking by numerical simulation, Hot Cracking in Welds II, Springer, Germany 2008.
7. Cross, C. E.: On the Origin of Weld Solidification Cracking, Hot Cracking Phenomena in Welds. Springer, Germany, pp. 3-18, 2005.
8. Kannengiesser, Th.; Cross, C. E.: Effect of Tack Placement on Local Weld Displacement and Solidification Cracking during Arc Welding of Aluminium Alloy 6083, The First South-East Asia IIW Congress, paper no. 52, pp. 480-491, Welding Institute of Thailand, 2006.
9. Kannengiesser, Th.; Kromm, A.: Design-Specific Influences in Local Weld Displacement and Hot Cracking, II International Conference on Welding and Joining of Materials, Peru 2007.
10. Cross, C. E.; Boellinghaus, T.: The Effect of Restraint on Weld Solidification Cracking in Aluminium, Welding in the World, Vol. 50, Issue 11-12, pp 51-54, Springer Verlag, 2006.
11. Kromm, A.; Kannengiesser, T.: Anwendung des neuen Controlled Tensile Weldability (CTW) Tests zur Untersuchung der Heißrissneigung. Fortschritte der Kennwertermittlung für Forschung und Praxis, Tagung-Werkstoffprüfung, pp. 451-456, 2006.
12. Matsuda, F.; Nakagawa, H.; Nakata, K.; Kohomoto, H.; Honda, Y.: Quantitative evaluation of solidification brittleness of weld metal during solidification by means of in-situ observation and measurement. Transactions of JWRI, vol. 12, no. 1, pp. 65-72, 1983.
13. Matsuda, F.; Nakagawa, H.; Matsubara, Y.: Quantitative evaluation of solidification brittleness of weld metal during solidification by means of in-situ observation and measurement (Report II). Transactions of JWRI, vol. 12, no. 1, pp. 73-80, 1983.
14. Pan, B.; Qian, K.; Xie, H.; Asundi, A.: Two-dimensional digital image correlation for in-plane displacement: a review, Measurement Science And Technology, 2009.
15. Gollnow, C.; Kannengießer, Th.: Analyse lokaler Verformungsfelder mittels DIC-Technik während der Heißrissbildung beim Schweißen. Schweißen und Schneiden, Heft 3, pp. 114-119, 2013.
16. Gollnow, C.; Kannengießer, Th.: Hot cracking analysis using in situ digital image correlation technique. Welding in the World, Springer Verlag, pp. 277-284, 2013.

17. Gollnow, C.; Kannengießer, Th.; Lauer, R.; Dong, M.; Riekers, R.: Experimental and Numerical Analysis of Local Thermomechanical Behavior for Hot Cracking Assessment in Welds. Mathematical Modelling of Weld Phenomena 10, pp. 357-370, 2013.
18. Kromm, A.; Kannengiesser, T.: Influence of Local Weld Deformation on the Solidification Crack Susceptibility of a Fully Austenitic Stainless Steel, Hot Cracking in Welds II, Springer, Germany 2008.
19. Herold, H.; Streitenberger, M.; Pshennikov: Prevention of Centreline Solidification Cracking during one-side welding, IIW-Doc. IX-2000-01, Online Dokument, 2001.
20. Wang, N., Mokadem, S., Rappaz, M., Kurz, W.: Solidification cracking of superalloy single and bi-crystals. Acta Materialia vol. 52, pp. 3173-3182, 2004.

# Prediction of Ductility-Dip Cracking in Narrow Groove Welds Using Computer Simulation of Strain Accumulation

**Steven L. McCracken and Jonathan K. Tatman**

**Abstract** Ductility-dip cracking (DDC) in high chromium nickel-base weld metals has been an issue during fabrication and repair of nuclear power plant components for many years. DDC is a solid-state cracking phenomenon, and several theories [1–10] have been proposed for the mechanism. Research conducted to develop these theories has primarily been performed using test methods involving small-scale specimens that may not replicate actual welding conditions (e.g., strain-to-fracture, hot-ductility, and varestraint). Due to the complexities of welding, there are potentially significant differences in the strain, strain-rates, stresses, and thermal cycles that can occur between these small-scale test methods and actual welding conditions. To eliminate this uncertainty, a high-restraint, narrow groove weld mockup was developed to assess DDC in this work. Filler metals 52 and 52M (AWS specifications ERNiCrFe-7 and ERNiCrFe-7A, respectively), compositions considered susceptible to DDC, are deposited with cold wire GTAW in a narrow groove with precise heat input and bead placement controls to isolate the occurrence of DDC to a known region of the weld deposit. Computer modeling using SysWeld™ with validated weld parameter inputs was also performed to simulate the narrow groove weld. Comparison of test specimens to computer simulations shows that the highest occurrence of DDC is in weld regions with multiple reheat cycles and high strain accumulation. This and future work is intended to develop a method to predict DDC susceptibility in multi-pass welds and to develop procedures and techniques that minimize the occurrence of DDC.

## Ductility-Dip Cracking: Established Theory Overview

Ductility-dip cracking occurs in austenitic weld metals within an elevated temperature range below the solidus, where there is an innate drop in ductility. This elevated temperature range (referred to as the ductility-dip temperature range) varies

S.L. McCracken (✉) · J.K. Tatman
EPRI Welding and Repair Technology Center, 1300 West W.T. Harris Blvd., Charlotte 28262, NC, USA
e-mail: smccracken@epri.com

 
Th. Boellinghaus et al. (eds.), *Cracking Phenomena in Welds IV*,
DOI 10.1007/978-3-319-28434-7_7

in welding literature; however it is well-known that recovery of this ductility drop occurs at a temperature above half the absolute solidus temperature of the material. This drop in ductility combined with the development of thermal strains during the weld cycle result in the formation of intergranular cracking.

A plot is provided in Fig. 1 which further describes the DDC phenomenon. As displayed in this plot, DDC occurs once the weld-induced strain reaches a threshold value ($E_{min}$) within the ductility-dip temperature range (DTR). Hot cracking occurs within the higher brittle temperature range (BTR). In many instances it is difficult to metallographically distinguish between hot cracking and DDC since they are both common in Ni-based weld deposits. However, the cause and mechanism of 52M hot cracking is considered to be better understood and typically occurs following dilution with stainless steel base metals where Nb, Si, S or P promote formation of low melting point constituents [11, 12].

Although the timing of DDC occurrence in Ni-base welds is generally well-understood, an agreement on the exact metallurgical cause has not been fully resolved. There are several predominant theories that have been proposed for the mechanism of DDC. One theory attributes DDC to the formation of excessive localized tensile stresses at grain boundaries due to intergranular formation of $M_{23}C_6$ precipitates [4]. Other research has refuted this theory [1–3], indicating that DDC results from plastic strain and grain boundary sliding (GBS) at elevated temperatures. In this case, it is suggested that intergranular precipitation of carbides are beneficial since they act to pin the grain boundaries at elevated temperatures, thus making them become jagged and tortuous. This tortuous nature of the grain boundaries prevents sliding and therefore increases resistance to DDC.

Grain boundary embrittlement due to migration of sulfur has also been proposed as a cause of DDC in Ni-base alloys [5–7]. Relatively high sulfur concentrations have been noted on DDC surfaces in these studies, however there has yet to be definitive evidence suggesting that the sulfur segregates to the grain boundaries prior to the occurrence of cracking. It is important to also note that the validity of this theory has been questioned further since DDC has been found in Ni-base alloys

**Fig. 1** Plot illustrating the DDC phenomenon [13]

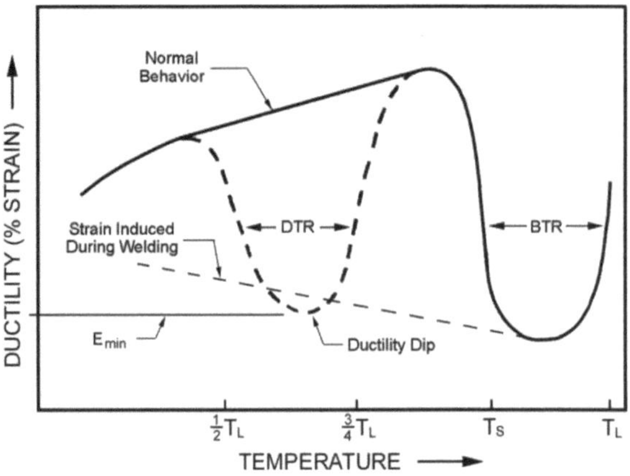

with extremely low levels of sulfur [8]. In addition to sulfur, DDC propensity has been shown to increase with the introduction of hydrogen [9] and oxygen [10] into the weld (e.g., delivered via shielding gas), although the exact reason is not fully understood.

A majority of the data used to support the above theories was gathered using test methods involving small-scale specimens. Data obtained from small-scale test methods (e.g., strain-to-fracture, hot-ductility, varestraint, PVR) can be deceiving since the augmented stresses and strains applied to these specimens may not replicate those which occur during actual high-restraint welding conditions. Uncertainty is compounded even further since the specimen designs and fabrication methods for these tests have not been standardized. Test methods have been developed and implemented (e.g., circular patch test) to replicate high restraint welding conditions [14, 15]; however attempts to discern between DDC and hot cracking in these welds have not been performed.

To better understand the mechanism of DDC, it is therefore necessary to develop a repeatable test method which replicates the high restraint conditions and multiple heat cycles that occur with multi-pass welding where DDC is most prevalent.

## Test Mockup Designed to Evaluate DDC Mechanism

Based on review of literature and industry experience, it was hypothesized in this work that DDC occurs in multi-pass weld regions that experience multiple elevated temperature cycles within the DTR, high grain boundary shear stress, and high plastic strain accumulation. This proposed hypothesis further supports the grain boundary sliding theory [1–3], and suggests the existence of a preferred grain boundary orientation for DDC to occur. The hypothesis also suggests that the grain size and straightness of the grain boundaries within the elevated temperature regions are important factors in determining whether or not DDC will occur. As an example, if the grains within the susceptible temperature region are long and straight, the likelihood of DDC will increase since there is minimal resistance to prevent the grain boundaries from sliding when exposed to weld shear stresses and plastic strain accumulation.

This proposed hypothesis implies a modified mechanism for DDC, which is the concept of multiple weld thermal cycles increasing the propensity for cracking. There are many regions in multi-pass welds that are exposed to multiple thermal cycles within the DTR. This multiple thermal cycle phenomenon, referred herein as "strain ratcheting", accumulates significant plastic strain in these regions which, when combined with the shear stresses developed from welding, results in grain boundary sliding and formation of DDC.

To confirm the proposed mechanism in this work, a narrow groove mockup design was developed to replicate a high restraint (i.e., high strain) welding condition. For this mockup design, the narrow groove was machined into a 1 in. (25.4 mm) thick gas metal arc weld (GMAW) nickel-base pad deposited on a 2 in.

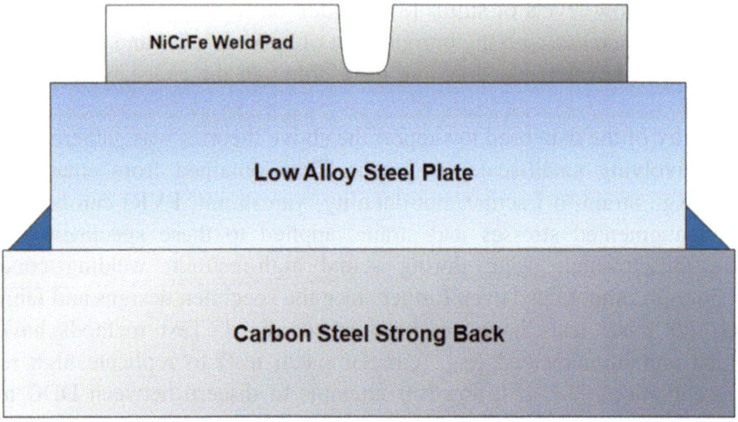


**Fig. 2** Cross-sectional schematic of narrow groove mockup. Note the NiCrFe GMAW pad is 1 in. (25.4 mm) thick and both steel plates are 2 in. (50.8 mm) thick

(50.8 mm) thick low alloy steel plate. A schematic of this mockup design is provided in Fig. 2. To eliminate impurities and the potential for hot cracking, the Ni-base filler material used to fabricate the weld pad on the low alloy steel was the same heat of weld metal used to make the narrow groove weld. The geometry of the narrow groove was designed for 12 weld layers and two passes per layer. The 2 in. (50.8 mm) thick carbon steel strong back was welded to a second 2 in. (50.8 mm) thick steel plate to provide added restraint.

To investigate the postulated theory that strain ratcheting promotes DDC, two controlled gas tungsten arc weld (GTAW) narrow groove bead sequences were tested in this study. These bead sequences involved the incorporation of multiple autogenous (i.e., no filler metal) remelt cycles on specific weld layers to concentrate the plastic strain accumulation and strain ratcheting within specific regions of the narrow groove weld.

The initial bead sequence tested is detailed in Fig. 3. For this bead sequence, there were 12 total weld layers with one, two, and three autogenous remelt cycles performed on layer four, seven, and ten weld beads respectively. The intention of the remelt cycles was to produce higher strain ratcheting and accumulation in the previously deposited weld metal regions directly below the remelt cycle layers (e.g., DDC would be expected to occur within weld layers eight and nine due to the remelt cycles imposed on layer ten). It is also important to note the pattern at which the remelt cycles were performed on each of the weld layers, which is detailed in Table 1.

Prior to performing the actual weld as specified in Fig. 3 and Table 1, a 2D proof-of-concept SysWeld™ finite element model (Fig. 4) was developed to examine the evolution of temperatures, shear stresses, and plastic strain accumulation when performing this bead sequence within the narrow groove. This SysWeld™ model was developed using ESI Group® Visual Environment version 9.5 software. Material databases readily available from ESI Group® for filler metal 52 and base

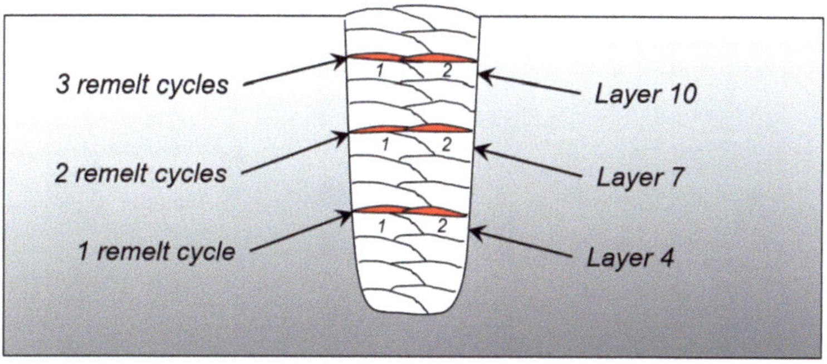


**Fig. 3** Autogenous remelt cycle descriptions for weld layers *four*, *seven*, and *ten*. The number of remelt cycles specified on the *left* were performed on each pass of the respective layers specified on the *right*. For example, three remelt cycles were performed on both weld *pass one* and *pass two* of *layer ten*

**Table 1** Bead deposition and autogenous remelt layer sequence pattern performed for the initial narrow groove weld test

| Weld pass number | Layer 4 | Layer 7 | Layer 10 |
|---|---|---|---|
| 1 | $1^a$ | $1^a$ | $1^a$ |
| 2 | $1^b$ | $1^b$ | $1^b$ |
| 3 | $2^a$ | $1^b$ | $1^b$ |
| 4 | $2^b$ | $2^a$ | $1^b$ |
| 5 | N/A | $2^b$ | $2^b$ |
| 6 | N/A | $2^b$ | $2^b$ |
| 7 | N/A | N/A | $2^b$ |
| 8 | N/A | N/A | $2^b$ |

See Fig. 3 for locations of layers 4, 7, and 10 for which 1, 2, and 3 remelt cycles were performed, respectively

[a] Original weld pass with filler wire deposition

[b] Autogenous (no filler wire) weld pass (i.e., remelt pass) using the same current, voltage, and travel speed as a weld pass with filler wire

metal ASTM A-516 Grade 70 were used for the NiCrFe and low alloy steel materials denoted in Fig. 4, respectively. The Goldak et al. [16] three-dimensional double-ellipsoid moving heat source calculation method was used to calculate the temperature fields of each simulated weld pass. Weld parameters intended to be used during fabrication of the actual weld (i.e., current, voltage, and travel speed) were used as the inputs to perform the moving heat source calculations.

As anticipated, evaluation of the proof-of-concept model confirmed the presence of high cumulative plastic strain directly below the remelt layers in weld metal regions that experience strain ratcheting and temperatures within the DTR (Fig. 5). Based on these results, it is expected that a higher density of DDC will occur within these high cumulative strain weld regions, with the highest density being below layer ten since this region experiences the greatest number of strain ratcheting

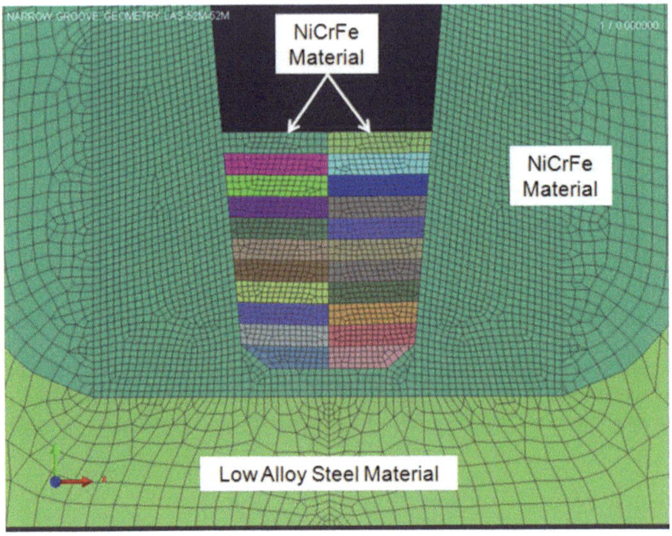


**Fig. 4** 2D finite element proof-of-concept SysWeld™ model

cycles and highest cumulative plastic strain. Note that although the remelt layers themselves experience multiple temperature cycles within the DTR, DDC is less likely to occur since these regions experience melting during each remelt cycle. Because of this melting, the cumulative strain within the melt zone of these layers resets to zero during each remelt cycle, and therefore does not experience the strain ratcheting or accumulation effect.

Further evaluation of the model was performed on nodes below each remelt layer which experience high cumulative strain (Fig. 6), and a plot was developed to compare the shear stresses, cumulative plastic strain, and thermal cycles of these node locations (Fig. 7). The data acquired and provided in this plot was taken from the final weld thermal cycle of each remelt layer. For ease of comparison, the time was normalized for each of the node locations. As shown in Fig. 7, the only significant difference in the data between the three node locations is the quantity of cumulative plastic strain, which was anticipated due to the difference in the number of strain ratcheting cycles. Several other details are also relevant to denote from this plot. The first is the presence of shear stress simultaneous to the occurrence of plastic strain accumulation, which provides an indication that grain boundaries are prone to shearing within these regions. It is also important to note that this dual combination of shear stress and plastic strain accumulation primarily occurs when temperature ranges are within the DTR. Finally, the cumulative plastic strain rate is higher during the on-heating portion of the thermal cycle, providing evidence that DDC may be more likely to initiate in these regions during the on-heating portion of the thermal cycle.

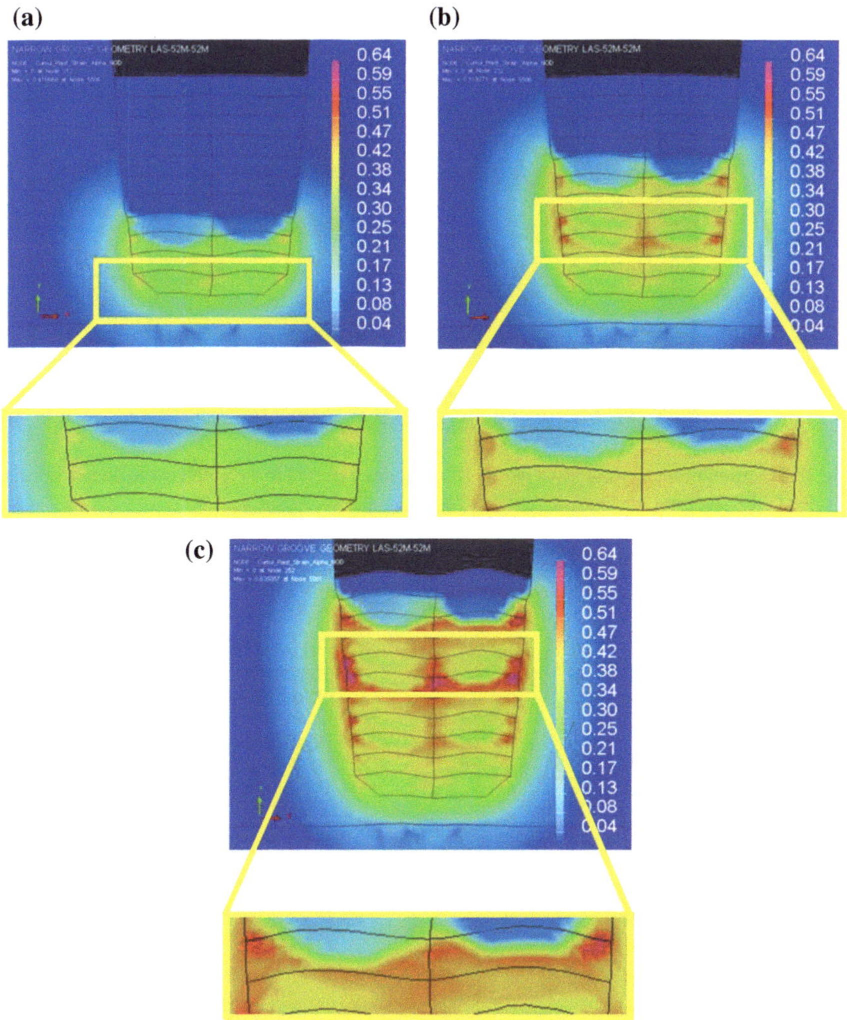


**Fig. 5** Cumulative plastic strain maps immediately following the final remelt cycle for weld layers **a** four, **b** seven, and **c** ten

## Confirmation of Strain Ratcheting Effect on DDC

To confirm the results of the proof-of-concept model and strain ratcheting effect on DDC susceptibility, a narrow groove test was performed using the proposed DDC mockup design. Filler metal 52M (ERNiCrFe-7A) was used to deposit the 1 in. (25.4 mm) thick GMAW pad over ASTM A-516 Grade 70 low alloy steel substrate. Following fabrication of the weld pad, a narrow groove was machined through the center of the GMAW weld pad transverse to the direction of the GMA weld beads

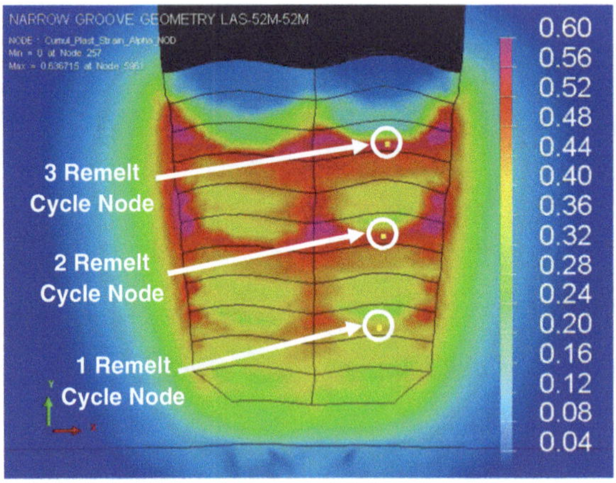


**Fig. 6** Location of high cumulative plastic strain nodes, referred to as remelt cycle nodes, selected for detailed evaluation

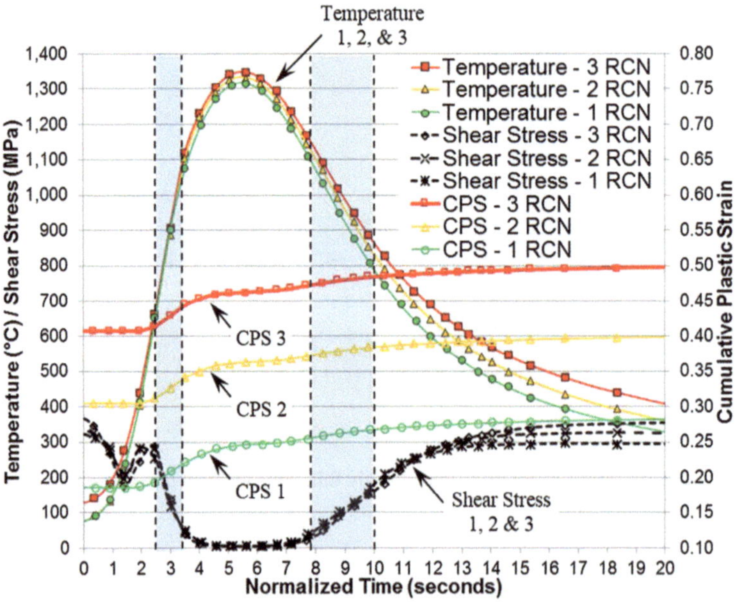


**Fig. 7** Plot showing thermal, shear stress, and cumulative plastic strain (*CPS*) data from the three remelt cycle nodes (*RCN*) selected for detailed examination (see Fig. 6). Note that the *blue shaded regions* indicate the approximate ductility-dip temperature range (*DTR*) during on-heating and on-cooling cycles

**Fig. 8** Image of narrow groove machined transverse to the GMA 52M weld pad

(Fig. 8). A 12 layer weld was performed within the narrow groove as specified in Fig. 3, and remelt passes were performed on weld layers four, seven, and ten using the sequence specified in Table 1.

Following completion of the narrow groove weld, two cross-sections were machined approximately 2 in. (50.8 mm) from the weld start and finish locations and examined for cracking. These locations were initially selected to determine if a difference in DDC susceptibility was present between the weld start and finish locations (it was later determined that no significant difference was present). Cracks less than 0.004 in. (0.10 mm) in size were not counted and multiple cracks along a single grain boundary were counted as a single crack if the length of the connecting ligament was less than the length of the smaller of the two cracks.

The weld start and finish cross-sections were polished and electrolytically etched using a 10 % Nital solution. Crack counts for both cross-sections were combined and the totals are provided in Table 2. The crack counts in Table 2 are separated based on the regions where the cracks occurred. As an example, the triple remelt pass region is considered to be layers seven, eight, and nine (i.e., the first three layers below triple remelt layer ten). Actual crack locations were also noted on the macrographs of both cross-sections, and are provided in Fig. 9.

The effect of strain ratcheting and cumulative plastic strain on DDC occurrence is evident based on the crack data in Table 2. As depicted in this table, crack totals increased significantly with increasing remelt cycles. To further confirm the effect

**Table 2** Crack count totals on the cross-sections removed from the weld start and finish locations of the 52M narrow groove DDC test mockup

| DDC susceptible weld region | Total number of cracks | Total crack length (in./mm) | Maximum crack length (in./mm) |
|---|---|---|---|
| Single remelt (layers 1, 2, 3) | 0 | 0 | 0 |
| Double remelt (layers 4, 5, 6) | 7 | 0.063/1.60 | 0.013/0.33 |
| Triple remelt (layers 7, 8, 9) | 10 | 0.108/2.74 | 0.020/0.51 |

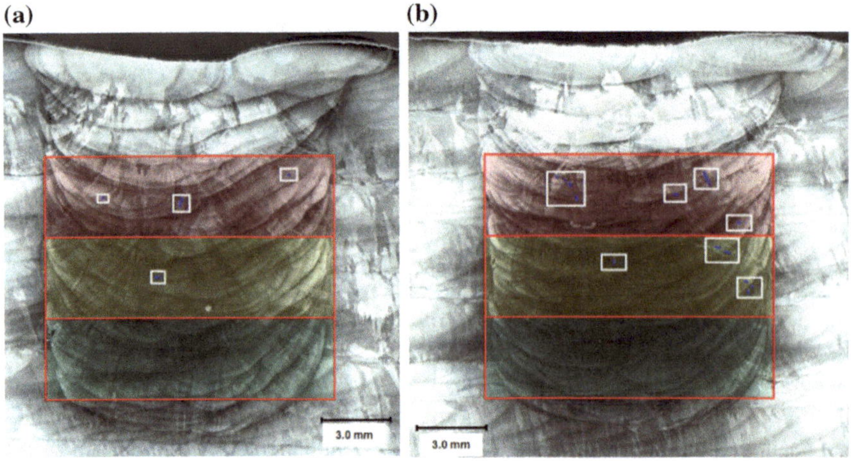


**Fig. 9** Cracks (identified by a *blue line* in the *small white boxes*) on the **a** weld start and **b** weld finish cross-sections taken from the 52M narrow groove DDC test mockup. The *three larger red rectangular boxes* indicate the approximate DDC susceptible regions

of cumulative plastic strain and strain ratcheting on DDC occurrence, the cumulative plastic strain map from the SysWeld™ simulation was superimposed on the weld finish cross-section (Fig. 10). The superimposed strain map reveals that the cracks are primarily located in the high cumulative plastic strain regions, providing further evidence of the strain ratcheting effect on DDC susceptibility.

Although the cracking was found to be located in the regions of high cumulative plastic strain, it should be noted that attempts were not performed to benchmark the SysWeld™ simulation results. Therefore, the exact location and magnitude of the cumulative plastic strains are qualitative, and should only be used for observing trends. A second narrow groove test was performed in this study to benchmark the simulation and more accurately predict the exact locations of cumulative plastic strain.

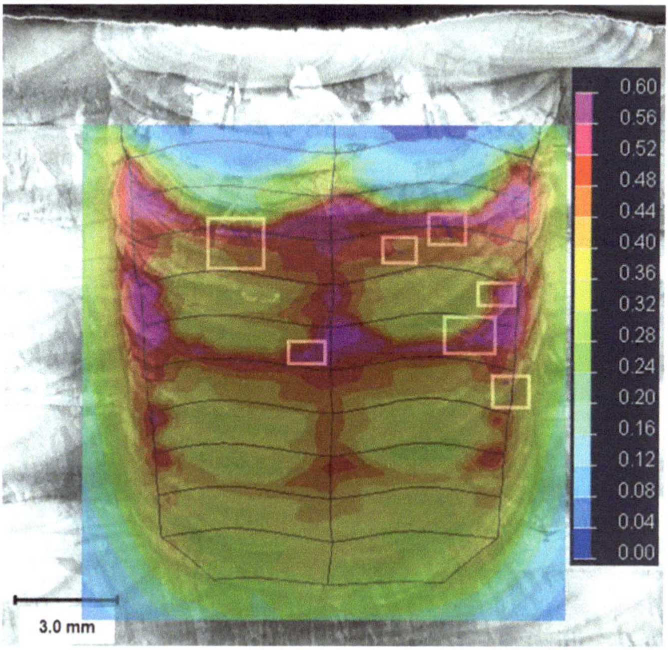


**Fig. 10** Crack locations (denoted by *white boxes*) on the weld finish narrow groove DDC test cross-section relative to the cumulative plastic strain map predicted from the proof-of-concept SysWeld™ simulation

## Comparison of First and Second Narrow Groove Test

Several differences were observed between the first and second narrow groove weld test performed in this study:

1. The second narrow groove test was performed using filler metal 52 (ERNiCrFe-7), a composition considered slightly more susceptible to DDC than the 52M filler material used for the first test.
2. The single remelt pass layer was eliminated since no cracking was present below this layer in the first weld test. In lieu of performing the remelt cycles on layers four, seven, and ten, the remelt cycles were performed on layers four and eight (Fig. 11).
3. The bead deposition order on the second test was inadvertently reversed on weld layer 6 (Fig. 11). This change in bead order was initially assumed to be insignificant and not to influence DDC susceptibility.
4. Rather than depositing the first bead and performing all corresponding remelt cycles on that bead prior to depositing the second weld bead, both beads were deposited prior to imposing the remelt cycles on layers four and eight.

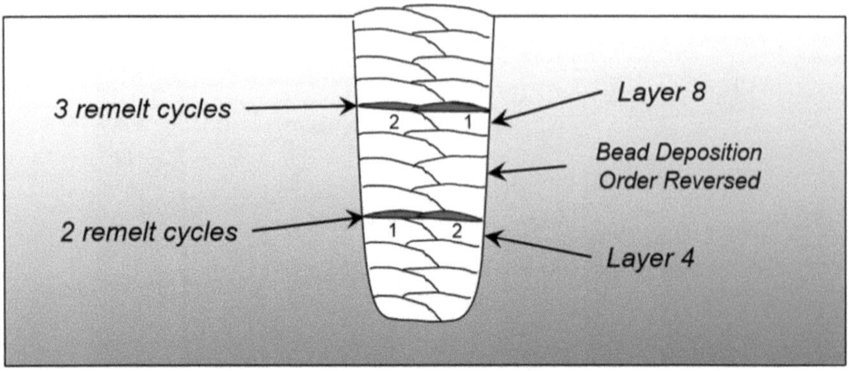


**Fig. 11** Bead deposition order and remelt cycle description for weld layers *four* and *eight*

**Table 3** Bead deposition and remelt layer sequence pattern performed for the second narrow groove weld test

| Weld pass number | Layer 4 | Layer 8 |
|---|---|---|
| 1 | 1[a] | 1[a] |
| 2 | 2[a] | 2[a] |
| 3 | 1[b] | 1[b] |
| 4 | 2[b] | 2[b] |
| 5 | 1[b] | 1[b] |
| 6 | 2[b] | 2[b] |
| 7 | N/A | 1[b] |
| 8 | N/A | 2[b] |

See Fig. 11 for the location of layer 4 with 2 remelt cycles and layer 8 with 3 remelt cycles

[a] Original weld pass with filler wire deposition

[b] Autogenous remelt pass (i.e., no filler wire) using the same current, voltage, and travel speed as a weld pass with filler wire

Additionally, the remelt cycles were alternated between beads one and two rather than being performed continuously on each bead. This different remelt layer sequence for the second test is described further in Table 3.

## Benchmarking of SysWeld™ Simulation

A number of items must be addressed to successfully benchmark the proof-of-concept SysWeld™ simulation developed for this study, which was attempted following fabrication of the second narrow groove DDC test described above. It is important to note that benchmarking the exact quantities of cumulative plastic strain throughout the narrow groove weld was outside the scope of this work, and attempts were only made to benchmark locations of high cumulative plastic strain occurrence within the narrow groove weld.

To properly benchmark the cumulative plastic strain regions within the narrow groove weld, the actual shapes of the weld deposit areas for each pass must be accurately simulated. The initial proof-of-concept simulation developed in this study contained non-representative weld beads that did not overlap and were rectangular in shape. For the actual narrow groove weld, significant bead overlap and slight differences in bead shape occur on each layer, and therefore a method to profile each weld pass is necessary to replicate the actual shapes. To accomplish this for the second narrow groove weld test, a Keyence LJ-V7080 laser profilometer was used (Fig. 12) to acquire profile data immediately following deposition of each weld pass.

Following completion of the narrow groove weld, the profile data was combined on a single plot to reveal the shapes of the deposited weld beads. As displayed in Fig. 13, an added benefit to obtaining this profile data is it can be overlapped onto the actual weld cross-section to reveal exact locations within the narrow groove weld.

As mentioned previously, the Goldak [16] three-dimensional double-ellipsoid moving heat source calculation method was used to calculate the temperature fields for the proof-of-concept simulation. Although this calculation method is considered accurate in terms of predicting the heat supplied by the welding arc, it does not take into account variations in substrate heat distribution due to changes in weld pool fluid flow, which can vary considerably depending on the composition of the

**Fig. 12** Keyence LJ-V7080 laser profilometer used for gathering bead shape data within narrow groove weld

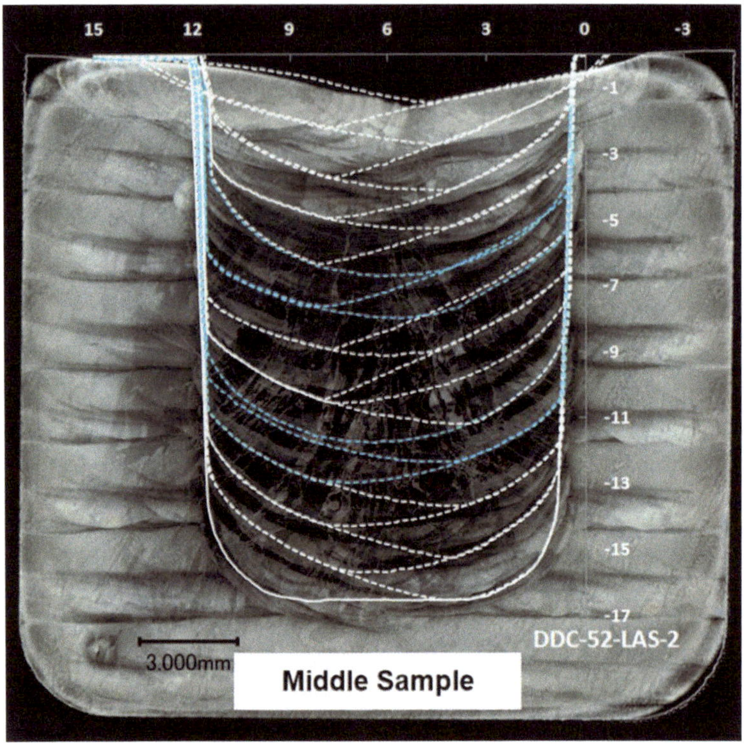


**Fig. 13** Bead profile data superimposed on macrograph of completed 52 narrow groove DDC mockup. Note the numbers on the *horizontal* and *vertical axes* represent millimeters

substrate. Changes in weld pool fluid flow can have an effect on the shape of the fusion zone. One element that has been shown to have a significant effect is sulfur [11, 12, 17–19]. Figure 14 shows macrographs of two GTAW 52M beads performed on two variants of low alloy steel substrate. Although identical weld conditions were used to fabricate these weld beads, they exhibited significant differences in bead shape (e.g., bead width, bead height, penetration). In this case, the difference in bead shape is likely attributed to the difference in sulfur content between the two substrates.

NiCrFe materials typically contain extremely low levels of sulfur to minimize the potential for hot cracking. Because of this, the shape of weld beads deposited on this material-type will favor a wide and shallow profile; similar to the bead-on-plate weld on the ASTM A-516 Grade 70 low alloy steel in Fig. 14.

To investigate the fusion zone shape for NiCrFe materials, a single GTAW bead-on-plate test was performed with 52M filler metal on Alloy 690 substrate

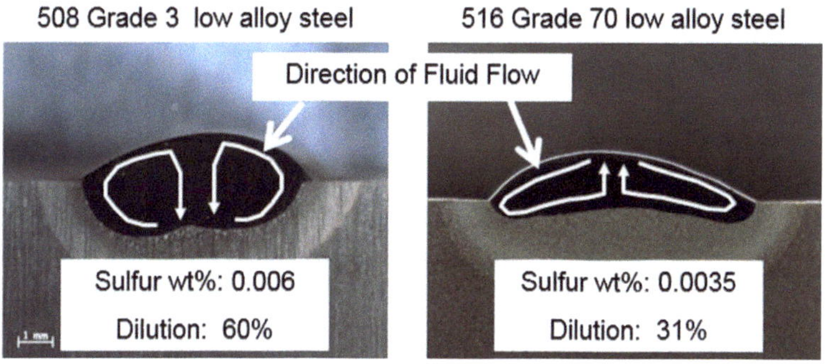


**Fig. 14** Macrographs of weld trials provided to show the significance of fluid flow on bead shape [19]. The 508 Grade 3 and 516 Grade 70 sulfur content was determined with a SpectroMax 5F arc spark OES metal analyzer

using the same parameter inputs used for the proof-of-concept simulation. The melt zone of the actual bead-on-plate weld was then compared to a simulated bead-on-plate weld using the default SysWeld™ heat source calculation method. Comparison of the simulated fusion zone using the default heat source and a cross section of the actual bead are shown in Fig. 15a and c. Assessment of Fig. 15 reveals significant differences between the actual and simulated fusion zones, indicating that the proof-of-concept simulation did not accurately replicate the fusion zone shape of the actual weld. The reason for this difference is attributed to the uniform Gaussian heat source profile used for distributing the calculated arc heat within the substrate. In actuality, distribution of the arc heat transferred to the substrate varies depending on the fluid flow of the weld pool, and therefore using the uniform Gaussian thermal distribution for simulations is likely not applicable for all cases.

Although the fluid flow of the weld pool significantly influences the final shape of the fusion zone, studies have been performed which show the depth and shape of the heat affected zone remains relatively unaffected [20]. This indicates the accumulation of plastic strain outside of the melt zone may not be significantly influenced by changes in fluid flow, providing a potential justification for not requiring replication of the actual fusion zone shape in the simulation. However as previously indicated, the cumulative plastic strains are reset in the fusion zone. Therefore, it may be necessary to modify the simulated heat source to reset the cumulative plastic strains in appropriate regions throughout the narrow groove weld. Due to this uncertainty, the heat source for the simulation was modified such that the simulated fusion zone matched the actual fusion zone shape (see Fig. 15b and c). This modified heat source distribution was used for simulating a second narrow groove DDC weld test performed in this study.

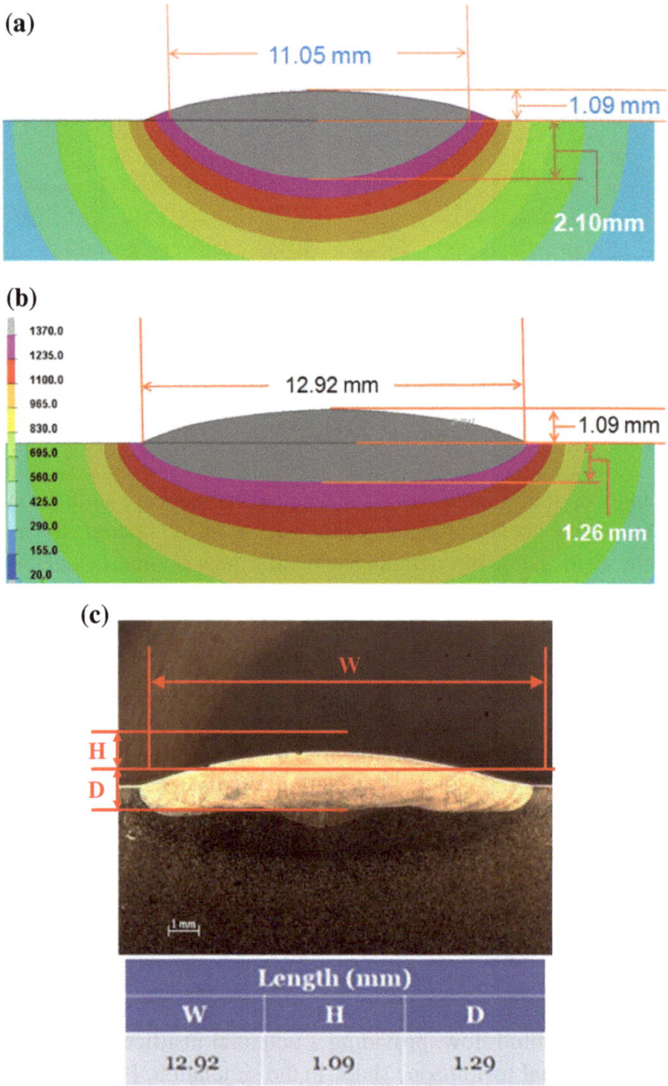


**Fig. 15** Comparison between **a** simulated fusion zone (*gray region*) using the default heat source distribution, **b** simulated fusion zone using the modified heat source distribution and **c** a cross section of the actual fusion zone. Simulation image **a** and **b** are courtesy of ESI Group®

## Crack Analysis of Second Narrow Groove DDC Test and Comparison with Benchmarked SysWeld™ Simulation

A second narrow groove DDC test was performed with filler metal 52 to benchmark the SysWeld™ simulation with the modified heat source. Three cross-sections were removed and examined for DDC. One cross-section was removed from the mid-location of the narrow groove weld; the same location as where the profile data was acquired. Similar to the first filler metal 52M narrow groove test, the other two cross-sections were removed 2 in. (50.8 mm) from the weld start and finish ends of the weld pad. The same minimum crack size limit of 0.004 in. (0.10 mm) was also used. Crack totals for the 52 narrow groove test are provided in Table 4.

Comparison of the crack totals between the first test with 52M and second test with 52 reveals significantly less total crack length (TCL) in the 52 weld metal. This was the case even though cracks were counted on three cross-sections from the 52 narrow groove test as opposed to only two cross-sections from the 52M test. This decrease in TCL was unexpected since filler metal 52 is considered to be slightly more susceptible to DDC than filler metal 52M. The unexpectedly lower TCL in the 52 weld metal indicates that the change in remelt pass sequence may have had a significant influence on DDC susceptibility. Close examination of the data shows that although less cracking was found in the 52 weld metal, the majority of cracking still occurred within the regions containing the most strain ratcheting cycles (i.e., below layer ten), further confirming the influence of strain ratcheting on DDC. Crack locations were noted on the mid-location cross-section containing the overlapped profile data (Fig. 16), revealing the exact weld layers where the cracking occurred.

As discussed previously, the primary reason for obtaining the bead profile data for the 52 narrow groove test was to replicate the exact bead shapes in the 2D finite element SysWeld™ simulation. The finite element mesh for this simulation is displayed in Fig. 17. The calibrated heat source distribution and Table 3 remelt sequence were imposed on the simulated weld beads.

Following simulation of the heat cycles, the cumulative plastic strains of the benchmark simulation were compared to the original proof-of-concept simulation and the crack locations on the 52 narrow groove test. The simulated cumulative plastic strain map for the filler metal 52 narrow groove test is presented in Fig. 18. The highest plastic strain accumulation was located in regions directly below the

**Table 4** Crack count totals from cross-sections removed from the weld start, middle, and finish locations of the 52 narrow groove DDC test

| DDC susceptible weld region | Total number of cracks | Total crack length (in./mm) | Maximum crack length (in./mm) |
|---|---|---|---|
| Double remelt (layers 1, 2, 3) | 4 | 0.034/0.864 | 0.018/0.457 |
| Triple remelt (layers 5, 6, 7) | 10 | 0.085/2.159 | 0.015/0.381 |

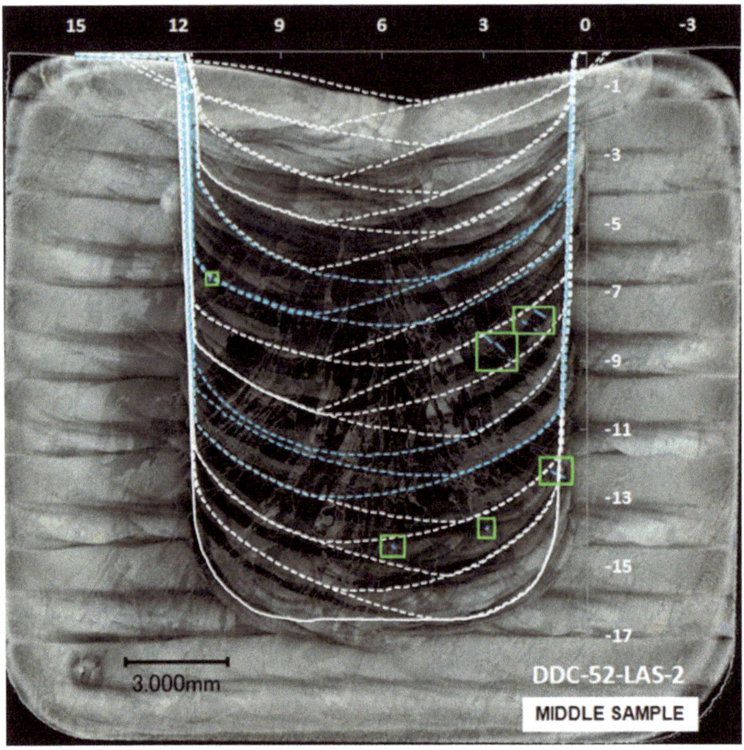


**Fig. 16** Exact crack regions on the mid-location cross-section taken from the 52 narrow groove DDC test. Note that the bead profile data is superimposed on the macrograph to reveal the weld layers where cracking occurred. The numbers on the *horizontal* and *vertical axes* represent millimeters, and the *blue dotted lines* represent the two remelt layers

weld layers with two and three remelt cycles, which agrees with the original proof-of-concept model. However, the notable difference between the benchmarked model and the proof-of-concept model is the high cumulative plastic strain regions are focused in the center of the weld layers, and are not evenly distributed across the entire transverse width of the narrow groove weld (compare Figs. 6 and 18). This difference is most likely the change in remelt bead sequence pattern between the proof-of-concept model and the benchmarked model. An additional contributing factor may be the overlapping of the weld beads in the benchmarked model as opposed to the rectangular, non-overlapped beads simulated in the proof-of-concept model.

Similar to Fig. 10, the cumulative plastic strain map from the benchmarked SysWeld™ simulation was superimposed on the middle cross-section containing the overlapped profile data (Fig. 19). Note that two separate strain maps are present on the macrograph. The top and bottom maps superimposed on this macrograph represent the cumulative plastic strains following completion of weld layers 11 and 7,

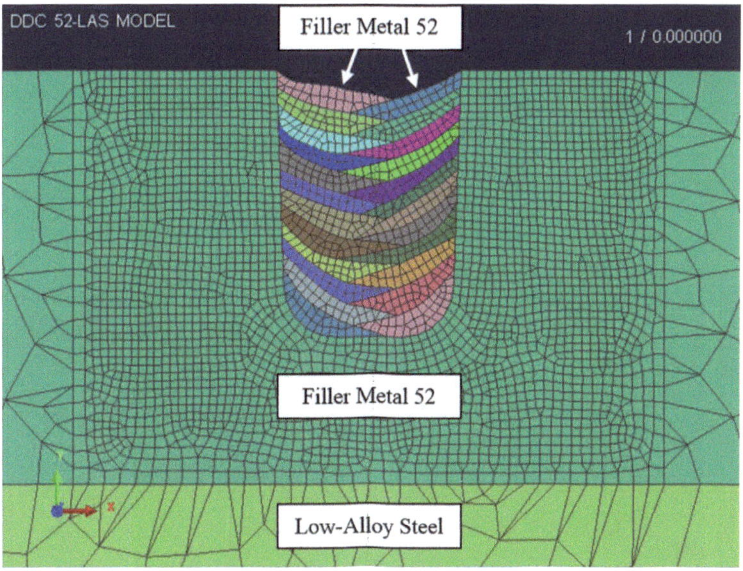


Fig. 17 Benchmarked 2D finite element mesh in SysWeld™

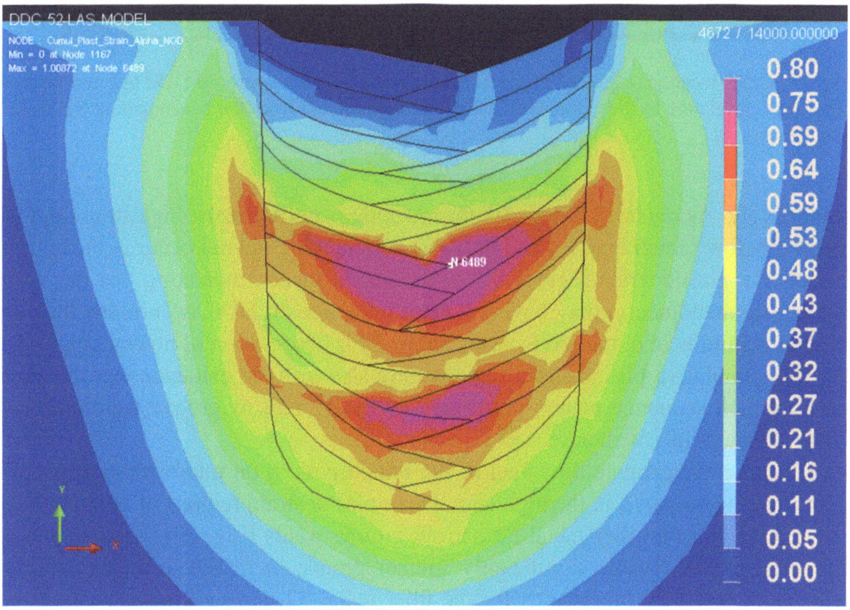


Fig. 18 Cumulative plastic strain map of the benchmarked SysWeld™ model following simulation of all weld passes

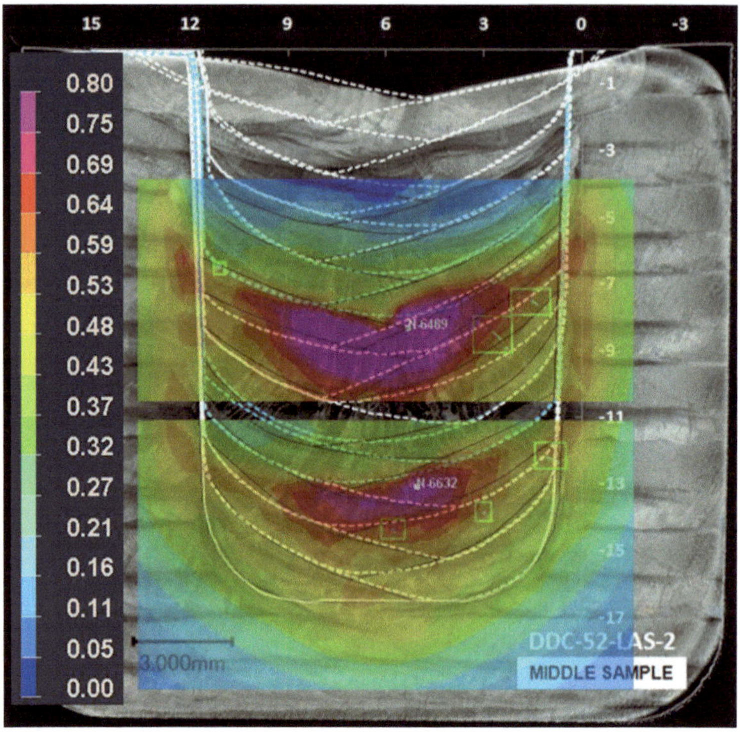


**Fig. 19** Crack locations (denoted by *green boxes*) on the 52 narrow groove DDC test mid-location cross-section relative to the cumulative plastic strain predicted from the benchmarked SysWeld™ simulation

respectively. As revealed in Fig. 19, the cracking occurred in the high strain regions directly below remelt layers 4 and 8. However, cracking did not occur in the highest cumulative plastic strain regions below the remelt layers, indicating that the strain ratcheting phenomenon may have a greater influence on DDC occurrence than the actual magnitude of cumulative plastic strain. This observation of no cracking in the highest cumulative plastic strain regions may also provide evidence that the alternating remelt sequence pattern performed on the 52 narrow groove weld creates a DDC resistant microstructure.

To further investigate the effect of the remelt pattern on microstructure, micrographs were compared in the high cumulative plastic strain regions of both the filler metal 52M and the filler metal 52 narrow groove welds (Fig. 20). Comparison of the micrographs reveals significantly smaller grains, increased tortuosity, recrystallization, and twin boundaries for the filler metal 52 narrow groove weld; all of which have been concluded to be beneficial factors for resisting DDC [21]. This observation is uncharacteristic of 52 weld metal, and implies that the alternating remelt pattern promoted a DDC resistant microstructure. This alternating remelt

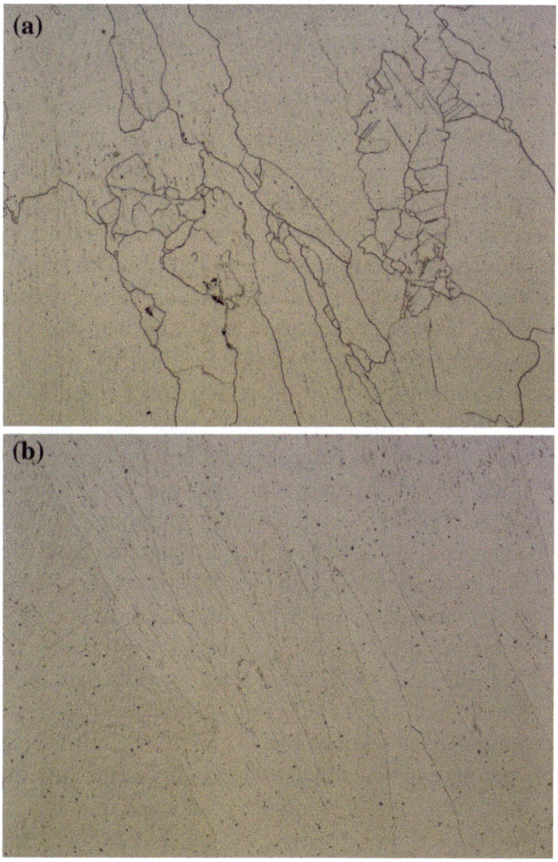


**Fig. 20** Comparison of micrographs within the high cumulative plastic strain regions of **a** filler metal 52 and **b** filler metal 52M narrow groove welds

pattern effect is further confirmed by the fact that less DDC occurred in the filler metal 52 narrow groove weld, which is considered slightly more susceptible to DDC than filler metal 52M. Additional narrow groove weld trials are required to further confirm this observation.

## Conclusions

1. A modification to the grain boundary sliding theory [1–3] was proposed in this work for the DDC mechanism in NiCrFe weld metals. This modified theory implies that multiple re-heat cycles in the ductility-dip temperature range accumulate plastic strains during both the on-heating and on-cooling portions of the re-heat cycles, resulting in a strain ratcheting effect. It is suggested that this strain ratcheting along with the accumulation of plastic strain exhausts the strength of the grain boundaries within these excessively reheated weld regions,

and the shear stress induced at the grain boundaries during these ratcheting events results in the formation of DDC.

2. A proof-of-concept SysWeld™ model was developed to investigate the effect of strain ratcheting in reheated weld regions. Results from the proof-of-concept SysWeld™ model confirmed the highest presence of cumulative plastic strains in regions experiencing the most reheat (strain ratcheting) cycles within the ductility-dip temperature range.
3. A new high restraint narrow groove test mockup was successfully developed in this study to confirm the proposed strain ratcheting effect on DDC susceptibility. Cracking results from test mockups fabricated using both filler metal 52 and 52M narrow groove weld tests revealed the highest density of DDC in weld metal regions experiencing the highest number of strain ratcheting cycles, confirming the strain ratcheting effect on DDC susceptibility.
4. A different remelt cycle pattern was used to fabricate the narrow groove weld test with filler metal 52, and cross-sections revealed less overall cracking than the filler metal 52M narrow groove weld test. This was unexpected since filler metal 52 is considered slightly more susceptible to DDC than filler metal 52M, and suggests the bead sequence pattern may have a significant effect on microstructure and DDC susceptibility.
5. A benchmarked SysWeld™ simulation was developed for comparison with the filler metal 52 narrow groove test performed in this study. The model results indicate that DDC was most prevalent in regions experiencing the highest number of strain ratcheting cycles and not the regions containing the highest cumulative plastic strain. Although further testing is required for confirmation, this may indicate that strain ratcheting has a greater effect on DDC occurrence than the actual magnitude of cumulative plastic strain.

**Acknowledgements** The authors would like to thank Takanori Kanehiro from IHI Corporation for assistance with conceptual development of the new narrow groove DDC test. The authors would also like to thank Brian Shula from ESI Group® for his assistance on development of the benchmarked SysWeld™ simulation used in this study.

## References

1. Ramirez A.J. and Lippold, J.C. 2004. High temperature cracking in nickel- base weld metal, Part 2– Insight into the mechanism, Materials Science and Engineering A, 380: 245-258.
2. Collins, M. G., Ramirez, A. J., and Lippold, J. C. 2004. An investigation of ductility dip cracking in nickel-based weld metals — Part 3. *Welding Journal* 83: 39-s to 49-s.
3. Noecker, F.F. II, and Dupont, J.N. 2009. Metallurgical Investigation into Ductility Dip Cracking in Ni-Based Alloys: Part 2. *Welding Journal* 88: 62-s to 77-s.
4. Young, G. A., Capobianco, T.E., Penik, M.A., Morris, B.W., and McGee, J.J. 2008. The Mechanism of Ductility Dip Cracking in Nickel-Chromium Alloys. *Welding Journal* 87: 31 (s)-43(s).

5. Nishimoto, K., Saida, K., Okauchi, H., and Ohta, K. 2006. Microcracking in multipass weld metal of Alloy 690. Part 2: Microcracking mechanism in reheated weld metal. *Science and Technology of Welding and Joining* 11(4): 462-470.
6. Nishimoto, K., Saida, K., Okauchi, H., and Ohta, K. 2006. Microcracking in multipass weld metal of Alloy 690. Part 3: Prevention of microcracking in reheated weld metal by addition of La to filler metal. *Science and Technology of Welding and Joining* 11(4): 471-479.
7. Nishimoto, K., Mori, H., and Hongoh, S. 1999. *Effect of sulfur and thermal cycles on reheat cracking susceptibility in multipass weld metal of Fe - 36% Ni alloy*, International Institute of Welding, IIW Doc. IX - 1934 - 99.
8. Dupont, J. N., Lippold, J. C., Kiser, S. D. 2009. *Welding Metallurgy and Weldability of Nickel-Base Alloys*, John Wiley & Sons Inc., 128-143.
9. Collins, M. G. and Lippold, J. C. 2003. An investigation of ductility – dip cracking in nickel - based filler metals — Part I , *Welding Journal* , 82 ( 10 ): 288s – 295s .
10. Nissley, N. 2006. *Intermediate temperature grain boundary embrittlement in Ni -base weld metal*. PhD Dissertation, The Ohio State University.
11. McCracken, S. L., Smith, R. E., *Evaluation of Filler Metal 52M (ERNiCrFe-7A) Hot Cracking When Welding on Cast Austenitic Stainless Steel Base Materials*, Proc. ASME PVP Conf., July 17-21, 2011, Baltimore, MD, Paper No. PVP2011-57703.
12. McCracken, S. L., Tatman, J. K., *Influence of Austenitic Stainless Steel Base Metals on Hot Cracking Susceptibility of High Chromium Nickel-Base Filler Metals*, Proc. ASME PVP Conf., July 15-19, 2012, Toronto, Ontario, Canada, Paper No. PVP2013-78614.
13. McCracken, S. L., *Review of Thermo-Mechanical Testing and Micro- Characterization Studies of Nickel-Base Alloy 52 Weld Filler Metal for Susceptibility to Ductility-Dip Cracking,* Proc. ASME IMECE Conf., November 15-21, 2003, Washington D.C., Paper No. IMECE2003-43342.
14. Chabenat, A., Thomas, A., Waskey, D. *Hot Cracking Susceptibility of 30% Cr Alloy Filler Metal in GTAW Deposits,* Welding & Repair Technology for Power Plants 6th International EPRI Conference, June 2004, Sandestin, Florida
15. Brown, C.M., Hicks, T.G., Tatman, J. K. June 2010, *Fatigue, Fracture Toughness, and Weldability Characteristics of Three Nickel-Chromium-Iron Alloy Welding Products,* EPRI WRT Conference.
16. Goldak, J., Chakravarti, A., and Bibby, M. 1985. *A double ellipsoid finite element model for welding heat sources*, IIW Document Number 212-603-85.
17. McCracken, S. L., Smith, R. E. *Hot Cracking Phenomena in Welds III*, 2011, pp. 333-352.
18. *Welding and Repair Technology Center: Measures to Minimize 52M Hot Cracking on Stainless Steel Base Materials.* EPRI, Palo Alto, CA: 2012. 1025167
19. Tatman, J. K., McCracken, S. L., and Hicks, T. G., Development of New Weld Heat Input and Dilution Equations for Gas Tungsten Arc Welding – Part 1, *Proc. ASME PVP Conf.,* July 14-18, 2013, Paris, France, Paper No. PVP2013-97358.
20. Glickstein, S. S. and Friedman, E. June 1981. Technical Note: Effect of Weld Pool Configuration on Heat-Affected Zone Shape, *Welding Journal*: 110s – 112s.
21. Nissley, N. E., and Lippold, J. C. "Ductility-Dip Cracking Susceptibility of Nickel-Based Weld Metals: Part 2—Microstructural Characterization." *Welding Journal* 88.6 (2009): 131s-140s.

# Part III
# Hot Cracking Phenomena—Materials

# Ductility-Dip Cracking Susceptibility of Commercially Pure Ni and Ni-Base Alloys Utilizing the Strain-to-Fracture Test

**Vern C. Kreuter V and John C. Lippold**

**Abstract** The ductility dip cracking (DDC) response of commercially pure Ni and two solid-solution strengthened Ni-base alloys was conducted using the strain-to-fracture (STF) test method that was developed at Ohio State University by Nissley and Lippold (Welding J 82(12):355–364, 2003 [1]). Alloys 200, 600, and 625 were tested over a range of strain at 950 °C to determine the threshold strain for fracture ($\varepsilon_{min}$) and the cracking response as a function of strain. Using established procedures, samples were prepared with an autogenous gas-tungsten arc (GTA) spot weld in order to provide consistent grain boundary character among samples and then tested using a Gleeble® thermo-mechanical simulator. In order to evaluate the effect of oxygen on DDC susceptibility, spot welds were made using Ar-2 % $O_2$ shielding gas. It was found that Ni-200 was most susceptible to DDC due to the large grain size and non-tortuous migrated grain boundaries (MGB's). Alloy 625 was extremely resistant to DDC, as reported by Zhang et al. (Trans JWRI 14(2):325–334, 1985 [2]), exhibiting a strain threshold greater than 15 %. The addition of oxygen reduced DDC resistance in alloys 600 and 625. Metallographic analysis showed that DDC initiates at grain boundary triple points and then propagates along straight migrated grain boundaries. SEM fractography revealed that crack surfaces exhibited smooth intergranular features with little evidence of ductile rupture.

## Introduction

Solid-solution strengthened Ni-base alloys are commonly used in high temperature applications that require excellent corrosion resistance due to their inherent strength and resistance to stress corrosion cracking (SCC) at temperatures up to 800 °C [3]. These alloys are strengthened by the additions of substitutional elements such as Cr,

V.C. Kreuter V (✉) · J.C. Lippold
Welding and Joining Metallurgy Group, The Ohio State University, Columbus, USA
e-mail: kreuter.8@buckeyemail.osu.edu



Th. Boellinghaus et al. (eds.), *Cracking Phenomena in Welds IV*,
DOI 10.1007/978-3-319-28434-7_8

Mo, Nb and Fe that provide a solid-solution strengthening effect to the austenite matrix. Specifically, nickel alloys 600 and 625 are commonly used in chemical processing and power generation applications, where high temperatures and corrosive environments are generally encountered. Due to high restraint and multipass weld conditions in these systems, DDC can be a significant weldability issue.

DDC is a solid-state cracking phenomenon that is associated with an abrupt drop in ductility within an intermediate temperature range ($T_s$–$0.5T_s$). Relaxation of residual stresses during reheating in multipass welding combined with the precipitous drop in ductility can lead to localized ductility exhaustion along migrated grain boundaries (MGBs), leading to DDC. This type of cracking phenomenon has been studied in a variety of Ni-base alloys and filler metals [4–13], and has proven to be a significant weldability issue in thick section, multipass weld metal exhibiting high restraint. While microcracks may be acceptable in some applications, DDC can be problematic in situations where defect tolerance is quite low, such as in nuclear power plant components.

DDC is an intergranular form of cracking observed most often in the weld metals of single-phase austenitic (FCC) materials, where MGBs may form readily during initial weld cooling and subsequent reheating (multipass welds). MGBs are especially susceptible to DDC in high purity materials, where grain boundary pinning effects are minimal. This leads to large grain size and non-tortuous grain boundaries which promote strain localization at the boundary. When strain is applied, grain boundary sliding occurs and localized strain concentrations at grain boundary triple points lead to void initiation. Grain boundary orientation relative to the applied strain seems to have a role in crack propagation [13].

Although the exact mechanism for DDC is still under debate, many factors have been identified which contribute to the development of DDC under specific conditions. Table 1 provides a summary of the mechanisms that have been proposed over the past 50 years. Weld metal composition is the dominant factor which affects elevated temperature ductility as it controls the weld metal microstructure as well as microstructural creep properties and precipitation behavior. The combination of a susceptible microstructure and high restraint are essential for the formation of a ductility dip crack.

The most noteworthy and logical of the factors promoting DDC is the concept of grain boundary sliding. Grain boundary sliding is a creep- like phenomenon that occurs at intermediate temperatures due to a drop in the cohesive strength of migrated grain boundaries. Impurity elements tend to reduce this cohesive energy even further which lowers the amount of strain in a material necessary to initiate grain boundary sliding, and therefore cracking. This suggests that a reduction in impurity concentrations (S, P, O, N) will result in an improvement in DDC resistance. In fact, this is not the case since high purity materials have been shown to be susceptible to DDC. This may be due to excessive grain growth and very straight boundaries that lead to high strain concentrations. For example, commercially pure Ni-200 shows a very high susceptibility to DDC.

Of the mechanisms listed in Table 1, only precipitation-induced cracking appears not to be applicable since DDC is observed in alloys where precipitation

**Table 1** Proposed DDC mechanism summaries: based on previous experimentation

| Year(s) | Author(s) | Alloys tested | Mechanism summary | Reference |
|---|---|---|---|---|
| 2003, 2006, 2013 | M. Collins, K. Nishimoto, K. Saida | FM-82, Alloy 690, high Ni–Cr SS | Grain boundary segregation of P and S in reheated weld metal | [12, 14, 15] |
| 1961, 1971, 1985, 2004 | F. Rhines, A. Dix, Y. Zhang, A. Ramirez J. Lippold | Monel®, Inconel X-750, Fe-36 % Ni, FM-52, | Grain boundary sliding, creep-like phenomena | [10, 16–18] |
| 2008 | G. Young | Alloy 690 and associated filler metals | Precipitation induced cracking | [4] |
| 1985 | Y. Zhang | Fe-36 % Ni (invar) | Synthesis of intergranular precipitation, serrated gb's, gb sliding, gb migration, flow stress, and dynamic recrystallization | [2] |
| 2005 | T. Capobianco | FM-52 | Sulfur embrittlement and local stresses from $Cr_{23}C_6$ carbides | [19] |
| 2004 | M. Collins | FM-52 and FM-82 | Complex interplay of alloy composition, impurity addition, gb segregation, gb sliding, precipitation, recrystallization, accumulated local strain, and gb misorientation | [9] |
| 1984 | F. Matsuda | Fe-36 % Ni (invar) | S, O, N, and Al embrittlement | [20] |

does not occur in the DDC temperature range. A number of factors may be involved and the importance of these factors may be different for different alloy systems.

The ductility trough that defines DDC susceptible materials is depicted in Fig. 1. The dotted line represents a solid-state dip in ductility which is necessary if DDC is

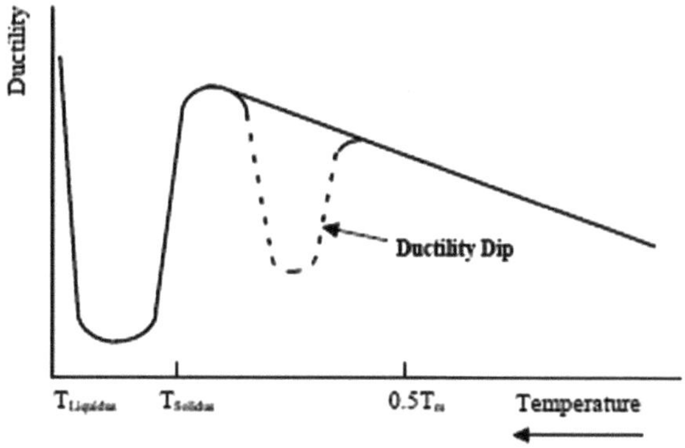


**Fig. 1** Ductility versus temperature trough resulting in the formation of DDC in multiple alloy systems

likely to occur. The bounds of this trough may also be used to help shed some light on the mechanism behind DDC. At the lower temperature bound, close to ½ of the effective solidus temperature, materials tend to have improved creep properties as well as lower diffusivities, which would make it very difficult for grain boundary sliding or impurity segregation to occur. At the upper bound, closer to the effective solidus temperature, recrystallization will occur at the boundary due to the combined effect of strain concentration and temperature. When recrystallization occurs, grain boundary area and tortuosity increases abruptly and initiation and crack propagation is arrested.

In order to quantify and categorize the DDC susceptibility of multiple alloy systems, the Strain-to-Fracture (STF) test was developed at The Ohio State University by Nissley and Lippold [1, 8, 21]. The STF test is a Gleeble®—based test method that isolates materials within the ductility-dip temperature range while applying a pre-programed strain to the sample. DDC quantification is determined by a threshold strain for fracture to occur ($\varepsilon_{min}$) and different alloy susceptibilities may be compared. The STF test allows the development of a strain-temperature envelope for a certain material that provides a unique "signature" of ductility dip behavior.

The purpose of this study is to quantify and categorize the DDC susceptibility of commercially pure Ni (Ni-200) and two common nickel- base alloys (Alloys 600 and 625) by determining the maximum strain achievable before cracking occurs. The large variation in composition among these alloys was intended to provide insight into compositional and microstructural effects on DDC. The effect of oxygen in the weld metal was also tested with the introduction of an Ar-2 % $O_2$ shielding gas in order to determine the relationship between impurity segregation and DDC susceptibility.

## Experimental Approach

### *Materials*

Two solid-solution strengthened Ni-base alloys (Alloys 600 and 625) and commercially pure Ni-200 were evaluated for susceptibility to DDC. The nominal chemical compositions of these alloys are provided in Table 2. All of these materials were in plate form of nominal thickness 0.25 in. (6.35 mm).

### *Sample Preparation*

STF samples were prepared by water-jet cutting the plate to the dog bone tensile dimensions recommended by Nissley [8]. Ten samples of each alloy were prepared for initial STF testing. After cutting the samples, an autogenous GTA spot weld is

**Table 2** Nominal base metal compositions for tested nickel alloys: 200, 625, and 600, respectively (wt%)

| | Alloy 200 | Alloy 625 | Alloy 600 |
|---|---|---|---|
| Ni | 99.00 | 58.00 | 72.00 |
| Cr | – | 21.50 | 15.50 |
| Fe | 0.40 | 5.00 | 8.00 |
| C | 0.15 | 0.10 | 0.15 |
| Mn | 0.35 | 0.50 | 1.00 |
| Mo | – | 9.00 | – |
| Si | 0.35 | 0.50 | 0.50 |
| Cu | 0.25 | – | 0.50 |
| S | 0.01 | – | 0.015 |
| Nb | – | 3.65 | – |
| Al | – | 0.40 | – |

made in the gauge section on both side of each sample. Spot weld parameters include a current downslope that allows for directional columnar growth in a radial array within the spot weld. Such a solidification pattern results in grain boundaries at various angles to the applied strain during STF testing. In some cases, the spot weld was only applied to one side of the sample in order to evaluate DDC susceptibility of the base metal. Spot weld parameters used in this study are shown in Table 3. Baseline testing was done with spot welds made with a 100 % Ar shielding gas. The second round of testing was conducted on samples where the spot weld on one side was made with Ar-2 % $O_2$ shielding gas in order to examine the effect that oxygen on DDC susceptibility.

Following the completion of the spot weld, sample gauge sections were ground flat through 800 grit silicon carbide in order to make crack detection easier. This surface preparation removes the surface roughness and oxide layers resulting from the spot weld, thus facilitating crack detection after testing. Indentation marks ~8 mm apart on the sides of the gauge section were applied in order to determine the strain induced in each sample during testing. Initial marks were measured with the use of an optical microscope and the change in length after testing divided by the original length was used to calculate the amount of strain in each sample.

**Table 3** Spot weld parameters

| | Time (S) | Value |
|---|---|---|
| Shielding gas preflow | 10 | 9.42 L/m |
| Initial current | 0.1 | 20 A |
| Upslope | 5 | 24 A/s |
| Weld current | 10 | 140 A |
| Downslope | 12.7 | 9.5 A/s |
| Final current | 0.1 | 20 A |
| Shielding gas postflow | 20 | 9.42 L/min |
| Voltage | – | 12.5 V |
| Shielding gas | Argon (99.98 %) | |

## *STF Testing*

Strain-to-fracture testing was conducted using a Gleeble 3800 thermo-mechanical simulator. Samples were fit with a type K thermocouple for temperature control. Stainless steel high force jaws system with knurled tool steel grips were used to maintain a uniform temperature profile in the gage section. A jaw spacing of 30 mm was used in these tests to allow for accurate positioning of each sample. Once samples are mounted and positioned appropriately, a roughing pump reduces the chamber pressure to $10^{-2}$ torr followed by a back purge of 99.998 % argon. This process is repeated twice in order to reduce oxidation of the sample during testing.

Samples were heated to a peak temperature of 950 °C at a heating rate of 100 °C/s, held at the peak temperature for 10 s, and then pulled in tension to a pre-determined stroke length. A stroke rate of 0.6 mm/s was used for all testing. The stroke length was varied using pre-determined data in order to generate the range of strain required. Upon completion of the stroke cycle, the samples were allowed to free cool to room temperature. Cooling was conducted under a zero load condition in order to prevent thermal contraction. Thermal and stroke cycles are shown graphically in Fig. 2.

Analysis of cracking on the surface of the samples was conducted using a Nikon SMZ1000 optical microscope at 30–70× magnification. The distance between indentation marks were also measured and strain was calculated.

## *Metallurgical Analysis*

In order to gain a better understanding on the mechanisms of DDC, metallurgical analysis was conducted on samples above and below the threshold strain for cracking. A plan view of the gauge section was used and samples were polished through 1 μm diamond paste. Electrolytic etching with 10 % chromic acid at 2.5 V for 10–15 s was used to reveal the microstructure.

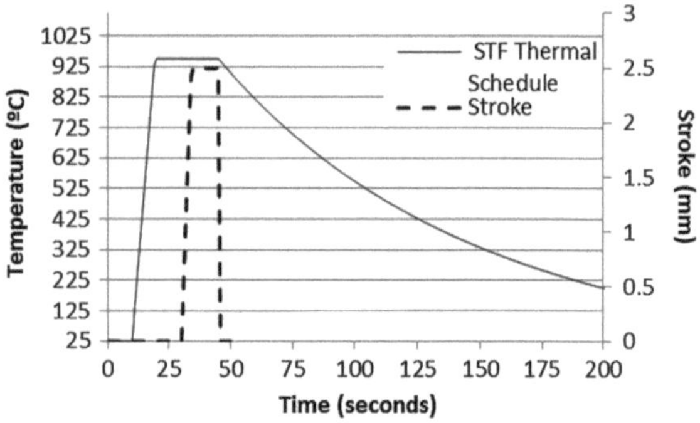


**Fig. 2** Thermal and stroke cycle of Gleeble for STF testing

Fractography was then conducted with the use of a scanning electron microscope on samples tested at the highest strains. The gauge sections of these sample were bent ~45° in order to open the fracture surface and facilitate observation in the SEM.

## Results and Discussion

### *STF Test Results*

A ductility dip in austenitic (FCC) materials has been shown by Nissley and others [1, 6–11] to occur over the temperature range from 700 to 1200 °C, with the minimum threshold strain for cracking in the range from 900 to 1000 °C. Based on work done by Nissley [1], a peak temperature of 950 °C was shown to adequately predict DDC behavior and use of a single temperature greatly reduces the number of samples required. In this study, this approach was taken and the DDC response of the three alloys at 950 °C is shown in Fig. 3.

### *Alloy 200*

Five Ni-200 samples were tested in this investigation and strained at various rates to determine threshold values for DDC. This alloy system proved to be the most susceptible to DDC, exhibiting a threshold value of ~1 % and a transition to gross cracking above 2 %. Gross cracking can be defined as the strain where cracking becomes so abundant that cracks begin to connect and grow [7]. This is likely due to the significant amount of long and non-tortuous MGBs in the weld microstructure of Ni-200.

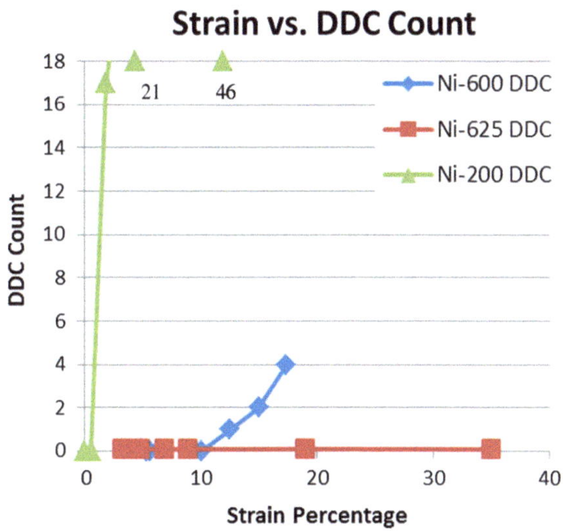


**Fig. 3** STF cracking data at 950 °C for Alloys 600 and 625, and Ni-200

## Alloy 600

Seven samples of Alloy 600 were tested in this investigation to determine the threshold strain for fracture in this system. It was determined that alloy 600 is resistant to DDC according to Dupont [3] by having a strain threshold greater than 6 %. Cracking was not observed until the applied strain exceeded 10 % and no transition to gross cracking was observed in samples tested.

## Alloy 625

Ten STF tests of Alloy 625 were conducted in this experiment and no cracking was observed in any of the samples tested, up to 35 % strain. The formation of a secondary phase (NbC) at the end of solidification resulted in very tortuous grain boundaries as well as a fine grained microstructure that is typically resistant to DDC.

## Effects of Oxygen Additions

The effect of impurity concentration on crack formation has been studied in many alloy systems. Specific to this study, Collins studied the effect of hydrogen and sulfur on DDC formation in two nickel-based filler metals [12] utilizing the STF testing method. It was found that both of these elements increased the susceptibility of DDC and lowered the effective $\varepsilon_{min}$. Hydrogen, which was added in through the shielding gas, had the most dramatic effect lowering the $\varepsilon_{min}$ value of Alloy 690 to less than 1 %.

In this study the effect of oxygen was investigated with the use of an Ar-2 % $O_2$ shielding gas in the spot weld. Spot welds were made on Alloys 600 and 625 to determine the effect that oxygen has on the strain threshold values. STF data with oxygen additions is shown in Fig. 4. The results showed a significant decrease in strain threshold required for DDC development when oxygen concentrations are increased. The strain threshold for Ni-600 dropped from $\sim$11 to $\sim$7 %, whereas the threshold for Ni-625 was established at $\sim$20 %. Like hydrogen, oxygen may have some factor in reducing grain boundary cohesion resulting in an increased susceptibility to DDC in materials exhibiting critical concentrations. Surface oxide formations tend to decrease the toughness and ductility of a material due to the brittle nature of the oxide. The extent to which oxygen effects DDC susceptibility and the amount required to shift threshold levels is still under investigation.

**Fig. 4** STF cracking data showing the effect of oxygen additions to the weld metal. **a** Alloy 600. **b** Alloy 625

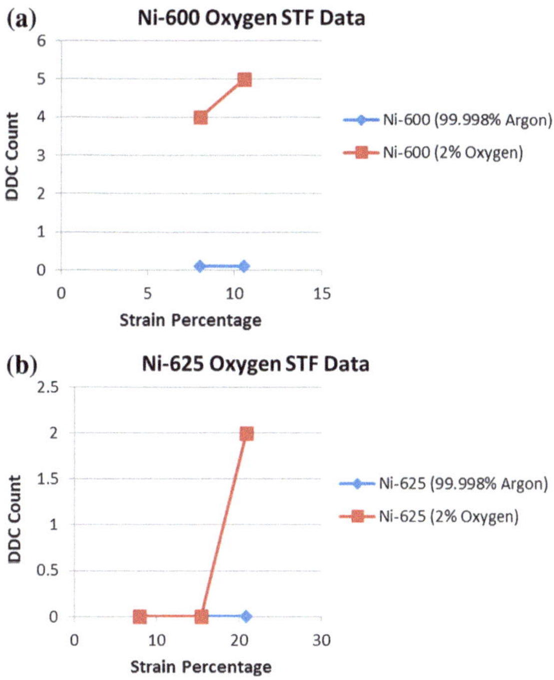


# Microstructure Characterization

## *Alloy 200*

Pure materials, such as Ni-200, can be susceptible to weld metal DDC due to their homogenous microstructure that is free of secondary phases. On cooling, the crystallographic portion of the solidification grain boundary will migrate via grain growth and boundary straightening mechanisms, leading to migrated grain boundaries that are long and non-tortuous. The lack of a secondary phase leads to free grain boundary motion within the weld metal; leading to large grain size and reduced MGB area. Since weld metal DDC forms preferentially along MGBs, Ni-200 has high susceptibility to DDC. Figure 5 shows DDC in the weld metal of a Ni-200.

STF sample tested at 1.8 % strain. A purely austenitic (FCC) microstructure is seen in both photomicrographs and ductility dip cracks can be viewed in the weld metal, initiating at the grain boundary triple point. Triple points may act as strain concentrators to initiate DDC formation by a grain boundary voiding mechanism. Once voids are formed, cracks tend to propagate along non-tortuous MGBs, leading to elongated cracks. DDC crack lengths in alloy 200 were on average 100 μm in length and sometimes exceeded 200 μm in samples subjected to higher strains.

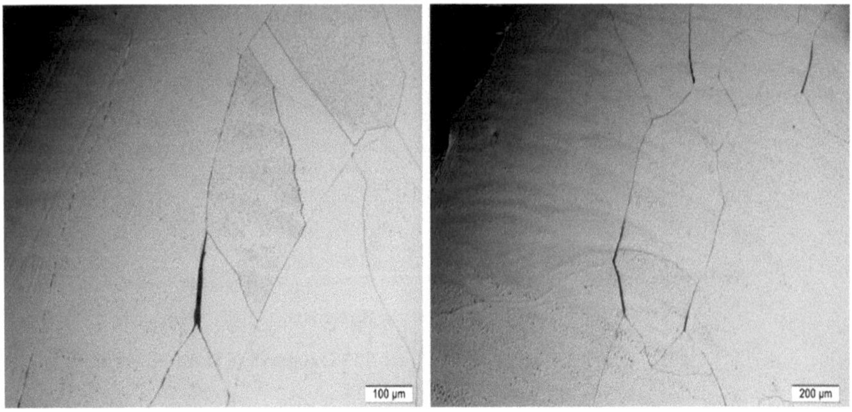


**Fig. 5** Alloy 200 weld metal at 100× and 50× magnification, respectively, showing elongated DDDC formations which tend to initiate at grain boundary

## *Alloy 600*

Phase identification software, JMatPro-Ni, suggests the formation of M7C3 carbides at ~1200 °C in Alloy 600. The formation of secondary phases within the austenitic matrix tend to pin grain boundaries via the grain boundary locking mechanism, leading to the formation of tortuous MGBs. Photomicrographs of Ni-600, as seen in Fig. 6, show a relatively fine grained austenitic microstructure with the presence of carbides along grain boundaries. Grain boundaries in Alloy 600 tend to be shorter and more tortuous, which result in the higher strain threshold exhibited by this material. DDC cracks also tend to be much shorter with an average crack length of ~40 μm.

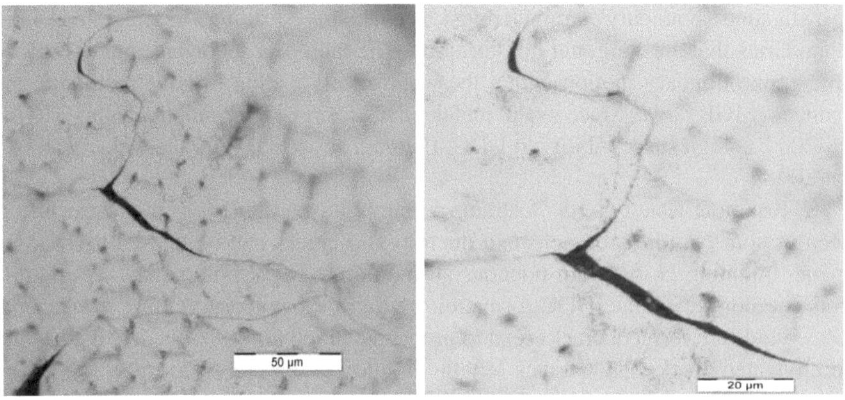


**Fig. 6** Photomicrographs of nickel alloy 600 weld metal. DDC initiates at grain boundary triple points and propagates along MGBs

## *Alloy 625*

Significant additions of Nb, Mo, and Cr in Alloy 625 leads to the formation of a complex, multi-phase microstructure that is resistant to DDC at all intermediate temperatures. Inter-dendritic regions, as seen in Fig. 7, contain large niobium carbides while adjacent regions contain Laves phase. Both secondary phases limit grain boundary motion resulting in a very fine grain structure with tortuous grain boundaries.

## *Fracture Morphologies*

The scanning electron microscope (SEM) was used in this study to determine the fracture morphology in Ni-200 and Alloy 600. Fracture surfaces were examined to confirm the existence of intergranular fracture at intermediate temperatures. Typically, DDC is categorized by the presence of macroscopically smooth intergranular crack surfaces, which represents complete exhaustion of ductility in the intermediate temperature range.

The fracture morphology of Ni-200 is shown in Fig. 8. Figure 8a shows the entire ductility-dip crack with small fractures stemming off of the perpendicular conjoining grain boundaries. From Fig. 8b–d, increasing magnification reveals a

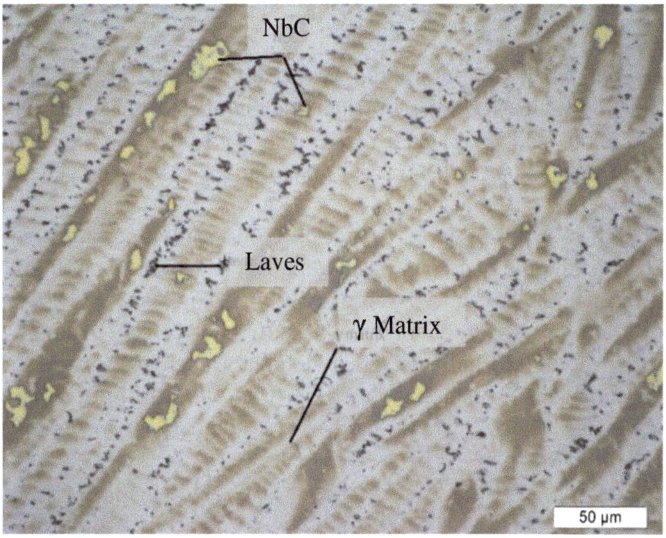


**Fig. 7** Nickel 625 weld metal at 500× magnification exhibiting complex secondary phase formation at inter-dendritic regions

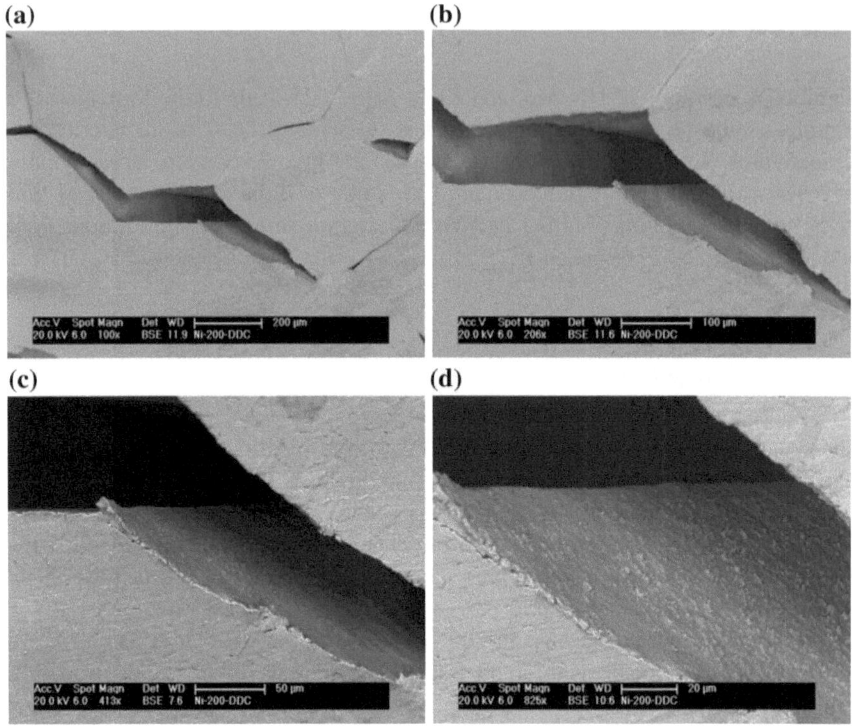


**Fig. 8 a–d** Ni-200 fracture morphology in a sample tested at 10 % strain. The smooth intergranular fracture morphology is indicative of DDC

smooth intergranular fracture morphology indicative of ductility-dip cracking. It is evident in this figure that the crack initiated at the grain boundary triple point in the center of the photo and propagated outward in both directions as strain concentrations increased. This particular crack is rather large and has a crack length of over 600 μm.

Figure 9 shows DDC observed in an Alloy 600 sample tested at 25 % strain. Figure 9a, b illustrate the degree of grain boundary cracking. Smooth intergranular fracture surfaces may be viewed more closely in Fig. 9c, which confirms the presence of DDC in this alloy. Figure 9b shows how DDC fractures propagate along the grain boundary until the grain boundary changes direction, where the crack propagation tends to stop. This is due to the tortuosity of the grain at this point, proving that DDC fractures tend to be abundant in materials exhibiting non-tortuous MGB's.

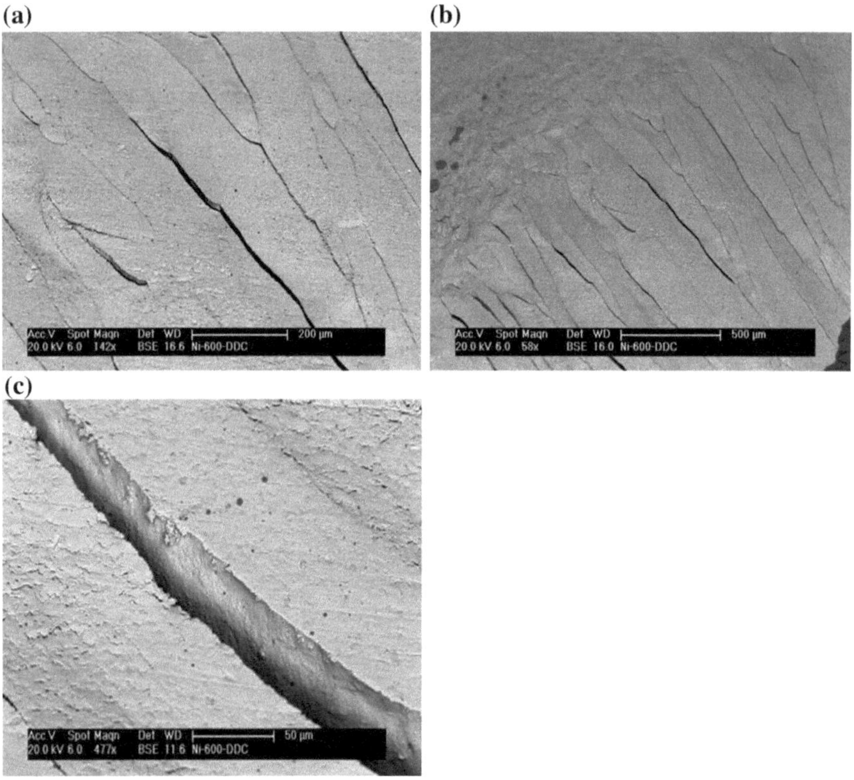


**Fig. 9 a–c** Alloy 600 fracture morphology in a sample tested at 25 % strain

## *Improving DDC Response*

Previous research conducted by Ramirez and Lippold proposed that grain boundary tortuosity has the greatest influence on DDC susceptibility [22]. Studies showed that increasing the level of precipitate-forming elements in nickel-based filler metals, such as niobium and titanium, may lead to grain boundaries which are more tortuous, and therefore less susceptible to DDC. Although tortuous grain boundaries may suppress the formation of DDC, it should be noted that the introduction of precipitate forming elements may widen the solidification temperature range making the alloy more susceptible to solidification cracking. Work conducted in this study supports the grain boundary tortuosity hypothesis and crack formation seems to be suppressed with the addition of precipitation forming elements in Ni-600 and Ni-625, as compared to commercially pure Ni-200. Continued work is being done to look at the effect of grain size and recrystallization behavior on DDC susceptibility.

## Conclusions

1. The STF test is a verifiable test in quantifying and categorizing DDC susceptibility in various alloy systems. Consistent results show that the STF test is highly repeatable in producing DDC and determining strain thresholds.
2. Commercially pure Ni-200 was most susceptible to DDC due to the absence of secondary phases. Large grain size and a low area fraction of non-tortuous MBGs allow for high strain concentration at the grain boundaries.
3. Alloy 625 showed very high resistance to DDC with no cracking observed at strains up to 35 %. This was the result of very tortuous grain boundaries arising from NbC and Laves precipitation at the end of solidification.
4. DDC propagates preferentially along non-tortuous MGBs.
5. Oxygen additions increased DDC susceptibility. Strain thresholds were lower with the introduction of oxygen to the weld metal resulting in an increased susceptibility to DDC formation.

**Acknowledgements** This work was supported by the Electric Power Research Institute. Special thanks to our program manager, Mr. Steve McCracken.

## References

1. Welding Journal, 2003. **82**(12): p. 355-364.
2. Zhang, Y.-C., H. NAKAGAWA, and F. MATSUDA, *Weldability of Fe-36 % Ni Alloy (Report VI): Further Investigation on Mechanism of Reheat Hot Cracking in Weld Metal (Materials, Metallurgy & Weldability).* Transactions of JWRI, 1985. **14**(2): p. 325-334.
3. DuPont, J.N.L.J.C.K.S.D., *Welding metallurgy and weldability of nickel- base alloys*. 2009, Hoboken, N.J.: John Wiley & Sons.
4. Young, G., Capobianco, T., Penik, M., *The mechanism of ductility dip cracking in nickel-chromium alloys.* Welding Journal, 2008. **87**(2): p. 31.
5. Dave, V., Cola, M., Schwartz, A., Hussen, A., *Grain boundary character in Alloy 690 and ductility-dip cracking susceptibility.* Welding Journal, 2004. **83**(1): p. 1-S.
6. Nissley, N. and J. Lippold, *Ductility-Dip Cracking Susceptibility of Nickel-Based Weld Metals Part 1: Strain-to-Fracture Testing.* Welding Journal, 2008. **87**(10): p. 257s-264s.
7. Nissley, N. and J. Lippold, *Ductility-Dip Cracking Susceptibility of Nickel-Based Weld Metals: Part 2—Microstructural Characterization.* Welding journal, 2009. **88**(6): p. 131s-140s.
8. Nissley, N.E., *Development of the strain-to-fracture test to study ductility-dip cracking in austenitic alloys*. 2002.
9. Collins, M., A. Ramirez, and J. Lippold, *An investigation of ductility-dip cracking in nickel-based weld metals-Part III.* Welding Journal, 2004. **83**(2): p. 39-S.
10. Ramirez, A. and J. Lippold, *High temperature behavior of Ni-base weld metal: Part II–Insight into the mechanism for ductility dip cracking.* Materials Science and Engineering: A, 2004. **380**(1): p. 245-258.
11. Ramirez, A. and J. Lippold, *High temperature behavior of Ni-base weld metal: Part I. Ductility and microstructural characterization.* Materials Science and Engineering: A, 2004. **380**(1): p. 259-271.

12. Collins, M. and J. Lippold, *An investigation of ductility dip cracking in nickel-based filler materials-Part I.* Welding Journal, 2003. **82**(10): p. 288.
13. Kikel, J. and D. Parker. *Ductility of dip cracking susceptibility of filler metal 52 and alloy 690.* in *5 th International Conference: Trends in Welding Research Conference Precedings*. 1998.
14. Nishimoto, K., Saida, K., Okauchi, H., Ohta, K., *Microcracking in multipass weld metal of alloy 690 Part 2–Microcracking mechanism in reheated weld metal.* Science and Technology of Welding & Joining, 2006. **11**(4): p. 462-470.
15. Saida, K., Nishijima, Y., Kiuchi, K., Nakayama, J., *Contribution of Phosphorus and Sulfur on Hot Cracking Susceptibility for Extra-High- Purity 25Cr-35Ni Stainless Steels.* Welding Journal, 2013. **92**(11): p. 322-s-331-s.
16. Rhines, F. and P. Wray, *Investigation of the intermediate temperature ductility minimum in metals.* ASM TRANS Q, 1961. **54**(2): p. 117-128.
17. *Inconel X-750.* Weld J, 1971. **50**(6): p. 247.
18. Zhang, Y.C., H. Nakagawa, and F. Matsuda, *Weldadility of Fe-36 % Ni Alloy (Report V): Behaviors of Grain-boundary Sliding and Cavity Formation Preceding Reheat Hot Cracking in Weld Metal (Materials, Metallurgy & Weldability).* Transactions of JWRI, 1985. **14**(2): p. 319-324.
19. Capobianco, T. and M. Hanson, *Auger Spectroscopy Results from Ductility Dip Cracks Opened Under Ultra-High Vacuum*. 2005. Medium: ED.
20. Matsuda, F., Nakagawa, H., Minehisa, S., *Weldability of Fe-36 % Ni alloy report II: Effect of chemical composition on reheated hot cracking in weld metal.* Trans. JWRI, 1984. **13**(2): p. 69-75.
21. Nissley, N., Collins, M., Guaytima, G., Lippold, J., *Development of the strain-to-fracture test for evaluating ductility-dip cracking in austenitic stainless steels and Ni-base alloys.* Welding in the World, 2002. **46**(7-8): p. 32-40.
22. Ramirez, A.J., J.W. Sowards, and J.C. Lippold, *Improving the ductility- dip cracking resistance of Ni-base alloys.* Journal of Materials Processing Technology, 2006. **179**(1): p. 212-218.

# Evaluation of Solidification Cracking Susceptibility in Austenitic Stainless Steel Welds Using Laser Beam Welding Transverse-Varestraint Test

**Eun-Joon Chun, Hayato Baba, Kazutoshi Nishimoto and Kazuyoshi Saida**

**Abstract** In order to quantitatively evaluate the solidification cracking susceptibility in laser welds of type 310S and type 316L stainless steels, the Varestraint testing system for laser beam welding (LBW transverse-Varestraint test) was newly constructed. The timing-synchronisation among the laser oscillator, welding robot and hydraulic pressure devices was established by employing high-speed camera observation together with electrical signal control among the three components. Moreover, the yoke-drop time measured by high-speed camera was compensated to prevent underestimation of the crack length. The LBW transverse-Varestraint test was conducted varying the welding speed from 10.0 to 40.0 mm/s, and the transverse-Varestraint test with gas tungsten arc welding was also performed varying the welding speed from 1.67 to 5.00 mm/s. As the welding speed increased from 1.67 to 40.0 mm/s, the solidification brittle temperature range (BTR) of type 310S stainless steel welds was reduced from 146 to 120 K, while the BTR enlarged from 36 to 49 K in type 316L stainless steel welds. A numerical simulation of the solid/liquid coexistence temperature range, using solidification segregation model combined with the Kurz–Giovanola–Trivedi model, explained the mechanism of the BTR shrinkage in type 310S stainless steel welds by reduction of the solid/liquid coexistence temperature range of the weld metal due to the inhibited solidification segregation of solute elements and promoted dendrite tip supercooling attributed to rapid solidification in the LBW process. The reason why the BTR enlarged in type 316L stainless steel welds could be clarified by the enhanced solidification segregation of solute elements (mainly P and S), corresponding to the decrement in

---

E.-J. Chun (✉) · H. Baba · K. Saida
Division of Materials and Manufacturing Science, Graduate School of Engineering, Osaka University, 2-1 Yamada-oka, Suita-shi, Osaka 565-0871, Japan
e-mail: ejchun@mapse.eng.osaka-u.ac.jp

K. Nishimoto
Department of the Application of Nuclear Technology, Fukui University of Technology, Gakuen 3-6-1, Fukui-shi, Fukui 910-8505, Japan



Th. Boellinghaus et al. (eds.), *Cracking Phenomena in Welds IV*,
DOI 10.1007/978-3-319-28434-7_9

δ-ferrite content at the solidification completion in the weld metal. It follows that the opposite tendencies on solidification cracking susceptibility with increasing the welding speed in LBW could be explained by the different solidification segregation behaviour of solute elements, closely related with the δ-ferrite content.

## Introduction

Austenitic stainless steels have been widely applied to various industrial fields, such as oil, gas industry as well as nuclear power plant, predominantly because of their superior corrosion resistance and mechanical properties under demanding service environments. The laser beam welding (LBW) offers various advantages in industrial fabrication in comparison with other conventional welding processes, such as highly automated and flexible production lines involving short production cycles. From a metallurgical viewpoint, the suppressed solidification segregation, microstructure refinement and minimisation of the heat-affected zone is attain through rapid solidification together with the low heat input [1, 2]. Furthermore, LBW also leads to low distortion in the welded structure even for alloys with a high thermal expansion coefficient and low thermal conductivity. These reasons explain why the LBW process has been frequently preferred for the welding fabrication of austenitic stainless steels. On the other hand, during welding of austenitic stainless steels with a laser beam, special attention should be paid to the weldability, which is commonly examined in terms of solidification cracking [3–9]. Several studies have already suggested that the solidification cracking susceptibility would be enhanced during LBW as compared with arc welding, especially at high welding speeds [1, 10].

The weld metal of stainless steels generally consists of four kinds of solidification modes, i.e. A mode: fully austenitic solidification, AF mode: primary austenite with secondary δ-ferrite solidification, FA mode: primary δ-ferrite with secondary austenite solidification, F mode: fully ferritic solidification [3, 4]. It is well known that the solidification mode of stainless steels has the predominant effect on the solidification cracking susceptibility, mainly through change in the constituent phases, solidification morphology and segregation behaviour of solute elements [3, 4, 11–15]. In particular, LBW generally involves rapid solidification that substantially deviates from the equilibrium state, such as a high temperature gradient, growth rate and supercooling. Furthermore, these non-equilibrium phenomena in LBW also can produce the partionless microstructure formation together with a transition of the solidification mode (changing from primary ferrite to primary austenite phase) in austenitic stainless steel weld metals [12–21]. Hence, the solidification cracking susceptibility in LBW must be different from that in arc welding, and deeper understanding of solidification cracking behaviour in LBW is required to clarify the metallurgical aspects of the solidification cracking. Although some studies on solidification cracking in LBW have been already reported using hot deformation rate, controlled tensile weldability and preloading tensile tests, and so on [1, 2, 10, 22–28], further investigation is still required, because these results explain only the cracking phenomena in terms of whether cracks occurred or not. In addition, there are no further discussions of the suppression mechanism of the cracking susceptibility. Only extremely limited studies have been made involving a quantitative comparison of the solidification cracking susceptibility in laser welds.

Even though high-temperature ductility curves (consisting of solidification cracking temperature range and augmented strain for crack occurrence) have been widely used to assess the solidification cracking susceptibility, the corresponding information based on actual testing in laser welds has not been reported yet except for a few cases [10, 27, 28], as compared to the numerous conventional studies on arc welding [29–47]. For assessment of the high-temperature ductility curve, the transverse-Varestraint test has been commonly used as one of the most practical candidates from the industrial standpoint, because of its reliable reproducibility and its ability to evaluate several kinds of hot cracks (solidification, liquation, ductility-dip cracks) simultaneously. However, evaluation of the ductility curve using the Varestraint test for LBW would be extremely difficult due to difficulties caused by the rapid travelling speed.

In the present study, the Varestraint testing setup assembled with the LBW apparatus (LBW transverse-Varestraint test) was newly constructed in order to quantitatively investigate the solidification cracking susceptibility in laser welds. Based on the developed procedures, the solidification cracking susceptibility in laser welds of type 310S and 316L stainless steels was evaluated. Furthermore, the effect of welding speed on the solidification brittle temperature range (BTR) was investigated, and the mechanism of the variations in the BTR was theoretically discussed based on numerical calculation of the solid/liquid coexistence temperature range using the Kurz–Giovanola–Trivedi (KGT) model combined with a solidification segregation model.

# Materials and Experimental Procedures

## *Materials*

The materials used in this study were commercial type 310S and 316L stainless steels. The chemical compositions of these steels are described in Table 1. According to the $Cr_{eq}/Ni_{eq}$ value of the steels, type 310S steel indicates the solidification with A mode, and type 316L steel indicates FA mode, respectively.

**Table 1** Chemical composition of the steel used (mass%)

| | C | P | S | Cr | Ni | Mo | Si | Mn | Fe |
|---|---|---|---|---|---|---|---|---|---|
| Type 310S stainless steel | 0.050 | 0.021 | 0.001 | 24.88 | 19.18 | – | 0.69 | 0.81 | Bal. |
| Type 316L stainless steel | 0.008 | 0.036 | 0.030 | 17.43 | 12.21 | 2.06 | 0.68 | 1.22 | Bal. |

### *Transverse-Varestraint Test Procedure with LBW (LBW Transverse-Varestraint Test)*

Figure 1 depicts the appearance of the transverse-Varestraint testing apparatus with LBW with schematic descriptions of the testing procedures (side and front view). The LBW transverse-Varestraint test setup consisted of a laser welding robot and the Varestraint test apparatus. As similar with the transverse-Varestraint test with conventional arc welding, a hydraulic pressure was applied after the melt-run (transverse) welding. And the whole testing sequence was observed by a high-speed camera in

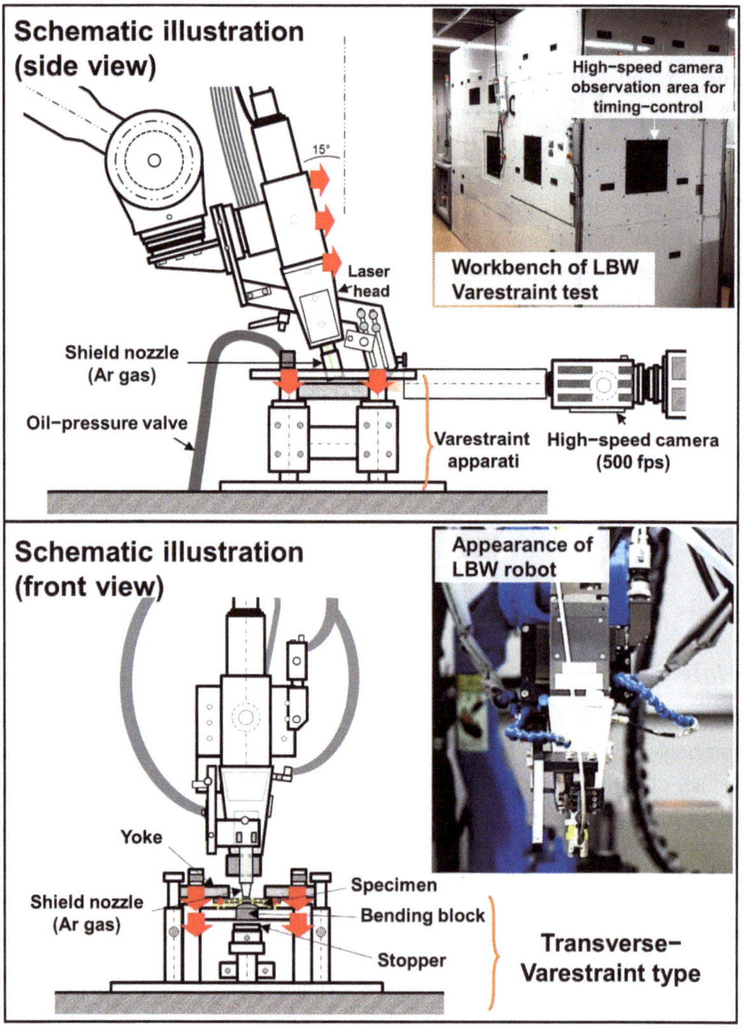


**Fig. 1** Appearance and schematic diagram of LBW transverse-Varestraint test

**Table 2** Conditions for LBW transverse-Varestraint test

| Test parameters | Values | | |
|---|---|---|---|
| Oscillator | Multi-mode fibre laser (maximum power: 6.5 kW) | | |
| Welding speed (mm/s) | 10.0 | 20.0 | 40.0 |
| Laser power (kW) | 1.5 | 2.0 | 2.5 |
| Defocus length (mm) | 0 | | |
| Beam radius (mm) | 0.4 | | |
| Beam irradiation angle (°) | 15 | | |
| Shield gas | Argon (99.99 % purity) | | |
| Gas flow rate (l/min) | 50 | | |
| Augmented strain (%) | 0.25–5.88 | | |

front of the test workbench through the protective filter as also indicated in Fig. 1 by arrow (in the workbench photograph). The test parameters of the LBW transverse-Varestraint test are listed in Table 2. Three levels of the welding speed (10.0, 20.0 and 40.0 mm/s) were employed, and the laser power was adjusted to obtain a half-penetration bead at each welding speed for the transverse welding. During the welding, argon was adopted as a shield gas to protect the bead surface from oxidation. The augmented strain was varied between 0.25 and 5.88 %. Figure 2 shows the dimensions of the test specimen used (thickness: 5 mm) together with the bending block arrangement. A length of the transverse weld bead was adjusted to 30 mm in every welding speed. The temperature history during welding solidification was measured using a thermocouple (W–Re) directly plunged into the weld pool during the melt-run welding to obtain the cracking temperature range. A thermocouple was inserted in different positions to obtain the cooling rate distribution in the bead width direction. The crack surface occurring in the transverse-Varestraint test was observed using a scanning electron microscope (SEM). The Specific test procedures are explained in the following section.

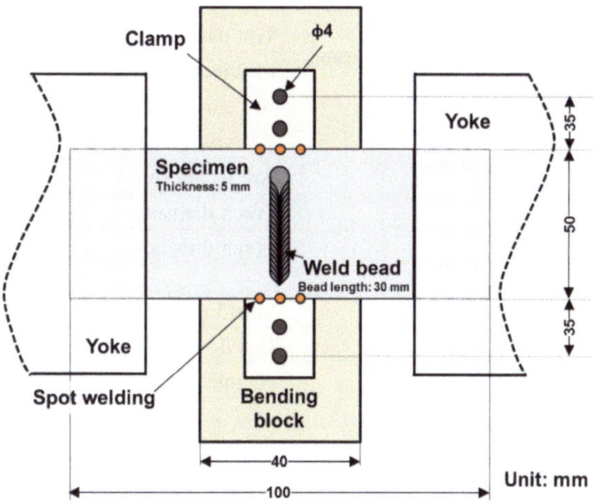


**Fig. 2** Dimension of specimen used for LBW transverse-Varestraint test

### *Transverse-Varestraint Test Procedure with GTAW*

The conventional transverse-Varestraint test using gas tungsten arc welding (GTAW) was also conducted using the test conditions listed in Table 3. Two levels of the welding speed (1.67 and 5.00 mm/s) were employed. The dimensions of the specimen used in the test were identical to those in the LBW transverse-Varestraint test (displayed in Fig. 2). The arc current was adjusted to obtain an one-third penetration bead at each welding speed for the transverse welding. The temperature distribution behind the molten pool in the bead was also measured by a thermocouple for conversion of the cracking temperature range from the crack length. The crack surface occurring in the transverse-Varestraint test was also observed by SEM.

### *Microstructural Analysis*

Solidification microstructure in the weld bead was observed by SEM after melt-run of LBW and GTAW under the identical conditions of the transverse welding in the transverse-Varestraint tests (as listed in Tables 2 and 3).

## Solidification Cracking Behaviour in LBW Transverse-Varestraint Test

### *Appearance and Characterisation of Hot Cracks*

Figures 3 and 4 show typical examples of the hot cracking appearance in the LBW transverse-Varestraint test at the welding speed of 20.0 mm/s for type 310S and 316L steels. Hot cracks occurred at the central region of the weld bead for both steels, and

**Table 3** Conditions for GTAW transverse-Varestraint test

| Test parameters | Values | |
|---|---|---|
| Electrode type | 2 % cerium–tungsten | |
| Arc current (A) | 1.67 | 5.00 |
| Arc voltage (V) | 120 | 200 |
| Arc length (mm) | 2 | |
| Torch diameter (mm) | 2 | |
| Torch direction | Perpendicular to specimen | |
| Shield gas | Argon (99.99 % purity) | |
| Gas flow rate (l/min) | 15 | |
| Augmented strain (%) | 0.25–4.00 | |

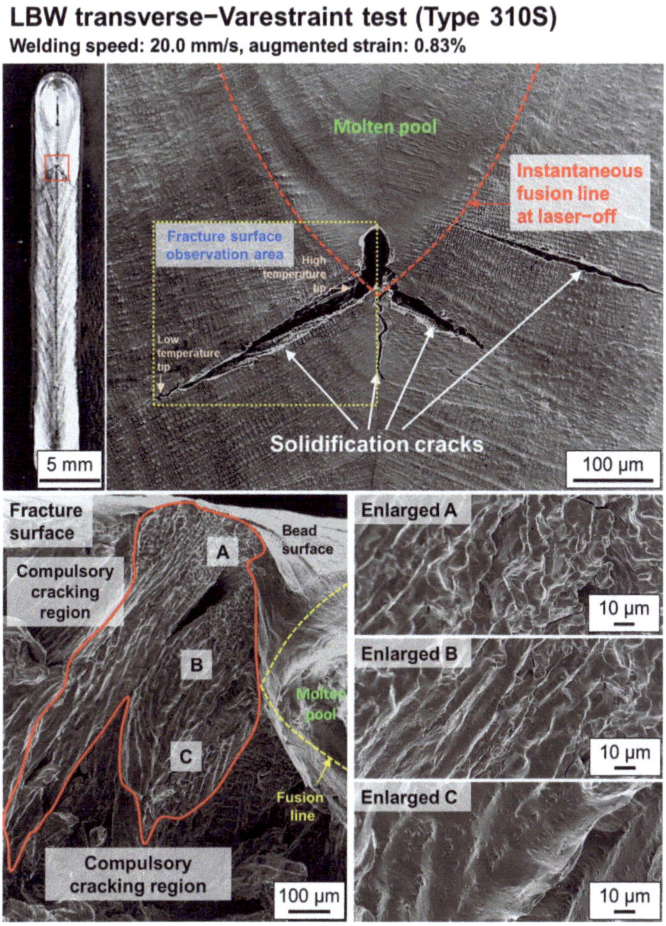


**Fig. 3** Appearance and surface microstructure (SEM) after LBW transverse-Varestraint test at a welding speed of 20.0 mm/s with corresponding fracture surface (type 310S)

the cracks were classified by a relatively short centreline crack and long transverse cracks. Figures 3 and 4 also show the fracture surface of the crack as indicated in each figure. The fracture surfaces at both the higher (under the fusion line) and lower temperature sides (away from the fusion line) indicate a dendritic structure with a trace of melting. Namely, the cracks formed in the LBW transverse-Varestraint test could be regarded as solidification cracks. Figures 5 and 6 show the appearance after the test at welding speeds of 10.0 and 40.0 mm/s for both steels. Similarly, solidification cracks could be also observed at the central region of the weld bead. Figure 7 shows typical solidification cracking appearance after the conventional GTAW Varestraint test at the welding speed of 1.67 mm/s. Different cracking propagation behaviours (i.e. propagation angle) could be observed between the LBW and GTAW tests.

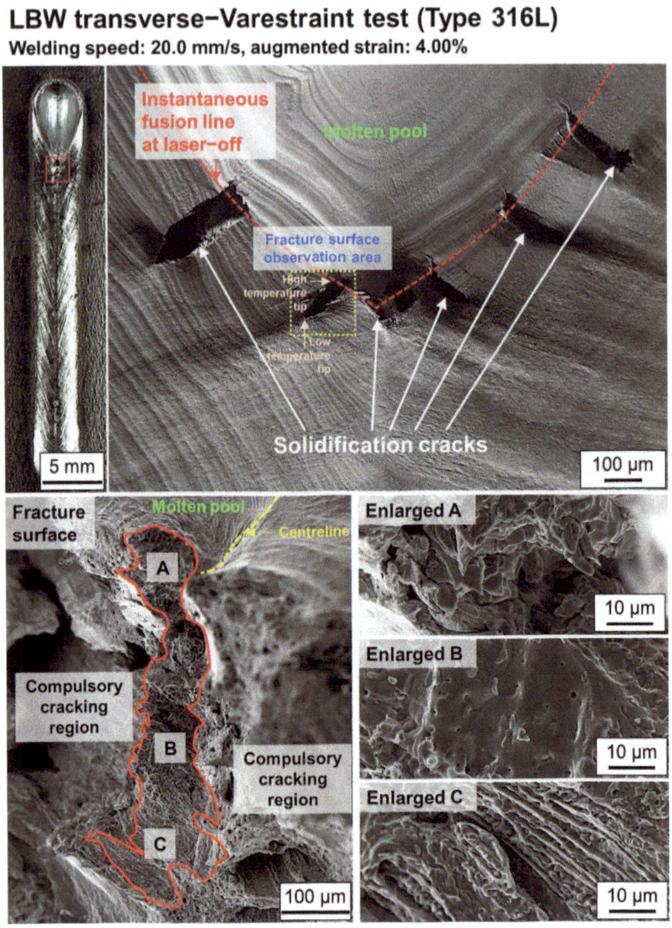


**Fig. 4** Appearance and surface microstructure (SEM) after LBW transverse-Varestraint test at a welding speed of 20.0 mm/s with corresponding fracture surface (type 316L)

## *Solidification Cracking Behaviour*

Figures 8 and 9 summarise the solidification cracking behaviour (crack length and number as a function of augmented strain) of type 310S and type 316L steel welds, respectively, together with the angle corresponds to the crack propagation from the bead centreline for both LBW (at 20.0 mm/s) and GTAW (at 1.67 mm/s) transverse-Varestraint tests. The crack occurrence was classified by three types, such as crack, back-fill and blank from the fusion line [54]. As the strain applied increased, the solidification cracking susceptibility increased (number of cracks and crack length), and the occurrence of solidification crack with the maximum length was also concentrated at the central region of the weld bead in both LBW and

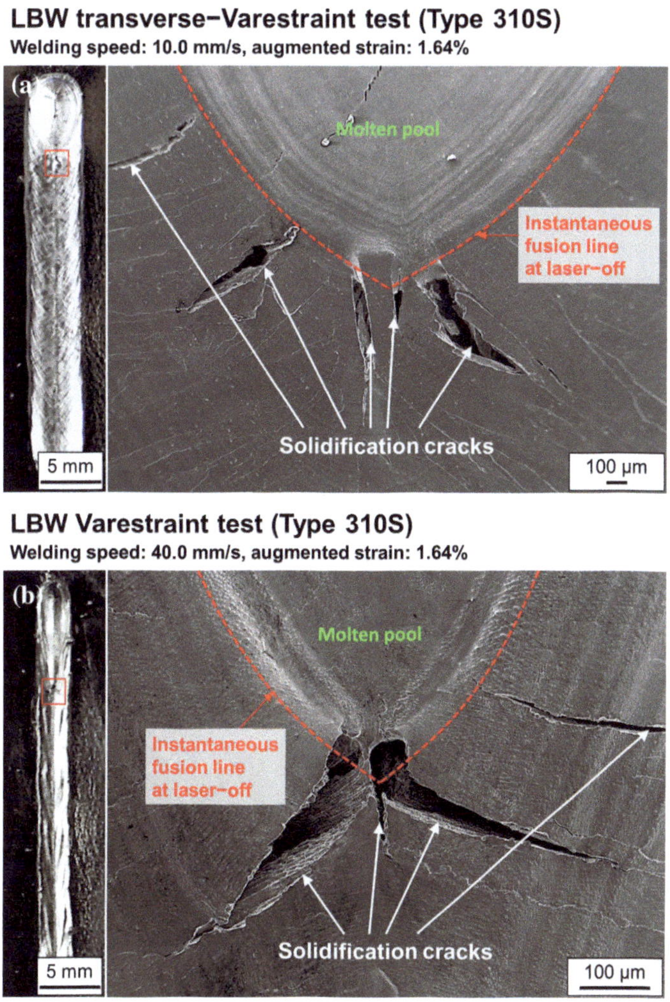


**Fig. 5** Appearance and surface microstructure (SEM) after LBW transverse-Varestraint test at welding speeds of **a** 10.0 and **b** 40.0 mm/s (type 310S)

GTAW transverse-Varestraint tests. On the other hand, transverse cracks with a certain angle (55°–70°) from the bead centreline occurred with maximum length in LBW transverse-Varestraint test, differently from that found in GTAW transverse-Varestraint test (0°–25°). It could be also confirmed that solidification cracking for both LBW and GTAW was much suppressed in type 316L steel welds compared with that of type 310S steel welds.

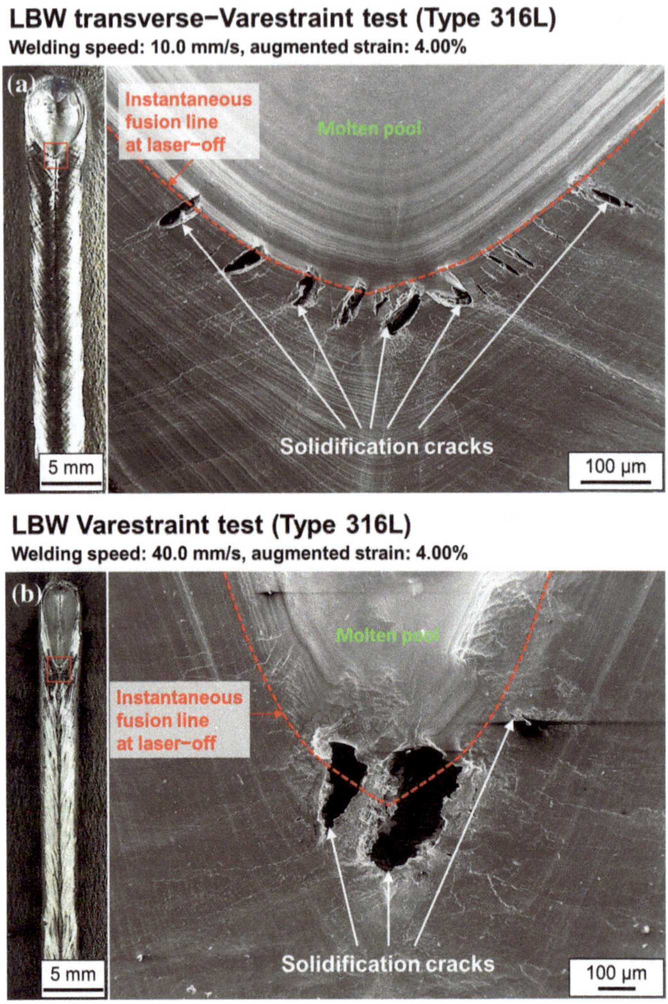


**Fig. 6** Appearance and surface microstructure (SEM) after LBW transverse-Varestraint test at welding speeds of **a** 10.0 and **b** 40.0 mm/s (type 316L)

## *Influence of Solidification Microstructure on Solidification Cracking Morphology*

Figure 10 summarises the surface microstructures at the bead central region after melt-run welding by GTAW and LBW at the welding speeds of 1.67 (GTAW), 10.0, 20.0 and 40.0 (LBW) mm/s for type 310S steel. As mentioned in sections "Appearance and Characterisation of Hot Cracks" and "Solidification Cracking Behaviour", the propagation angle of the solidification cracks in the LBW transverse-Varestraint test quite differed from that in the GTAW transverse-Varestraint test (i.e. higher

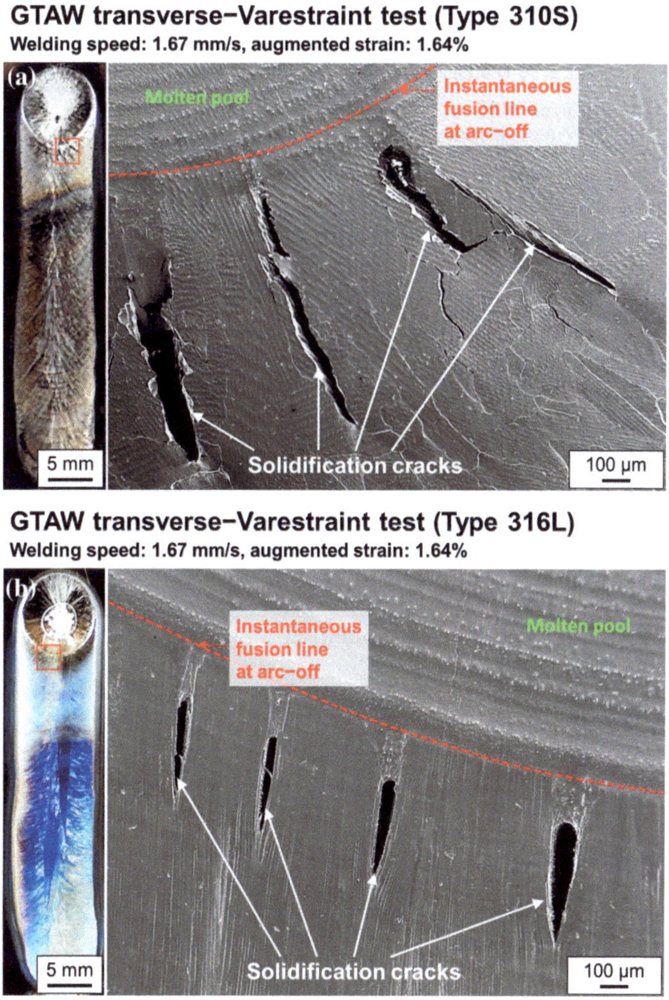


**Fig. 7** Appearance and surface microstructure (SEM) after GTAW transverse-Varestraint test at a welding speed of 1.67 mm/s for **a** type 310S and **b** type 316L steels

propagation angle of the solidification crack with maximum length in the transverse-LBW Varestraint test as compared with that in the transverse-GTAW Varestraint test). As the welding speed increased from GTAW to LBW, quite different solidification morphology in the vicinity of the bead centreline, where the solidification crack mainly occurred in the Varestraint test, could be clearly observed. At the welding speed of 1.67 mm/s (GTAW), the bead centre region composed of straight dendrite boundary with the same growth direction. Similarly, at the welding speed of 10.0 mm/s (LBW), each columnar dendritic zone composed of the same dendrite growth direction. However, at the welding speeds of 20.0 and 40.0 mm/s (LBW), complex

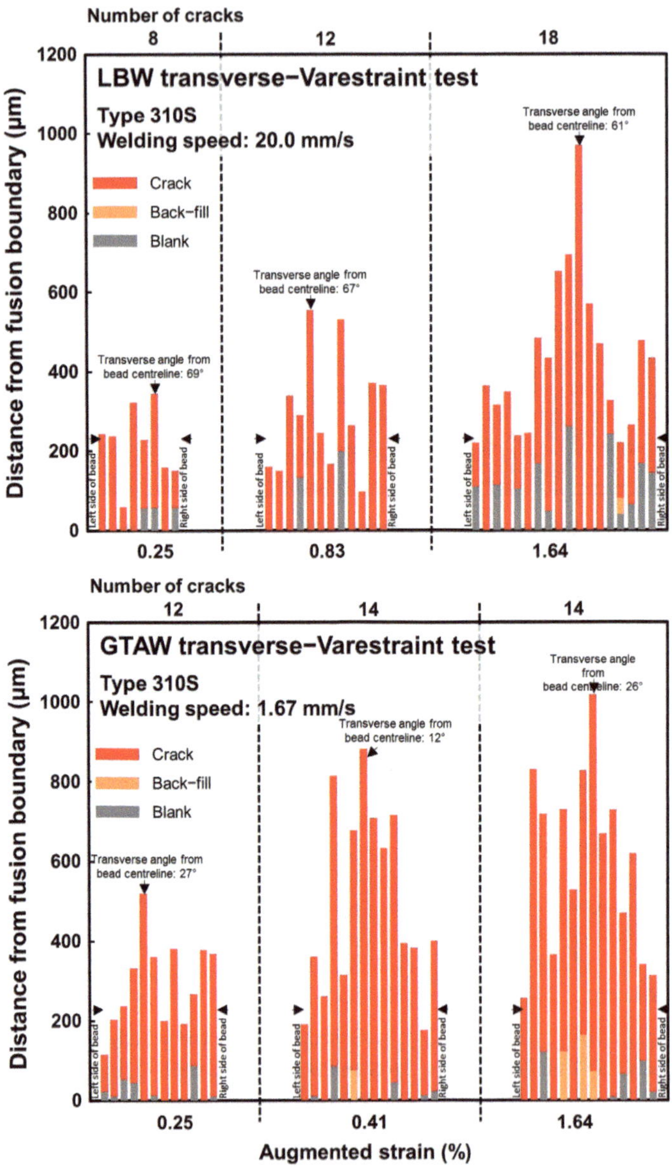


**Fig. 8** Quantitative analysis of solidification cracking susceptibility: location and length of the cracks in LBW and GTAW transverse-Varestraint tests (type 310S)

microstructural morphology composed of each columnar dendritic zone with the different growth directions was emerged in the vicinity of the bead centreline (as indicated by a red line surrounded in the schematic description). Namely, as compared with the straight dendrite boundary within the columnar dendritic zone from both ends of the

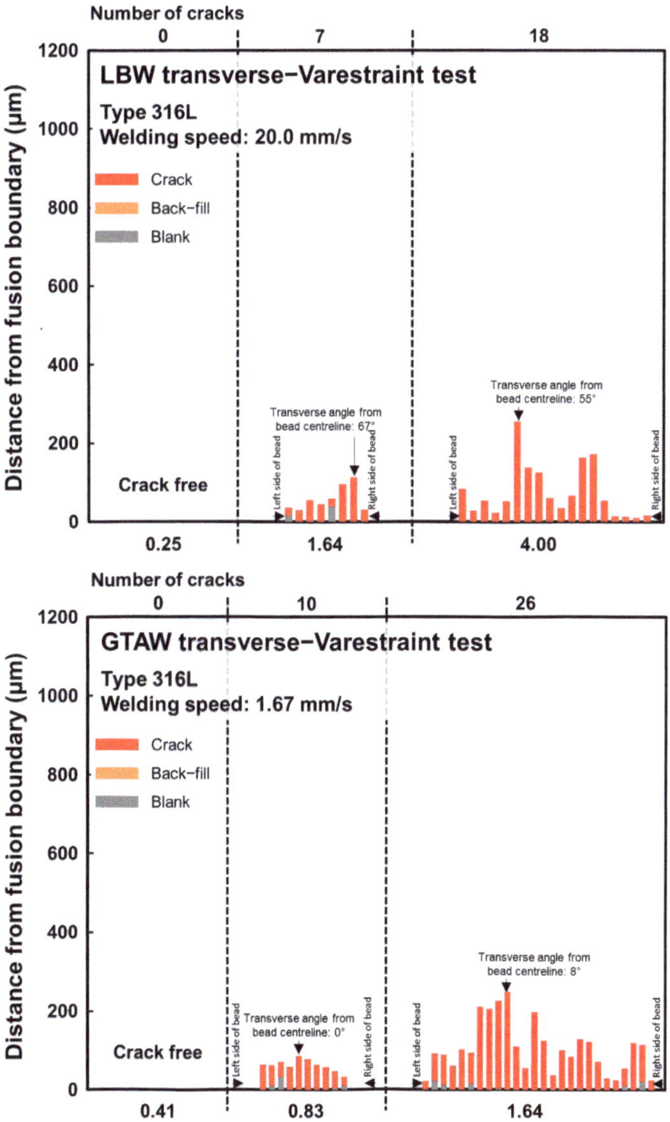


**Fig. 9** Quantitative analysis of solidification cracking susceptibility: location and length of the cracks in LBW and GTAW transverse-Varestraint tests (type 316L)

weld bead, the complex morphology consisted of the serrated dendrite boundaries was formed in the vicinity of the centreline. Therefore, it could be deduced that the transverse propagation behaviour of solidification crack with the maximum length in the LBW transverse-Varestraint test (especially at the welding speeds of 20.0 and 40.0 mm/s) caused from complex morphology at the centreline (difficult to propagate

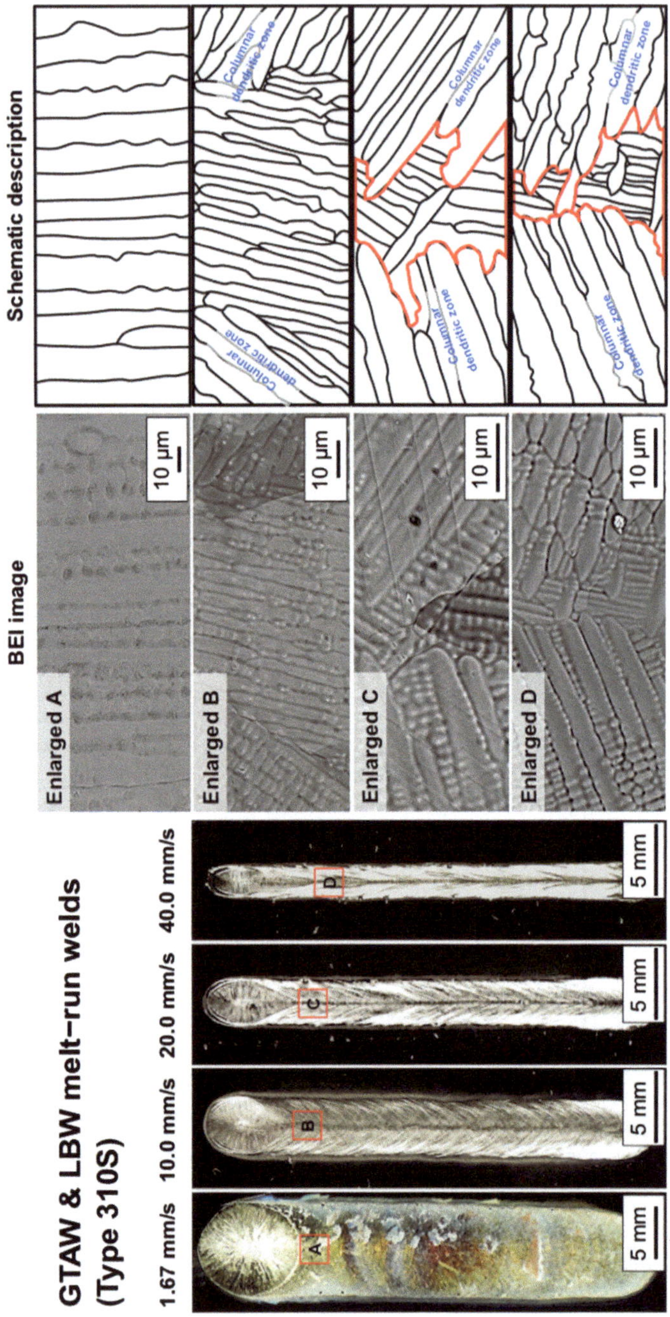


**Fig. 10** Appearance and surface microstructure (SEM) after LBW and GTAW (type 310S)

through the centreline) as compared with the straight dendrite boundary within the columnar dendritic zone. It follows that the cracking morphology in the Varestraint test is strongly affected by the solidification microstructure.

## Establishment of LBW Transverse-Varestraint Testing Methodology

### *Features in Test Procedures*

#### Timing-Synchronisation with High-Speed Camera Observation

Generally, the concept of the transverse-Varestraint test is to apply a hydraulic pressure immediately after the heat source-off to fix the temperature fields backward the molten pool. However, a minute time gap between the yoke-drop and the laser beam-off in the LBW transverse-Varestraint test has the potential for causing the under or overestimation of the solidification cracking susceptibility (especially the cracking temperature range) due to the extremely rapid cooling rate in LBW compared with GTAW. Therefore, strict timing-synchronisation between them should be essentially achieved. Therefore, in order to synchronise the timing between laser beam-off and yoke-drop, the electric signal among the components was systematically controlled. Furthermore, high-speed camera observation was simultaneously employed to prevent the inaccurate timing control. A schematic illustration of the timing-synchronisation procedure is displayed in Fig. 11. A beam from a pilot laser lamp, which was directly attached on the yoke, and the laser beam during the melt-run welding were simultaneously observed during the LBW transverse-Varestraint test by a high-speed camera [as indicated by arrow in Fig. 1 (side view)]. Figure 11 also shows the recorded images from a high-speed camera. A laser beam from the pilot laser lamp, indicating the yoke position and its movement during testing together with the laser beam during the melt-run welding could be successfully detected. Namely, based on this observation procedure, laser-off (Fig. 11b) and yoke-drop (Fig. 11c) were perfectly synchronised in the LBW transverse-Varestraint test.

#### Compensation of Yoke-Drop Time by High-Speed Camera Observation

Figure 12 shows recorded images (observed from the front of the Varestraint apparatus) by a high-speed camera during the yoke-drop, the measurement results on the yoke-drop time at certain strain levels together with a schematic description for compensation method of the cracking temperature range. As shown in the recorded images, a time-gap between the initiation and completion of yoke-drop is inevitable during strain loading in the transverse-Varestraint test. Therefore, during

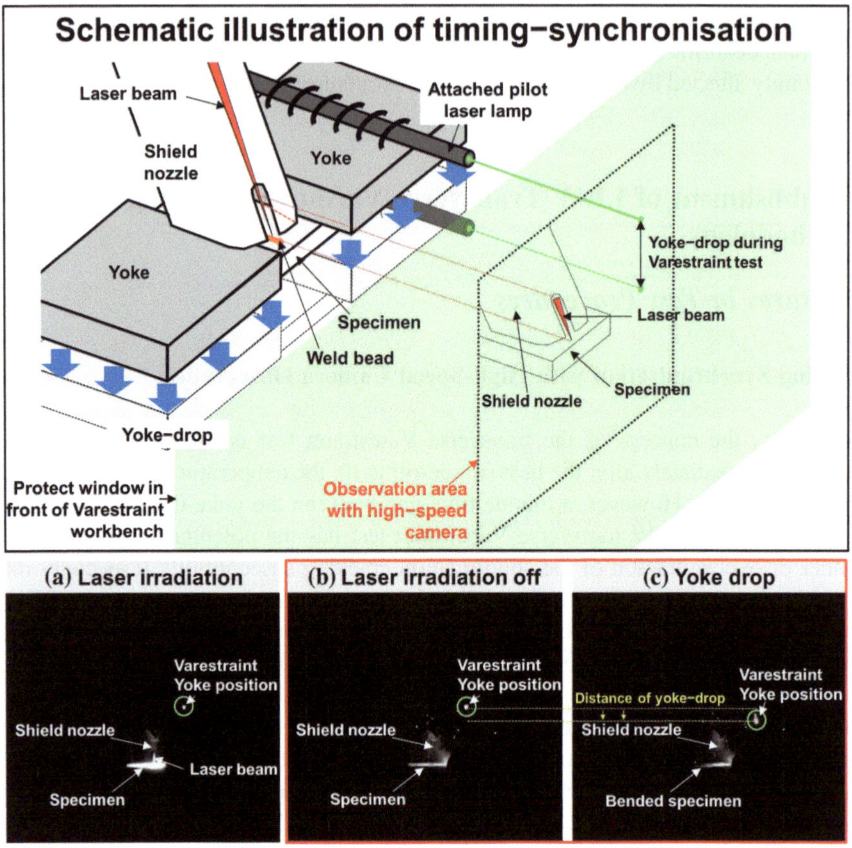


**Fig. 11** Schematic illustration of timing-synchronisation and recorded image from high-speed camera during LBW transverse-Varestraint test

yoke-drop, the resulting crack would partially escape the cracking temperature range, and the possibility of underestimation of the crack length is an inevitable issue. For the GTAW transverse-Varestraint test, this underestimation is likely to be negligible because of relatively low temperature gradient (i.e. slow cooling rate). Hence, using high-speed camera observation, the yoke-drop time was measured in all strain conditions used for the LBW transverse-Varestraint test. Figure 12 also lists the yoke-drop time under representative strain conditions. As the augmented strain increased from 0.25 to 4.00 %, the yoke-drop time also increased from 3.6 to 13.3 ms. Based on the crack length (i.e. actual solidification cracking temperature range) observed after the transverse-Varestraint test, the temperature range during the yoke-drop was compensated by the additional cracking temperature range in every strain condition used.

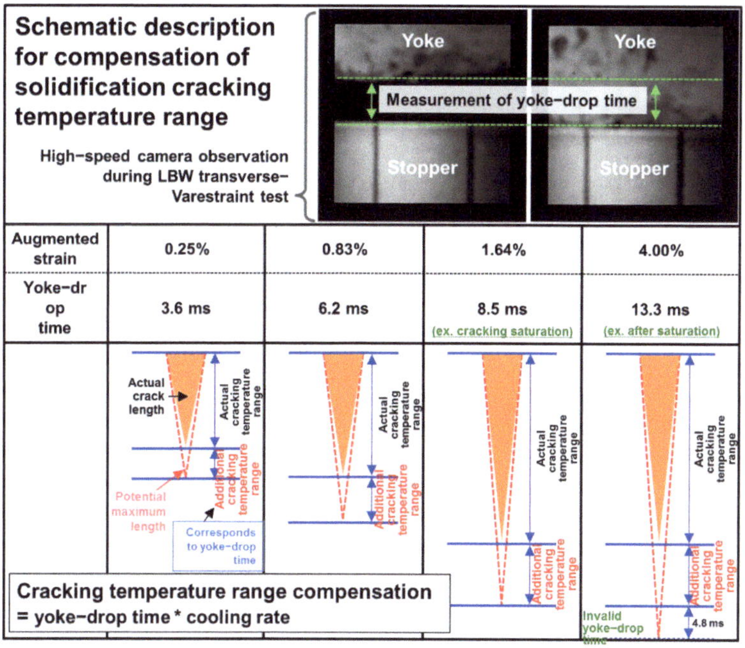


**Fig. 12** Schematic description of BTR compensation during yoke-drop with recorded image from high-speed camera

## *Determination of High-Temperature Ductility Curve for Solidification Cracking*

### Conversion of Solidification Cracking Temperature Range from Crack Length

Figure 13 shows a schematic description for conversion of the solidification cracking temperature range. The upper schematic diagram shows the solidification cracking appearance after the LBW transverse-Varestraint test based on the results in Fig. 3. The lower schematic is the cooling rate distribution actually measured at the welding speed of 20.0 mm/s for type 310S steel welds. For each centreline and transverse crack that occurred after the test, an individual cooling rate was adopted for the conversion of the respective cracking temperature range (i.e. the corresponding temperature gradient was selected for each crack at different positions). For this purpose, the cooling rate should be measured at different positions in the bead width direction. The result is shown in lower part of Fig. 13. The overall distribution of cooling rate could be obtained using inter-or extrapolation based on the actual measured values. The cooling rate was ranging from 1080 to 1215 K/s from the bead centreline to the weld toe line. Consequently, for the transverse crack, the cracking temperature range was converted using the cooling rate (A′), and

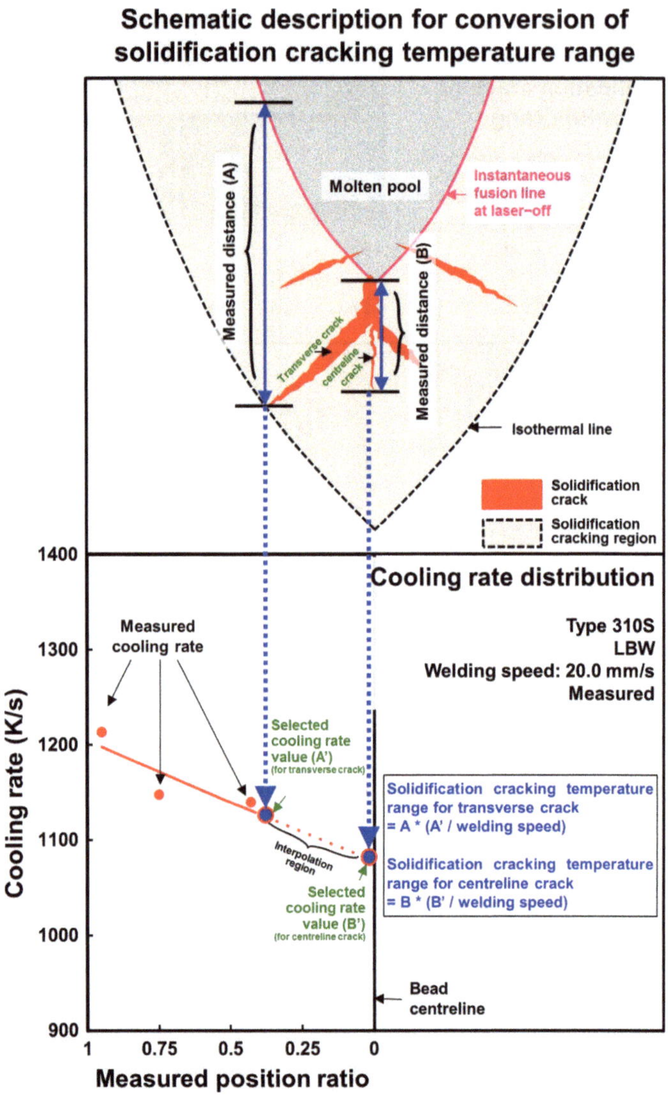


**Fig. 13** Schematic description of conversion method for solidification cracking temperature range and cooling rate distribution in bead width direction

cooling rate (B′) adopted for the centreline crack using the measured distance (A) and (B), respectively. In other words, this procedure could reflect the angle component for the cracking temperature range of the transverse crack. Figure 14a shows a typical example of the measured temperature history at the welding speed of 20.0 mm/s for the conversion procedure of the cracking temperature range. During the melt-run welding, the peak temperature extended over the equilibrium

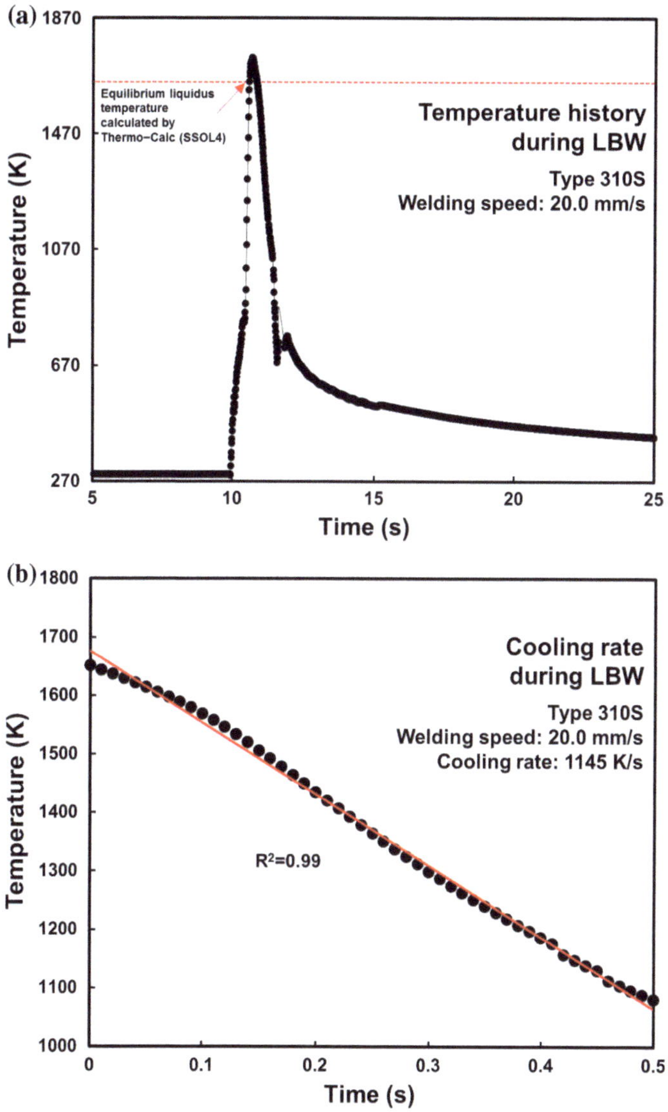


**Fig. 14** Measured **a** temperature history and **b** cooling rate during LBW

liquidus temperature (calculated from Thermo-Calc SSOL4 database). Also, the temperature decrement from the liquidus temperature for 0.5 s was adopted as the cooling rate as indicated in Fig. 14b. Mentioned procedures above were identically applied to other welding speeds.

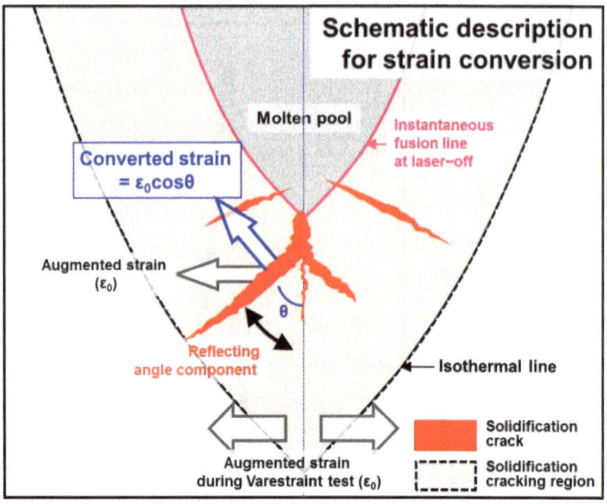


**Fig. 15** Schematic description of augmented strain conversion for high-temperature ductility curve

### Compensation of Augmented Strain

Figure 15 shows a description of the augmented strain compensation for transverse crack formation with solidification cracking appearance schematically depicted as in Fig. 13. As a general rule, the direction of applied strain during the transverse-Varestraint test is perpendicular to welding direction. In other words, the transverse cracks (especially with maximum length) formed with a large transverse angle (55°–70°) as compared with those in the GTAW transverse-Varestraint test and should be compensated to reflect the propagation angle. Consequently, based on the augmented strain during the transverse-Varestraint test ($\varepsilon_0$) and the propagation angle of a crack ($\theta$), the effective augmented strain was compensated as $\varepsilon_0 \cos\theta$ for the transverse cracks.

## High-Temperature Ductility Curve for Solidification Cracking in Laser Welds

### *Difference Between LBW and GTAW*

Figures 16 and 17 show the ductility curves of type 310S and 316L steel welds for welding speeds of 20.0 (LBW) and 1.67 (GTAW) mm/s. When the applied strain reached to about 1.00 % (for type 310S steel welds) and 2.00 % (for type 316L steel welds), the temperature range of solidification cracking was saturated, respectively. Therefore, the BTR could be estimated as 123 K for LBW (at the welding speed of 20.0 mm/s) and 146 K for GTAW (at the welding speed of 1.67 mm/s) in type 310S steel welds. Moreover, the BTR of type 316L steel welds was 51 K for LBW (at the welding speed of 20.0 mm/s) and 36 K for GTAW (at the welding speed of

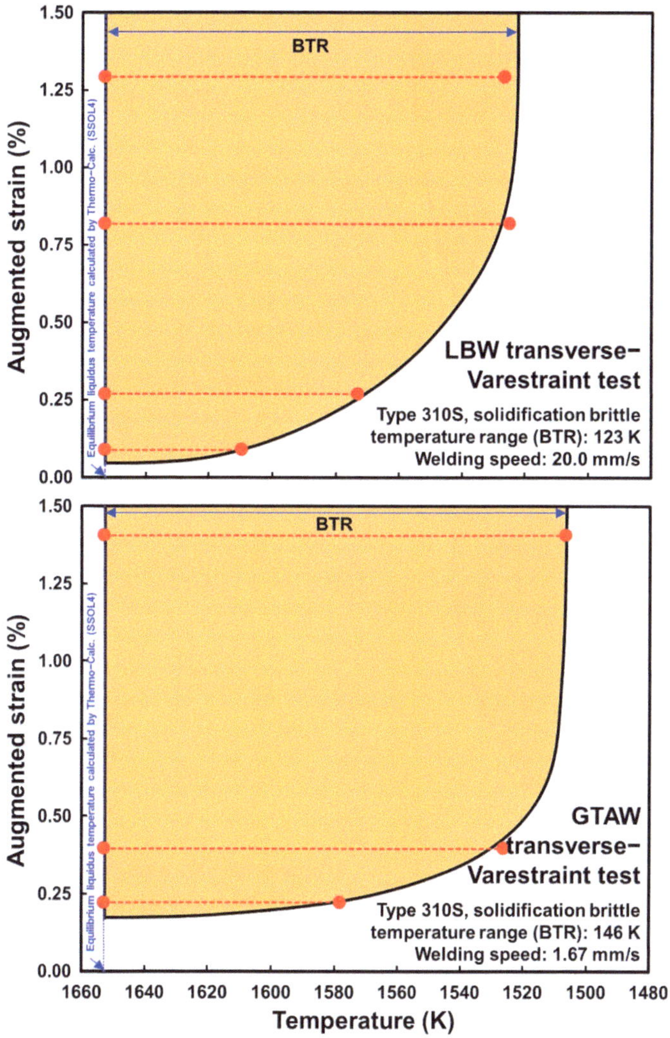


**Fig. 16** High-temperature ductility curve for solidification cracking and BTR obtained from LBW and GTAW transverse-Varestraint tests (type 310S)

1.67 mm/s). For type 310S steel welds, the BTR in LBW (at the welding speed of 20.0 mm/s) was quite narrower than that in GTAW (at the welding speed of 1.67 mm/s), and vice versa for type 316L steel welds. Furthermore, it was also confirmed that the critical strain for occurrence of solidification cracking (i.e. $\varepsilon_{min}$) in the laser welds was slightly lower than that for the arc welds in both type 310S and type 316L steel welds. Namely, it could be clearly confirmed that the solidification cracking susceptibility was different between the LBW and GTAW.

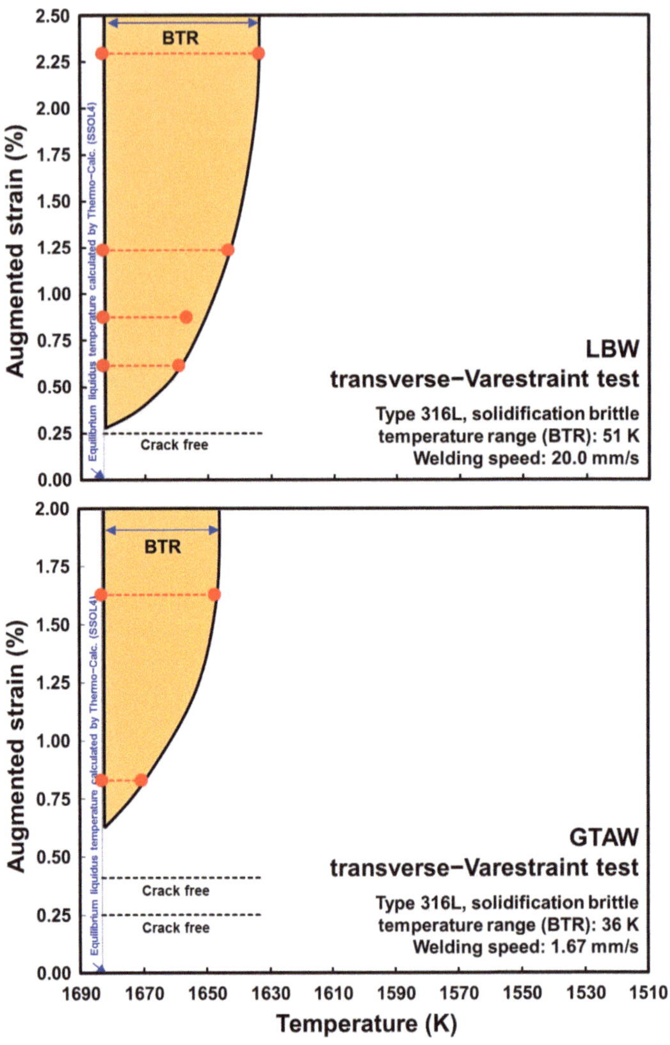


**Fig. 17** High-temperature ductility curve for solidification cracking and BTR obtained from LBW and GTAW transverse-Varestraint tests (type 316L)

## *Change in BTR with Welding Speed*

Figure 18 summarises the change in BTR as a function of the welding speed from 1.67 (GTAW) to 40.0 mm/s (LBW). In the figure, the solid mark indicates the BTR in GTAW transverse-Varestraint test (at 1.67 and 5.00 mm/s) and the open mark is for the BTR in LBW transverse-Varestraint test (at 10.0, 20.0 and 40.0 mm/s), respectively. The BTR shrank with an increase in the welding speed, from 146 to 120 K (i.e. 26 K BTR shrinkage) for type 310S steel welds. However, for type 316L

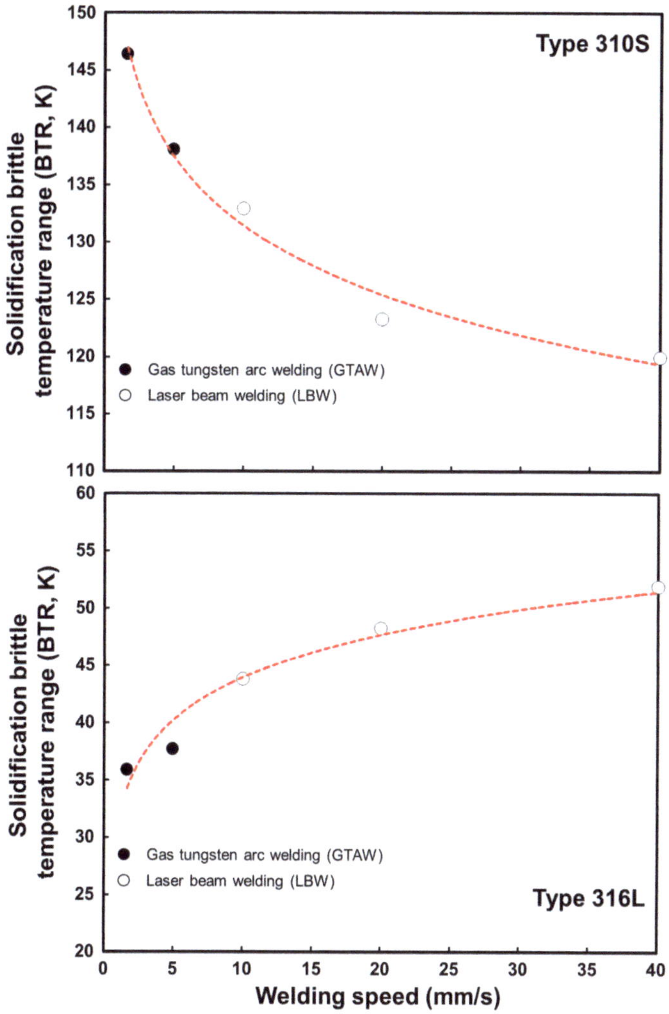


**Fig. 18** Relationship between welding speed and BTR measured by using LBW and GTAW transverse-Varestraint tests

steel welds, the BTR slightly enlarged with an increase in the welding speed from 36 to 52 K. In addition, the transition of solidification mode could not be detected within the employed welding speed through microstructure analysis (i.e. FA mode solidification in all welding speed employed) for type 316L steel welds. It follows that the opposite tendency could be clearly confirmed between type 310S and 316L steel welds.

## Mechanism of BTR Variation in LBW

Hereinafter, the mechanism of the opposite BTR variation with the welding speed (i.e. the BTR shrinkage in type 310S steel welds and the BTR enlargement in type 316L steel welds) is discussed by employing a numerical calculation of the solid/liquid coexistence temperature range. Figure 19 summarises the calculation concept. Generally, the BTR is characterised by upper and lower temperature limits. The upper temperature limit of the BTR approximately corresponds to the solidification initiation temperature, which can be also expressed by the liquidus temperature ($T_L$). And the solidification completion temperature approximates the lower temperature limit of the BTR. During the welding process, segregation of solute elements occurs along with dendrite growth because a solid phase exhausts solute elements into the remaining liquid phase. Consequently, the solute content of the liquid continuously increases during the solidification. Thus, the true solidification completion temperature ($T_C$) deviates from the nominal solidus temperature ($T_S$). A difference between solidification initiation ($T_L$) and the completion ($T_C$) temperatures could be regarded as the solid/liquid coexistence temperature range during the welding solidification. Therefore, the variation mechanism of the solid/liquid coexistence temperature range as a function of the welding speed would govern the BTR change mechanism. The changes in solidification initiation ($T_L$) and completion ($T_C$) temperatures were computed using the KGT and a solidification segregation model, respectively.

### *Theoretical Basis of Solidification Initiation Temperature ($T_L$) Calculation*

The solidification initiation temperature ($T_L$) can be approximated by the dendrite tip temperature ($T_{tip}$). In order to calculate the dendrite tip temperature, the KGT

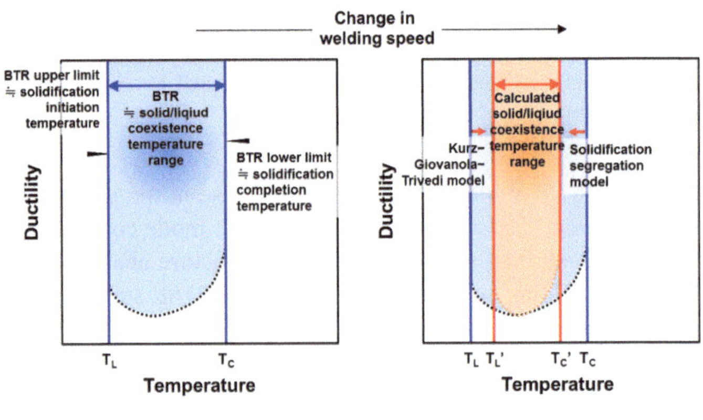


**Fig. 19** Schematic illustration of the numerical simulation concept for calculation of solid/liquid coexistence temperature range

model extended for Fe–Cr–Ni ternary system has been applied [21, 48–50]. The calculation procedure is similar to that used in several studies already reported [21, 24, 26, 50]. Below is a brief description of the theoretical basis.

The solidification initiation temperature ($T_L$) [=dendrite tip temperature ($T_{tip}$)], which could be deduced from the difference between the equilibrium liquidus temperature ($T_L^{eq}$) and the actual undercooling degree ($\Delta T$), can be expressed as follows;

$$T_L = T_L^{eq} - \Delta T \tag{1}$$

$$\Delta T = \Delta T_C + \Delta T_R + \Delta T_K + \Delta T_{cell} \tag{2}$$

where $\Delta T_C$ is the constitutional undercooling, $\Delta T_R$ is the curvature undercooling, $\Delta T_K$ is the kinetic undercooling, and $\Delta T_{cell}$ is the cellular undercooling.

Constitutional undercooling ($\Delta T_C$) is defined as;

$$\Delta T_C = m_i^{eq} C_i^0 - m_i^V C_i^L \tag{3}$$

where $m_i^{eq}$ is the equilibrium liquidus gradient of each alloying element, $m_i^V$ is the velocity dependent liquidus gradient, $C_i^0$ is the initial composition of alloying elements, and $C_i^L$ is the liquid concentration at the dendrite tip. Because equilibrium partitioning cannot be assumed in a rapid solidification process, the theory that considers the deviation from the equilibrium partitioning discussed by Boettinger et al. [51] as well as the variation of the partition coefficient with growth velocity suggested by Aziz [52] was introduced. Thus,

$$m_i^V = m_i^{eq} \left\{ 1 + \frac{k_i^{eq} - k_i^V \left(1 - \ln \frac{k_i^V}{k_i^{eq}}\right)}{1 - k_i^{eq}} \right\} \tag{4}$$

$$k_i^V = \frac{k_i^{eq} + \frac{a_0 V}{D_i^L}}{1 + \frac{a_0 V}{D_i^L}} \tag{5}$$

where $V$ is the dendrite growth velocity, $k_i^V$ is the velocity-dependent partitioning coefficient, $k_i^{eq}$ is the equilibrium partitioning coefficient of each alloying element, $D_i^L$ is the temperature dependent diffusion coefficient of the elements in the liquid, $a_0$ is the length scale related to the interatomic distance. Accordingly the stability criterion, $C_i^L$ is determined by the following equations;

$$C_i^L = \frac{C_i^0}{1 - (1 - k_i^V) I_V(P_i^C)} \tag{6}$$

$$I_V\left(P_i^C\right) = P_i^C \exp\left(P_i^C\right) \int_{P_i^C}^{\infty} \frac{\exp(-z)}{z} dz \tag{7}$$

$$GR^2 + 2R \sum_i m_i^V P_i^C \left(1 - k_i^V\right) C_i^L \xi_i^C + 4\pi^2 \Gamma = 0 \tag{8}$$

where $P_i^C$ is the solute Peclet number, $I_V\left(P_i^C\right)$ is the Ivantsov's solution, $G$ is the temperature gradient, $V'$ is the welding velocity, $\varepsilon$ is the cooling rate, and $\xi_i^C$ is the absolute stability coefficient determined as;

$$\xi_i^C = \begin{cases} 1 - \dfrac{2k_i^V}{\sqrt{1 + \left(\frac{2\pi}{P_i^C}\right)^2} - 1 + 2k_i^V} & \left(P_i^C \le \dfrac{\pi^2}{\sqrt{k_{V,i}}}\right) \\ \dfrac{\pi^2}{k_i^V P_i^C} & \left(P_i^C > \dfrac{\pi^2}{\sqrt{k_{V,i}}}\right) \end{cases} \tag{9}$$

The curvature undercooling ($\Delta T_R$) is expressed as [24, 26];

$$\Delta T_R = -\frac{2\Gamma}{R} \tag{10}$$

where $\Gamma$ is the Gibbs–Thompson parameter.

The kinetic under cooling ($\Delta T_K$) can be described as [24, 26];

$$\Delta T_K = \frac{V}{\mu_0} = \frac{R_g T_m^2}{\Delta H_f V_0} V \tag{11}$$

where $\mu_0$ is the linear interface kinetic coefficient, $V_0$ is the sonic velocity in the liquid, $\Delta H_f$ is the enthalpy of fusion, and $T_m$ is the liquidus temperature of pure iron.

Burden et al. defined $\Delta T_{cell}$ as [24, 26, 53];

$$\Delta T_{cell} = \frac{GD}{V} \tag{12}$$

On the basis of the above, the solidification initiation temperature ($T_L$) could be summarised as Eq. 13. In type 316L steel welds, the dendrite tip temperature ($T_{tip}$) was calculated for each δ-ferrite and austenite phase respectively, and the higher of them was selected as the solidification initiation temperature ($T_L$). For the calculation of constitutional undercooling component ($\Delta T_C$), each binary system for alloying elements (i.e. Fe–Cr, Ni, Mo, Si, Mn, P, S and C) was employed, and the contribution of each system for the constitutional undercooling ($\Delta T_C$) was summed as also indicated in Eq. 13;

$$T_L = T_L^{eq} + \sum_i m_i^{eq} C_i^0 \left\{ 1 - \frac{1 + \frac{k_i^{eq} - k_i^V \left(1 - \ln \frac{k_i^V}{k_i^{eq}}\right)}{1 - k_i^{eq}}}{1 - (1 - k_i^V) I_V(P_i^C)} \right\} - \frac{2\Gamma}{R} - \frac{R_g T_m^2}{\Delta H_f V_0} V - \frac{GD}{V} \quad (13)$$

## *Theoretical Basis of Solidification Completion Temperature ($T_C$) Calculation*

In order to calculate the solidification completion temperature ($T_C$), the solidification segregation behaviours of all solute elements to the cell boundary were calculated for each binary system (i.e. Fe–Cr, Ni, Mo, Si, Mn, P, S and C) using the finite differential method. The single A mode solidification segregation model was employed for type 310S steel welds and the divorced-eutectic FA mode solidification model (based on Ref. [55]) was adopted for type 316L steel welds to reflect the respective solidification mode. The models are schematically depicted in Fig. 20. The cross-section shape of a dendrite was basically assumed to be a hexagonal prism. A one-dimensional regular triangle model for type 310S steel welds and half-quadrangle model for type 316L steel welds were employed, respectively. In a particular, for type 316L steel welds, δ-ferrite was solidified from the cell core to the boundary as a primary phase, and then austenite phase subsequently solidified from cell boundary to cell core as a secondary phase. Consequently, residual liquid phase at solidification completion was located between austenite and cell boundary for type 310S steel welds, and positioned between δ-ferrite and austenite phases for type 316 steel welds, respectively. The principle of the calculation procedure is similar to those previously reported [24, 36, 46, 54]. Below is a brief description of this segregation model.

The distribution of the solute elements during the solidification process was determined by solving the diffusion equation for both solid and liquid phases. Symmetrical boundary conditions were applied to the both end segments. The diffusive flux $J_i$ from a segment $i$ to a segment $i + 1$ is given by Fick's first law;

$$J_i = D \frac{C_{i+1} - C_i}{\Delta x} \quad (14)$$

where $D$ is the diffusion coefficient of the solute, $C_i$ and $C_{i+1}$ are the concentrations in segment $i$ and $i + 1$ respectively, and $\Delta x$ is the segment width. The change in solute concentration $\Delta C_i$ during a minute time interval $\Delta t$ at segment $i$ is expressed by;

$$\frac{\Delta C_i}{\Delta t} \frac{S_i + S_{i-1}}{2} \Delta x_i = J_i S_i - J_{i-1} S_{i-1} \quad (15)$$

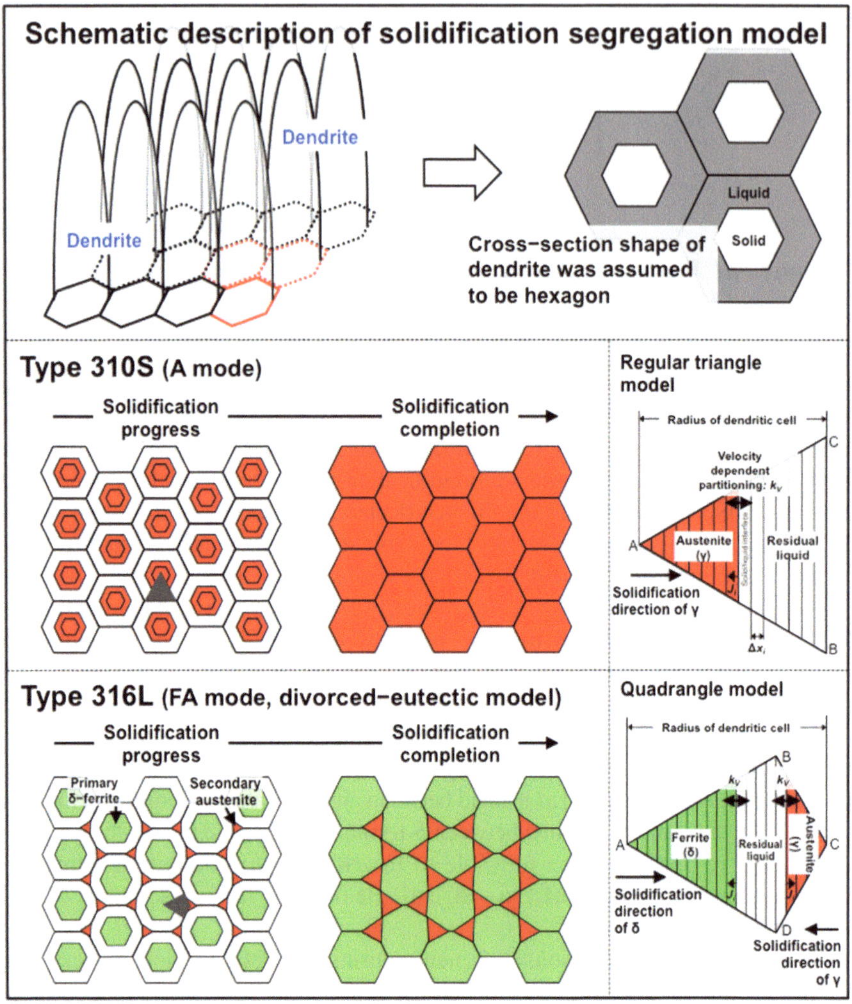


**Fig. 20** Schematic illustration of the analysis model for calculation of solidification completion temperature (solidification segregation model)

where $S_i$ and $S_{i-1}$ are the sectional areas of segments $i$ and $i - 1$ respectively. $S_i$ is expressed as Eqs. 16 (single A mode solidification segregation model for type 310S steel) and 17 (divorced-eutectic FA mode solidification segregation model for type 316L steel);

$$S_i = \frac{1}{\sqrt{3}} \left( \sum_{k=1}^{i} \Delta x_k \right)^2 \quad (16)$$

$$S_i = \frac{1}{2\sqrt{3}}\left(\sum_{k=1}^{i} \Delta x_k\right)^2 \tag{17}$$

From Eqs. (15) to (17), $\Delta C_i$ can be found from;

$$\Delta C_i = \frac{2D\Delta t}{\Delta x_i(S_i + S_{i-1})}\left\{S_i \frac{C_{i+1}^B - C_i^B}{\Delta x_i} - S_{i-1}\frac{C_i^B - C_{i-1}^B}{\Delta x_{i-1}}\right\} \tag{18}$$

where $C_{i+1}^B$, $C_i^B$ and $C_{i-1}^B$ are the solute concentrations in segment $i + 1$, $i$ and $i - 1$ at the previous $\Delta t$. Assuming the mass conservation law of solute in a dendrite, Eq. (19) can be obtained.

$$\left(\sum_{k=1}^{N} \Delta x_k\right)^2 C_0 = \sum_{i=1}^{N}\left\{\left(\sum_{k=1}^{i} \Delta x_k + \sum_{k=1}^{i-1} \Delta x_k\right)\Delta x_i C_i^S\right\} + \sum_{i=j+1}^{N}\left\{\left(\sum_{k=1}^{i} \Delta x_k + \sum_{k=1}^{i-1} \Delta x_k\right)\Delta x_i C_i^L\right\} \tag{19}$$

where $N$ is the total number of segments, $C_0$ is the initial solute concentration, $C_i^S$ is the solute concentration in solid phase segment $i$, $C_i^L$ is the solute concentration in the liquid phase and $j$ is the segment number of the solid phase at the solid/liquid interface.

During the solidification process, the solute concentration at the solid/liquid interface is determined by the non-equilibrium concept as follows;

$$C_j^S = k_V C_{j+1}^L \tag{20}$$

where $k_V$ is the non-equilibrium distribution coefficient defined by Eq. (5), $C_j^S$ and $C_{j+1}^L$ are the solute concentrations of solid and liquid phases at the solid/liquid interface respectively.

In the solidification segregation calculation mentioned above, the solidification was initiated from the solidification initiation temperature [$T_L$ = dendrite tip temperature ($T_{tip}$)] for both welds, linked with the result in Eq. 13. Especially, for type 316L steel welds, it was assumed that austenite solidification as a secondary phase was initiated when the liquidus temperature of residual liquid phase is reduced less than the dendrite tip temperature ($T_{tip}$) of austenite phase also calculated from Eq. 13. After that, the mentioned above segregation computation was progressed for respective phases until solidification completion. The liquidus temperature in the residual liquid phase during the solidification segregation ($T_m$) is assumed by decrement of liquidus temperature corresponds to segregated concentration from the initial solute composition, reflecting the velocity dependent liquidus gradient as follows;

$$T_m = T_{tip} + \sum m^V\left(C_{j+1}^L - C^0\right) \tag{21}$$

where $T_{tip}$ is the calculated dendrite tip temperature from Eq. 13, $m^V$ is the velocity dependent liquidus gradient from Eq. 4. And the decrement of the liquidus temperature was summed by contribution of each binary system for alloying elements (i.e. Fe–Cr, Ni, Mo, Si, Mn, P, S and C).

The solidification segregation was computed by 95 % solidification completion (i.e. fraction of residual liquid phase was 5 %), based on tin-quenched solidification microstructure observed by Ref. [46]. The segregated concentration of all solute elements in the residual liquid phase at solidification completion (=95 % solidification completion) was inputted to Thermo-Calc software (SSOL4 database), and the equilibrium solidus temperature computed by the Thermo-Calc was adopted as the solidification completion temperature ($T_C$) in the welding solidification as similar in Ref. [46].

## *Conditions of Numerical Calculation*

Tables 4 and 5 summarise the material constants used in the calculation for both the solidification initiation and completion temperatures [56–58]. The equilibrium liquidus temperatures ($T_L^{eq}$) of type 310S and 316L steels used were obtained from the Thermo-Calc software with the SSOL4 database. For simplicity, all the calculations were carried out under the following assumptions; the dendrite solidification velocity is equal to the welding speed, the cooling rate remains constant during solidification, and interactions between solute elements (i.e. cosegregation) are negligible. The calculation conditions are also listed in Table 6. The dendrite radius employed was 10.0 μm, and the mesh was divided into 100 segments. The welding speed was varied as 1.67, 5.00 mm/s (GTAW) and 10.0, 20.0, 40.0 mm/s

**Table 4** Material constants used for numerical calculation of solidification initiation temperature (for the KGT model)

| Constants | Value | |
|---|---|---|
| Equilibrium liquidus temperature (K) | Type 310S: 1654 (Thermo-Calc, SSOL4) | |
| | Type 316L | δ: 1683 (Thermo-Calc, SSOL4) |
| | | γ: 1678 (Thermo-Calc, SSOL4) |
| Length scale for solute trapping (m) | $5.0 \times 10^{-9}$ | |
| Gas constant (J/mol K) | 8.31446 | |
| Liquidus temperature of pure Fe (K) | 1803 | |
| Sonic velocity in liquid (m/s) | 2000 | |
| Gibbs–Thomson coefficient (km) | γ: $3.2 \times 10^{-7}$ | |
| | δ: $2.6 \times 10^{-7}$ | |
| Enthalpy of fusion (J/kg) | γ: 237201 | |
| | δ: 221381 | |

δ: δ-ferrite, γ: austenite

**Table 5** Material constants used for numerical calculation of solidification initiation and completion temperatures in binary systems of Fe–C, S, P, Cr, Ni, Si, Mn and Mo (for KGT and solidification segregation models)

| Constants | C | S | P | Cr | Ni | Si | Mn | Mo |
|---|---|---|---|---|---|---|---|---|
| Equilibrium partition coefficient in γ | 0.40 | 0.02 | 0.11 | 0.88 | 0.95 | 0.61 | 0.85 | 0.61 |
| Equilibrium partition coefficient in δ | 0.20 | 0.05 | 0.14 | 1.02 | 0.77 | 0.62 | 0.74 | 0.86 |
| Equilibrium liquidus gradient in γ (K/mass%) | −39.6 | −21.0 | −20.6 | −1.96 | −2.20 | −12.1 | −3.71 | −2.76 |
| Equilibrium liquidus gradient in δ (K/mass%) | −59.0 | −26.5 | −24.6 | 0.34 | −3.47 | −12.0 | −4.19 | −1.29 |
| Diffusion coefficient in liquid ($10^{-7}$m$^2$/s) | 5.20 | 41.4 | 0.43 | 2.67 | 4.92 | 0.51 | 4.6 | 0.61 |
| Activation energy for diffusion in liquid (kJ/mol) | 19.1 | 87.9 | 33.5 | 67.2 | 68.0 | 38.4 | 70.6 | 42.0 |
| Diffusion coefficient in γ ($10^{-4}$m$^2$/s) | 2.30 | 0.75 | 8.70 | 0.17 | 0.77 | 0.30 | 0.16 | 0.036 |
| Diffusion coefficient in δ ($10^{-3}$m$^2$/s) | 0.002 | 0.14 | 29.0 | 3.70 | 0.97 | 0.17 | 0.15 | 15.0 |
| Activation energy for diffusion in γ (kJ/mol) | 148 | 236 | 273 | 264 | 281 | 252 | 262 | 240 |
| Activation energy for diffusion in δ (kJ/mol) | 83.9 | 203 | 271 | 267 | 262 | 229 | 234 | 283 |

δ: δ-ferrite, γ: austenite

(LBW). For both steels, the cooling rate during solidification was assumed to be 240 K/s for 1.67 mm/s (GTAW), 540 K/s for 5.00 mm/s (GTAW), 1000 K/s for 10.0 mm/s (LBW), 1200 K/s for 20.0 mm/s (LBW), and 1950 K/s for 40.0 mm/s (LBW) based on the measured thermal cycle during the melt-run welding as explained in Fig. 14.

**Table 6** Conditions for numerical calculations

| Analysis conditions | |
|---|---|
| Dendrite radius (μm) | 10.0 |
| Number of mesh division | 100 |
| Welding speed (mm/s) | 1.67, 5.00 (GTAW), 10.0, 20.0, 40.0 (LBW) |
| Cooling rate (K/s) | Welding speed 1.67 mm/s: 240 |
| | Welding speed 5.00 mm/s: 540 |
| | Welding speed 10.0 mm/s: 1000 |
| | Welding speed 20.0 mm/s: 1200 |
| | Welding speed 40.0 mm/s: 1940 |

## *Calculation Results*

### Solidification Segregation Behaviour

Among various alloying elements in stainless steels, sulphur (S) and phosphorous (P) have been generally regarded as the most governing elements of the solidification cracking susceptibility (especially for cracking temperature range), segregating at the remaining liquid phase between dendrites during the welding solidification [3–7, 22, 24, 36, 46, 54]. Figures 21 and 22 representatively show the segregation

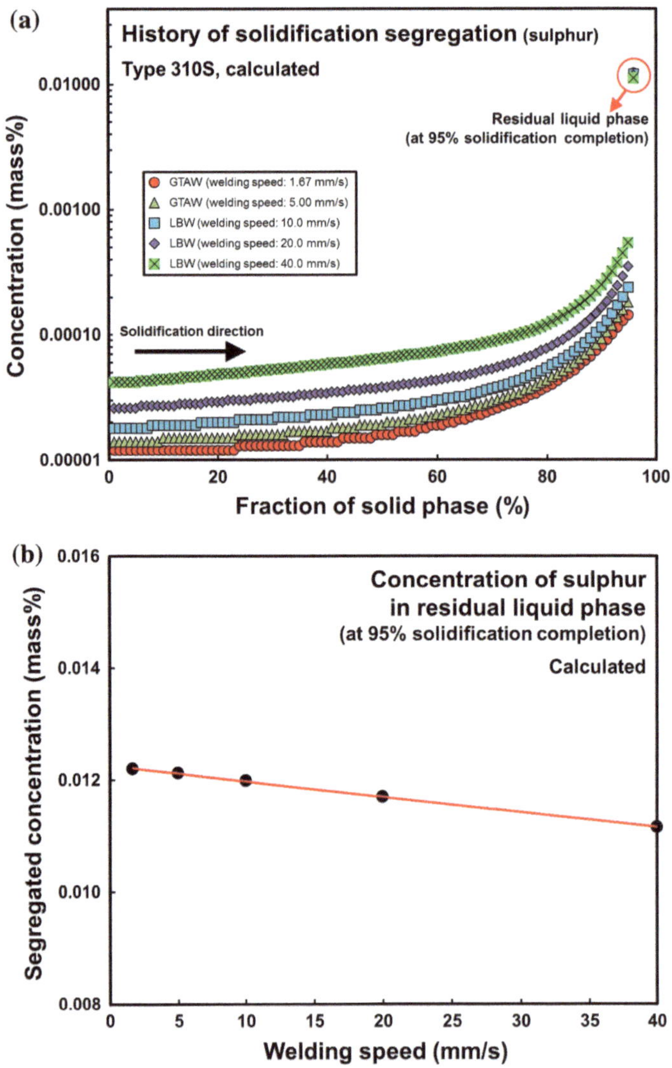


**Fig. 21** Calculation results of solidification segregation for S: **a** history of solidification segregation during welding solidification and **b** concentration of S in residual liquid phase (type 310S)

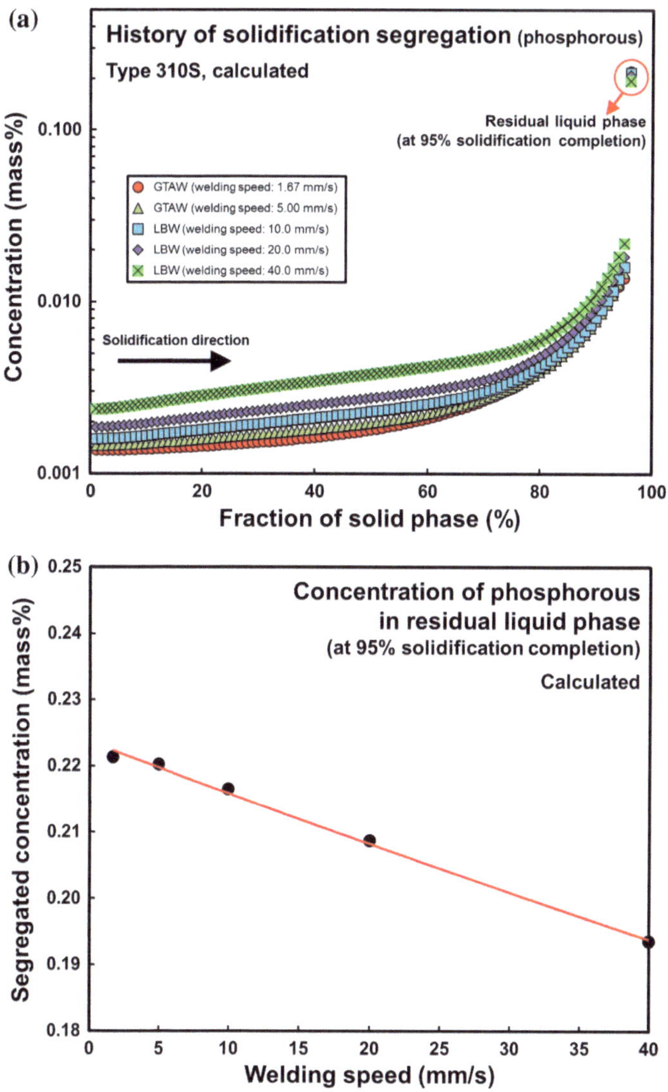


**Fig. 22** Calculation results of solidification segregation for P: **a** history of solidification segregation during welding solidification and **b** concentration of P in residual liquid phase (type 310S)

behaviours of S and P for type 310S steel welds. During the welding solidification, the solute concentration increased with the progress of the solidification, and high amounts of S and P were segregated to the remaining liquid phase at solidification completion (i.e. solidified fraction: 95 %) as described in Figs. 21a and 22a. As the welding speed increases, however, S and P were easily distributed in the solid phase, and the concentration of them in the residual liquid phase at the solidification completion was different in accordance with the welding speed. Figures 21b and 22b show change of the segregated concentrations in remaining liquid phase at solidification completion as a function of the welding speed. The segregated concentrations for both elements indicated the decrement tendency with an increase in the welding speed. Similar results could be also confirmed for other solute elements. Namely, it could be confirmed that employing LBW process alleviated the solidification segregation behaviour of the impurity and solute elements.

Figures 23a and 24a show the solidification segregation behaviours of S and P for type 316L steel welds. As indicated by arrows, solidification of δ-ferrite was

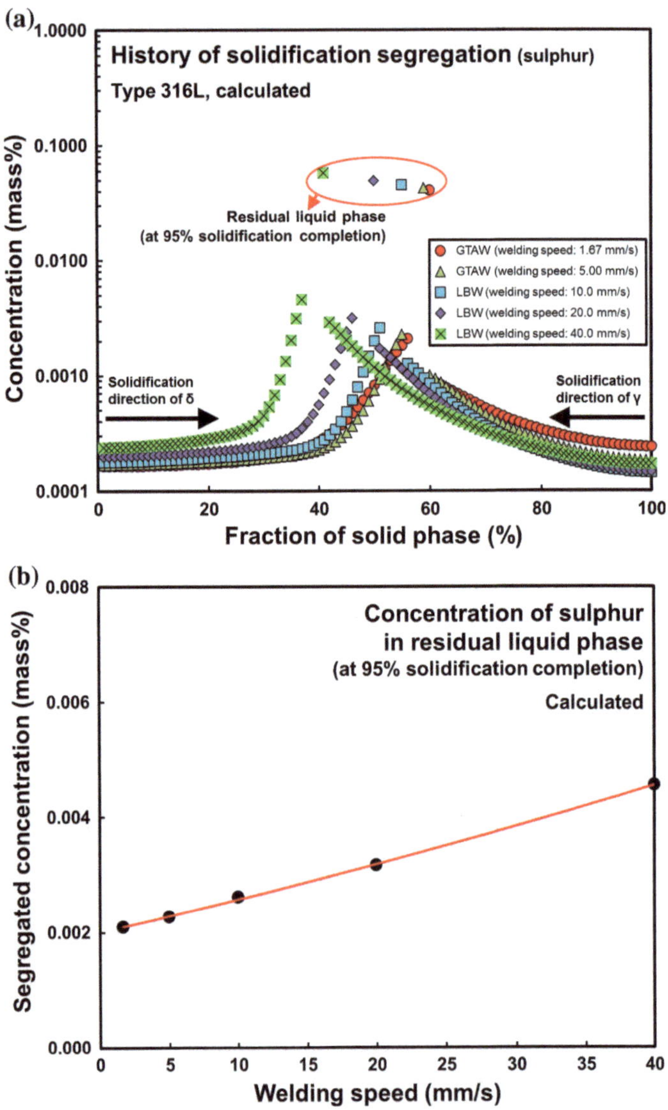


**Fig. 23** Calculation results of solidification segregation for S: **a** history of solidification segregation during welding solidification and **b** concentration of S in residual liquid phase (type 316L)

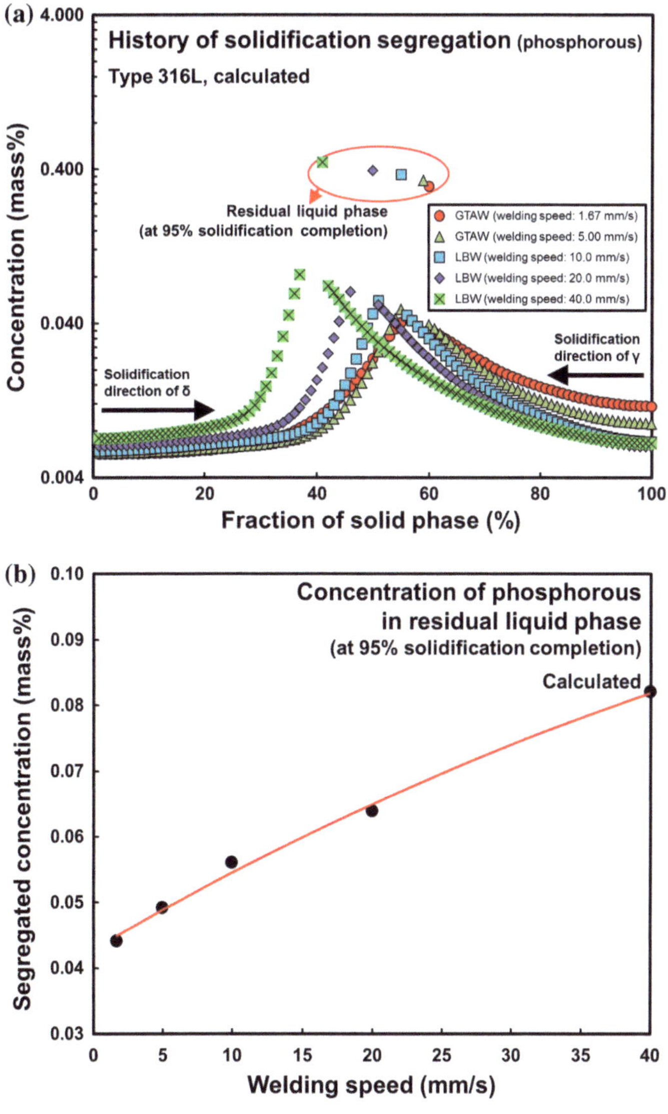


**Fig. 24** Calculation results of solidification segregation for P: **a** history of solidification segregation during welding solidification and **b** concentration of P in residual liquid phase (type 316L)

progressed from left ends and from the right ends for that of austenite phase in each welding speed. And the concentration in residual liquid phase at the solidification completion (i.e. solidified fraction: 95 %) was located between the respective peak concentration of δ-ferrite and austenite phases in each welding speed. During the solidification, the element concentrations of S and P increased with the progress of the solidification for both δ-ferrite and austenite phases, and high amounts of the S and P were also segregated to the remaining liquid phase after the solidification completion in every welding speed. As similar with type 310S steel welds in Figs. 21a and 22a, the elements were easily distributed in the solid phase (especially in δ-ferrite phase) with an increase in welding speed. Figures 23b and 24b also show change in segregated S and P concentrations in the residual liquid phase at solidification completion. As opposed to type 310S steel welds, the segregated concentrations increased with an increase in the welding speed. Namely, it could be also confirmed that the solidification segregation behaviour of the impurity elements could not be mitigated in LBW for type 316L steel welds with an increase in the welding speed. Therefore, the opposite tendency of solidification segregation behaviour could be detected between type 310S and 316L steel welds.

## Effect of Welding Speed on Solid/Liquid Coexistence Temperature Range

Figure 25 shows calculation results of the solidification initiation ($T_L$), the completion temperatures ($T_C$) and the solid/liquid coexistence temperature range ($T_L - T_C$) as a function of welding speed for type 310S steel welds. The solidification initiation temperature ($T_L$) decreased with an increase in the welding speed, because of the increased supercooling (predominantly constitutional undercooling behaviour). And the solidification completion temperature ($T_C$) increased with an increase in the welding speed due to the mitigated solidification segregation. As a result, the calculated solid/liquid coexistence temperature range ($T_L - T_C$) shrank with an increase in the welding speed. Figure 26 shows the result for type 316L steel welds. The solidification initiation temperature ($T_L$ = dendrite tip temperature ($T_{tip}$) for δ-ferrite phase) decreased with an increase in the welding speed same as type 310S steel welds. However, the solidification completion temperature ($T_C$) also decreased with an increase in the welding speed. As a result, the calculated solid/liquid coexistence temperature range ($T_L - T_C$) enlarged with an increase in the welding speed. Consequently, it could be recognised that the solid/liquid coexistence temperature range ($T_L - T_C$) for both welds was oppositely changed as a function of the welding speed, and the coexistence temperature range was strongly governed by variation of the solidification completion temperature ($T_C$), which was dominated by the solidification segregation behaviour. Namely, it could be deduced that the solid/liquid coexistence temperature range ($T_L - T_C$) was predominately governed by solidification segregation behaviour.

Figure 27 shows the comparison of the experimental BTR with the calculated solid/liquid coexistence temperature range ($T_L - T_C$). In the figure, a solid bar indicates for the experimental BTR, and a open bar is for the calculated solid/liquid coexistence temperature range ($T_L - T_C$), respectively. And Fig. 28 additionally shows relationship between experimental BTR and the calculated solid/liquid coexistence temperature range ($T_L - T_C$). Although absolute value of the calculated solid/liquid coexistence temperature range ($T_L - T_C$) deviated from the experimental BTR, the BTR obtained from the Varestraint tests was positively correlated with the calculated coexistence temperature ranges ($T_L - T_C$) for both welds. Consequently,

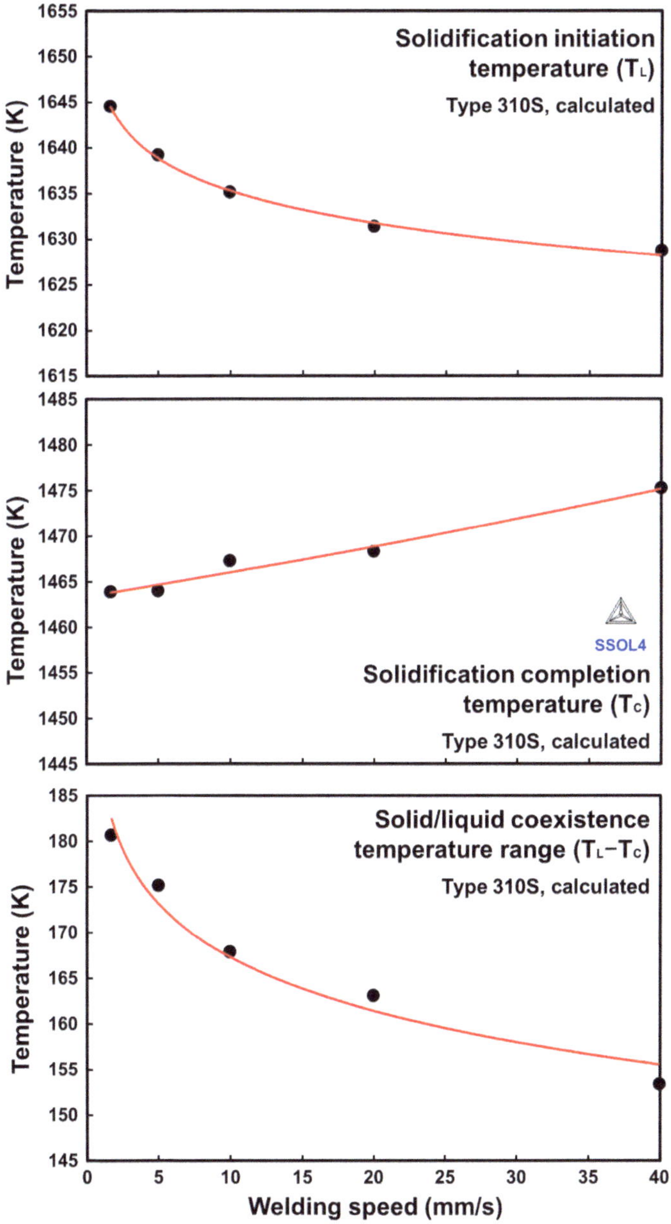


**Fig. 25** Calculation results: solidification initiation temperature, solidification completion temperature, solid/liquid coexistence temperature range as a function of welding speed (type 310S)

the mechanism of the opposite variation in the BTR between type 310S and 316L steel welds could be explained by the variation of the solid/liquid coexistence temperature range ($T_L - T_C$) (i.e. the different solidification segregation behaviour).

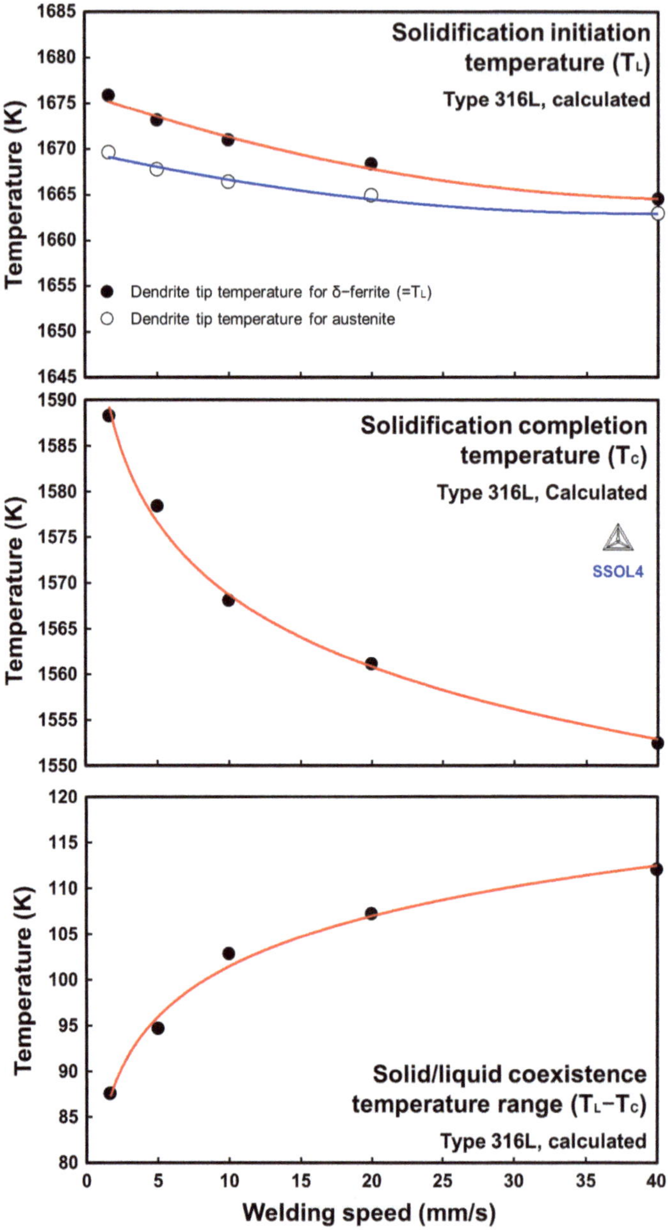


**Fig. 26** Calculation results: solidification initiation temperature, solidification completion temperature, solid/liquid coexistence temperature range as a function of welding speed (type 316L)

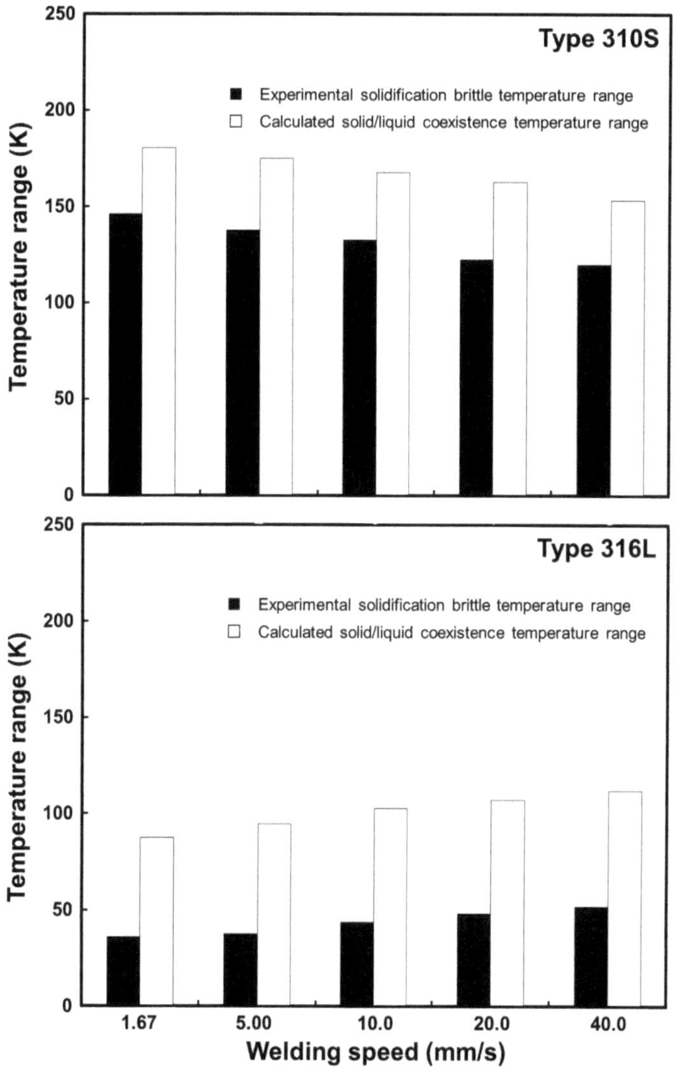


**Fig. 27** Comparison of the calculated solid/liquid coexistence temperature range with the experimental BTR

## *Variation Mechanism of Solidification Cracking Susceptibility in LBW*

The change in calculated partitioning coefficients (using Eq. 5) of S and P for δ-ferrite and austenite phases as a function of the welding speed is shown in Fig. 29. In the figure, a solid mark corresponds to the coefficient in δ-ferrite phase, and a

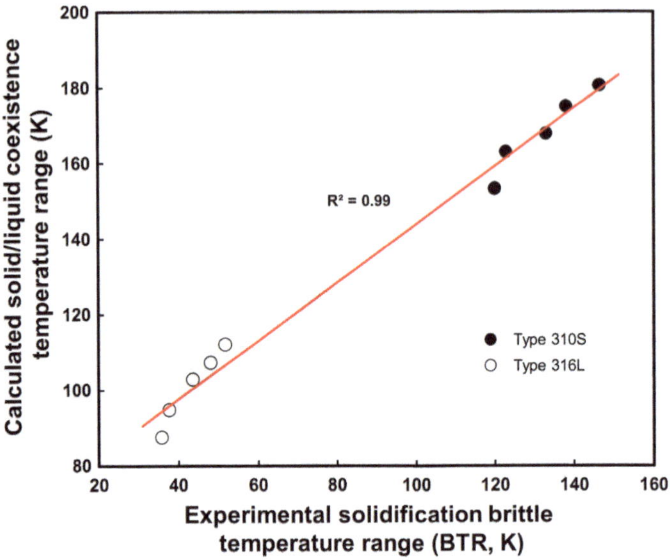


**Fig. 28** Relationship between the calculated solid/liquid coexistence temperature range and the experimental BTR

open mark is for that in austenite phase for each element. The partitioning coefficients for the elements increased with an increase in the welding speed in both austenite and δ-ferrite phases. Therefore, for type 310S steel welds, the inhibited

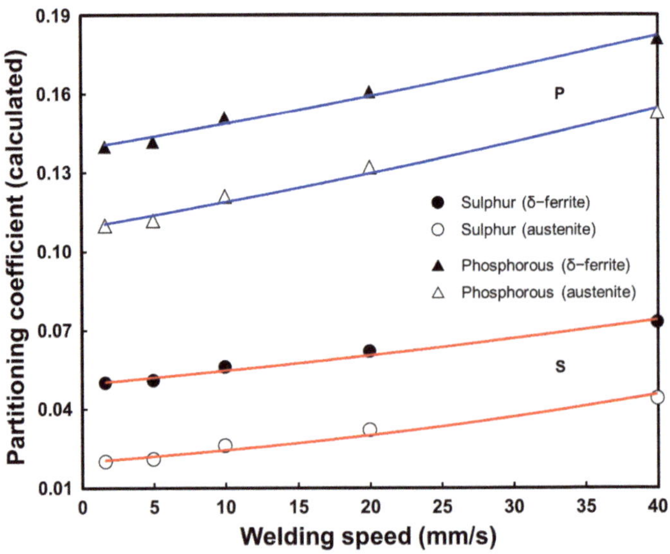


**Fig. 29** Partitioning coefficients for S and P as a function of welding speed

solidification segregation as shown in Figs. 21 and 22 could be attributed to the increased partitioning coefficients in austenite phase during the welding solidification (i.e. solute trapping). Namely, the mechanism of mitigated solidification cracking susceptibility (i.e. BTR shrinkage) in type 310S steel welds with an increase in the welding speed could be regarded as combined effects of the dendrite tip supercooling with the inhibited segregation of solute and impurity elements during the welding solidification, shrinking the solid/liquid coexistence temperature range ($T_L - T_C$).

Figure 30 shows calculated result for the δ-ferrite amount at solidification completion as a function of the welding speed for type 316L steel welds. The amount of δ-ferrite phase at solidification completion was calculated based on the model as depicted in Fig. 20 [i.e. area fraction of δ-ferrite phase = area of δ-ferrite phase/total area of the half-quadrangle (ABC)]. The ferrite content at solidification completion decreased with an increase in the welding speed from 41.8 to 18.3 %. Consequently, the enhanced solidification segregation for type 316L steel welds with an increase in welding speed (in Figs. 23 and 24) could be related with decrement of the δ-ferrite amount. It has been commonly regarded that δ-ferrite formation during the welding solidification can play a positive role in the solidification cracking susceptibility, because the ferrite phase possesses the higher solubility of impurity elements (such as S and P) compared with that in austenite phase [3, 4]. Moreover, the ferrite content during the welding solidification must be also varied as a function of the welding speed as already reported by several studies [12, 14, 15, 20]. Namely, the enhanced solidification cracking susceptibility (i.e. BTR enlargement) with an increase in the welding speed for type 316L steel

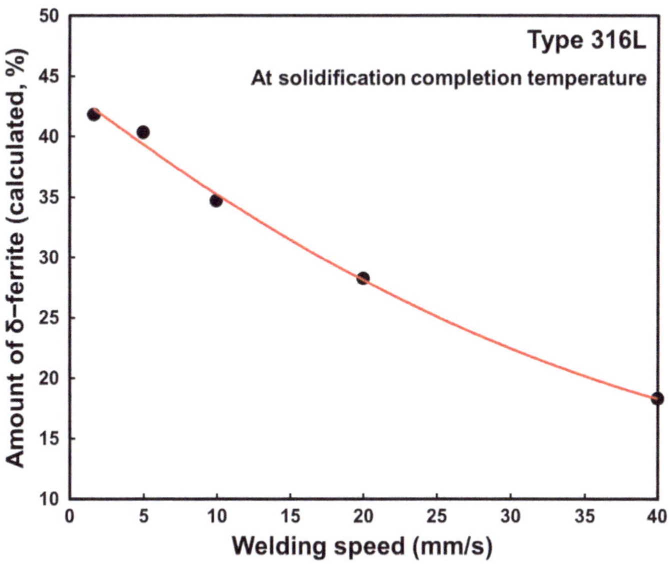


**Fig. 30** Calculated amount of δ-ferrite at solidification completion as a function of welding speed

welds could be attributed to the enhanced solidification segregation of solute elements (mainly S and P) highly correlated with a decrease in the δ-ferrite amount, which could be regarded as the more significant contributing factor than partitioning coefficient increment in the ferrite phase (in Fig. 29) for type 316L steel welds, enlarging the solid/liquid coexistence temperature range ($T_L - T_C$).

## Conclusions

In the present study, a new evaluation procedure for solidification cracking susceptibility in laser welds using a laser beam welding (LBW) transverse-Varestraint test was developed. The solidification cracking susceptibility of type 310S and 316L stainless steel laser welds was quantitatively evaluated. The main conclusions are summarised below;

1. The testing procedure consisted of electrical signal control and high-speed camera observation, timing-synchronisation among the laser oscillator, welding robot and hydraulic pressure. In order to prevent underestimation of the BTR, the yoke-drop time was also compensated as an additional component of the BTR.
2. Solidification cracks occurred in the LBW transverse-Varestraint test for type 310S and 316L stainless steels. Transverse cracks with a certain angle (55°–70°) occurred from the bead centreline differently from that found in gas tungsten arc welding (GTAW) transverse-Varestraint test (0°–25°). This difference in the solidification cracking was dominantly dependent on the solidification morphology of weld bead.
3. In order to accurately determine the high-temperature ductility curve for solidification cracking in the LBW transverse-Varestraint test, the corresponding temperature gradient was measured for each crack at different positions. For the transverse cracks, the effective augmented strain was also compensated based on the augmented strain during the test, considering the crack propagation angle.
4. As the welding speed increased from 1.67 mm/s (GTAW) to 40.0 mm/s (LBW), the BTR was reduced from 146 to 120 K (26 K shrinkage) for type 310S stainless steel welds. In other words, employing LBW could mitigate the solidification cracking susceptibility. However, for type 316L stainless steel welds, the BTR enlarged from 36 to 52 K (16 K enlargement). Namely, the opposite tendency of the BTR variation could be confirmed as a function of the welding speed.
5. The mechanism of the variation in the BTR as a function of the welding speed was probed by a numerical calculation of the solid/liquid coexistence temperature range combined with the Kurz–Giovanola–Trivedi (for calculation of solidification initiation temperature) and solidification segregation models (for calculation of solidification completion temperature) respectively.

6. As the welding speed increased, the solidification initiation temperature decreased, while the solidification completion temperature increased for type 310S stainless steel welds. As a result, the solid/liquid coexistence temperature range shrank with an increase in the welding speed. However, for type 316L stainless steel welds, solidification completion temperature decreased with an increase in the welding speed, and the solid/liquid coexistence temperature range enlarged with an increase in the welding speed.
7. The BTR reduction behaviour for type 310S stainless steel welds with an increase in welding speed could be explained by the dendrite tip supercooling together with the suppressed solute elements segregation in the remaining liquid phase during welding solidification due to the increased partitioning coefficients, thereby shrinking the solid/liquid coexistence temperature range. Furthermore, the BTR enlargement behaviour for type 316L stainless steel welds with an increase in welding speed could be also clarified by the enhanced solidification segregation behaviour (mainly impurity elements: P and S) in the remaining liquid phase during the solidification, caused by the decreased δ-ferrite amount despite the dendrite tip supercooling and the solute trapping phenomena, thereby expanding the solid/liquid coexistence temperature range.

## References

1. Boellinghaus T, Gumenyuk A and Quiroz V (2011) Short term metallurgy and hot cracking during laser beam welding of austenitic stainless steels. Hot Cracking Phenomena in Welds III Springer, Berlin, pp. 103–129.
2. Gittos MF, Birch SMI and Pargeter RJ (2011) Solidification cracking susceptibility in C−Mn steel $CO_2$ laser welds. Hot Cracking Phenomena in Welds III Springer, Berlin, pp. 225−262.
3. Lippold JC and Kotecki DJ (2005) Welding metallurgy and weldability of stainless steels. A John Wiley & Sons, Inc., pp. 141–229.
4. Kou S (2003) Welding metallurgy. A John Wiley & Sons Inc., pp. 216−300.
5. Ogawa T and Tsunetomi E (1982) Hot cracking susceptibility of austenitic stainless steels. Welding Journal 61(3):82s−93s.
6. Lippold JC and Savage WF (1982) Solidification of austenitic stainless steel weldments: Part 3 −the effect of solidification behaviour on hot cracking susceptibility. Welding Journal 61 (12):388s−396s.
7. Katayama S (2001) Solidification phenomena of weld metals: solidification cracking mechanism and cracking susceptibility (3rd report). Welding International 15(8):627−636.
8. Cross CE (2005) On the origin of weld solidification cracking. Hot Cracking Phenomena in Welds Springer, Berlin, pp. 3−18.
9. Cross CE and Coniglio N (2008) Weld solidification cracking: critical conditions for crack initiation and growth. Hot Cracking Phenomena in Welds II Springer, Berlin, pp. 39−58.
10. Kadoi K, Fujinaga A, Yamamoto M and Shinozaki K (2013) The effect of welding conditions on solidification cracking susceptibility of type 310S stainless steel during laser welding using an in−situ observation technique. Welding in the World 57(3):383−390.
11. Lippold JC (1994) Solidification behavior and cracking susceptibility of pulsed−laser welds in austenitic stainless steels. Welding Journal 73(6):129s−139s.

12. David SA, Vitek JM and Hebble TL (1987) Effect of rapid solidification on stainless steel weld metal microstructure and its implications on the schaeffler diagram. Welding Journal 66(10): 289s–300s.
13. Weeter LA, Albright CE and Jones WH (1986) Development of weldability test for pulsed laser beam welding. Welding Journal 65(3):51s–62s.
14. Vitek JM, Dasgupta A and David SA (1983) Microstructure modification of austenitic stainless steels by rapid solidification. Metallurgical Transactions A 14A(9):1833–1841.
15. Elmer JW, Allen SM and Eagar TW (1989) Microstructural development during solidification of stainless steel alloys. Metallurgical Transactions A 20A(10): 2117–2131.
16. Nakao Y, Nishimoto K and Zhang WP (1989) Effects of rapid solidification by laser surface melting on solidification modes–Study on laser surface modification oh stainless steel (Report 1). Quarterly Journal of Japan Welding Society 7(3):414–421.
17. Nakao Y, Nishimoto K and Zhang WP (1990) Mechanism of formation of rod–like eutectic structure in rapidly solidified stainless steels by laser surface melting–Study on laser surface modification oh stainless steel (Report 2). Quarterly Journal of Japan Welding Society 8 (3):365–370.
18. Nakao Y, Nishimoto K and Zhang WP (1990) Mechanism of formation of massively solidified structure in rapidly solidified stainless steels by laser surface melting–Study on laser surface modification oh stainless steel (Report 3). Quarterly Journal of Japan Welding Society 8 (3):371–377.
19. Nakao Y, Nishimoto K and Zhang WP (1991) Effect of rapid solidification on the microstructures and delta–ferrite morphologies–Study on laser surface modification oh stainless steel (Report 4). Quarterly Journal of Japan Welding Society 9(1):104–111.
20. Nakao Y, Nishimoto K and Zhang WP (1991) Effect of rapid solidification on the delta–ferrite content in microstructure and its implications on the Schaeffler diagram–Study on laser surface modification oh stainless steel (Report 5). Quarterly Journal of Japan Welding Society 9(1):111–116.
21. Fukumoto S and Kruz W (1997) The δ to γ transition in Fe–Cr–Ni alloys during laser treatment. ISIJ International 37(7):677–684.
22. Matsuda F, Nakagawa H and Ueyama T (1987) Solidification crack susceptibility in laser beam weld metal of 0.2C–low alloy steels–effect of bead configuration and S and P contents. Transactions of JWRI 16(2):103–114.
23. Cross CE, Coniglio N, Westin EM and Gumenyuk A (2011) Laser weldability of stainless steel. Hot Cracking Phenomena in Welds III Springer, Berlin, pp. 103–129.
24. Saida K, Nishijima Y, Ogiwara H and Nishimoto K (2013) Prediction of solidification cracking in laser welds of type 310 stainless steels. Quarterly Journal of Japan Welding Society 31(2):157– 166.
25. Quiroz V, Gumenyuk A and Rethmeier M (2012) Investigation of the hot cracking susceptibility of laser welds with the controlled tensile weldability test. The Journal of Strain Analysis for Engineering Design 47(8):587–599.
26. Nishimoto K and Mori H (2004) Hot cracking susceptibility in laser weld metal of high nitrogen stainless steels. Science and Technology of Advanced Materials 5(1–2): 231–240.
27. Wen P, Shinozaki K, Yamamoto M, Senda Y, Tamura T and Nemoto N (2009) In–situ observation of solidification cracking of laser dissimilar welded joints. Quarterly Journal of Japan Welding Society 27(2):134–138.
28. Shinozaki K, Weng P, Yamamoto M and Kadoi K (2010) Effect of grain size on solidification cracking susceptibility of type 347 stainless steel during laser welding. Transactions of JWRI 39(2):136–138.
29. Savage WF and Lundin CD (1965) The Varestraint test. Welding Journal 44(10):433s–442s.
30. Savage WF and Lundin CD (1966) Application of the Varestraint technique to the study of weldability. Welding Journal 45(11):497s–503s.
31. Lingenfelter AC (1972) Varestraint testing of nickel alloys. Welding Journal 51(9):430s–436s.

32. Senda T, Matsuda F, Takano G, Watanabe K, Kobayashi T and Matsuzaka T (1972) Studies on solidification crack susceptibility for weld metals with trans−Varestraint test (1). Quarterly Journal of Japan Welding Society 41 (6):709−723.
33. Senda T, Matsuda F and Takano G (1973) Studies on solidification crack susceptibility for weld metals with trans−Varestraint test (2). Quarterly Journal of Japan Welding Society 42 (1):48−56.
34. Arata Y, Matsuda F and Saruwatari S (1974) Varestraint test for solidification cracking susceptibility in weld metal of austenitic stainless steels. Transactions of JWRI 3(1):79−88.
35. Matsuda F, Nakagawa H, Ogata S and Katayama S (1978) Fractographic investigation on solidification crack in the Varestraint test of fully austenitic stainless steel. Transactions of JWRI 7(1):59–70.
36. Saida K, Matsushita H, Nishimoto K, Kiuch K and Nakayama J (2013) Quantitative influence of minor and impurity elements of solidification cracking susceptibility of extra high purity type 310 stainless steel. Science and Technology of Welding and joining 18(7):616–623.
37. Destructive tests on welds in metallic materials–hot cracking tests for weldments–arc welding processes. Draft European International Standard. PrEN ISO 17641.
38. Farrar JCM (2005) Hot cracking tests–the route to international standardization. Hot Cracking Phenomena in Welds Springer, Berlin, pp. 291−304.
39. Lippold JC (2005) Recent development in weldability testing. Hot Cracking Phenomena in Welds Springer, Berlin, pp. 271−290.
40. Herold H, Pchennikov A and Streitenberger M (2005) Influence of deformation rate of different tests on hot cracking formation. Hot Cracking Phenomena in Welds Springer, Berlin, pp. 328−346.
41. Matsuda F, Nakagawa H, Nakata K and Okada H (1979) The VDR cracking test for solidification crack susceptibility on weld metals and its application to aluminum alloys. Transactions of JWRI 8(1):85−95.
42. Matsuda F, Nakagawa H, Nakata K, Kohmoto H and Honda Y (1983) Quantitative evaluation of solidification brittleness of weld metal during solidification by means of in−situ observation and measurement (report 1)−development of the MISO technique. Transactions of JWRI 12 (1): 65−72.
43. Wolf M, Schobbert H and Boellinghaus T (2005) Influence of the weld pool geometry on solidification crack formation. Hot Cracking Phenomena in Welds Springer, Berlin, pp. 245−268.
44. Kromm K and Kannengieber T (2008) Influence of local weld deformation on the solidification cracking susceptibility of a fully austenitic stainless steel. Hot Cracking Phenomena in Welds II Springer, Berlin, pp. 127−145.
45. Kruz W, Giovanola B and Trivedi R (1986) Theory of microstructure development during rapid solidification. Acta Metallurgica 34(5):823−930.
46. Nishimoto K, Saida K, Kiuchi K and Nakayama J (2008) Influence of minor and impurity elements on hot cracking susceptibility of extra high−purity type 310 stainless steels. Hot Cracking Phenomena in Welds III Springer, Berlin, pp. 183−208.
47. Kannengiesser T and Boellinghaus T (2014) Hot cracking test–an overview of present technologies and applications. Welding in the World 58(3):397−421.
48. Lipton J, Kurz W and Trivedi R (1987) Rapid dendrite growth in undercooled alloys. Acta Metallurgica 35(4):957−964.
49. Takahiro O, Hiroyuki H and Kazuhiro O (2007) Numerical analysis of solidification process of austenitic stainless steel weld metal with dependence of welding velocity−Study of solidification structure on austenitic stainless steel weld metal. Quarterly Journal of Japan Welding Society 25(1):140−148.
50. Mori H and Nishimoto K (2004) Theoretical analysis on hot cracking behavior in laser surface melted region of austenitic stainless steel. Proceedings of the Finnishi−German−Japanese Joint International Seminar on Mechanical and Metallurgical Approaches to New Joining Process. pp. 221−231.

51. Boettinger WJ, Shechtman D, Schaefer RJ and Biancaniello FS (1984) The effect of rapid solidification velocity on the microstructure of Ag–Cu alloys. Metallurgical Transactions A 15A(1):55–66.
52. Aziz MJ (1982) Model for solute redistribution during rapid solidification. Journal of Applied Physics 53(2):1158–1168.
53. Burden MH and Hunt JD (1974) Cellular and dendritic growth–II. Journal of Crystal Growth 22(2):109–116.
54. Saida K, Nishijima Y, Nishimoto K, Kiuchi K and Nakayama J (2013) Contribution of Phosphorus and sulfur on hot cracking susceptibility of extra–high–purity–25Cr–35Ni stainless steels. Welding Journal 92(11):322s–331s.
55. Morishita M, Ishida H and Yoshida M (2010) Modeling of microsegregation and precipitation of iron metallic compounds in Al–Fe–Si ternary alloy. Journal of Japan Institute of Light Metals 60(4):157–163.
56. The Japan Institute of Metals (2004) Kinzoku data book. Maruzen, Tokyo, pp. 20–25.
57. Morita Z and Tanaka T (1988) Thermodynamics on the equilibrium distribution coefficients of solute elements between solid and liquid phases in iron alloys. Tetsu–to–Hagane 74(7):543–552.
58. Ono Y (1977) Diffusion in liquid iron and its alloys. Tetsu–to–Hagane 63(8):1350–1361.

# Comparative Evaluation of Mesoscale Sensitivity to Crack Formation in Multi-pass Welds Using Wires IN52 and IN52MSS

**K.A. Yushchenko, V.S. Savchenko, A.V. Zvyagintseva, N.O. Chervyakov and L.I. Markashova**

**Abstract** Assessment of susceptibility of multipass welds, produced by using wires ERNiCrFe-7 (IN52) and ERNiCrFe-13(IN52MSS), to crack formation was carried out. The PVR test was used to determine susceptibility to cracking. This test uses forced deformation of specimens being welded at a variable rate where the criterion of sensitivity to crack formation is the presence of cracks and their number at the specimen surface as a function of the rate of forced deformation. Welds made using IN52 have high sensitivity to the formation of ductility-dip cracks (DDC) in the weld metal heat-affected zone (HAZ) as compared with IN52MSS, which were not sensitive to the formation of cracks of this type at the same range of deformation rates. The characteristics of plastic deformation of both the weld metals were determined using transmission microscopy and electron backscattered diffraction (EBSD).

## Introduction

High sensitivity to the formation of ductility-dip cracks (DDC) in welded joints of nickel alloys of IN 690 type is well known. The cracks are formed in the temperature interval of about 700–1000 °C along the high-angle migrated boundaries of austenite grains, mainly in multi-pass welds [1–3]. Mechanical tests in laboratory machine of Gleeble type showed the differences in susceptibility to the formation of ductility dip. Welded joints, made by wire IN52, unlike the welded joints, made by wire IN52MSS, have the lowered values of elongation $\delta$ in the temperature interval of 600–1000 °C. In the metal of welded joints of IN52MSS the ductility dip in the same interval of temperatures is negligible [2].

K.A. Yushchenko (✉) · V.S. Savchenko · A.V. Zvyagintseva
N.O. Chervyakov · L.I. Markashova
The E.O. Paton Electric Welding Institute, Kiev, Ukraine
e-mail: yu.kon@paton.kiev.ua



Th. Boellinghaus et al. (eds.), *Cracking Phenomena in Welds IV*,
DOI 10.1007/978-3-319-28434-7_10

Sensitivity to crack formation was investigated on nickel alloy IN690 welded joints made by wires IN52 and IN52MMS. To evaluate the weldability, the destructive methods of testing with applying load were used in accordance with ISO/TR 17641-3:2005 and ISO/TR 17641-1-2011. The procedure of PVR-Test with a programmable deformation (Programmierter Verformungs Riss Test) was used [4].

## Materials and Experimental Procedures

To model the conditions of fulfillment of multi-pass welds in welded structures, the six-pass deposits were made on plates of alloy IN 690 of 8 × 40 × 200 mm sizes into preliminary cut-out grooves. Scheme of edge preparation and sequence of welds layout are given in Fig. 1a.

Chemical composition of welding wires and base metal are given Table 1.

Regime of the automatic groove welding using 0.9 mm filler wire was performed by GTAW in argon under conditions: $U_a = 10.5$ V; $v_f = 75$ m/h; $f_{osc} = 60$ osc/min; $A_{osc} = 2.5$ mm; $v_w = 6$ m/h. is shown in Fig. 2. The selected welding conditions provided a good formation of each bead of the multilayer weld, absence of undercuts and lack-of-fusion between the beads and minimum penetration of previous weld.

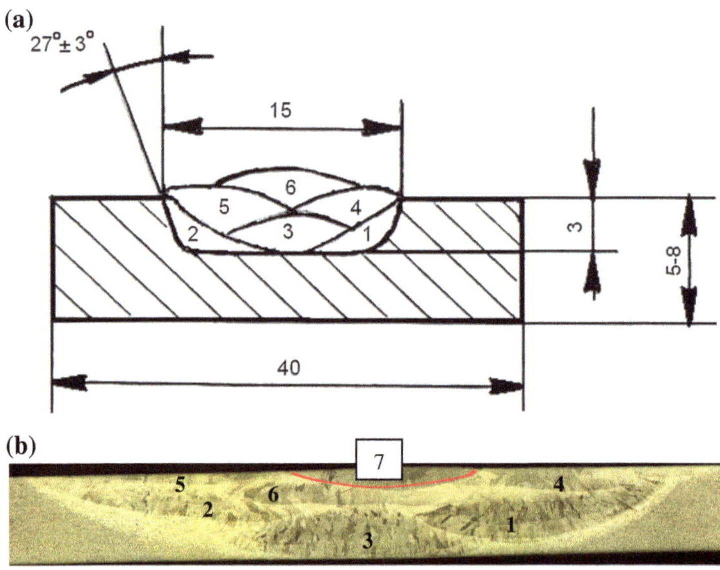


**Fig. 1** Scheme of welds layout using wires IN52 or IN52MSS (**a**), a *circle* shows the place of sampling for transmission microscopy; macrostructure of multi-layer deposition (*1–6*) and reference weld of GTAW without filler (*7*) in machine PVR-test (**b**)

**Table 1** Chemical composition of welding wires and base metal

| | Elements (wt%) | | | | | | | | | | | |
|---|---|---|---|---|---|---|---|---|---|---|---|---|
| | C | Mn | Ni | Cr | Fe | Nb | Mo | Ti | S | P | Al | Si |
| IN690 | 0.025 | 0.24 | base | 29.72 | 10.3 | – | – | 0.28 | 0.002 | 0.005 | 0.87 | 0.32 |
| IN52 | 0.026 | 0.31 | base | 28.8 | 8.5 | 0.03 | 0.03 | 0.51 | 0.001 | 0.004 | 0.72 | 0.12 |
| IN52MSS | 0.024 | 0.29 | base | 30.3 | 7.2 | 2.52 | 3.51 | 0.25 | 0.0008 | 0.0006 | 0.22 | 0.15 |

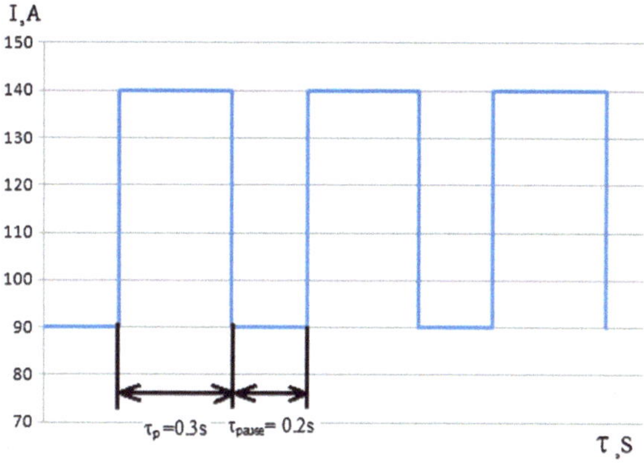


**Fig. 2** Regime of the automatic groove welding

Specimens were manufactured from deposited plates (gouging and grinding up to 340 × 180 mm sizes) for testing in machine PVR-Test at a variable rate of deformation during welding in the ranges of 0–12 mm/min. The specimen prepared for the tests is fixed between the upper (1) and lower (2) grips (Fig. 3). Welding is performed with the welding torch (4), which is moved vertically up along the specimen by the movement mechanism (3). The force and movement are registered

**Fig. 3** Loading device for testing the specimens during welding: *1*, *2* grips; *3* mechanism for vertical movement of the torch; *4* welding torch

with the bridge-circuit strain sensors. The sensors are powered from the stabilized local power supply.

## Results and Discussion

Number of cracks at the preset rate of deformation was taken as a criterion of crack resistance. Number of cracks in HAZ was calculated by using the optical microscope at 50× magnification. The generalized data about sensitivity to formation of ductility dip cracks in welds, made by wires IN52 and IN52MSS, is given in Fig. 4 [5], the nature of crack propagation at tests is shown in Fig. 5a.

Analysis of data shows that in welding with wire In52MSS (Fig. 5b) the welds at forced variable rate of deformation of less than 6 mm/min are almost not sensitive to the formation of ductility dip cracks.

One of the possible causes of intergranular fracture in welding of nickel alloys is the non-equilibrium segregation of some impurity elements, first of all sulphur and oxygen, dissolved on defects of crystalline lattice, to the boundaries [6]. These data are confirmed by the Auger-spectroscopy of grain boundaries [7]. The intragranular plastic deformation in welds under the effect of welding cycle also contributes to the presence of segregation in welds made on alloys of IN 690 type [8]. The intensification of plastic deformation in near-boundary regions of grains in welds of such type is shown elsewhere [9].

At the E.O.Paton Electric Welding Institute the characteristics of plastic deformation of welds made by wires IN52 and IN52MSS were examined using the

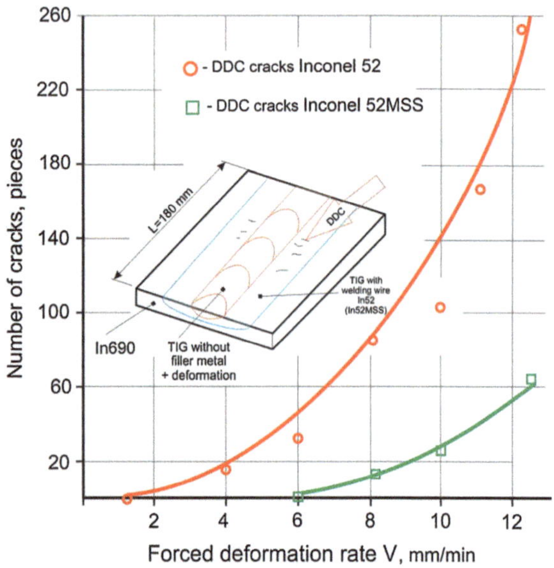


**Fig. 4** Effect of speed of displacement of the PVR-test machine grips in making test GTA weld on the amount of cracks in the welded joint made with wire IN52 and IN52MSS

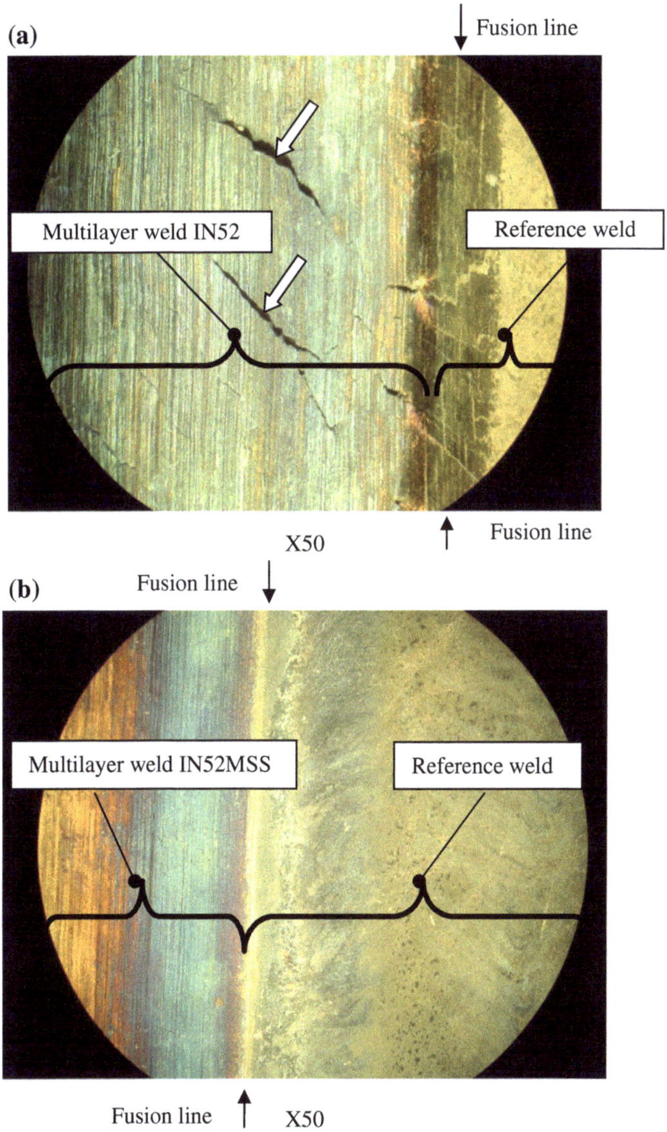


**Fig. 5** Appearance of surface of welds of IN52 (**a**) and IN52MSS (**b**) after tests by procedure PVR-Test at section with deformation rate 6 mm/min. *Arrows* show the DDC cracks in IN52

transmission microscopy (TEM), and also by electron back scattered diffraction (EBSD).

To examine the fine structure, the multilayer welds (see Fig. 1a), made by two wires IN52 and IN52MSS, were selected. Direct transmission of fine structure was performed in microscope JEM-200 CX at accelerating voltage of 200 kV.

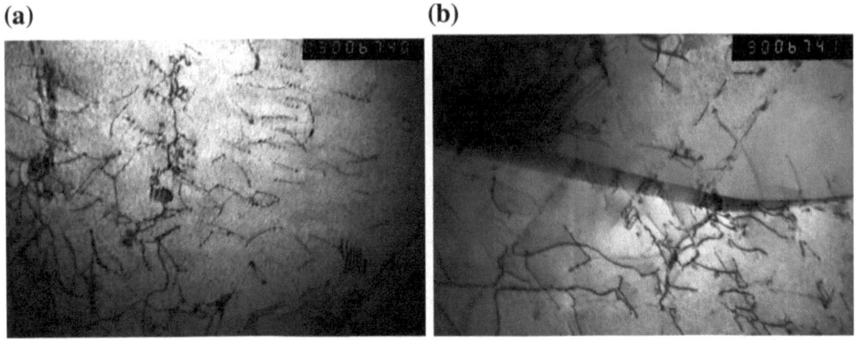

**Fig. 6** Fine structure and distribution of density of dislocations in volume of grains (**a**—×30,000) and near intergranular boundaries of weld metal, made by wire IN52MSS (**b**—×30,000)

Results of examinations showed that in structure of weld metal made by wire IN52MSS the uniform distribution of dislocations of low density $\rho \sim 1 \times 10^8 \ldots 1 \times 10^9$ cm$^{-2}$ in grain volumes takes place (Fig. 6a). Along the intergranular boundaries the negligible increase in density of dislocation is observed, which is $\rho \sim 6 \times 10^9 \ldots 7 \times 10^{10}$ cm$^{-2}$ (Fig. 6b).

In weld metal, made by wire In52, the increased density of dislocations up to $\rho \sim 2 \times 10^9 \ldots 8 \times 10^{10}$ cm$^{-2}$ is observed (Fig. 7a). Density of dislocations along the sub-boundaries is increased up to $\rho \sim 9 \times 10^{10}$ cm$^{-2}$, the significant increase in density of dislocations was observed near intergranular boundaries, which amounts to $\rho \sim 1 \times 10^{11} \ldots 2.2 \times 10^{11}$ cm$^{-2}$ (Fig. 7b).

Measurement of scalar density of dislocations was performed by using the secant method in accordance with procedures described elsewhere [9, 10, 11].

Calculation of the density of dislocations was performed by microphotos at magnification of not less than ×30000. Results on estimation of density of

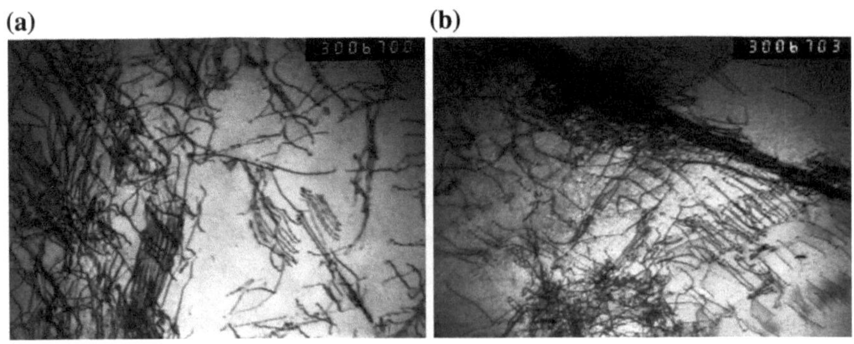

**Fig. 7** Fine structure and distribution of density of dislocations in weld metal, made by wire IN52 in volume of grains (**a**—×30,000) and near intergranular boundaries (**b**—×30,000)

**Table 2** Values of density of dislocations (ρ) in inner volumes of grains and along intergranular and subgranular boundaries of weld metal in use of wires of different alloying

| | IN52MSS | IN52 |
|---|---|---|
| | ρ ($cm^{-2}$) | ρ ($cm^{-2}$) |
| Grain volume | $1 \times 10^8 \ldots 1 \times 10^9$ | $2 \times 10^9 \ldots 8 \times 10^{10}$ |
| Grain subboundary | $1 \times 10^9$ | $9 \times 10^{10}$ |
| Grain boundary | $6 \times 10^9 \ldots 7 \times 10^{10}$ | $1 \times 10^{11} \ldots 2.2 \times 10^{11}$ |

dislocations in weld metal in applying wires IN52 and IN52MSS in different regions of grain are given in Table 2.

Value of localized deformation ($\varepsilon_{loc}$) was determined according to the relation of Konrad [12–15]. Evaluation of localized deformation ($\varepsilon_{loc}$) in different investigated regions of weld metal (volume of grain, near-boundary zone) showed that in the case of applying wire IN52MSS the localized deformation in grain volumes is negligible and amounts to about 1 %. Near the intergranular boundaries the value of localized deformation is somewhat increased up to $\varepsilon_{loc} = 2 \ldots 6$ %. In case of use of wire IN52 the localized deformation is increased and equal to 10 % in grain volumes, while near the intergranular boundaries it reaches up 30…40 %. Zones of localized deformation are given in Fig. 8.

Value of local inner stresses ($\tau_{l/in}$) was determined by a formula for inner stresses [16–19]. Analytical estimation valuation of local inner stresses ($\tau_{l/in}$) showed that in case of use of wire IN52MSS the uniform distribution of $\tau_{l/in}$ is observed without abrupt gradients. In volumes of grains $\tau_{l/in} = 1.7 \ldots 17.34$ MPa that in comparison with the theoretical strength ($\tau_{theor.} = 46{,}556$ MPa) is $0.0003 \ldots 0.0037\tau_{theor}$. Near subbounadries $\tau_{l/in}$ is not almost changed and equal to 17.34 MPa ($0.0037\tau_{theor}$). Near intergranular boundaries a little increase in density of dislocations is observed that leads to little increase in value of local inner stresses $\tau_{l/in} \sim 104.2 \ldots 73.7$ MPa, that is $0.002 \ldots 0.037\tau_{theor}$. Data are given in Fig. 9 and Table 3.

In case of applying wire IN52 in welding, the non-uniform distribution of density of dislocations is observed in grain volume/boundary zones, thus leading to an abrupt gradient of local inner stresses. The analytical estimation showed that in

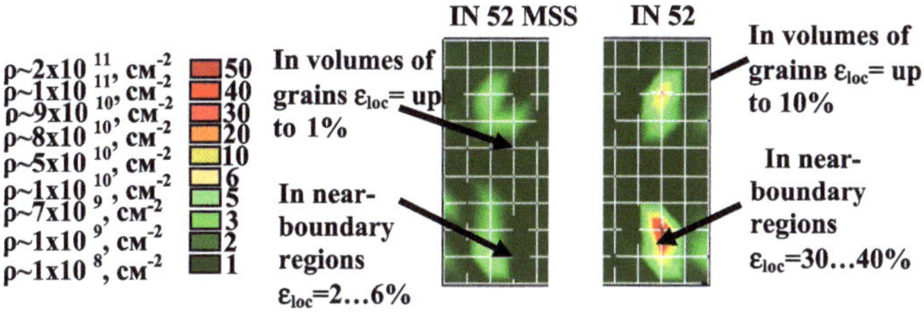


**Fig. 8** Zones of localized deformation ($\varepsilon_{loc}$) in different areas of weld metal in use of wires IN52MSS and IN52

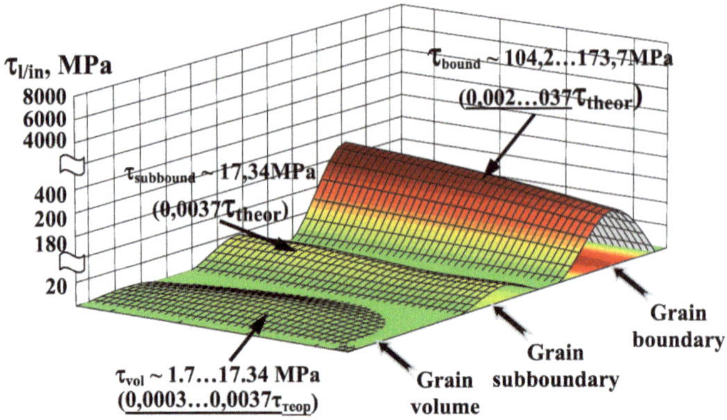


**Fig. 9** Distribution of local inner stresses ($\tau_{l/in}$) in weld in use of wire IN52MSS

**Table 3** Values of local inner stresses ($\tau_{l/in}$): in inner volumes of grains and along intergranular and subgranular boundaries in metal weld of different alloying

| | IN52MSS | IN52 |
|---|---|---|
| | ($\tau_{l/in}$) MPa<br>($x \cdot \tau_{theor}$) | ($\tau_{l/in}$) MPa<br>($x \cdot \tau_{theor}$) |
| Grain volume | 1.74…17.3<br>(0.0003…0.0037$\tau_{theor}$) | 347.5…1390<br>(0.074…0.3$\tau_{theor}$) |
| Grain subboundary | 17.34<br>(0.0037$\tau_{theor}$) | 1564<br>(0.33$\tau_{theor}$) |
| Grain boundary | 104.2…173.7<br>(0.022…0.037$\tau_{theor}$) | 1737.6…3822.7<br>(0.37…0.82$\tau_{theor}$) |

grain volume $\tau_{l/in}$ = 347.5…1390 MPa that is (0.074…0.3$\tau_{theor}$) of theoretical strength, increase in $\tau_{l/in}$ is observed along the subboundaries and equal to 1564 MPa (0.33$\tau_{theor}$). Abrupt increase in density of dislocations ($10^{11}$…$2.2 \times 10^{11}$) is observed in regions near the intergranular boundaries that gives $\tau_{l/in}$ = 1737.6…3822.7 MPa and this amounts to 0.37…0.82$\tau_{theor}$ of the theoretical strength (Fig. 10; Table 3).

One of the parameters of defect formation is the stacking fault energy (SFE) which can be measured by width of splitting the dislocations [10, 20].

In weld metal, when IN52MSS wire is used, the width of splitting the dislocations amounts on average to 0.045…0.07 μm, both in volume of grains and also near the intergranular boundaries (Fig. 6). This width gives the energy of stacking fault of about $\gamma_{SFE}$ = 0.091 J/m$^2$, Fig. 11.

In weld metal, when IN52 wire is used, the width of splitting the dislocations in grain body is decreased on average to 0.022…0.04 μm, that gives $\gamma_{SFE}$ = 0.19 J/m$^2$ (Fig. 7). The values SFE are given in Fig. 11.

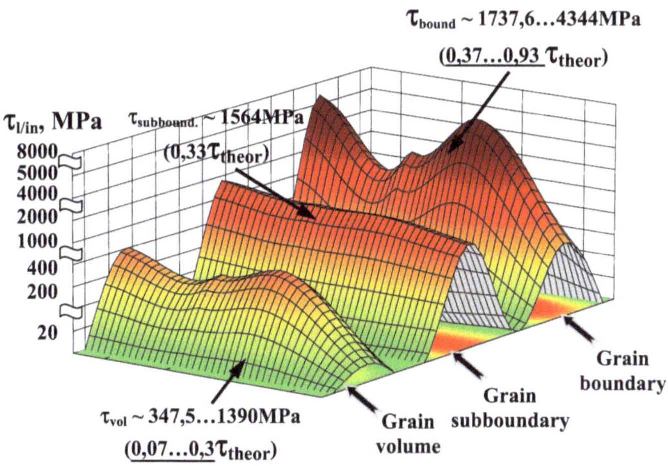


**Fig. 10** Distribution of local inner stresses ($\tau_{l/in}$) in weld metal in use of wire IN52

**Fig. 11** Change in $\gamma_{SFE}$, J/m$^2$

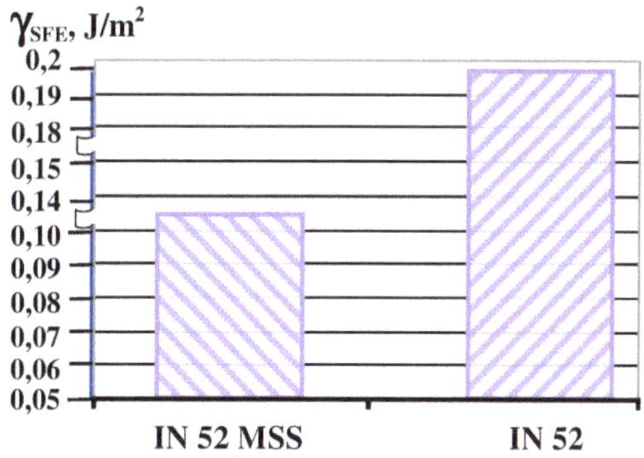


To analyze the nature of propagation of DDC cracks in welds, made by wires IN52MSS and IN52, the experimental investigations of grain structure near crack in HAZ of weld were carried by scanning electron microscope of EVO-50 of "Zeiss" company using a CCD-detector. Maps of crystallographic orientation of grains (Fig. 12), maps of diffraction of back-scattered electrons (Kikuchi patterns) (Fig. 13) in local regions for each specimen being investigated to prove that ductility dip cracks in HAZ of welded joint are propagated along the large-angle boundaries. This is also confirmed by others [9].

Experimental Kikuchi patterns [21, 22] from different regions of the same grain is characterized by the degree of the blurring effect of diffraction fringes with removal from the crack, positions 1, 2, 3, 4, Fig. 12. Here, the local deformation (relative change of inter-plane distance $d_{hkl}$) was evaluated by area under the profile of intensity of Kikuchi line [23, 24]

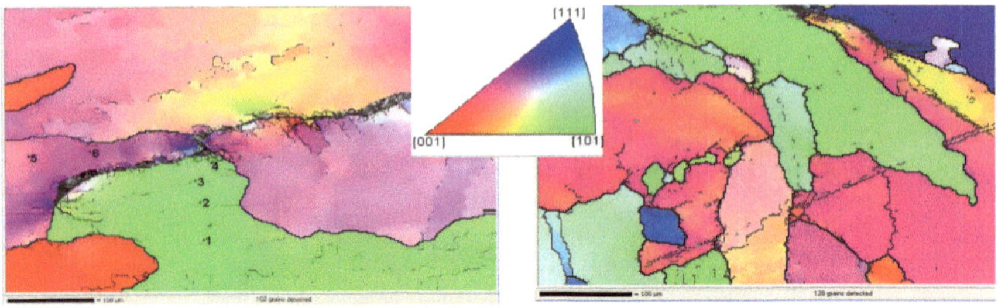


**Fig. 12** Fragments (580 × 440 μm) of maps of crystallographic orientation (inverse pole figure) of HAZ of welds made by wires IN52 (**a** near the DDC crack) and IN52MSS (**b**)

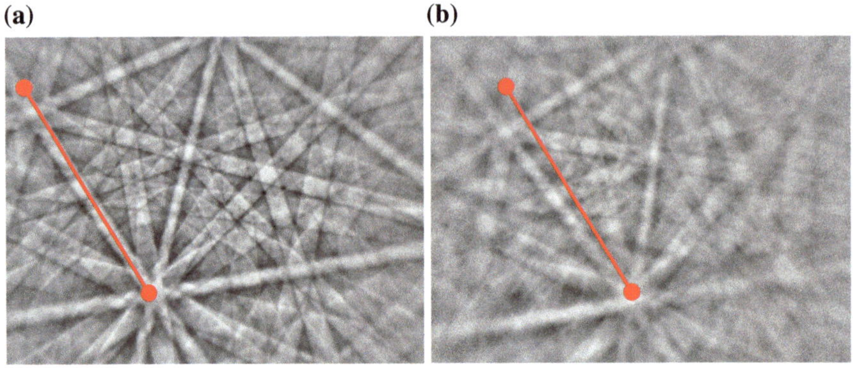


**Fig. 13** Patterns of Kikuchi, obtained from region No.1 (**a**); from region No.4 (**b**); two similar lengths on different Kikuchi patterns demonstrate relative change of scale and turn of patterns for each region

$$\varepsilon_{hkl} = \frac{\Delta d_{hkl}}{d_{hkl}}$$

The more intensive deformation in definite region of specimen (region 4, where $\varepsilon = 3.5 \times 10^{-4}$ %, the more blurred is the proper Kikuchi pattern (as compared with region 1, where $\varepsilon = 10^{-5}$ %).

## Conclusions

1. Welds made by wire IN52, as compared with welds made by wire IN52MSS, have higher sensitivity to the formation of ductility dip cracks (DDC) in the weld metal heat-affected zone (HAZ). In IN52 cracks initiate at the rate of deformation of less than 2 mm/min while with IN52MSS, the cracks of this type initiate at the rate of deformation of more than 6 mm/min.

2. It was found that thin structure of weld in use of wire IN52 is characterized by high density of dislocations, in particular near weld metal migrated grain boundaries (about up to $\rho \sim 1 \times 10^{11} \ldots 2.2 \times 10^{10}$ $cm^{-2}$); high level of localized deformation (up to $\varepsilon_L$ = 30…40 %) and significant local inner stresses ($\tau_{l/in}$ = 1737.6…4344 MPa, 0.37…0.93$\tau_{theor}$), in particular near grain boundaries. An increase stacking fault energy (about $\gamma_{SFE}$ = 0.19 J/m$^2$) is also observed. In case of IN52MSS, the uniform (without gradients) distribution of dislocation density, low and uniform level of localized deformation (to $\varepsilon_L$ = 6 %) and local inner stresses (to $\tau_{l/in}$ = 173.7 MPa, 0.37$\tau_{theor}$) are observed, and also the energy of stacking fault energy (about $\gamma_{SFE}$ = 0.091 J/m$^2$) is decreased.
3. Maps of crystallographic orientation, local deformation anomalies and distortion of FCC lattice in sites of formation of ductility dip cracks in HAZ of welded joints were investigated. Cracks propagated along high-angle boundaries. EBSD patterns were obtained in different regions of the same grain that abutted a crack. They were characterized by different degrees of blurring of the diffraction fringes with the highest degree of blurring at the highest deformation.
4. It can be assumed that formation of ductility dip cracks is the limiting state of degree of degradation of grain boundaries, at which cracking takes place.

## References


1. Torres EA, Peternella FG, Caram R, Ramirez AJ, (2010) In situ scanning electron microscopy high temperature deformation experiments to study ductility dip cracking of Ni-Cr-Fe alloys. Ed. by T. Kannengiesser et al. In: In-situ studies with phonons, neutrons and electrons scattering. Berlin Heidelberg: Springer-Verlag, P. 28.
2. Yushchenko K, Savchenko V, Chervyakov N, Zvyagintseva A. and Guyot E (2011) Comparative hot cracking evaluation of welded joints of alloy 690 using filler metals Inconel ® 52 and 52 MSS. Welding in the World, Vol. 55, n° 9/10.
3. Lippold JC and Kotecki DJ (2005) Welding metallurgy and Weldability of stainless steels. Copyright by John Wiley&Sons.
4. ISO/TR 17641-3:2005. Destructive tests on welds and metallic materials – Hot cracking tests for weldments - Arc welding processes - Part 3: Externally loaded tests
5. Kiser D, Zhang R, Baker BA. A new welding material for improved resistance to ductility dip cracking. www.Specialmetals Welding Products Company, Newton, North Carolina, USA.
6. Song S-H, Zhang O, Weng L-Q, (2008) Deformation-induced non-equilibrium boundary segregation in dilute alloys. Materials Sci. and Eng., A473, P. 226-232.
7. Savchenko VS, Yushchenko KA, Zvyagintseva AV, David SA and Vitek JM (2007) Investigation of structure and crack formation in welded joints of single crystal Ni-base alloys. Welding in the World, Vol. 51, n° 11/12, P. 76-81.
8. Saida K, Nishimoto K (2010) Regulatory criteria of minor and impurity elements for welding integrity of alloy 690 and alloy 690 to type 316L stainless steel multipass welds – evaluation. In: Proc. of Int. Symp. on the Ageing Management & Maintenance of Nuclear Power Plants, P. 207-220.
9. Collins MG, Ramirez AI, Lippold JC (2004) An investigation of ductility-dip cracking in nickel-based weld metals. Pt. 3. Welding J., n° 2, P. 39 s-49 s.

10. Hirsh P, Hovi A, Nickolson R et al. (1968) Electron microscopy of thin crystals. Moscow: Mir, 574 p.
11. Tomas G (1963) Electron microscopy of metals. Moscow: Inostr. Lit-ra,, 351 p.
12. Suzuki HO (1967) On yield strength of polycrystalline metals and alloys. In: Structure and mechanical properties of metals. Moscow: Metallurgiya, P. 255-260.
13. Trefilov VI, Moiseev VF, Pechkovsky EP et al. (1987) Strain hardening and fracture of polycrystalline metals. Kyiv: Naukova Dumka,, 248 p.
14. Conrad H (1963) Effect of grain size on the lower yield and flow stress of iron and steel. Acta Metallurgica, 11, P. 75-77.
15. Markashova LI Kushnareva OS (2013)- Influence of structure on the mechanical properties of the weld metal of welded joints in aluminum alloys of Al-Cu-Li, Physico-chemical mechanics of materials. № .5. - P. 112-118.
16. Ke AS (1964) Dislocation arrangement in α-iron at strain and return. In: Direct observation of crystal imperfections. Moscow: Metallurgiya,, P. 160-178.
17. Koneva NA, Lychagin DV, Teplyakova LA et al. (1986) Dislocation-disclination substructures and hardening. In: Theoretical and experimental study of disclinations. Leningrad: Izd-vo LFTI, P. 116-126.
18. Markashova LI, Poznyakov VD, Gaivoronskii AA, Berdnikova EN, Alekseenko TA (May, 2012) Estimation of the strength and crack resistance of the metal of railway wheels after long-term operation // Materials Sciense,. - Vol. 47, №. 6. – P.799-806.
19. Markashova LI, Poznyakov VD, Gaivoronskii AA, Berdnikova EN, Alekseenko TA (May 2012) Structural Performance Evaluation properties of welded joints of high strength steels under repair works / / Proceedings of the 6th International Conference “Mathematical modeling and information technologies in welding and allied processes”,, Katsively Crimea. - Kiev: PWI. Paton NASU, 2012. - P.40-43
20. Gorelik SS, Rostarguev LN, Skakov Yu.A (1979) X-ray and electrooptic analysis. Moscow: Metallurgiya.
21. Davidson DL (1982) The effect of deformation on selected area electron channeling patterns. J. Mater. Sci. Lett.,, Vol. 1, n° 6, P. 236-238.
22. Yoshitomi Y, Ohta K, Harase J, SugaY (1994) A method for measurement by analyzing sharpness of ECP with image analysis. Textures and Microstructures, n° 22, 199-218.
23. Borcha MD, Balovsyak SV, Fodchuk IM, Khomenko VYu, Tkach VN (2013) Distribution of local deformations in diamond crystals according to the analysis of Kikuchi lines profile intensities. J. of Superhard Materials, Issue 4, P.34-42.
24. Sasaki Y, Igushi M and Hino M (2006) Measuring strains for hematite phase in sinter ore by electron backscattering diffraction method. Key Engineering Materials, 326-328, P. 237-240.

# Welding Optimization of Dissimilar Copper-Aluminum Thin Sheets with High Brightness Lasers

**Fidel Zubiri, María del Mar Petite, Jaime Ochoa and María San Sebastian**

**Abstract** Dissimilar welding of Al and Cu represents a big challenge due to the formation of brittle intermetallic phases. The present study is focused on the laser beam welding optimization of Al-Cu thin sheets using high brightness lasers. Joints were produced with welding speeds as fast as 12 m/min, characterized both mechanically and electrically, with microstructure analysis. The results exhibit that when Al is on top of Cu, the weld volume is large and several microcracks show up due to the formation of $CuAl_2$ ($\theta$), CuAl ($\eta_2$) and $Cu_9Al_4$ ($\alpha$) phases. When Cu is on top of Al instead, the weld volume is small and just $CuAl_2$ ($\theta$) phase is formed; no cracks are observed. This behavior is explained through the correlation between the weld volume, the different interaction of Al and Cu with the laser beam and the intermetallic phases generated in the weld.

**Keywords** Aluminum-copper · Dissimilar welding · Laser welding · Intermetallic phases · Electrical conductivity · Lithium ion battery

## Introduction

Al-Cu joints have become a matter of great interest during the last years since they can be used as a cheap and reliable replacement for Cu current carrying components in newly envisaged applications like vehicle batteries or energy storage systems [1, 2]. However, the achievement of sound Al-Cu dissimilar joints is a very challenging task as several difficulties arise when joining these metals by fusion welding technologies. Their different chemical and physical properties (thermal conductivity, thermal expansion coefficient, melting point) as well as their limited mutual solubility leads to the formation of several brittle intermetallic phases that reduce the

F. Zubiri · M. del Mar Petite · J. Ochoa · M.S. Sebastian (✉)
IK4-LORTEK Technological Centre, Arranomendia Kalea 4A, 20240
Ordizia, Gipuzkoa, Spain
e-mail: msansebastian@lortek.es



Th. Boellinghaus et al. (eds.), *Cracking Phenomena in Welds IV*,
DOI 10.1007/978-3-319-28434-7_11

strength and ductility of the joints and giving rise to extensive cracking [3]. In this respect, high power lasers have been introduced to perform the joints, given that laser welding has the advantage of significantly reducing the formation of these phases due to the associated high cooling and solidification rates [4, 5]. However, Cu and Al are difficult to melt with lasers due to their high reflectivity and high thermal conductivity. Consequently, only a small amount of laser energy is absorbed (less than 5 %), making very difficult to generate and maintain a molten pool. Moreover, if very thin sheets must be laser welded, it is required a very precise control of the heat input on the weld in order to avoid distortion and weld defaults. This results in a narrow process window [6].

To overcome this problem, we have used the most recent developments in high brightness lasers, which are able to focus their light on a very small focus diameter (10–20 μm) and have excellent beam qualities, to improve laser welding of Al and Cu at high welding speeds. The goal of the present study is therefore to evaluate and optimize the mechanical and electric performance of dissimilar Al-Cu lap welds carried out at high speeds with high brightness and high focusability lasers. The paper is organized in two main sections, one in which we will describe the laser properties, welding configuration and experimental conditions employed in our study and other where we will discuss the suitability of welding through a complete analysis of the weld microstructure and weld hardness.

## Experimental Procedure

To carry out the experiments, the latest generation of fiber lasers was employed. In particular, a commercial TruFiber with a maximum average output power of 400 W, which generates a gaussian beam of $M^2 < 1.1$ at $\lambda = 1064$ nm. The beam was guided to the focusing head through a 15 μm diameter fiber. The optical head is composed of two lenses of 200 mm for collimation and focusing, thus producing a focus spot size on the metal surface of the same diameter of the fiber, 15 μm. A lateral nozzle with an internal filter was placed close to the focused spot, to generate a controlled laminar flow of Ar (15 l/min), and thus avoiding as much as possible oxidation and internal defects on the weld.

Al 3003 and Cu 101 sheets with 0.3 mm thicknesses were employed for the laser welding tests. Table 1 summarizes their chemical composition and physical properties. Two flat sheets of each material were positioned and fixed with a clamping device in lap configuration (see Fig. 1 top) after cleaning their surface with acetone. The entire device was set over a CNC stage to control speed and position of the weld bead over samples.

In order to establish the parameter window, several lap welding trials were carried out both in Al/Cu (aluminum on top) and Cu/Al (copper on top) configurations, setting the focus point just on the surface of the uppermost sheet. Welding trials with two parallel welds tracks were performed (see Fig. 1). The optimum welding parameters, corresponding to the highest realizable welding speed, were

**Table 1** Specifications of materials used for the Al-Cu dissimilar welding tests

| Material | Cu 101 | Al 3003 |
|---|---|---|
| Composition | Cu (min 99.99 %)<br>$O_2$ (max 5 ppm) | Al balance<br>Cu (0.05–0.2 %)<br>Fe (0.7 max)<br>Mn (1–1.5 %)<br>Si (0.6 % max)<br>Zn (0.1 % max) |
| Density (kg/m$^3$) | $8.94 \times 10^3$ | $2.73 \times 10^3$ |
| Melting point (°C) | 1084 | 655 |
| Boiling point (°C) | 2562 | 2467 |
| Specific heat (J/kg K) | 385 | 896 |
| Thermal conductivity (RT) (W/K m) | 401 | 198 |

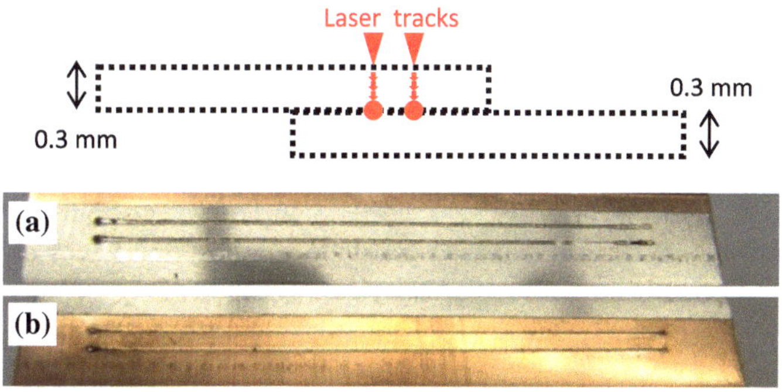


**Fig. 1** Lap welding scheme employed in tests. Welded samples with two parallel weld tracks and two different configurations: **a** Al/Cu: aluminum on top; **b** Cu/Al: copper on top

determined by checking that both sheets were adequately joined by microstructural analysis and mechanical testing. Keeping constant laser power, Al/Cu and Cu/Al configurations needed speeds of 8 and 12 m/min respectively to perform a proper weld. In other words, Cu/Al configuration needed of a larger energy input (lower speed) to perform the weld. This is explained by the particular reflectivity of copper which is much higher than that of aluminum at this wavelength. Because of this reason, the optimum parameter window for Cu/Al configuration was found to be much smaller with respect to Al/Cu configuration [7, 8].

Once manufactured the best welds with these parameters, they were cut to perform a metallographic analysis. Keller and Klem reagents were utilized to etch Al and Cu surfaces of the cross sections obtained, respectively. The samples were subsequently characterized by optical and SEM microscopy. In addition, its composition was analyzed by performing EDX measurements. Micro-hardness (HV0.05) testing as well as tensile shear testing was also carried out in order to obtain mechanical properties. Finally, electrical resistance of the Al/Cu and Cu/Al laser welds were measured using a digital ohmmeter.

## Results and Discussion

Figure 2 shows cross sections of the two Al/Cu and Cu/Al welds. At this magnification neither cracking nor porosity were observed in the joints. The interconnecting surfaces were small and reached 150–200 μm for the Cu/Al configuration and 200–250 μm for the Al/Cu. It can be observed in both images that the melted volume of Al is much larger compared to that of Cu. Comparing in between configurations, the melted volume of Al is considerably larger when it is on top of the lap stack than when it is at the bottom. This is not equivalent in the case of Cu, where the amount of melted volume is approximately similar. The reason for this behavior can be found in the different conductivities and melting temperatures of Cu and Al, affecting the way heat reaches the bottommost layer. With the Cu/Al configuration, only partial penetration was achieved due to Cu screening, and the underlying material was melted by heat conduction form the surface. In the opposite configuration, the bottom sheet was melted directly from the interaction with the laser beam given that Al absorbed heat more easily. As a consequence, a high Al enrichment and a low Cu dilution in the weld seam are expected in the Al/Cu configuration, which may impact the presence of undesired intermetallic phases and consequently the quality of the weld.

It is worth noticing here that, despite the low laser power and high welding speed, a keyhole mode was achieved avoiding lack of fusion issues in both configurations. Recently, these kind of problems were reported when using pulsed Nd: YAG lasers to weld Al and Cu, and Al being the top sheet [9]. Under these circumstances, lack of fusion was detected in the underlying Cu sample even though the top Al sheet was fully melted down to the interface. The origin for this was related to the lower melting temperature of Al, insufficient heat transfer to Cu (with a higher melting temperature) and a lack of surface wetting. To solve this, the energy density and heat input were increased by employing higher pulse power and reducing welding speed [10]. Nonetheless, in our study lack of fusion in the bottom sample was not registered at the high welding speeds thanks to the increased energy density provided by the high brightness laser.

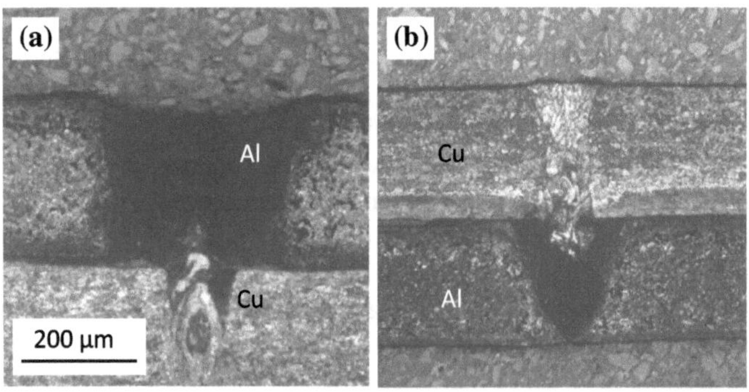


**Fig. 2** Macrographs of **a** Al/Cu and **b** Cu/Al laser welds

To complement these measurements, higher magnification inspection of the previous cross sections was accomplished by SEM microscopy. Figure 3 reveals in detail the welding zone and interface of each weld. According to these images, there are clear differences in the mixing patterns of Al/Cu and Cu/Al weld samples. In the Al/Cu sample there are two distinctive melting zones, one central Cu-rich brighter zone and a surrounding Al-rich darker zone. The Cu-rich zone rises up from the bottom Cu sheet up to the top Al sheet but the mixing is very limited. In the Cu/Al sample instead, there is a smaller and more homogeneous mixing zone. It mainly extends down to the bottom of the Al sheet. These mixing patterns were confirmed by EDX mapping analysis, which have been depicted in Fig. 3b, c, e, f for each one of the configurations.

Despite no cracks observed in optical macrographs, several microcracks were observed at higher magnifications in the melted zone of the Al/Cu sample. An example of them can be appreciated in Fig. 4a. These microcracks were observed in different cross-sections. On the other hand, only some of the inspected cross sections of Cu/Al samples showed these (and much smaller) cracks (see Fig. 4b).

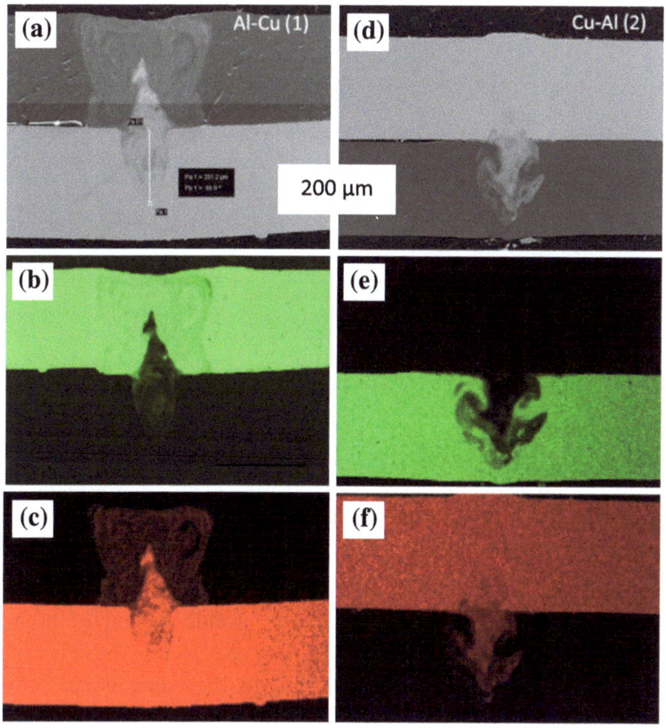


**Fig. 3** SEM images of **a** Al/Cu and **d** Cu/Al welds. Al appears darker than Cu in the images. **b**, **c**, **e**, **f** EDX mappings depicted for each one of the configurations. *Green* and *red* colors represent Al and Cu materials respectively (color figure online)

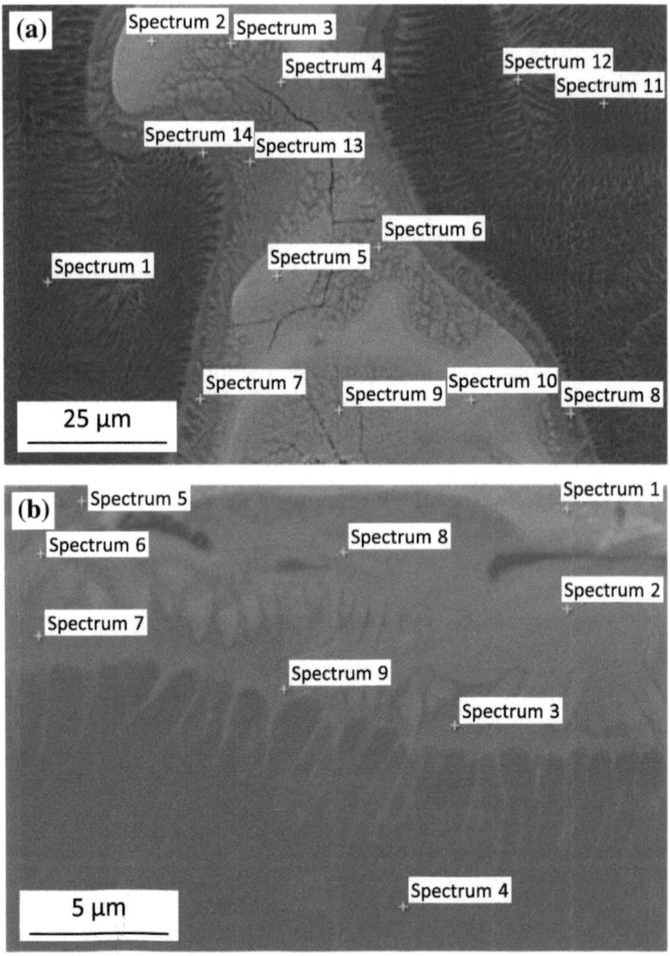


**Fig. 4** SEM images of **a** Al/Cu and **b** Cu/Al sample showing microcracks inside the melting zone

The origin of the cracks in Al/Cu samples is associated with the presence of hard and brittle Al-Cu intermetallic phases. To confirm this assumption, we deeply analyzed the microstructure and phases of the welds. First, we performed a quantitative EDX analysis of Al and Cu contents in the melted zone, including the heat affected zone and the base metals in all samples. Results of this analysis are pointed out in Table 2, where we have presented the atomic weight of both metals at different points illustrated in Fig. 4a, b. Recalling the binary Al-Cu binary diagram, and after analyzing EDX results for Al/Cu sample, one can conclude that spectra 1 and 11, located outside the melting zone, are very rich in Al with minimal presence of Cu. The microstructure of this zone is consequently composed by Al(α) phase. The layer surrounding the melted zone is the eutectic phase, according to results of

**Table 2** Quantitative EDX analysis of points 1–14 shown in Fig. 4 (SEM images of Al/Cu samples with microcracks)

| Configuration | at% | 1 | 2 | 3 | 4 | 5 | 6 | 7 | 8 | 9 | 10 | 11 | 12 | 13 | 14 |
|---|---|---|---|---|---|---|---|---|---|---|---|---|---|---|---|
| Al/Cu | Al | 95.2 | 27.2 | 60.4 | 66.4 | 29.1 | 60.9 | 72.2 | 81.2 | 52.1 | 24.2 | 93.5 | 88.0 | 76.3 | 84.1 |
| | Cu | 4.8 | 72.3 | 39.6 | 33.6 | 70.9 | 39.1 | 23.8 | 17.1 | 47.9 | 75.8 | 6.3 | 11.8 | 23.5 | 15.9 |
| | | Al(α) | $AlCu_4$ | | $CuAl_2(\theta)$ | $Cu_9Al_4(\gamma)$ | | | Eutectic | CuAl | $AlCu_4$ | Al(α) | | | Eutectic |
| Cu/Al | Al | 0.7 | 66.4 | 80.5 | 99.0 | 61.7 | 75.5 | 80.5 | 67.8 | 81.6 | | | | | |
| | Cu | 99.3 | 33.3 | 19.2 | 0.8 | 35.9 | 24.3 | 19.1 | 30.2 | 18.1 | | | | | |
| | | Cu | $CuAl_2$ | Eutectic | Al(α) | $CuAl_2$ | | Eutectic | $CuAl_2$ | Eutectic | | | | | |

spectra 8 and 14, and can be also considered as part of melted zone. Focusing on the melted zone, there is a mixture of complex intermetallic phases of different stoichiometry including ($Cu_9Al_4$, $AlCu_4$ or CuAl) and θ ($CuAl_2$), which is eventually the origin of the cracks as these phases are very hard and fragile. The same analysis was then done for the Cu/Al sample as it is shown in Fig. 4b. In this case, the melted zone (spectra 2, 5 and 8) is totally surrounded by eutectic phase (spectra 3, 7 and 9) and contains only one type of intermetallic phase ($CuAl_2$).

This was confirmed after obtaining a micro-hardness (HV0.05) survey in the cross section of the samples. Hardness measurements were achieved along several horizontal lines in steps of 0.05 mm at different depths (y = 0, ±0.05, ±0.1 and ±0.2 mm), where z = 0 mm represents the interface between Al and Cu. The results (see Table 3) show that the highest hardness values are concentrated in the melted zone, being coincident with the region where intermetallic phases are concentrated. In the Al/Cu samples, hardness values above 500 HV0.05 were consistently measured in the region y = ±0.1, while in the Cu/Al samples, these high values are similar but are concentrated in the y = ±0.05 region. Far from these regions, hardness decreases to the nominal values for Al and Cu, but not in similar ways. In the case of Al/Cu, hardness of Aluminum is recovered almost 0.4 mm away from the interface, while it is recovered at 0.2 mm in the case of Cu/Al, pointing out the influence of the different laser-matter interaction of each material on top of the lap weld.

After characterization, the samples were analyzed to determine their mechanical properties through tensile shear testing. Rectangular weld samples containing two weld seams were employed in the study. Three samples were tested in each configuration. The measured shear strength showed very similar values for the Cu/Al sample, while the variation for the Al/Cu sample was a little bit larger (see Table 4). In any case, the results demonstrate that the welds are highly reproducible and close to the typical values obtained in this type of welds (around 65 MPa [10]). The electrical resistance of the dissimilar welds was measured as well by passing a

**Table 3** Hardness survey of Al/Cu and Cu/Al samples at different depths (distance in mm)

| | Al | ← | Al/Cu | → | | Cu |
|---|---|---|---|---|---|---|
| x\y | 0.20 | 0.10 | 0.05 | 0.00 | -0.05 | -0.10 | -0.20 |
| -0.25 | 38.7 | 38.4 | | | | 109.0 | 108.0 |
| -0.15 | 144.0 | 167.0 | 120.0 | | 102.0 | 95.4 | 104.0 |
| -0.10 | 142.0 | 192.0 | 150.0 | | 82.2 | 90.9 | 99.0 |
| -0.05 | 150.0 | 149.0 | 165.0 | 615.0 | 357.0 | 596.0 | 92.5 |
| 0.00 | 149.0 | 645.0 | 596.0 | 635.0 | 231.0 | 818.0 | 160.0 |
| 0.05 | 145.0 | 173.0 | 448.0 | 833.0 | 268.0 | 500.0 | 115.0 |
| 0.10 | 140.0 | 141.0 | 127.0 | | 114.0 | 99.0 | 99.6 |
| 0.15 | 137.0 | 159.0 | 146.0 | | 121.0 | 99.0 | 90.8 |
| 0.25 | 41.1 | 40.8 | | | | 110.0 | 96.0 |

| | Cu | ← | Cu/Al | → | | Al |
|---|---|---|---|---|---|---|
| x\y | 0.20 | 0.10 | 0.05 | 0.00 | -0.05 | -0.10 | -0.20 |
| -0.25 | 89.3 | 87.2 | | | | 46.2 | 51.6 |
| -0.15 | 82.7 | 81.3 | | | | 36.7 | 38.6 |
| -0.10 | 88.2 | 80.0 | 79.8 | | 249.0 | 47.2 | 35.1 |
| -0.05 | 87.2 | 80.7 | 100.0 | 315.0 | 472.0 | 276.0 | 64.6 |
| 0.00 | 82.7 | 179.0 | 247.0 | 839.0 | 5780 | 384.0 | 116.0 |
| 0.05 | 86.7 | 349.0 | 749.0 | 657.0 | 486.0 | 578.0 | 115.0 |
| 0.10 | 87.2 | 82.7 | 84.7 | | 66.7 | 107.1 | 38.6 |
| 0.15 | 92.5 | 89.3 | | | | 36.1 | 36.0 |
| 0.25 | 99.0 | 88.7 | | | | 45.4 | 46.0 |

Hardness values higher than 450 HV0.05 are highlighted

**Table 4** Tensile strength and electrical resistance measured in lap welding samples of Al/Cu and Cu/Al

| Configuration | Shear strength (MPa) | Electrical resistance (mΩ/m) |
|---|---|---|
| Al/Cu | 65.9 | 2.93 |
| | 47.6 | 2.90 |
| | 76.4 | 2.93 |
| Cu/Al | 62.5 | 3.00 |
| | 76.4 | 2.97 |
| | 70.5 | 2.93 |

controlled electrical current through the melted zone from one terminal to the other and measuring the voltage. Measured values are lower than those obtained in resistance spot welded samples, fact that emphasizes, in combination with the rest of results, the superior approach of utilizing high brightness lasers to achieve suitable Al/Cu welds.

## Conclusions

In view of the results presented in this work the feasibility for achieving sound dissimilar Al-Cu joints at high speeds with the new generation of high brightness lasers has been proven. The advanced laser beam features avoided lack of fusion issues both in Al/Cu and Cu/Al configurations. However, the lap weld configuration still greatly influences the weld aspect, melting zone volume and mutual mixing. The volume of the mixing zone, where detrimental Al-Cu intermetallic phases are concentrated, is much reduced in Cu/Al configuration, thus reducing the cracking susceptibility. The nature (stoichiometry) and size of the intermetallic phases are also different depending on how the sheets are placed. Current results showed that Cu/Al configuration is the optimum one in order to reduce the number of micro-cracks along the whole weld track, as expected.

Additionally, shear tensile test and electrical resistance of Cu/Al samples welded at high speed were consistent and meet the requirements that are defined for current applications like lithium ion battery manufacturing. Therefore, great advantages can be made from high brightness lasers compared with current joining technologies (resistance spot welding, ultrasonic welding, …) included quicker process, precise, contactless, small heat affected zone and big automation capacity.

**Acknowledgements** The authors would like to thank R&D centre CARTIF for the technical support in the project. In addition, the authors acknowledge Dr. Juan B. González-Díaz for fruitful discussions and his valuable contribution to this work.

## References

1. Ream, S.: Laser welding for battery. Industrial Laser Solutions, March/April 2010, pp. 12-14 (2010).
2. Speth, D., Eastman, J.: Laser welding batteries for electric vehicles. Industrial Laser Solutions, March/April 2012, pp. 27-32 (2012).
3. Schmidt, M., Weigl, M.: Laser-Bonding in High Power Electronics - Influencing the Connection Quality of Copper-Aluminum Welds by a Lateral Beam Displacement and Laser Power Modulations. J Laser Micro Nanoen. 5 (3), 242-247 (2010).
4. Jokiel, M., Holtz, R.: Experiences with laser beam welding of dissimilar materials. The Industrial Laser User, 43, pp. 26-29 (2006)
5. Katayama, S.: Laser welding of aluminium alloys and dissimilar metals. Welding International, 18 (8), 618-625 (2004)
6. Theron, M., Van Rooyen, C., Ivanchev, L.H.: CW ND:YAG laser welding of dissimilar sheet metals, 26th International Congress on Applications of Lasers & Electro-Optics(ICALEO), 29, October-November 2007, pp. 8 (2007).
7. Xue, Z.Q., Hu, S.S., Zuo, D., Cai, W., Lee, D., Kannatey-Asibu, E.: Molten pool characterization of laser lap welded copper and aluminum, J. Phys. D. 46, 495501 (2013)
8. Balu, P., Carlson, B., Kovacevic, R.: An Investigation into the Laser Micro-Welding of Aluminum and Copper in Lap Joint Configuration, TMS2011 Supplemental Proceedings 3, 295 (2011)
9. Metsios, I.: QCOALA (Quality Control of Aluminium Laser-welded Assemblies) project technical report, October 2012.
10. Hailat, M.M., Mian, A., Chaudhury, Z.A., Newaz, G., Patwa, R., Herfurth, H.J.: Laser micro-welding of aluminum and copper with and without tin foil alloy, Micros system Technologies, 18 (1), 103-112 (2012).

# Elevated Temperature, Solid-State Cracking in Welds

**John C. Lippold**

**Abstract** A number of elevated-temperature, solid-state cracking phenomena are associated with welded fabrication. These include ductility dip cracking which generally occurs during multipass welding, and reheat cracking which is usually associated with postweld heat treatment. Reheat cracking includes stress relief cracking of steels, strain-age cracking of Ni-base alloys, and relaxation cracking of stainless steels and Ni-base alloys. This paper describes the mechanism associated with each of these forms of cracking and methods to avoid such cracking. Weldability tests that can be used to quantify susceptibility to the various forms of cracking will also be described.

## Introduction

Elevated temperature, solid-state cracking has been observed in welds in a number of materials including low alloy steels, stainless steels, and Ni-base alloys. In most cases, cracking occurs when the material is reheated either during intentional postweld heat treatment (PWHT) or in multipass welding when the weld metal or base metal is reheated by sequential weld passes. The most common forms of cracking are known as ductility-dip cracking, stress relief cracking, PWHT cracking, and strain-age cracking.

Ductility dip cracking is manifested as a severe drop in ductility at intermediate temperatures and is most common in materials that have a face-centered cubic (fcc) crystal structure, including Ni-base alloys and austenitic stainless steels. Stress relief and PWHT cracking are associated with steels that must be heat treated after welding to temper the microstructure and/or relieve the residual stresses resulting from welding. In the low alloy steels, this heat treatment is conducted below the lower critical temperature ($A_1$). A related cracking phenomenon, known as

J.C. Lippold (✉)
Welding Engineering Program, Ohio State University, Columbus, OH, USA
e-mail: lippold.1@osu.edu

 
Th. Boellinghaus et al. (eds.), *Cracking Phenomena in Welds IV*,
DOI 10.1007/978-3-319-28434-7_12

relaxation cracking, occurs after extended exposure at the service temperature due to the slow relaxation of residual stresses. Strain-age cracking is normally associated with Ni-base superalloys where postweld solution annealing and aging are required to restore mechanical properties following welding.

All of these forms of cracking involve the relaxation of tensile residual stresses at temperatures where changes to the microstructure are also occurring. The superposition of these phenomena results in high stresses at grain boundaries that lead to cracking. Virtually all of the elevated temperature cracking that occurs in welded structures due to stress relaxation is intergranular, so understanding the nature of grain boundaries is critical to describing the mechanisms involved.

There are other forms of solid-state cracking that will not be discussed here. These include lamellar cracking which is associated with the heat-affected zone (HAZ) of welds made on steel plate, and so called Type IV cracking which is a creep-related failure mechanism that occurs in the fine grained HAZ of low alloy steels after elevated temperature service exposure.

## Ductility Dip Cracking

This form of solid-state cracking occurs in a number of engineering materials. The presence of a ductility dip in austenitic (fcc) alloys was reported as early as 1912 by Bengough [1]. In 1961, Rhines and Wray [2] reported that a ductility dip occurs in copper alloys, nickel alloys, austenitic stainless steels, titanium, and aluminum. Ductility dip cracking (DDC) was originally identified as a problem associated with the hot working of materials—such as the hot deformation of cast ingots to form wrought bar or plate products [2, 3]. DDC was identified as a problem during welding in the 1970s and 1980s. Early papers documented DDC in austenitic stainless steels, although this form of cracking was deemed "anomalous" [4]. In fact, DDC in welds has been around for quite some time and was often misinterpreted as some form of hot cracking, often termed "microfissuring" [5, 6]. Since the late 1990s, severe DDC has been encountered in Ni-base weld metals used in the nuclear power industry [7–9]. This has resulted in considerable study of the DDC phenomenon and a desire to develop DDC-resistant filler metals.

DDC may occur in both the weld metal and HAZ but, is generally considered a weld metal problem. In weld metals, DDC is always intergranular and generally occurs in reheated weld metal during multipass welding. The term ductility dip refers to the sharp reduction in ductility which susceptible materials exhibit in a temperature range between the alloy solidus and approximately half the solidus temperature. The mechanism responsible for DDC is still the subject of considerable debate. Oddly enough, this type of cracking is frequently observed in austenitic (fcc) materials which have very low levels of impurities (sulfur, phosphorus, and boron). Because grain boundary liquation does not play a role in DDC, the importance of impurity segregation to grain boundaries is reduced.

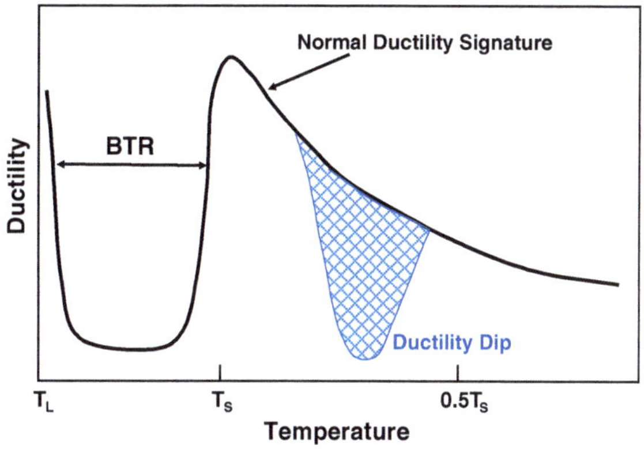


**Fig. 1** Ductility-temperature relationship for a material exhibiting a solid-state ductility dip

A schematic representation of the "ductility signature" of a material that exhibits a ductility dip is shown in Fig. 1. In many materials, ductility will be very high at temperatures just below the solidus temperature ($T_S$) and then gradually decrease upon cooling to room temperature. In materials susceptible to DDC, a sharp drop in ductility occurs between approximately $0.9T_S$ and $0.6T_S$ (for stainless steels and Ni-base alloys the range is between 800 and 1150 °C). Both the width and depth of this "dip" define the DDC susceptibility of the material.

Failure due to ductility dip cracking usually occurs at elevated temperatures where rapid grain growth occurs. Most theories suggest that cracking is associated with large grain size and the behavior of the grain boundaries (such as grain boundary precipitation and boundary sliding). The absence of impurities and second phase particles removes any obstacles for grain boundary motion and the resultant grain size in susceptible weld metals can be extremely large. As a result, the structure contains very little grain boundary area and it is hypothesized that strain is concentrated at these boundaries [10]. In support of this hypothesis, fracture due to the ductility dip is intergranular, but sometimes with ductile features (usually termed ductile intergranular).

From a welding standpoint, the two material systems that show the highest susceptibility to DDC are the austenitic stainless steels and Ni-base alloys. DDC has also been reported in copper and titanium alloys, but this is unusual in practice.

The micrograph in Fig. 2 shows a solidification grain boundary (SGB) and migrated grain boundary (MGB) pair in Ni-base alloy weld metal. The MGB represents the crystallographic component of the SGB that has migrated during cooling from the solidification temperature or during reheating. Note that the MGB is very straight. In the weld metal, DDC *always* occurs along the MGB.

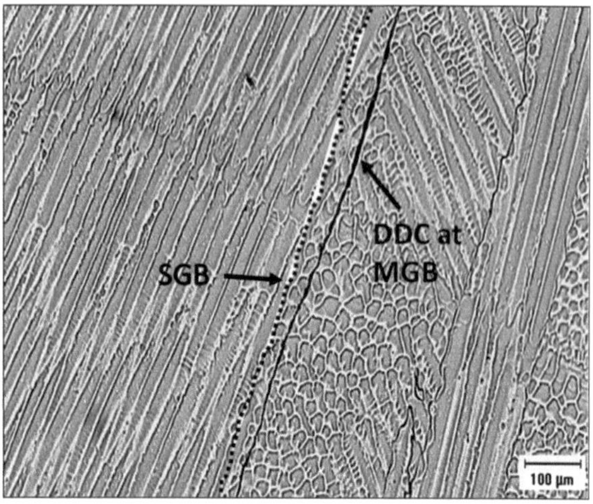


**Fig. 2** Ductility-dip cracking along a migrated grain boundary (MGB) in a Ni-base alloy weld metal. The dotted line represents the solidification grain boundary (SGB)

## Proposed Mechanisms

A number of theories have been proposed to describe the mechanism of DDC, as summarized in Table 1. The first theory of ductility dip cracking was proposed by Rhines and Wray [2] based on studies of the hot working behavior of base metals. They believed that the loss of intermediate temperature ductility was caused by grain boundary shearing, similar to creep-rupture failures that occur at lower temperatures over longer exposure times. According to their theory, at temperatures below the recrystallization temperature, grain boundary voids have time to join by grain boundary shearing and cause a fracture. Above the recrystallization temperature where recrystallization occurs because of mechanical working, the formation of new grain boundaries makes the joining of the voids difficult, presumably due to the creation of more grain boundary area. This mechanism is in general agreement

**Table 1** Ductility dip cracking theories

| Name | Description | Year |
|---|---|---|
| Rhines and Wray [2] | Grain boundary shearing up to recrystallization temperature | 1961 |
| Yamaguchi et al. [14] | Sulfur segregation and embrittlement | 1979 |
| Zhang et al. [20] | Combination of effects up to recrystallization temperature | 1985 |
| Ramirez and Lippold [11] | Grain boundary sliding, microvoid formation, boundary tortuosity | 2004 |
| Nishimoto et al. [16–18] | Impurity segregation | 2006 |
| Noecker and DuPont [12, 13] | Grain boundary sliding, carbide distribution and morphology | 2007 |
| Young et al. [22] | Precipitation-induced cracking | 2008 |

with that recently proposed by Ramirez and Lippold [11], and Noecker and DuPont [12, 13] for weld metals, as discussed in more detail later.

Yamaguchi et al. [14] proposed that higher sulfur levels increased the tendency for a ductility dip in Ni-base superalloys at temperatures between 950 and 1150 °C (1740 and 2100 °F), where sulfur segregated to and embrittled the grain boundaries and which cracked under an applied stress. Similar sulfur segregation mechanisms have also been proposed by Matsuda [15] and more recently by Nishimoto et al. for Ni-base Alloy 690 weld metals [16–18]. Recent work by Collins et al. [19] with Ni-base filler metal 82 (ERNiCr-3) also showed that sulfur additions increase susceptibility to DDC. While sulfur and other impurities may contribute to DDC, the work by Ramirez and Lippold concluded that differences in susceptibility could not be simply explained by impurity (S and P) content since many materials with very low impurity content were also susceptible [11].

Zhang et al. [20] reported that the combined effects of grain-boundary precipitation, grain-boundary sliding, grain boundary migration, and grain boundary serration affect the DDC performance of the low expansion alloy Invar (Fe-36Ni). They also suggested that recrystallization and decreased flow stress were factors in the recovery of ductility at elevated temperatures [21].

Young et al. [22] have suggested that DDC in high-Cr, Ni-base filler metals results from grain boundary carbide precipitation and associated "precipitation induced cracking." While this mechanism may have some relevance to high-Cr, Ni-base alloys that form grain boundary carbides, it cannot explain DDC that occurs in materials where grain boundary precipitation does not occur.

Work at Ohio State University on a variety of austenitic stainless steels and Ni-base alloys has shown that precipitation behavior and grain boundary "tortuosity" have a major influence on DDC susceptibility [11, 23–31]. It was concluded that DDC is essentially an elevated temperature, grain boundary sliding phenomenon. They also found that impurity (P, S, O and H) segregation, grain boundary precipitation, and boundary tortuosity affect DDC susceptibility and that controlling the nature of grain boundaries is the key to avoiding DDC in Ni-base weld metals.

The DDC theory proposed by Ramirez and Lippold is shown schematically in Fig. 3. When there is no grain boundary precipitation, the grains are able to grow and the grain boundaries tend to be very straight. When strain is applied to this microstructure within the ductility dip temperature range, grain boundary sliding occurs and stresses are high at the grain boundary triple points leading to void formation at these locations. Eventually cracks propagate along the grain boundaries.

When precipitates form along the grain boundaries in the solid state, stresses will now concentrate at the triple points and also at the precipitate-grain boundary interface. This results in void formation along the grain boundary where precipitates are present. In the situation where precipitates form at the end of solidification, the grain boundaries are pinned by these precipitates resulting in a "tortuous" grain boundary. Because of this tortuosity, the boundary resists sliding due to the mechanical locking effect of the tortuous boundary.

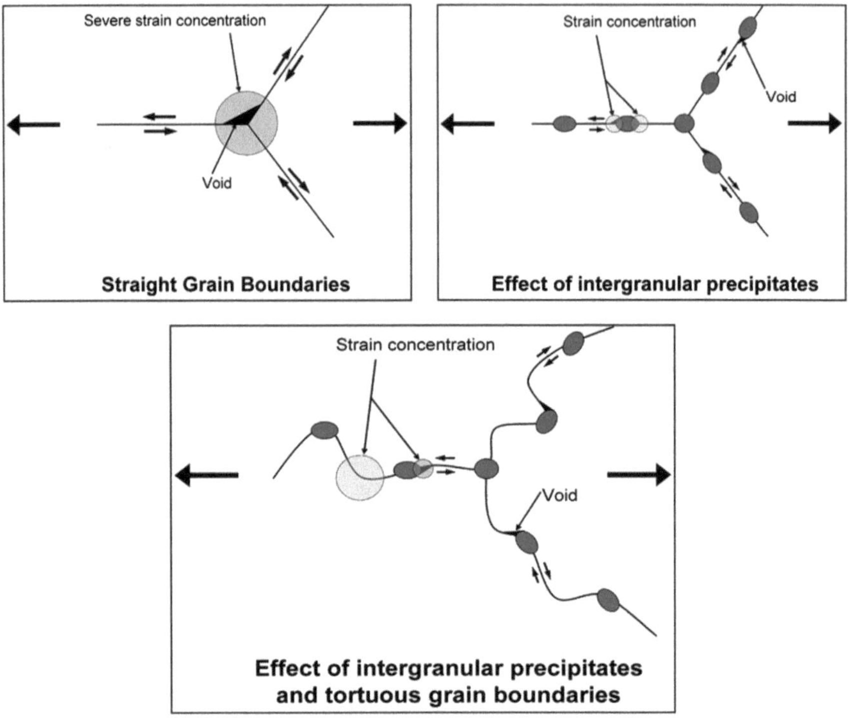


**Fig. 3** Schematic of grain boundary character as a function of precipitation

It has been shown that weld metals that exhibit "clean" and straight grain boundaries are most susceptible to DDC, while those that contain tortuous grain boundaries are the most resistant. Obviously, designing the microstructure in order to control grain boundary character is a key factor in preventing DDC.

The photomicrographs shown in Fig. 4 are from strain-to-fracture test samples tested at approximately 5 % total strain within the ductility-dip temperature range.

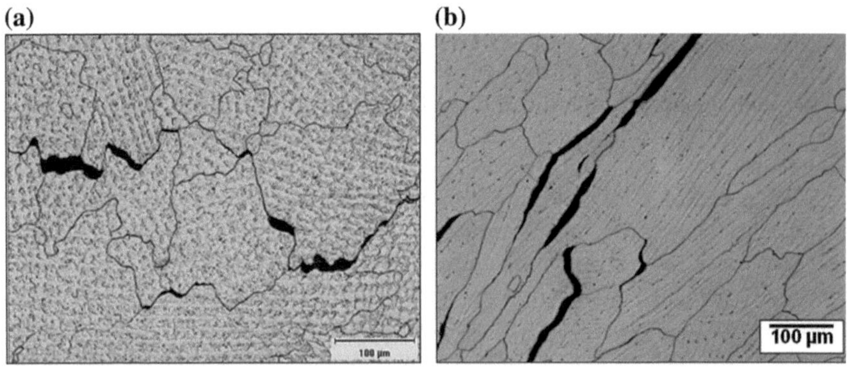


**Fig. 4** Ductility dip cracks in strain-to-fracture samples tested at 950 °C and 5 % strain, **a** Filler Metal 82, **b** Filler Metal 52

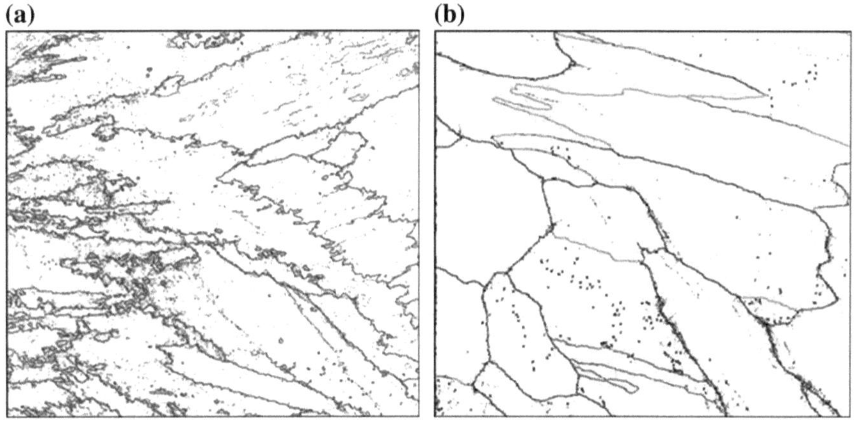


**Fig. 5** EBSD maps showing migrated grain boundaries in Ni-base weld metal, **a** Filler Metal 52 MSS, **b** Filler Metal 52M

This test will be described later in this section. Note that the FM 82 grain boundaries are somewhat tortuous and the cracks are quite short. In contrast, the FM 52 grain boundaries are very straight allowing more grain boundary sliding and promoting larger cracks. The strain threshold to produce DDC is much lower for FM 52 than for FM 82.

A technique for evaluating grain boundary orientation is called Electron Backscatter Diffraction (EBSD). It is also commonly referred to as Orientation Imaging Microscopy (OIM). This technique uses diffraction patterns generated in the scanning electron microscope to identify crystal orientation differences in a crystalline material. Thus, the high angle grain boundaries in the weld metal or HAZ can be easily delineated.

The EBSD patterns shown in Fig. 5 are from two Ni-base weld metals that exhibit a large difference in DDC susceptibility. Note that the DDC-resistant weld metal in Fig. 5a exhibits very tortuous migrated grain boundaries. This boundary type was developed by the formation of precipitates (Nb-rich carbides) at the end of solidification that promote boundary pinning. In contrast, the weld metal in Fig. 5b forms no precipitates at the end of solidification and pinning of the MGBs does not occur. This results in large grains with very straight boundaries.

One of the interesting features of austenitic (fcc) materials in the DDC temperature range is that the strain tends to concentrate at the grain boundary, rather than being distributed uniformly throughout the microstructure. This strain localization is what gives rise to grain boundary sliding and preferential void formation and cracking at the grain boundaries.

## Quantifying Ductility Dip Cracking

A number of tests have been used to determine susceptibility to DDC. Most prevalent are the hot ductility test, the spot Varestraint test, the double-spot Varestraint test, and the strain-to-fracture test. The slow bending Varestraint test used by Matsuda [32] to develop BTR curves for solidification cracking also yielded ductility-dip temperature curves for DDC. The spot Varestraint test is normally used to evaluate HAZ liquation cracking susceptibility in base metals, but can also be used to evaluate weld metal liquation and ductility dip cracking.

The double-spot Varestraint test was originally developed by Lippold and Lin at Edison Welding Institute [33] in order to isolate DDC from solidification cracking during standard transverse Varestraint testing. With this test an initial gas tungsten arc spot weld was applied to a sample and then a second spot weld applied within this initial weld upon performing a standard spot Varestraint test. Weld metal DDC could then be generated in the initial spot weld upon the application of sufficient strain. While this was an improvement over the standard Varestraint test, there were still three major complicating issues, (1) weld metal liquation cracking and DDC could not be separated, (2) it was difficult to determine the temperature range over which cracking occurred, and (3) some materials only exhibited DDC at the highest achievable strain ($\sim$10 %) so determining a strain threshold was difficult. Despite these issues, the double-spot Varestraint test did provide some useful DDC susceptibility data for a number of Ni-base filler metals. For example, Kikel and Parker [34] were able to compare the DDC susceptibility of Filler Metal 52 and Alloy 690 with that of Filler Metals 82 and 625.

The Gleeble hot ductility test was used by both Noecker and Dupont [12, 13], and Young et al. [22] to quantify DDC in high-Cr, Ni-base weld metals. Since this is essentially a hot tensile test where the sample is pulled to failure, no information on the critical threshold strain to initiate DDC is obtained. In the case of Young et al. [22], susceptibility to DDC was determined by the drop in on-cooling ductility after heating to an arbitrary peak temperature.

The strain-to-fracture (STF) test was developed by Nissley and Lippold [35] to avoid the shortcomings of the Varestraint and hot ductility tests for determining susceptibility to DDC. It is a Gleeble-based test that allows weld metals or base metals to be evaluated using microstructure, strain and temperature as the principal variables. Using this test, strain-temperature envelopes can be developed that show the regime over which DDC occurs for a given material. A schematic representation of strain-temperature DDC behavior for susceptible, intermediate, and resistant materials is shown in Fig. 6. Actual data for STF testing of stainless steels and Ni-base alloys can be found elsewhere [23, 28–31].

Using the STF test, relative susceptibility to DDC can be determined by the minimum threshold strain for cracking across the entire temperature range and the

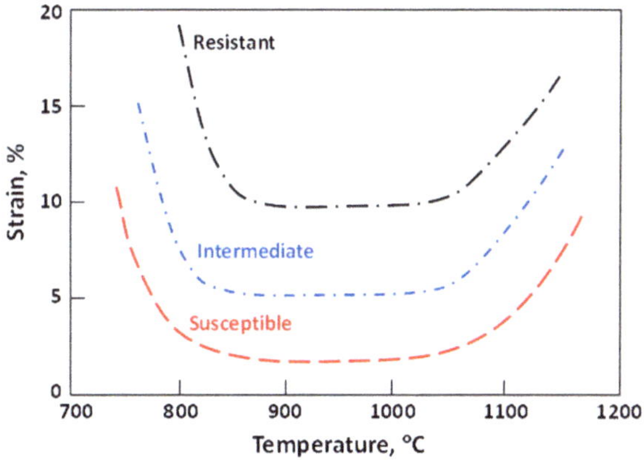


**Fig. 6** Schematic strain-to fracture behavior for predicting ductility dip cracking susceptibility of austenitic (fcc) materials

severity of cracking above the threshold strain. For example, in Fig. 6 the minimum threshold strain for susceptible materials is on the order of 2 % while that of intermediate materials is approximately 5 %. Although this difference in minimum threshold strain may seem small, actual practice has shown that weld metals with threshold strains of approximately 5 % are moderately susceptible to DDC, while those that approach or exceed 10 % tend to be quite resistant.

The minimum strain for cracking in the STF test tends to be in the range from 900 to 1000 °C. A more efficient method for comparing DDC susceptibility among materials has been to simply test over a range of strain at 950 °C to determine the threshold strain for cracking and the increase in cracking as a function of strain above the threshold. This yields data as shown schematically in Fig. 7 for susceptible, intermediate, and resistant materials, where the number of cracks on the sample surface is plotted as a function of applied strain.

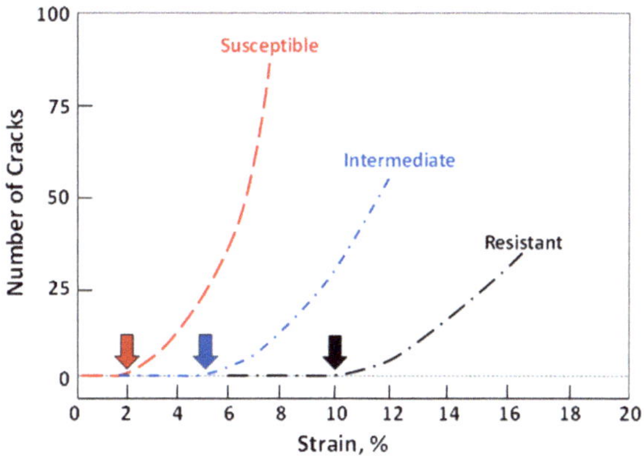


**Fig. 7** Schematic of ductility dip cracking as a function of applied strain at 950 °C for austenitic (fcc) materials. *Arrows* indicate threshold strain for cracking

## Preventing Ductility Dip Cracking

From a welding standpoint, DDC is most frequently found in the fusion zone, so the discussion here focuses primarily on weld metals. The most effective method for avoiding weld metal DDC is to select filler metals that are resistant to this form of cracking. For austenitic stainless steels, these include filler metals such as Types 308 and 309 that contain sufficient ferrite in the deposit to pin migrated grain boundaries and considerably increase boundary tortuosity. Weld metals of this type will usually exhibit threshold strain for cracking levels above 10 % (see Fig. 7). These weld metals are also very resistant to solidification cracking. For austenitic stainless steel weld deposits that are fully austenitic, control of DDC is more difficult and the general approach used is to reduce the restraint (residual stress) in the deposited weld metal, as discussed below.

Ni-base weld metals can also exhibit a range of DDC behavior, depending on composition and solidification behavior. Alloys that form no second phase at the end of solidification to control the character of the MGBs are very susceptible to DDC. This includes Filler Metal 52 which exhibits large grains and very straight MGBs, as shown in Figs. 4b and 5b. The addition of elements that promote the formation of a carbide at the end of solidification via a eutectic reaction will greatly improve resistance to DDC. For example, Filler Metals 82 and 625 which contain additions of Nb and form NbC at the end of solidification are quite resistant to DDC. The NbC that is present at the end of solidification is effective in pinning the MGBs and making these boundaries very tortuous and resistant to sliding in the DDC temperature range.

In general, weld metals that form a second phase at the end of solidification are effective in pinning the MGBs and show good resistance to DDC. This is a form of "grain boundary engineering" that uses the mechanical locking effect of a tortuous grain boundary to oppose boundary sliding. Similar advantages may be gained by reducing grain size, but this is often difficult in fusion welding. Impurities (particularly S) have been shown to have an adverse effect on DDC, but reducing impurities alone will not prevent DDC under high restraint conditions. In general, impurities have a secondary additive effect on susceptibility to DDC. If possible, control of the grain boundary character (tortuosity) is more effective in avoiding DDC. There is also some evidence that oxygen segregation to grain boundaries may have a negative effect, but this has not been well documented [32]. A better understanding of oxygen effects is recommended since oxygen may in fact be added to shielding gases, particularly for gas metal arc welding. There is also the possibility of oxygen pickup during the welding process. Finally, the use of shielding gases that contain H, such as 98Ar-2$H_2$, can lead to increased susceptibility to DDC, as shown by Collins et al. [23]. Such shielding gases are sometimes recommended since they improve the wetting characteristics of the molten pool and help avoid lack-of-fusion defects. $H_2$-bearing shielding gases should be avoided in situations where DDC is a potential problem.

In situations where DDC cannot be avoided by filler metal selection and grain boundary engineering, susceptibility must be reduced by reducing restraint. If the residual stress in the solidified weld metal can be reduced, relaxation of these stresses that occurs by reheating during multipass welding does not lead to DDC. Reduction of weld heat input and the use of small weld beads for thick section, highly restrained joints can be effective, although at a penalty to productivity. It has been shown that even highly susceptible filler metals such as Filler Metal 52 can be free of DDC if restraint is controlled through optimization of heat input, and bead size and placement.

## Reheat and Stress-Relief Cracking

Reheat cracking and stress-relief cracking, as the terms imply, are associated with postweld heat treatment (PWHT) or stress relief heat treatment following welding. These heat treatments are designed to temper martensitic microstructures, reduce residual stresses, or both. Reheat cracking also describes solid-state cracking in multipass welds where subsequent weld passes provide the "reheating". This form of cracking also includes "underbead cracking" that is associated with cladding of some pressure vessel steels. In general, reheat cracking is usually associated with steels that must be subjected to stress relief following welding due to code requirements. Another form of postweld heat treatment cracking, called "strain-age" cracking, is associated with precipitation strengthened Ni-base alloys. This form of cracking will be described in a separate section.

The subjects of reheat, stress-relief, and underbead cracking have received considerable attention since the 1950s and a number of excellent review papers are available [36–38]. Much of the interest in reheat cracking was driven by postweld heat treatment cracking in austenitic stainless steels and Cr-Mo-V pressure vessel steels. This was generally associated with large forgings or thick-section components fabricated for the power generation industries. Pressure vessel steels that were clad with stainless steels for use in steam generators and large pressure vessels in the nuclear power industry were susceptible to "underclad" cracking.

Cracking via the reheat cracking mechanism is most frequently observed in low alloy steels containing secondary carbide formers (Cr, Mo, and V) and which form untempered martensite in the HAZ. Cracking in these steels normally occurs along prior austenite grain boundaries in the coarse-grained region of the HAZ. Austenitic stainless steels containing carbide formers (Nb, Ti) and/or high carbon content for elevated temperature service can also be susceptible to reheat cracking. Another form of cracking, called "relaxation" cracking, occurs in these same steels via a similar mechanism but at lower temperatures and longer times. Relaxation cracking can occur after months or years in susceptible steels used for elevated temperature service.

Other material classes may also be susceptible. Technically, any material which exhibits a strong precipitation reaction in the stress relief temperature range will be

**Table 2** Steels susceptible to reheat cracking

| Alloy designation | Type/form | Composition |
|---|---|---|
| A508, grade 2 | Forging | 0.6Mo, 0.40Cr, 0.25C |
| A514, grade F | Plate | 0.5Mo, 0.5Cr, 0.05 V, 0.15C |
| A517, grade F | Plate | 0.50Mo, 0.5Cr, 0.15C, 0.003B |
| A533, grade B | Plate | 0.5Mo, 0.20C |
| A710 (HSLA-80/100) | Plate | 2.0Ni, 0.5Cr, 0.5Mo, 1.0Cu, 0.05C |
| F/P22 | Forging/plate | 2.25Cr-1.0Mo |
| Type 347 | Stainless steel | 18Cr-11Ni-0.6Nb-0.04C |
| Type 321 | Stainless steel | 18Cr-10.5Ni-0.4Ti-0.04C |
| Alloy 800H | High alloy SS | 21Cr-32Ni-0.4Ti-0.4Al-0.1C |

susceptible to this form of cracking. Reheat cracking has not been reported in aluminum or titanium alloys. Materials in which reheat cracking, including stress relief, PWHT, underbead, or relaxation types, have been observed are listed in Table 2.

## Low Alloy Steels

In low alloy steels, there are five necessary conditions for reheat cracking to occur.

1. *Elevated temperature thermal excursion*. In the HAZ, regions susceptible to reheat cracking are heated into the austenite phase field (above the $A_3$ temperature) and into a temperature range where coarsening of the austenite grains occurs.
2. *Carbide dissolution*. The time-temperature relationship in the austenite phase field must be such that carbide dissolution is relatively complete. It is particularly critical that the alloy carbides (Cr, Mo, V) at least partially dissolve.
3. *Residual stress*. As the weldment cools to room temperature, considerable residual stress accumulates in the structure.
4. *Reheating into the critical temperature range*. Reheating into the temperature range between 300 and 675 °C (570–1250 °F) is particularly damaging. This range represents the regime in which carbides re-precipitate.
5. *Creep or stress relaxation during reheating*. This requires that sufficient residual stress is present in the as-welded structure and that the reheat temperature is high enough to promote stress relaxation.

For austenitic stainless steels, such as Type 347, the HAZ thermal cycle results in some grain growth and carbide dissolution. There is no on-cooling transformation, at room temperature there is simply an austenitic (fcc) microstructure with residual stress.

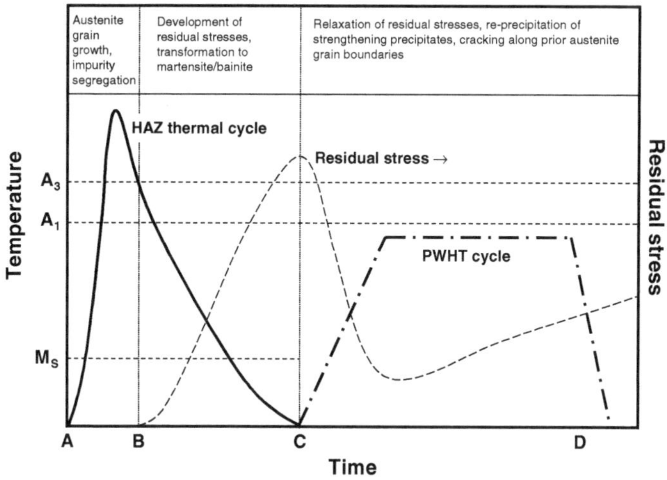


**Fig. 8** Schematic of thermal cycle associated with reheat cracking in low alloy steels, where *A–D* represent different stages of the HAZ and PWHT thermal cycles

A schematic of the weld and PWHT thermal cycle that can give rise to reheat cracking in low alloy steels is shown in Fig. 8. During the weld thermal cycle, regions of the HAZ are heated to a temperature where coarsening of the austenite grains occurs and there is sufficient thermal driving force for carbide dissolution (A to B in Fig. 8). Upon cooling, this region (typically the CGHAZ) transforms to martensite and/or bainite (B–C in Fig. 8). Since the carbides were completely (or partially) solutionized, the hardness of this region will be in the range from 40 to 50 HRC, requiring a PWHT to temper the martensite. For most low alloy steels susceptible to reheat cracking, the PWHT temperature will be in the range from 595 to 705 °C (1100–1300 °F). During PWHT, the alloy carbides re-precipitate and residual stresses are relaxed.

It is the combination of stress relaxation and precipitation hardening that leads to cracking along the prior austenite grain boundaries in the CGHAZ. The precise mechanism for reheat cracking is the subject of some debate. Some have suggested that *intragranular* precipitation of carbides during reheating strengthens the grain interiors relative to the boundaries and promotes localized deformation along the boundaries as simultaneous stress relaxation occurs. In support of this hypothesis, it has been shown that steels which contain molybdenum and vanadium and form intragranular carbides with these elements are particularly susceptible to reheat cracking [39–44].

Impurity segregation to grain boundaries may also contribute. Several studies have shown that the segregation of impurity elements to the austenite grain boundaries during the HAZ thermal cycle results in boundary decohesion during PWHT [39, 45, 46]. These impurities include the typical phosphorus and sulfur, but also others that are unique to this form of cracking, including copper (Cu), tin (Sn), antimony (Sb), and arsenic (As).

## Stainless Steels

As indicated above, a form of reheat cracking has also been observed in austenitic stainless steels. In particular, the HAZ and fusion zone of Type 347 stainless steel has been reported to be susceptible, particularly in thick sections [47–53]. This is a "stabilized" grade of stainless steel that contains niobium to reduce susceptibility to intergranular corrosion. Reheat cracking can also occur in other stainless steels, including Ti-stabilized Type 321 and high carbon grades such as 304H and 316H. There have also been reports of reheat cracking in high alloy stainless steels used for elevated temperature service, in particular Alloy 800[1] and its high carbon variant Alloy 800H [48, 54, 55].

Cracking normally occurs during the postweld stress relief that is often required for thick section stainless steel weldments. Many of these alloys have also been shown to be susceptible to relaxation cracking, as discussed below. Reheat cracking has also been observed in austenitic stainless steel weld metals, such as Type 308, that are deposited using the flux-cored arc welding (FCAW) process. Cracking in these weld deposits occurs due to the presence of bismuth, which is added to facilitate slag separation from the weld metal [56].

The mechanism for cracking in Type 347 is associated with the precipitation of NbC during the reheating cycle. Since intragranular precipitation occurs in the same temperature range as stress relaxation, locally high strains concentrate at grain boundaries and promote cracking. Grain growth in the HAZ increases cracking susceptibility. It is not clear whether there is any effect of impurity segregation in this alloy. The fracture mode is typically ductile intergranular, exhibiting extremely fine ductile dimples. It is interesting that this form of cracking can also occur in the weld metal of the austenitic stainless steels. In low alloy steels, it is almost always in the HAZ. The presence of delta ferrite in the weld deposit does not seem to influence susceptibility to this form of cracking in Type 347.

An example of reheat cracking in Type 347 weld metal is shown in Fig. 9. Cracking occurred after postweld stress relief of a large structure constructed from Type 347 stainless stainless and a matching filler metal. The ASME code required that this structure be stress relieved in the temperature range from 850 to 900 °C (1560–1650°F) before it could be put into service. This resulted in severe reheat cracking in the weld metal. The fracture occurs along migrated grain boundaries in the weld metal, as indicated by the intergranular features of the fracture surface.

[1]Note that Alloy 800 is often considered a Ni-base alloy, even though the nominal Fe content is higher than the Ni content.

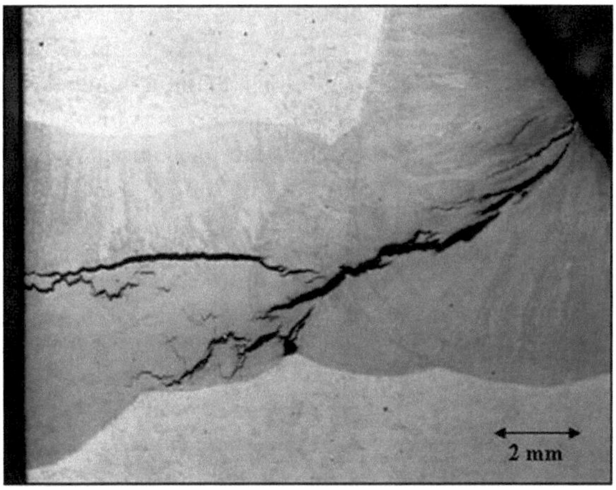


**Fig. 9** Reheat (stress relief) cracking in Type 347 stainless steel weldment

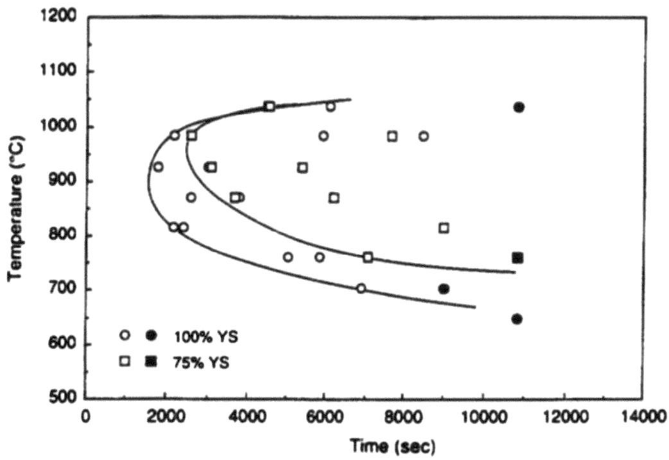


**Fig. 10** Temperature-time cracking envelope for Type 347 weld metal under two stress conditions

The cracking susceptibility of Type 347 weld metal exhibits a C-curve cracking response, as shown in Fig. 10. The two curves shown represent the onset of cracking when a weld metal sample of Type 347 was loaded to either 75 or 100 % of its yield strength at a given temperature and held at that temperature until failure occurs.

By plotting the fracture time at a given temperature, the reheat cracking envelope can be determined. For example, at 900 °C (1650 °F) reheat cracking occurs within 2000 s when yield strength level stresses are present. The cracking "envelope" described by these two C-curves represents the precipitation temperature range of NbC in stainless steel.

Another manifestation of reheat cracking has been observed in thick section welds made using Type 308 filler metal applied using the FCAW process [56]. Many flux-cored electrodes contain additions that facilitate the removal of the flux/slag from the weld surface. One of these elements is bismuth (Bi). In the weld metal, Bi apparently segregates to grain and interphase boundaries and reduces the ductility of the weld metal at temperatures above 700 °C (1290 °F). Reheat cracking occurs along the austenite-delta ferrite interface. At temperatures above 850 °C (1560 °F), the bismuth segregation promotes local melting and weld metal liquation cracking is possible.

## Relaxation Cracking

Relaxation cracking is another manifestation of reheat cracking that is related to elevated temperature service exposure. It typically occurs in austenitic stainless steels that are used at service temperatures in the range from 550 to 750 °C (1020–1380 °F). Susceptible materials are the same as indicated for susceptibility to reheat cracking: Types 321, 347, 304H, and 316H, and Alloy 800H. The cracking mechanism is the same as for reheat cracking, but occurs at much longer times. It is a particularly insidious form of cracking because it normally occurs after hundreds or thousands of hours of service. Work by van Wortel at the TNO Metals Research Institute in the 1990s, supported through a large Joint Industrial Program, evaluated a number of these materials and developed a test methodology to quantify susceptibility to relaxation cracking [57].

Relaxation cracks were found to have slightly different features than reheat cracks, primarily because they form over much longer exposure times. In Alloy 800H, the crack along the HAZ grain boundary exhibited a Ni-rich filament surrounded by a Cr-rich oxide layer. In advance of the crack, small voids (or cavities) were present along the grain boundaries. Large grain boundary carbides were present, surrounded by a precipitate-free zone and then a dense distribution of matrix carbides. The formation of matrix carbides results in an increase in hardness and, combined with stress relaxation, leads to grain boundary failure.

The relaxation cracking test developed at TNO identified specific temperature ranges over which relaxation cracking was most prevalent. For the 300-series austenitic stainless steels, this range was 525–600 °C (980–1110 °F) and for Alloy 800H, 550–650 °C (1020–1200 °F). Higher residuals stresses and large grain size were both found to accelerate failure. The use of a stress relief heat treatment prior to service exposure is an obvious method to avoid relaxation cracking, but in heavy section weldments the use of such a heat treatment usually results in reheat (stress relief) cracking.

## Quantifying Reheat Cracking Susceptibility

There has been considerable effort over the years to develop test techniques that quantify susceptibility to reheat cracking. These tests can be separated into two groups, those that use self-restraint and those that are simulative in nature and use externally applied load (or strain). Among the self-restraint tests, the Lehigh restraint test [58], Y-groove test [56], modified implant test [59], and the BWRA test [60] are most often cited.

These tests do not typically provided quantitative results, rather they tend to be of the "go-no go" type. Sectioning is required after testing to determine if reheat cracks are present. These tests can be useful in identifying welding procedures and postweld heat treatment conditions for avoiding reheat cracking. For example, Meitzner and Pense [53] used the Lehigh restraint test to develop a C-curve cracking envelope for stress relief cracking in A517, and Nishimoto et al. [56] used the Y-groove test to develop similar curve for Type 308 FCAW deposits containing bismuth, as shown in Fig. 11.

In order to better quantify susceptibility to reheat cracking, tests have been developed that simulate the HAZ microstructure and then apply a stress to promote cracking. Most of the simulation tests use a thermo-mechanical simulator, such as the Gleeble™ [61] to apply stress (and/or strain) to a sample after an appropriate HAZ thermal cycle. Tests developed by Balaguer et al. at RPI [62], Nawrocki et al. at Lehigh [63], and Norton and Lippold at Ohio State [64] all used the Gleeble to develop a representative HAZ microstructure in a small tensile sample and then a constant load was applied at various PWHT temperatures. Ideally, this test can develop a "C-curve" cracking response that indicates susceptibility to reheat cracking.

Such a response curve developed by Nawrocki et al. [63] is shown in Fig. 12. Typically these tests are subject to large scatter in terms of failure time. In order to reduce scatter and better define material ductility as a function of PWHT temperature and time, Norton and Lippold [64] used an approach where the sample was pulled to failure after up to 4 h of exposure at the PWHT temperature. Although developed for studying strain-age cracking in Ni-base superalloys, the test works equally well for steels. This approach will be described in the section on Strain Age Cracking.

## Preventing Reheat Cracking

Just as the mechanism for reheat cracking is quite complicated and varies as a function of the material composition, methods to avoid reheat cracking can also be complicated. The discussion here refers primarily to reheat and stress relief cracking that occurs during fabrication. Some of the same methods for prevention can apply to relaxation cracking that occurs during service, but this is not always the case.

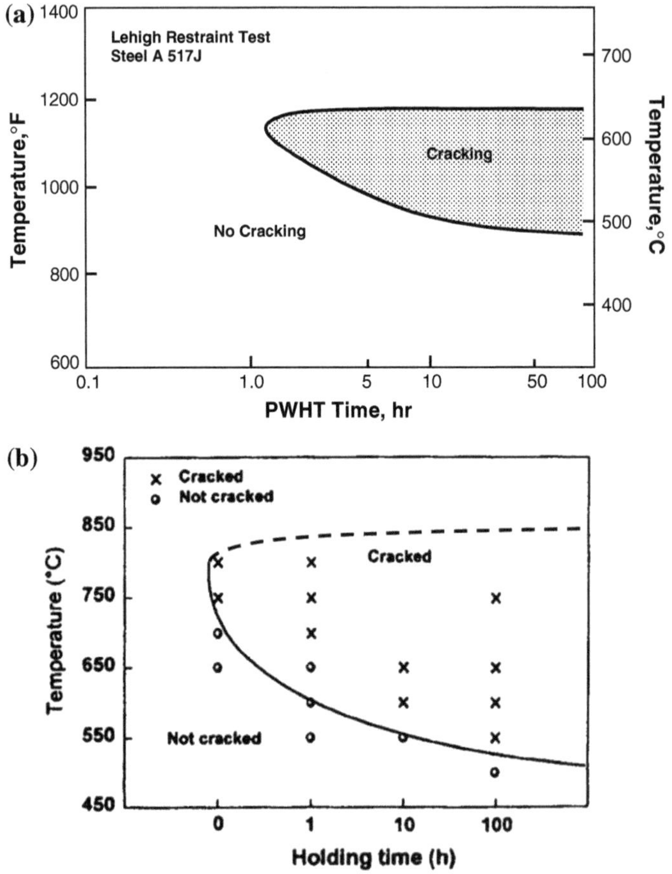


**Fig. 11** Effect of stress relief temperature on cracking susceptibility, **a** Lehigh restraint test for the HAZ of A517, **b** Y-groove test for Type 308 FCAW weld metal. From [53, 56]

Prevention methods are discussed in terms of (1) composition control, (2) effect of welding conditions, (3) control of residual stresses, (4) control of stress relaxation during reheating, (5) effect of stress concentration, and (6) "buttering" of the substrate.

**Composition Control**. The steels listed in Table 2 are inherently susceptible to reheat cracking because they contain secondary carbide formers. Reheat cracking can be reduced or eliminated by choosing steels that have reduced susceptibility, for example by using the relationship developed by Haure and Bocquet [65] or others as listed in Table 3. Often selection of an alternate material is not an option, since the base material has been specified for use in a certain application.

In general, as the secondary carbide former (Cr, Mo, V, Nb, Ti) content and carbon content increases, the material becomes more susceptible to reheat cracking. Impurities have been shown to contribute to the reheat cracking mechanism by

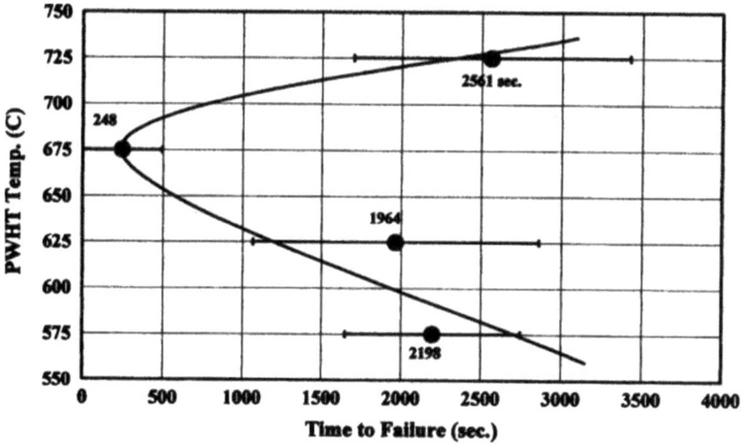


**Fig. 12** C-curve cracking response for 2.25Cr-1Mo steel developed using the Gleeble™. From [63]

**Table 3** Empirical relationships for reheat cracking susceptibility based on composition

| | Relationship | Constraints | Source |
|---|---|---|---|
| $\Delta G$ | Cr + 3.3Mo + 8.1 V − 2 | Susceptible if $\Delta G > 0$ | [82] |
| | | $C \leq 0.18$ %, $Cr \leq 1.5$ % | |
| $\Delta G_1$ | 10C + Cr + 3.3Mo + 8.1 V − 2 | Susceptible if $\Delta G > 2$ | [83] |
| | | $C \leq 0.18$ %, $Cr \leq 1.5$ % | |
| $P_{SR}$ | Cr + Cu + 2Mo + 10 V + 7Nb + 5Ti − 2 | Susceptible if $P_{SR} > 0$ | [84] |
| | | 0.1–0.25 % C, 0–1.5 % Cr, | |
| | | 0–0.2 % Mo, 0–1.0 % Cu, | |
| | | 0–0.15 % V, Nb, Ti | |
| CERL | 0.2Cu + 0.44S + 1.0P + 1.8As + 1.9Sn + 2.7Sb | Susceptibility increases | [85] |
| | | with value. Valid for A533, Type B steels | |
| X | 10P + 5Sb + 4Sn + As + Cu | Susceptibility increases with value | [86] |
| R | P + 2.43As + 3.57Sn + 8.16Sb | Susceptibility increases with value. Valid for 0.5Cr-Mo-V steels | [86] |

diffusing to grain boundaries and lowering the boundary cohesive strength. Reducing impurity content can improve cracking resistance in most of the susceptible materials, but impurity control alone cannot insure resistance to reheat cracking. Many of the impurity elements that promote reheat cracking in low alloy steels (in particular As, Sn, and Sb) do not appear on composition analysis reports that are provided by the material supplier, so it may be necessary to conduct (or specify) additional analysis.

**Effect of Welding Conditions**. The effect of welding conditions on susceptibility to reheat cracking can be profound and much research has been conducted on a number of steels to understand the effect. There are two basic, and conflicting, approaches. The first approach is to minimize weld heat input in order to reduce HAZ grain size. In both low alloy and stainless steels, larger grains in the HAZ have been shown to increase susceptibility, since less grain boundary area results in higher strain localization at the boundaries. Finer grains help to better distribute the strain and minimize void formation and/or boundary sliding. This approach works well with the stainless steels, such as Type 347 and Alloy 800H. The alternative approach is to use preheat and higher heat input in order to affect the transformation behavior in the CGHAZ. This approach does not apply to the stainless steels. For low alloy steels with intermediate hardenability, the use of preheat and high heat input will slow the cooling rate in the CGHAZ and reduce the hardness by limiting the formation of martensite. As the CGHAZ hardness decreases, relaxation of residual stress during reheating will be more uniformly distributed. It has also been argued that shallower temperature gradients in the HAZ resulting from higher heat input lead to a wider CGHAZ and more grain boundary area over which stress relaxation can be accommodated.

**Control of Residual Stresses**. Reheat cracking is almost always associated with thick-section weldments where the level of residual stress following welding is quite high—approaching the yield strength of the base or weld metal. Prediction and measurement of residual stresses in welds has been the subject of considerable research and the reader is referred to other authoritative texts that address these issues. It is well known, however, that the use of preheat can be a simple and effective method to reduce residual stress. Other techniques, including the control of bead size and placement, and welding sequence, can also be effective at reducing residual stresses. Another effective approach is to select filler metals whose strength undermatches that of the base metal. In this manner, restraint resulting from weld shrinkage during cooling concentrates in the weld metal rather than the CGHAZ. This assumes that the weld metal is of a composition that is not susceptible to reheat cracking. This approach may not be effective with the austenitic stainless steels, such as Type 347, since these steels are usually welded with a matching filler metal. The example shown in Fig. 9 demonstrates how weld metals can be as susceptible as the base metal to reheat cracking.

**Control of Stress Relaxation during Reheating**. Even in the presence of high residual stresses, cracking can be avoided if stress relaxation can be managed during the reheating cycle. This is a difficult problem in the low alloy steels, since the temperature range in which significant stress relaxation occurs overlaps the range where secondary carbides begin to reform in the microstructure. Postweld heat treatment cannot typically be conducted above the precipitation temperature range because the lower critical temperature ($A_1$) is an effective upper bound for postweld heat treatment, i.e. PWHT above the $A_1$ results in reformation of austenite. For the austenitic stainless steels, this is not the case and possible methods exist for relaxing residual stresses. Since the nose of the carbide precipitation curve is at relatively high temperatures (850–950 °C) it may be possible to hold the weldment at temperatures below this range to allow partial relaxation of residual stresses before heating to the normal stress relief temperature. Even fractional reduction of the residual stress may be successful in avoiding reheat cracking in some situations. The other possibility is to heat the structure rapidly to a temperature above the carbide precipitation temperature range. This will allow rapid relaxation of the stresses while avoiding the formation of intragranular precipitates. This solution may have limited usefulness for large structures, since the ability to heat rapidly enough to avoid the nose of the precipitation curve may be limited. As illustrated in Fig. 10, heating rates exceeding 30 °C/min may be required to avoid cracking in heavy section Type 347 weldments.

**Effect of Stress Concentration**. An obvious approach for reducing susceptibility to reheat cracking (and hydrogen induced cracking) is to eliminate stress concentration in the weldment. This may include elimination of slag intrusions at the weld toe, grinding or blending the weld toe, or use of other material removal techniques to eliminate stress concentrations. Since cracking usually initiates in the HAZ very close to the fusion boundary, attention to this area of the weldment is very important. Welds with partial penetration, lack-of-fusion, or other process-related defects can also greatly increase stress concentration. Defects of this type that are open to the surface are the most damaging. Some attempts to locally alter stress concentration and residual stress have used various "peening" techniques. Peening can generate local compressive stresses on the weld surface and potentially mitigate initiation of reheat cracks. Peening probably has little effect in large, thick-section weldments since peening only affects the structure within a few millimeters from the surface. Many of the austenitic stainless steels that are susceptible to reheat cracking are also susceptible to HAZ liquation cracking, including Type 321 and 347, and Alloy 800. Liquation cracks can also increase stress concentration in the coarse-grained region of the HAZ. Careful inspection prior to PWHT is advised to assure that these types of defects are not present.

**"Buttering" of the substrate**. A technique that has been used to avoid reheat cracking and other forms of base metal HAZ cracking (such as lamellar cracking) is often called "buttering". With this approach, a layer of resistant weld metal is applied to the base metal substrate. This layer (or layers) is applied at low heat input and subsequently stress-relieved to eliminate residual stresses in the susceptible CGHAZ of the base metal. When welding is conducted after application of the butter layer, the CGHAZ is contained within the butter layer which is a composition resistant to reheat cracking. While this approach is usually very effective, it adds another fabrication step and increases fabrication costs significantly.

## Strain-Age Cracking

Strain-age cracking (SAC) is a form of reheat, or postweld heat treatment, cracking that is specific to the precipitation-strengthened, Ni-base alloys. It is a solid-state cracking phenomenon that is most often observed in the HAZ just adjacent to the

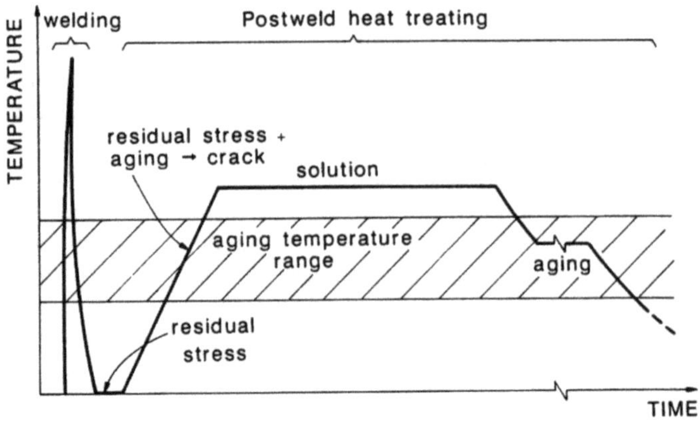


**Fig. 13** Schematic of the thermal history during welding and postweld heat treatment of Ni-base superalloys [66]

fusion boundary, although it is possible for SAC to occur in the weld metal of these alloys. In most cases, it occurs during postweld heat treatment, but is also possible (although unlikely) during reheating in multipass welds. This form of cracking is most prevalent with the “gamma-prime”, $Ni_3(Al,Ti)$, strengthened alloys and many of these alloys are considered “unweldable” because of this cracking phenomenon. The term “strain-age” refers to the simultaneous effect of relaxation of stresses causing high local strain and the age hardening of the structure by precipitate formation. Although SAC is similar to the reheat cracking mechanism that occurs in steels, there are some important differences that are unique to the Ni-base superalloys.

A schematic illustration of a representative thermal history for welded and postweld heat treated Ni-base superalloys is shown in Fig. 13 [66]. During the weld thermal cycle, strengthening precipitates (and other constituents) that are present in the base metal dissolve in the austenite matrix and some grain growth will occur, depending on weld heat input. Since the as-welded fusion zone and HAZ are effectively solutionized, significant softening occurs. A postweld heat treatment consisting of solution annealing and aging must be applied to strengthen the weldment and base metal to the original base metal strength level. The solution heat treatment also serves to relieve residual stresses resulting from the welding process. Ideally, the weldment is heated to an appropriate solution annealing temperature where alloying additions go back in solution (or homogenize in the weld metal due to solidification segregation) and residual stresses relax, and then cooled to an aging temperature where precipitation is controlled such that the required mechanical properties are achieved. A more detailed explanation of the SAC mechanism is provided in the next section.

In practice, it may be difficult (or impossible) to prevent the precipitation of $\gamma'$ during heating to the solution annealing temperature. This a function of the alloy composition and the differences in hardening rates. The relationship between the

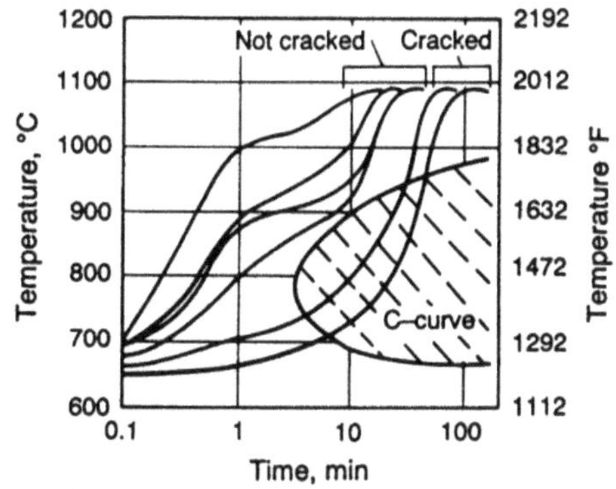


**Fig. 14** Schematic illustration of the effect of heating rate and precipitation behavior on susceptibility to strain-age cracking [67]

heating cycle and precipitation is illustrated in Fig. 14 [67]. Strengthening precipitates, such as $\gamma'$ and $\gamma''$, exhibit a distinct "C-curve" temperature-time regime where precipitation is possible. If the weldment can be heated rapidly enough to avoid intersecting the precipitation curve, then precipitation will not occur and solution annealing of the weldment can be achieved. If the weldment cannot be heated rapidly enough (or if the C-curve is shifted to the left in Fig. 14), precipitation will occur and the alloy will begin to harden. For Ni-base superalloys, precipitation during heating overlaps the temperature range where significant stress relaxation occurs and this can lead to locally high strains at the grain boundaries. If these strains are sufficiently high, grain boundary failure will occur and a strain-age crack will form.

## Mechanism for Strain-Age Cracking

As described above, the term strain-age cracking (SAC) is derived from the fact that both local strain and aging must occur nearly simultaneously. The term should not be confused with the metallurgical phenomenon of "strain aging" observed in carbon steels. SAC in welds has been studied extensively in various Ni-base superalloys and the severity of this problem has led to the development of SAC resistant alloys, such as Alloys 718 and 706.

It is generally agreed that SAC in Ni-base superalloys results from low ductility in the HAZ accompanied by high strain accumulation in the same region [68–71]. Such reduction of ductility is associated with the development of grain "stiffening" and/or grain boundary weakening during PWHT. Most investigators attribute this to intragranular precipitation hardening combined with precipitation-free zones at grain boundaries, or intergranular carbide precipitation. If the decrease in ductility during PWHT occurs before or at a faster rate than stress relief, the "embrittled" region in the HAZ may crack due to its inability to accommodate the redistribution of strain associated with the stress-relieving process. The following general observations have been made regarding SAC.

- It is always intergranular.
- It is most prevalent in the HAZ adjacent to the fusion line and in some cases is associated with the partially melted zone (PMZ).
- It occurs during postweld heating to the solution annealing temperature due to simultaneous precipitation and local strain accumulation at grain boundaries.

The stresses which cause cracking may have three origins, (1) weld residual stress, (2) thermally-induced stresses arising from the difference in the coefficient of thermal expansion between the base material and weld metal, and (3) stresses from dimensional changes caused by precipitation. In general, the precipitates have a different lattice parameter than the matrix and their formation will lead to a local grain boundary stress. Based on the published literature the metallurgical contributors to SAC are the following.

- The rate of hardening (strengthening). Materials which harden more slowly allow better accommodation of the stresses.
- Intragranular precipitation resulting in hardening of the grain interior leading to stress concentration at the grain boundaries. This mechanism was originally proposed by Prager and Shira [72] based on the work of Younger et al. [73] on austenitic steels.
- "Transient embrittlement" of the HAZ due to the precipitation of *intergranular* carbides. According to this theory [74–76] the embrittling reaction is thought to result from dissolution of carbides during the weld thermal cycle and subsequent re-precipitation in continuous "films" of $M_{23}C_6$-type carbides along grain boundaries during heat treatment. These carbide "films" are not capable of resisting the stresses caused by the precipitation of $\gamma'$, and presumably failure occurs at the carbide/matrix interface at the grain boundary.
- Partial melting along grain boundaries adjacent to the fusion line. This may be due to impurity segregation or constitutional liquation. Note that Ti which is added as a strengthening agent can also promote constitutional liquation if Ti-rich, MC carbides are present. Boron which is added to improve creep resistance in these alloys also promotes grain boundary liquation.

The mechanism for SAC in Ni-base alloys is still not precisely defined, although it is clear that both compositional and restraint factors play a role. It is well known, for instance, that certain alloys are more resistant to SAC than others. This resistance is generally attributed to the rate and nature of the precipitation reaction(s) that promote strengthening. The $\gamma'$ strengthened alloys are the most susceptible and the influence of Ti and Al content has been well documented. The relationship of Ti and Al content to SAC was originally proposed by Prager and Shira [72] and a diagram based on their work which includes additional modern alloys is shown in Fig. 15.

The original diagram contained a band running from approximately 6–7 at.% Al to 6–7 at.% Ti, separating alloys that are resistant (below) from those that are

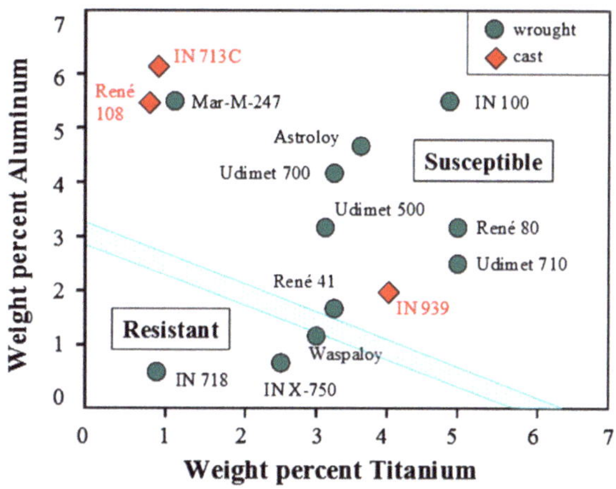


**Fig. 15** Effect of Al and Ti content on susceptibility to strain-age cracking in Ni-base superalloys (modified from Prager and Shira [72])

susceptible (above). It is not clear how this diagram was developed, but probably represented the results of different weldability tests and practical experience. Some diagrams of this type have replaced the band with a line, but the use of a transition band from resistant to susceptible is most appropriate since susceptibility to SAC is a strong function of restraint. For example, in alloys that are marginally susceptible to SAC (such as Waspaloy and René 41), it is known that minimizing the level of residual stress prior to PWHT is very effective in preventing SAC. Based on this diagram, it can be seen that the higher Ti + Al contents promote a stronger and more rapid precipitation of $\gamma'$. This, in effect, shifts the nose of the precipitation curve to much shorter times making it difficult to suppress precipitation during postweld heating to the solution annealing temperature range.

Duvall and Owzarski [77] demonstrated the effect of Ti + Al content in a study of the postweld heat treatment cracking susceptibility of Waspaloy and Alloy 718. They showed that HAZ cracking obeyed a C-curve behavior, as shown in Fig. 16, and that the C-curve for Alloy 718 was displaced to longer times. The C-curve for Waspaloy (containing 3 wt% Ti and 1.4 wt% Al) represents the $\gamma'$ precipitation regime while the Alloy 718 (containing 0.9 wt% Ti, 0.5 wt% Al, and 5 wt% Nb) C-curve represents the precipitation regime for $\gamma''$.

These results again demonstrate the beneficial effect of the sluggish precipitation reaction of $\gamma''$ for avoiding SAC during postweld heat treatment of Ni-base superalloys. In this same study, they found no indication of the "transient embrittlement" phenomenon associated with $M_{23}C_6$ precipitation at grain boundaries. Instead, the ductility was kept at a moderately low level during aging in the cracking temperature range by a combination of microstructural interactions produced during welding and heat treatment. Changes in cracking susceptibility between different heats of Waspaloy resulted from changes in ductility, which were caused by $\gamma'$ precipitation and intergranular carbide precipitation. Within groups of susceptible and non-susceptible microstructures, wide differences in the amount and morphology of carbides were observed.

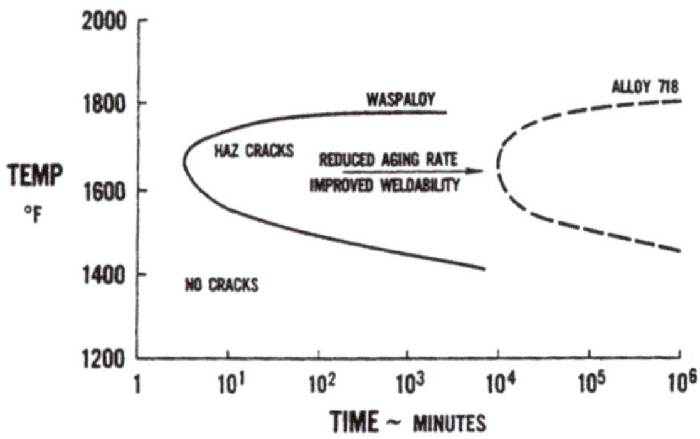


**Fig. 16** C-curve strain age cracking behavior for Waspaloy and Alloy 718 showing much greater tolerance for PWHT cracking for Alloy 718 due to slower aging rate [77]

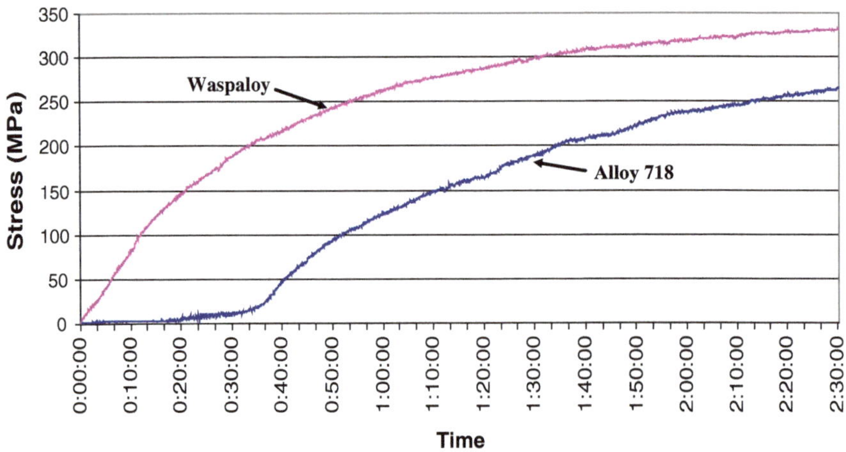


**Fig. 17** Effect of postweld aging time on stress in the simulated HAZ of Waspaloy and Alloy 718 [64]

Norton and Lippold [64] used a Gleeble-based test to study the SAC susceptibility of Waspaloy and Alloy 718. In this test, samples are initially subjected to a HAZ thermal cycle and then cooled to room temperature under restraint, so that considerable room temperature residual stress was present. The sample was then immediately heated into the aging temperature range where the stresses were allowed to relax and precipitation of $\gamma'$ (Waspaloy) or $\gamma''$ (Alloy 718) occurred with hold time. Since the sample was fixed, precipitation resulted in increased stress in the sample, as shown in Fig. 17. Note that the starting residual stress has been subtracted from this data to allow for easier comparison of the aging behavior.

After a predetermined time (up to 4 h) the samples were then pulled to failure at the test temperature and their ductility measured. These tests resulted in the

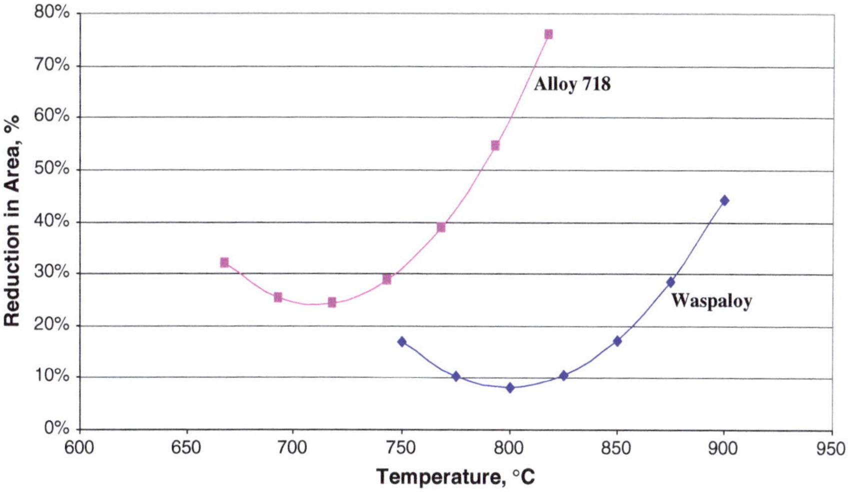


**Fig. 18** Postweld heat treatment ductility curves for Waspaloy and Alloy 718 after 3 h of aging following a simulated HAZ thermal cycle [64]

development of 3-dimensional C-curves based on test temperature, time, and strength/ductility. This data could then be used to generate 2-dimensional, ductility C-curves for specific time-temperature conditions. An example of this for both alloys after 3 h of PWHT is shown in Fig. 18. This data again shows the beneficial effect of $\gamma''$ versus $\gamma'$ precipitation with respect to SAC. Although both alloys show a drop in ductility in the PWHT temperature range, the minimum ductility for Waspaloy is much lower. As a result, much lower strains resulting from stress relaxation are required to promote SAC in Waspaloy relative to Alloy 718.

Analysis of these samples also clearly reveals the nature of SAC in Ni-base superalloys. The photomicrograph in Fig. 19 shows the region near the fracture in a Waspaloy sample. Cracks are intergranular and tend to initiate at grain boundary triple points. Higher magnification examination of the grain boundaries in the SEM revealed no evidence of carbides or continuous carbide precipitation, suggesting that the "transient embrittlement" phenomenon is not occurring in Waspaloy. This is in agreement with the conclusion of Duval and Owczarski [78]. Examination of the fracture surfaces of Waspaloy and Alloy 718 samples indicated that the fracture morphology is either smooth or ductile intergranular. An example of both of these fracture morphologies can be seen on the fracture surface of an Alloy 718 sample in Fig. 20.

Based on previous research, a mechanism for strain-age cracking in Ni-base superalloys is shown schematically in Fig. 21. It can be described in 4 stages, as shown in the schematics A through D in the figure. In ***Stage A*** the HAZ is heated to sub-solidus temperatures. If strengthening precipitates are present in the base metal, most of these will dissolve on heating to the peak temperature. At the highest temperatures in the HAZ, many of the Ni-base superalloys will undergo some grain

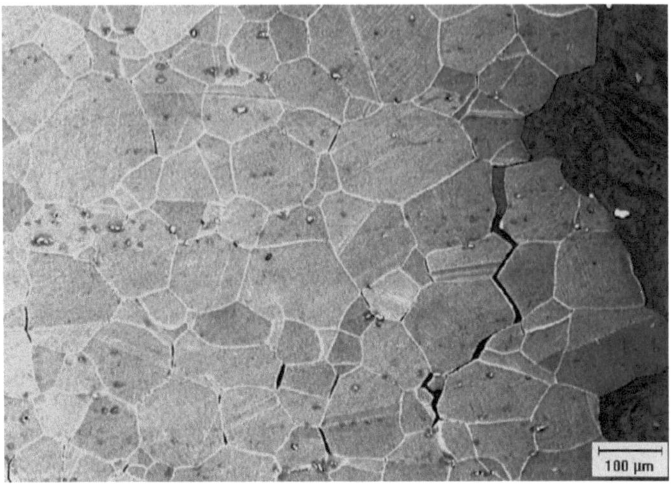


**Fig. 19** Intergranular strain-age cracking in the simulated HAZ of Waspaloy [64]

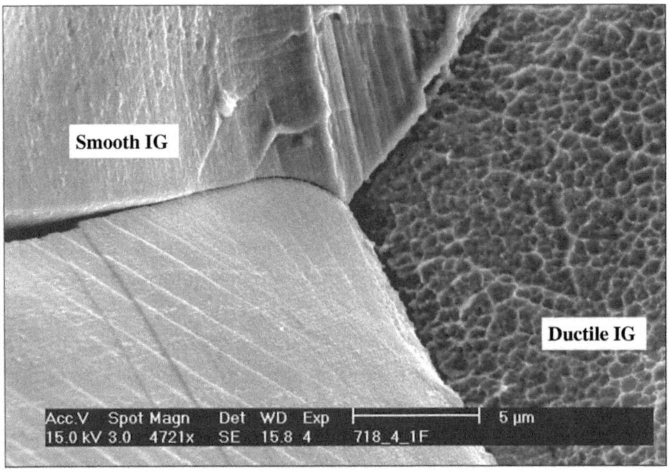


**Fig. 20** Fracture morphology of strain-age cracking in Alloy 718 [64]

boundary liquation just adjacent to the fusion boundary. There will also be some grain growth in the HAZ, the degree of which is determined by the starting base metal microstructure and the HAZ thermal cycle. As the weld cools to room temperature during ***Stage B***, the liquid films solidify (there could also be possible HAZ liquation cracking) and residual stress starts to accumulate in the system. Typically, there is little on-cooling re-precipitation and the HAZ is essentially in the solution annealed condition at room temperature. Upon reheating to the solution annealing temperature, some stress relaxation will occur as the weld is heated above

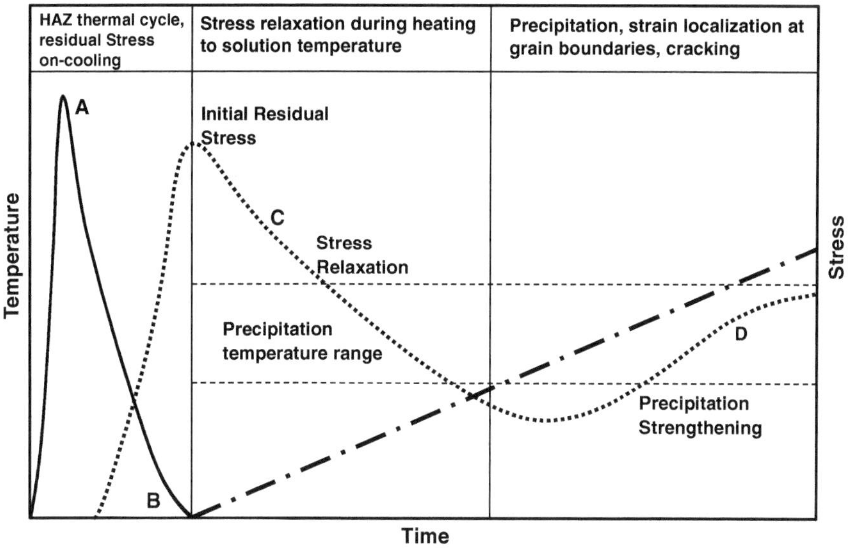


**Fig. 21** Schematic illustration of the strain-age cracking mechanism in Ni-base superalloys

approximately $0.5T_S$, as shown in ***Stage C***. Presumably, the residual stress does not completely relax and upon heating into the precipitation temperature range, additional contraction stresses accumulate due to precipitation. The combination of precipitation and relaxation stresses results in localized stress at the grain boundaries in ***Stage D*** that can lead to intergranular cracking.

## Quantifying Susceptibility to Strain-Age Cracking

There are a number of tests that have been used to quantify susceptibility to strain-age cracking. As with tests developed for other cracking phenomena, the techniques are of either the "self-restraint" or "simulative" (externally loaded) type. Generally speaking, the same techniques used for quantifying reheat cracking and discussed previously are used for strain-age cracking.

Prager and Shira [72] discuss a number of tests for evaluating susceptibility to SAC including the circular patch test, a "plug-weld" test, and a controlled heating rate test. The circular patch test has been widely used to study susceptibility to SAC, but it is a "go-no go" test. The test sample either cracks or it does not. In tests with René 41, they found that this test only identified the most susceptible heats of material. Heats with intermediate susceptibility, which may in fact be crack-susceptible in practice, could not be readily identified. Similar problems are encountered with other self-restraint tests. Another disadvantage of the self-restraint tests is that the samples must often be sectioned in order to determine if cracking has occurred since the cracking may not be evident on the surface.

Because of the disadvantages of the self-restraint tests, a number of tests that either apply a stress or allow stress relaxation to occur have been developed. The controlled heating rate test is essentially a high temperature tensile test where a sample is heated into the aging temperature range and then pulled to failure. This test shows a minimum in ductility over a narrow temperature range. Materials with higher minimum ductility were deemed to be more resistant to SAC. Note that this test was used to evaluate base materials of different composition (and presumably grain size), but did not specifically test the weld HAZ.

Researchers at Haynes International developed a modified version of the controlled heating rate test using a Gleeble™ thermo-mechanical simulator [79, 80]. Thin sheet material machined into reduced gauge tensile samples are first heated to 593 °C at 56 °C/s (1100 °F at 100 °F/s), then heated to 788 °C at 17 °C/s (1450 °F at 30 °F/s), and finally pulled to failure at 788 °C at a rate of 1.6 mm/min (0.063 in./min). Total elongation is used to determine ductility. The test temperature of 788 °C was selected as a compromise among a number of Ni-base superalloys based on minimum ductility in earlier testing. As such, it does not necessarily represent the absolute minimum for any of the materials tested. Typical results from this test are shown in Fig. 22 for two different sample geometries [80]. A solid-solution strengthened alloy (Hastelloy X) was added to the test matrix for comparison purposes.

The Welding Institute in the UK developed a constant load rupture test to evaluate samples in a variety of microstructure conditions, including simulated HAZs [70]. This test used a notched tensile sample that was heated to a predetermined temperature and held under constant load until failure occurs or sample run out (10,000 min). The use of a notch in the sample allowed a critical stress intensity to be determined that could be used to predict failure.

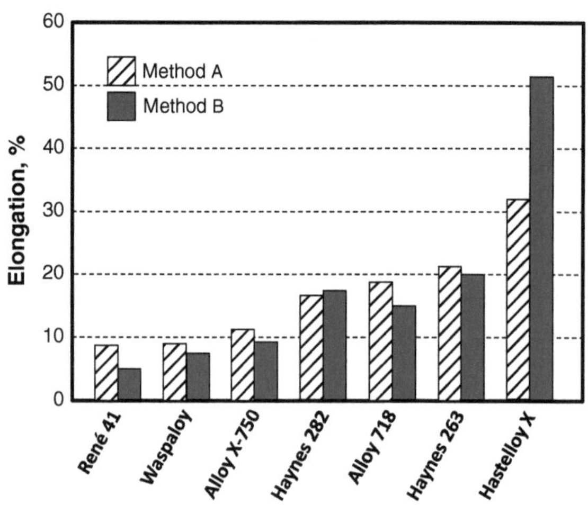


**Fig. 22** Results from the Haynes International controlled heating rate tensile test at 788 °C (1450 °F). Methods *A* and *B* refer to different sample geometries. From Metzler [80]

Many of the other tests for SAC susceptibility used a Gleeble™ thermo-mechanical simulator [61]. Wu and Herfert [74] used the Gleeble to first simulate the HAZ in small René 41 tensile bars. A notch was then machined in the sample and successive samples loaded in constant tension at increasing stress levels (24–33 ksi) until failure occurred. The stress to cause failure was used as a measure of SAC susceptibility. Note that this test did not allow for any stress relaxation since the stress was held constant during the test. Duvall and Owczarski performed a HAZ simulation (including heating in the partially-melted zone) and then used constant displacement at different PWHT temperatures to load the sample [77]. The sample was periodically cooled to room temperature to check for cracking. Using this technique, a C-curve cracking response was developed. Because this test used constant displacement, rather than constant stress, it allows for stress relaxation after the initial stress (displacement) is applied.

Franklin and Savage [69] used a technique similar to Duvall and Owczarski, but included constant displacement control to study the effect of base metal pre-treatment on SAC susceptibility in René 41. Their results on the base metal in the solution annealed condition showed the effect of initial stress relaxation and then precipitation strengthening at longer hold times. If the base metal was tested in the overaged condition, only stress relaxation occurs with no precipitation hardening at longer hold times. They concluded that since precipitates would be dissolved in the high temperature region of the HAZ, there was no advantage to overaging the base metal prior to welding.

Dix and Savage [81] developed a Gleeble-based test to evaluate the ductility response of the base metal as a function of displacement rate over a range of PWHT temperatures. These tests simply heated the sample to a given temperature and then pulled the sample to failure at a fixed displacement rate. For Inconel X-750, a ductility minimum is observed around 1600 °F (870 °C) at both displacement rates and lower displacement rates result in lower ductility over a range of temperature. This reflects the fact that precipitation of $\gamma'$ occurs much more rapidly at 870 °C than at 700 °C and that at the lower displacement rates there is more time for precipitation to occur.

There are a number of problems with the simulative tests for SAC that are described above. Many of them do not include an initial HAZ simulation thermal cycle. Since SAC almost always occurs in the high temperature HAZ, it is necessary to precondition the sample microstructure using a thermal cycle of that experienced in the HAZ. This is a critical consideration since the effect of the base metal microstructure condition is effectively removed in this region and the HAZ prior to PWHT is essentially in the solution annealed condition. Even if the base material is in the solution annealed condition, the HAZ thermal cycle can still lead to grain growth and impurity segregation that can influence SAC susceptibility.

Another problem with some of these tests is that they do not allow for stress relaxation which is a key element in the SAC mechanism. Samples tested under constant load (or stress) or constant extension rate do not allow relaxation to occur. Thus, the fixed displacement tests after a HAZ simulation thermal cycle are the most appropriate in terms of approximating actual conditions. Thus, of the simulative tests described above, that of Duvall and Owczarski [77] most closely simulates the actual conditions under which SAC occurs.

However, even this test ignores one key element critical to SAC. The tests that impose a HAZ thermal cycle to condition the microstructure prior to postweld heat treatment, typically cool the sample back to room temperature under no load. This ignores the fact that under ordinary welding conditions, residual stress accumulates in the HAZ during cooling. Recognizing this, Gleeble-based tests developed by Balaguer et al. [62], Nawrocki et al. [63], and Norton and Lippold [64] have incorporated a displacement during cooling that results in a tensile residual stress at room temperature. The test used by Nawrocki et al. to evaluate stress relief cracking in low alloy steels, maintains this tensile stress on the sample throughout the PWHT cycle. In practice, the tensile residual stress relaxes when the sample is heated to the PWHT temperature. This is the reason why tests using this constant load technique normally fail in relatively short times.

Recognizing this deficiency, the PWHT cracking technique developed by Norton and Lippold at OSU [64] maintains the sample in fixed displacement after cooling to room temperature. Upon heating to the PWHT temperature, the tensile stresses initially relax and with time and then the stress increases as contraction due to precipitation occurs. The rate and magnitude of this increase is indicative of SAC susceptibility, as shown in Fig. 17. Note that the stress vs. time curves shown in this figure represent the stress increase above the minimum relaxation stress after heating to the PWHT temperature.

Another problem encountered with Gleeble-based PWHT cracking tests is the large variation in time to failure when testing duplicate samples. In order to eliminate this variability and reduce testing time, Norton would terminate the test after a pre-determined time (up to 4 h) and pull the sample to failure at the test temperature. Using this approach, ductility curves at specific times and temperatures can be obtained, as shown previously in Fig. 18.

## Preventing Strain-Age Cracking

Strain-age cracking in the precipitation strengthened Ni-base alloys occurs due to the local accumulation of strain and the concomitant hardening of the microstructure due to precipitation. Local strains develop due to both the relaxation of weld residual stresses and precipitation-induced stresses. Non-uniform heating during postweld heat treatment may also contribute some thermally-induced stresses. The combination of straining and hardening due to precipitation is what leads to SAC. To reduce or eliminate SAC the following steps may be taken. A combination of these approaches may be necessary to avoid cracking.

- Minimize residual and thermally-induced stresses by appropriate joint design and choice of welding process and materials. The use of lower weld heat input and smaller weld beads to minimize residual stress is often effective.

- Reduce the strength of the weld metal and/or base metal. The use of a filler metal with lower strength than the base metal and/or a solution annealed base metal reduce residual stress in the HAZ. It has also been shown that overaging the base metal in $\gamma'$-strengthened alloys can reduce cracking susceptibility [78].
- During PWHT, heat as rapidly as possible through the precipitation temperature range (see Fig. 14). This suppresses intragranular precipitation and allows residual stress to relax uniformly in the microstructure. This approach is limited by two factors, (1) the ability to heat rapidly based on component size, and (2) the kinetics of precipitation. As the Ti + Al content increases, the nose of the precipitation curve occurs at much shorter times) which does not allow the suppression of precipitation.
- Avoid partial melting along grain boundaries adjacent to the fusion boundary. Although, it has not been conclusively shown that the presence of a PMZ increases susceptibility, the formation of HAZ liquation cracks clearly contribute to SAC since they can act as initiation point for cracking. Many of the Ni-base alloys are inherently susceptible to HAZ grain boundary liquation/cracking, so the only practical approach is to minimize the weld heat input and promote steeper HAZ temperature gradients which reduce the region over which liquation occurs.
- Minimize HAZ grain size. Since SAC occurs due to strain accumulation at the grain boundary during the period of stress relaxation, a finer grain size in the HAZ will reduce the local strain on individual grain boundaries. This can also be accomplished by minimizing the grain size of the base metal, reducing heat input, or using high energy density (HED) processes that create steep temperature gradients in the HAZ. It should be noted that fine grain size in the base metal may not be desirable in the application of many Ni-base superalloys, since this can compromise the elevated temperature creep properties.
- Select alloys with lower Ti + Al content. This is an obvious solution that is often not practical since the alloy was selected to meet elevated temperature properties that require high Ti + Al content. Lower Ti + Al content will allow slower heating to the solution annealing temperature without the onset of precipitation. The use of alloys that substitute Nb as a strengthening agent (such as Alloy 718) and form $\gamma''$, $Ni_3Nb$, are known to be resistant to SAC.

## Summary

A number of elevated-temperature, solid-state cracking phenomena that can occur in either the weld metal or HAZ have been reviewed. In general, the evolution of microstructure during welding and subsequent reheating, and the level of residual stress are important to all these mechanisms. Avoiding elevated temperature cracking is possible in most instances, either through control of composition or microstructure. Reducing residual stresses or controlling stress relaxation during reheating is also critical, but control of these factors is more difficult.

**Acknowledgements** This paper reflects the contributions of many graduate students and postdocs who I have had the pleasure of interacting with at Ohio State University. These individuals include Dr. Antonio Ramirez, Dr. Nathan Nissley, Mr. Matt Collins, and Dr. Seth Norton. Much of this paper was excerpted from a book entitled **Welding Metallurgy and Weldability** published by John Wiley and Sons, Inc.

## References

1. Bengough, G.D.1912. A study of the properties of alloys at high temperatures, *Journal of the Institute of Metals*, VII: pp. 123-174.
2. Rhines, F.N. and P.J. Wray. 1961. Investigation of the intermediate temperature ductility minimum in metals, *Transactions of the ASM*, 54:117-128.
3. Mintz, B., Abu-Shosa, R., and Shaker, M. 1993. Influence of deformation induced ferrite, grain boundary sliding, and dynamic recrystallization on hot ductility of 0.1-0.75 wt% c steels, *Mat. Sci. and Tech.*, 9:907-914
4. Hemsworth, B., T. Boniszewski, and N.F. Eaton. 1969. Classification and definition of high temperature welding cracks in alloys. *Metal Construction & British Welding Journal,* 16 (1):5-16.
5. Cordea, J.N., Kammer, P.A., and Martin, D.C. 1964. Causes of fissuring in Ni-base and stainless steel alloy weld metals, *Welding Journal*, 43(11):481s-491s.
6. Haddrill, D.M. and Baker, R.G. 1965. Microcracking in austenitic weld metal, *Brit. Weld. Jour.*, 12(8):411-418.
7. Lippold, J.C. and Ramirez, A.J. 2003. Elevated temperature grain boundary embrittlement and ductility-dip cracking of Ni-base weld metals, *Proc. Conf. on Vessel Penetration, Inspection, Cracking, and Repairs*, NRC, Gaithersburg, MD, Sept. 29-Oct 2, 2003.
8. Nissley, N.E, Collins, M.G., Guaytima, G., and Lippold, J.C. 2002. Development of the Strain-to-Fracture Test for Evaluating Ductility-Dip Cracking in Austenitic Stainless Steels and Ni-base Alloys, *Welding in the World*, 46, 7/8:32-40.
9. Lippold, J.C. and Nissley, N.E. 2007. Further investigations of ductility-dip cracking in high chromium, Ni-base Filler Metals, *Welding in the World*, 51, 9/10: 24-30.
10. DuPont, J.N., Lippold, J.C. and Kiser, S.D. 2009. *Welding Metallurgy and Weldability of Nickel Base Alloys*, pub. by Wiley and Sons, Inc. Hoboken, NJ, ISBN 978-0-470-08714-5, pp. 138-142.
11. Ramirez, A.J. and Lippold, J.C. 2004. High temperature cracking in nickel-base weld metal, Part 2– Insight into the mechanism, *Materials Science and Engineering A*, 380: 245-258.
12. Noecker II, F.F. and DuPont, J.N. 2009. Metallurgical investigation into ductility dip cracking in Ni-based alloys, Part I, *Welding Journal*, 88(1):7s-20s.
13. Noecker II, F.F. and DuPont, J.N. 2009. Metallurgical investigation into ductility dip cracking in Ni-based alloys, Part II, *Welding Journal*, 88(3):62s-77s.
14. Yamaguchi, S. 1979. Effect of Minor Elements on Hot Workability of Nickel-Base Superalloys, *Met. Technol.*,6(5):170-175.
15. Matsuda, F.1984. Weldability of Fe-36 % Ni alloy, II.-effect of chemical composition on reheated hot cracking in weld metal, *Trans. JWRI*, 13(2):241-247.
16. Nishimoto, K., Saida, K., and Okauchi, H. 2006. Microcracking susceptibility in reheated weld metal – Microcracking in multipass weld metal of Alloy 690 (Part 1), *Sci. and Tech. of Welding and Joining*, 11(4):455-461

17. Nishimoto, K., Saida, K., Okauchi, H. and Ohta, K. 2006. Microcracking susceptibility in reheated weld metal – Microcracking in multipass weld metal of Alloy 690 (Part 2), *Sci. and Tech. of Welding and Joining*, 11(4):462-470.
18. Nishimoto, K., Saida, K., Okauchi, H. and Ohta, K. 2006. Microcracking susceptibility in reheated weld metal – Microcracking in multipass weld metal of Alloy 690 (Part 3), *Sci. and Tech. of Welding and Joining*, 11(4):471-479
19. Collins, M.G. and Lippold, J.C. 2003. An investigation of ductility-dip cracking in nickel-based filler metals - Part I, *Welding Journal*, 82(10):288s-295s.
20. Zhang, Y.C., H. Nakagawa, and F. Matsuda. 1985. Weldability of Fe-36 %Ni Alloy (Report VI), *Trans. JWRI*, 14(5):125-134.
21. Zhang, Y.C., H. Nakagawa, and F. Matsuda. 1985. Weldability of Fe-36 %Ni Alloy (Report V), *Trans. JWRI*, 14(2):119-124.
22. Young, G.A, Capobianco, T.E, Penik, M.A, Morris, B.W., and McGee, J.J., 2008. The mechanism for ductility dip cracking in nickel-chromium alloys, *Welding Journal*, 87 (2):31s-43s.
23. Collins, M.G. and Lippold, J.C. 2003. An investigation of ductility-dip cracking in Ni-base filler metals-Part 1, *Welding Journal*, 82(10):288s-295s.
24. Collins, M.G. and Lippold, J.C. 2003. An investigation of ductility-dip cracking in Ni-base filler metals-Part 2, *Welding Journal*, 82(12):348s-354s.
25. Collins, M.G., Ramirez, A.J. and Lippold, J.C. 2004. An investigation of ductility-dip cracking in Ni-base filler metals-Part 3, *Welding Journal*, 83(2):39s-49s.
26. Ramirez, A.J. and Lippold, J.C. 2004. High temperature cracking in nickel-base weld metal, Part 1– Ductility and Fracture Behavior, *Materials Science and Engineering A*, 380:259-271
27. Ramirez, A.J. and Lippold, J.C. 2005. New Insight into the Mechanism of Ductility Dip Cracking in Ni-base Weld Metals, *Hot Cracking Phenomena in Welds*, ISBN 978-3-540-22332-0, publ. by Springer, pp.19-41
28. Lippold, J.C. and Nissley, N.E. 2008. Ductility dip cracking in high-Cr Ni-base filler metals, *Hot Cracking Phenomena in Welds II*, ISBN 978-3-540-78627-6, publ. by Springer, pp. 409-426.
29. Nissley, N.E. and Lippold, J.C. 2008. Ductility-dip cracking susceptibility of Ni-based weld metals-Part 1, *Welding Journal*, 87(10):257s-264s.
30. Nissley, N.E. and Lippold, J.C. 2009. Ductility-dip cracking susceptibility of Ni-based weld metals, Part 2 – Microstructural Characterization, *Welding Journal*, 88(6):131s-140s.
31. Alexandrov, B.T.,Hope, A.T, Sowards, J.W., McCracken, S. and Lippold, J.C. 2011. Weldability studies of high-Cr, Ni-base filler metals for nuclear applications, *Welding in the World*, Vol. 55, No. 3/4, pp. 65-76.
32. Matsuda, F. 1990. Hot crack susceptibility of weld metal, *Proc. of 1st US-Japan Symposium on Advances in Welding Metallurgy*, published by the American Welding Society, pp. 19-36.
33. Lippold, J.C. and Lin, W. 1994. Unpublished research performed at Edison Welding Institute.
34. Kikel, J.M. and D.M. Parker. 1999. Ductility dip cracking susceptibility of filler metal 52 and Alloy 690, *Trends in Welding Research V*, ASM International, pp. 757-762.
35. Nissley, N.E. and Lippold, J.C. 2003. Development of the strain-to-fracture test for evaluating ductility-dip cracking in austenitic alloys, *Welding Journal*, 82(12):355s-364s.
36. Meitzner, C.F. 1975. Stress relief cracking in steel weldments, *WRC Bulletin*, No. 211.
37. Dhooge, A. and Vinckier, A. 1986. Reheat cracking – a review of recent studies, *Welding in the World*, 24, 5/6:2-18.
38. Dhooge, A. and Vinckier, A. 1992. Reheat cracking – a review of recent studies (1984-1990), *Welding in the World*, 30, 3/4:44-71.
39. Tamaki, K., Suzuki, J., Nakaseka, Y., and Tajiri, M. 1984. Effects of carbides on reheat cracking sensitivity, *Trans. JWS*, 15(1).
40. Tamaki, K. and Suzuki, J. 1983. Effect of chromium and molybdenum on reheat cracking sensitivity of steels, *Trans. JWS*, 14(2).
41. Tamaki, K. and Suzuki, J. 1984. Combined influences of phosphorus, chromium and molybdenum on reheat cracking of steels, *Trans. JWS*, 15(2).

42. Tamaki, K., Suzuki, J. and Tajiri, M. 1984. Effect of vanadium and titanium on reheat cracking sensitivity, *Trans. JWS*, 15(1).
43. Tamaki, K. and Suzuki, J. 1983. Reheat cracking test on high strength steels by modified implant test, *Trans. JWS*, 14(2).
44. McMahon, Jr., C.J. and Shin, J. 1984. Comparison of stress relief cracking in A508, C12 and A533B pressure vessel steels, *Met. Sci.*, 8(18).
45. Watanabe, T. and Savage, W.F. 1984. A study of reheat cracking in weld heat affected zone of high strength steel, *Trans. Natn. Res. Inst. Metals Japan*, 26(4).
46. Kikuchi, T. and Nakao, Y. 1982. Effect of impurity elements on the reheat cracking in the weld zones of steel, *Proc. 4th Int. JWS Symposium*, Osaka, Japan, September 1982.
47. Thomas, Jr., R.D. 1984. HAZ cracking in thick sections of austenitic stainless steels-Part 1, *Welding Journal*, 63(9):24-32.
48. Thomas, Jr., R.D. 1984. HAZ cracking in thick sections of austenitic stainless steels-Part 2, *Welding Journal*, 63(9):355s-368s.
49. Curran, R.M., Rankin, A.W. 1955. Welding Type 347 stainless steel for 1100°F turbine operation, *Welding Journal*, 34(3 ):205-213.
50. Christoffel, R.J. 1962. Cracking in Type 347 heat-affected zone during stress relaxation, *Welding Journal,* 41(6):251s-256s.
51. Christoffel, R.J. 1960. Notch-rupture strength of Type 347 heat-affected zone, *Welding Journal*, 39(7):315s-320s.
52. Younger, R.N, Haddrill, d.M. and Baker, R.G. 1963. Post-weld heat treatment of high-temperature austenitic steels, *JISI*, 201:693-698.
53. Thomas, Jr., R.D. and Messler, Jr., R.W. 1997. Welding Type 347 stainless steel – an interpretive report, *WRC Bulletin*, No. 421.
54. Van Wortel, H. 2007. Control of relaxation cracking in austenitic high temperature components, *NACE 2007*, pp.2216-2228.
55. Kiso, T., Seshimo, I. and Okazaki, T. 2011. Cracking in welds of heavy wall nickel alloy piping during fabrication, *NACE 2011*.
56. Nishimoto, K., Matsunaga, T., Tanaka, T., and Okazaki, T. 1998. Effect of bismuth on reheat cracking susceptibility in Type 308 FCAW weld metal, *Welding in the World*, 41:220-235.
57. Van Wortel, H. 2007. Control of relaxation cracking in austenitic high temperature components, *NACE 2007*, pp.2216-2228.
58. Pense, A.W., Galda, E.J, and Powell, G.T. 1971. Stress relief cracking in pressure vessel steels, *Welding Journal*, 50(8):374s-378s.
59. Tamaki, K., and Suzuki, J. 1983. RHC test on high strength steel by a modified implant test, *Trans. JWS*, 14(2);33-38.
60. Younger, R.N. and Baker, R.G. 1960. Heat-affected zone cracking in welded high-temperature austenitic steels, *JISI*, 196:188-194.
61. Gleeble is the registered trademark of Dynamic Systems Inc., Poestenkill, NY., http://gleeble.com.
62. Balaguer, J.P., Wang, Z. and Nippes, E.F. 1989. Stress relief cracking of copper containing HSLA steel, *Welding Journal*, 68(4):121s-131s.
63. Nawrocki, J.G., Dupont, J.N., Robino, C.V., Puskar, J.D. and Marder, A.R. 2003. The mechanism of stress-relief cracking in a ferritic alloy steel, *Welding Journal*, 82(2):25s-35s.
64. Norton, S.J. and Lippold, J.C. 2003. Development of a Gleeble-based Test for Postweld Heat Treatment Cracking Susceptibility, *Trends in Welding Research VI*, publ. by ASM International, pp. 609-614.
65. Haure, J. and Bocquet, P. 1975. Fissuration sous les revêtments inoxyables des pièces pour cuvées sous pression (Cracking below stainless steel cladding under tension), Convention No. 6210-75/3/303, Creusot Loire, Sept. 1975.
66. S. Kou, Welding Metallurgy, 1st Edition, published by Wiley Interscience, Inc.
67. Berry, T.F. and Hughes, W.P. 1969.A study of the strain-age cracking characteristics in welded René 41 – Phase II, *Welding Journal*, 46(11):505s-513s.

68. Baker, R.G. and Newman, R.P. 1969. Cracking in welds, *Metal Construction and Br. Weld. J.*, 1:1-4.
69. Franklin, J.G. and Savage, W.F. 1974. Stress Relaxation and Strain-Age Cracking in René 41 Weldments, *Welding Journal*, 53(9):380s-387s.
70. McKeon, D. 1971. Re-heat cracking in high nickel alloy heat-affected zones, *Welding Journal*, 50(5):201s-205s.
71. Nakao, Y. 1988. Study on reheat cracking of Ni-base superalloy, Waspaloy, *Trans. JWS,* 19 (1):66-74.
72. Prager, M. and Shira, C. S. 1968. Welding of precipitation-hardening nickel-base alloys, Welding Research Council Bulletin 128.
73. Younger, R.N. and Barker, R.G. 1961. Heat-affected zone cracking in welded austenitic steels during heat treatment, *Brit. Weld. Jour.* 8(12):579-587.
74. Wu, K. C., and Herfert, R. E. 1967. Microstructural studies of René 41 simulated weld heat-affected zones, *Welding Journal* 46(1): 32s-38s.
75. Weiss, S., Hughes, W. P., and Macke, H. J. 1962. Welding evaluation of high temperature sheet materials by restraint patch testing, *Welding Journal* 41(1): 17s-22s.
76. Hughes, W. P., and Berry, T. F. 1967. A study of the strain-age cracking characteristics in welded Rene 41-Phase I, *Welding Journal* 46(8): 361s-370s.
77. Duvall, D. S., and Owczarski, W. A. 1969. Studies of postweld heat-treatment cracking in nickel-base alloys, *Welding Journal* 48(1): 10s-22s.
78. Duvall, D. S., and Owczarski, W. A. 1971. Heat treatments for improving the weldability and formability of Udimet 700, *Welding Journal* 50(9): 401s-409s.
79. Rowe, M.D. 2006. Ranking the resistance of wrought superalloys to strain-age cracking, *Welding Journal*, 85(2)27s-34s.
80. Metzler, D.A. 2008. A Gleeble®-based methodology for ranking the strain-age cracking susceptibility of Ni-based superalloys, *Welding Journal*, 87(10):249s-256s.
81. Dix, A.W. and Savage, W.F. 1971. Factors influencing strain-age cracking in Inconel X-750, *Welding Journal*, 50(6):247s-252s.
82. Nakamura, H., Naiki, T., and Okabayashi, H. 1970. Relation between stress-relief cracking and metallurgical properties of low alloy steels, *Trans. JWS*, 1(2):60-71.
83. Haure, J. and Bocquet, P. 1975. Fissuration sous les revêtments inoxyables des pièces pour cuvées sous pression (Cracking below stainless steel cladding under tension), Convention No. 6210-75/3/303, Creusot Loire, Sept. 1975.
84. Ito, Y. and Nakanishi, M. 1972. Study on stress relief cracking in welded low alloy steels, IIW Doc. X-668-72.
85. Brear, J.M. and King, B.L. 1980. An assessment of the embrittling effects of certain residual elements in two nuclear pressure vessel steels (A533B, A508), *Phil. Trans. Royal Soc. London*, A 295:291.
86. Hrivnak, I., Magula, A., Zajac, J., and Smida, T., 1985. Mathematical evaluation of steel resistance to reheat cracking, IIW Doc. IX-1346-85.

# Part IV
# Hot Cracking Susceptibility of Ni-Base Alloys

# Weldability Evaluation of High Chromium, Ni-Base Filler Metals Using the Cast Pin Tear Test

**Eric Przybylowicz, Boian Alexandrov, John Lippold and Steven McCracken**

**Abstract** High chromium, nickel-base filler metals have been commonly used throughout the nuclear power industry for the weld overlay repair of dissimilar metal welds. These alloys provide optimum resistance to primary water stress corrosion cracking in nuclear power plant cooling systems. However some of these nickel alloys present weldability challenges including susceptibility to solidification cracking and ductility dip cracking. ERNiCrFe-7A (52M) and ERNiCrFe-13 (52MSS) filler metals, including two heats of 52M and one heat of both 52MSS and 690Nb, have been evaluated in this study. The susceptibility to solidification cracking was evaluated using the cast pin tear test (CPTT). The CPTT was also used to evaluate the effect of dilution between two heats of ERNiCr-3 (FM82) on the solidification cracking behavior. Metallurgical characterization using light optical microscopy, scanning electron microscopy (SEM), and energy dispersive spectroscopy (EDS) in the SEM has been performed in order to identify solidification cracking mechanisms, and to study the effect of liquid film formation and backfilling on cracking susceptibility.

## Introduction

### *Motivation*

Ni-base alloy 600 filler metals, Alloy 182 and 82, have been commonly used in pressurized water reactors (PWRs) to join stainless steel piping to low alloy steel (LAS) components [1]. The LAS components, typically vessel nozzles or valve

E. Przybylowicz (✉) · B. Alexandrov · J. Lippold
Welding Engineering Program, Department of Materials Science and Engineering,
The Ohio State University, Columbus, USA
e-mail: eric.przybylowicz@gmail.com

S. McCracken
Electric Power Research Institute, Palo Alto, USA

 
Th. Boellinghaus et al. (eds.), *Cracking Phenomena in Welds IV*,
DOI 10.1007/978-3-319-28434-7_13

bodies, have a coefficient of thermal expansion (CTE) which differs significantly from that of stainless steel; Alloys 182 and 82 provide a CTE which bridges this mismatch [2, 3]. In addition to its thermo-mechanical properties, Alloy 600 was selected for this application due to its inherent corrosion resistance, although over the last 30 years primary water stress corrosion cracking (PWSCC) has be observed in reactor cooling system components welded with this alloy.

Stress corrosion cracking of metals requires three factors to occur: environment, material conditions, and stress [2]. The susceptibility of Alloy 600 and its filler metals are dependent on factors which include: chemical composition, metallurgical condition, and heat treatment. The element which appears to influence SCC most significantly is Chromium, which effects the formation of a protective Cr oxide layer and carbide precipitation during material processing [4, 5]. Other factors such as grain size, degree of cold work, and residual and operating stresses play a role in Alloy 600 PWSCC susceptibility. In order to eliminate PWSCC one of the three controlling factors must be addressed. The stress component can be addressed through the application of a weld overlay covering the length of the nozzle connection. An overlay such as this would apply a compressive stress at the crack tips which already exist in the connection due to thermal contraction upon cooling; inhibiting further crack propagation and initiation. Another approach to mitigating this problem is to look at the alloy which is used to overlay the connection. By increasing the Chromium from 22 % (wt) to 30 % (wt) the susceptibility of the overlay to PWSCC can be significantly reduced if not eliminated. This increase in Cr content has been shown to produce oxides with better mechanical properties and higher passivation rates as well as reduce Cr depletion after carbide precipitation [6]. Alloy 600 requires a 10–15 h period at elevated temperature in order to replenish the chromium depleted regions created during carbide precipitation whereas Alloy 690, having higher chromium content requires significantly less time, and is virtually PWSCC immune if the carbon content is maintained below 0.02 % (wt) [3, 7, 8].

Repair and replacement of these connections since the 1980s have mainly utilized Alloy 690 and its filler metals [7]. These high-Cr Ni-base filler metals, 52 (ERNiCrFe-7) and 52M (ERNiCrFe-7A), have been implemented in nuclear applications where PWSCC resistance is required. The repair is done by depositing a multi-pass overlay across the entire connection from the low alloy steel nozzle to the connecting stainless steel pipe, Fig. 1.

Although PWSCC resistant, weld filler metals 52 and 52M have been found to exhibit a higher susceptibility to ductility-dip cracking (DDC) compared to filler metal 82 [9].

Filler metal 52MSS is the next generation of high-Cr Ni-base filler metal containing additions of 2.5 % (wt) Niobium and 3.5 % (wt) Molybdenumn. Strain-to-fracture testing of this alloy has shown that these elemental additions significantly improve the material's resistance to DDC [9]. Single sensor differential thermal analysis (SSDTA) as well as previous work performed on the cast pin tear test (CPTT) indicates that the additions of niobium result in a widened solidification temperature range and an increased susceptibility to solidification cracking. This

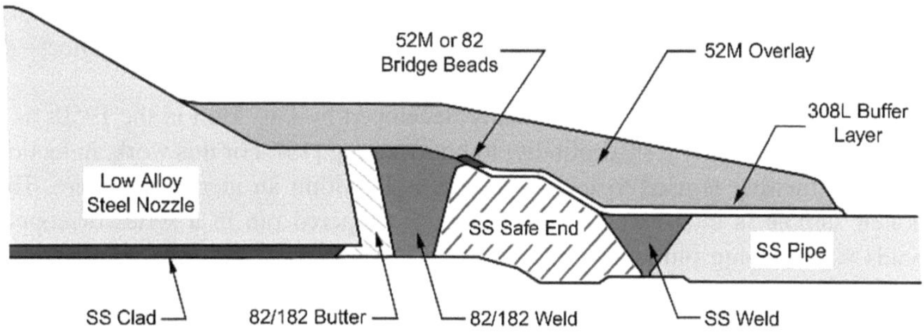


**Fig. 1** Cross section of typical 52M weld overlay (WOL) on a vessel nozzle-to-pipe dissimilar metal connection [1]

can be attributed to low melting point NbC eutectic and laves phase which forms in high Niobium bearing Ni-base alloys [10–12].

## *Background*

A variety of tests which evaluate a material's susceptibility to solidification cracking exist today. These test methods create strain conditions by self-restraint or external-restraint. Self-restrained solidification cracking tests use the thermo-mechanical reaction of the material to induce strain accumulation during solidification. In other words the shrinkage of the specimen itself results in strain during solidification. One example of a self-restrained solidification cracking test is the Cast Pin Tear Test. Externally-restrained tests utilize external mechanical restraint such as bending or applying a tensile load in addition to any intrinsic strain in the material. Some examples of externally-restrained solidification cracking tests are: Hot Tensile test, Programmable Deformation Rate test (PVR), and Transvarstraint test [13, 14].

The essence of the Cast Pin Tear Test is that as molten material is poured into the room temperature mold the heating of the mold material results in expansion while the cooling of the charge material results in contraction. As a result tensile strain accumulates in the sample as it solidifies. If this strain exceeds the ductility of liquid films present along grain boundaries during solidification, cracking will result. In other words longer pin lengths result in larger accumulations of tensile strain which result in larger cracking responses. Alloys are tested over a series of pin lengths and the cracking response at each length is observed and recorded. Material ranking is done according to the maximum pin length at which 0 % cracking consistently occurs and the minimum pin length at which 100 % cracking occurs. Using these criteria alloys can be ranked and their relative susceptibilities to solidification cracking compared. This test method offers a quick and economical means for determining solidification cracking susceptibilities of different material compositions. The speed and small amount of material required to rank an alloy makes the Cast Pin Tear Test a viable evaluation method for ranking existing materials, studying the effects of dilution, and the development of new materials.

The Cast Pin Tear Test was originally developed by F.C. Hull in the 1950s as a means to rank an alloy's susceptibility to hot cracking [15]. For this work, induction levitation melting is used to melt a 19 g sample within an inert atmosphere. The molten sample is then cast into the shape of a tapered pin in a series of copper molds with varying pin length, Fig. 2.

Several other versions of this test have been developed and used to rank alloy susceptibilities to hot cracking: such as the Threaded Mold Weldability Test [16], the Grooved Copper Mold Test [17], and two iterations of the Cast Pin Tear Test developed at The Ohio State University [11, 18–20]. The second generation Cast Pin Tear Test was developed at The Ohio State University and bears a resemblance to Hull's method of testing. Rather than induction levitation melting, the original second generation test utilized a retractable gate in the bottom of a cooled copper hearth on which a sample was melted using a gas-tungsten arc torch and released into the mold below. The newest generation of the test incorporates induction levitation melting of the sample in a cooled copper coil. The samples are prepared using a gas-tungsten arc button melting device within an argon purged chamber. The "buttons" are then induction levitation melted into a series of molds ranging in

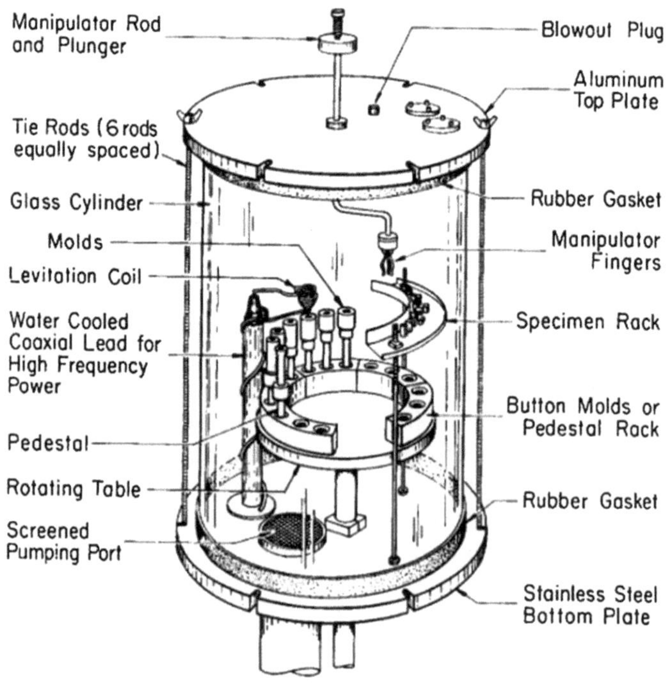


**Fig. 2** CPTT apparatus developed by Hull [15]

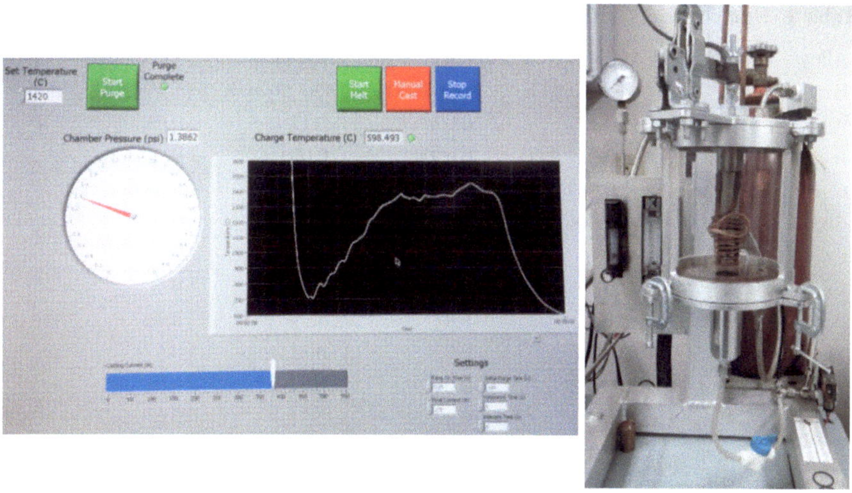


**Fig. 3** New Generation CPTT control screen (*left*), New Generation CPTT apparatus (*right*)

length from 0.5 to 2.5 in. (12.7–63.5 mm) with a 0.125 in. (3.2 mm) step between each length. From the cracking response of an alloy over the range of pin lengths, hot cracking susceptibility rankings are established (Fig. 3).

The objective of this study is to evaluate the repeatability of the new generation Cast Pin Tear Test and use it to develop solidification cracking susceptibility rankings for two heats of 52M, one heat of 52MSS-C, one heat of TG-SN690Nb, and one heat of filler metal 82. The Cast Pin Tear Test was also used to evaluate the effect of dilution on hot cracking susceptibility between two heats of filler metal 82. Filler metal $82_{(A)}$, which has been found in practice to be resistant to solidification cracking, is diluted at two different levels with filler metal $82_{(B)}$, which has been found to be more susceptible to solidification cracking.

## Materials and Procedures

### *Materials*

Six high-Cr Ni-base filler metals were evaluated in this study: two heats of (ERNiCr-3) Alloy 82, two heats of (ERNiCrFe-7A) Alloy 52M, Alloy (ERNiCrFe-13) 52MSS, (ERNiCrFe-7A) Alloy TG-SN690Nb. The compositions of these alloys are provided in Table 1. The composition for Alloy 82 is representative of AWS 5.14 ERNiCr-3 due to an unknown nominal composition for each heat.

**Table 1** Heat composition of tested alloys (wt%)

| Alloy | Alloy 82[a] | 52M | 52M | 52MSS-C | TG-SN690Nb |
|---|---|---|---|---|---|
| Heat | | NX7206TK | NX0T85TK | NX77W3UK | FBH2280 |
| Al | – | 0.12 | 0.09 | 0.13 | 0.01 |
| B | – | 0.001 | 0.0004 | 0.001 | 0 |
| C | 0.10 max | 0.017 | 0.02 | 0.023 | 0.03 |
| Co | – | 0.006 | 0.011 | 0.01 | 0 |
| Cr | 18–22 | 30.01 | 30.11 | 29.24 | 29.7 |
| Cu | – | 0.03 | 0.03 | 0.05 | 0 |
| Fe | 3 max | 8.51 | 8.87 | 8.79 | 8.4 |
| Mn | 2.5–3.5 | 0.72 | 0.72 | 0.31 | 0.5 |
| Mo | – | 0.01 | 0.05 | 3.51 | 0.01 |
| Nb | 2–3 | 0.81 | 0.87 | 2.51 | 0.6 |
| Ni | 67 min | 59.41 | 58.95 | 55.11 | 59.84 |
| P | 0.03 max | 0.003 | 0.002 | 0.004 | 0 |
| S | 0.015 max | 0.001 | 0.0005 | 0.00005 | 0.001 |
| Si | – | 0.12 | 0.11 | 0.11 | 0.21 |
| Ta | – | 0.01 | 0.01 | 0.01 | 0 |
| Ti | 0.75 max | 0.21 | 0.16 | 0.18 | 0.7 |
| Zr | – | 0.01 | 0.001 | 0.007 | 0 |

[a]Composition from AWS 5.14 ERNiCr-3

## *Procedure*

Solidification cracking susceptibility rankings were developed using the new generation Cast Pin Tear Test developed at The Ohio State University. The materials to be tested are cleaned and weighed to masses corresponding to different pin lengths, Table 2 [19].

Buttons of the test alloy are melted in an argon purged chamber, using a gas-tungsten arc torch over a cooled copper hearth. The buttons are placed inside the copper coil of the new generation cast pin tear tester, induction levitation melted, and cast into rigid molds of varying length (Fig. 4).

The cooling rate of the samples is controlled by the mold material; the molds in this study are made of an Al-bronze alloy (C630000) which provides a cooling rate comparable to conventional GTAW [20]. The stress level which accumulates in the sample is controlled by the pin length, and at a threshold pin length solidification cracking occurs (typically near the head of the pin). Upon further increase in pin length, cracking will increase until 100 % circumferential cracking is observed. The pins are examined using a binocular microscope at a magnification of up to 70×.

**Table 2** Mass of Ni-base material required to fill various mold lengths

| | | | | | | | | | |
|---|---|---|---|---|---|---|---|---|---|
| Length (in./mm) | 0.5/12.7 | 0.625/15.875 | 0.75/19.05 | 0.875/22.225 | 1/25.4 | 1.125/28.575 | 1.25/31.75 | 1.375/34.925 | 1.5/38.1 |
| Mass (g) | 9.5 | 10 | 10.5 | 11 | 11.5 | 12 | 12.5 | 13 | 13.5 |
| Length (in./mm) | 1.625/41.275 | 1.75/44.45 | 1.875/47.625 | 2/50.8 | 2.125/53.975 | 2.25/57.15 | 2.375/60.325 | 2.5/63.5 | |
| Mass (g) | 14 | 14.5 | 15 | 15.5 | 16 | 16.5 | 17 | 17.5 | |

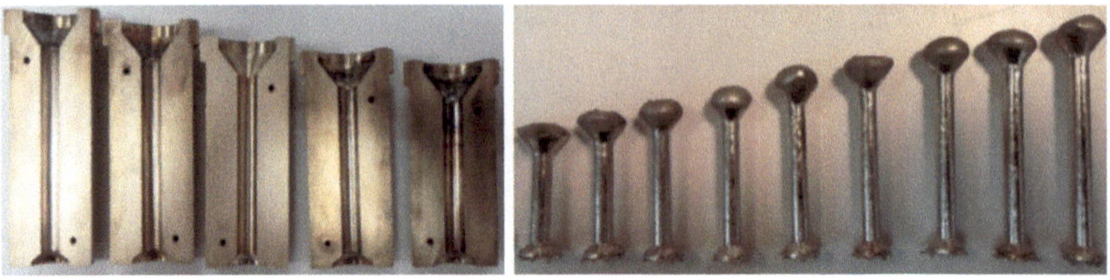

**Fig. 4** Al-Bronze molds (C630000) and 52M cast pins

Crack length is measured circumferentially around the pin in degrees and calculated using the equation below.

$$\% \, Cracking = \frac{L_T^0}{360^\circ} \times 100$$

$L_T$: Total length of all cracks measured on pin surface

Alloy susceptibility to solidification cracking is then characterized by the maximum pin length with 0 % cracking, or a maximum cracking of under 10 %, and minimum pin length with 100 % cracking. Rankings between alloys and their dilutions are generated by plotting the maximum circumferential cracking (MCC) response versus pin length. The maximum circumferential cracking response is the largest measured circumferential cracking value in at least four samples at a given pin length. The parameters used during button melting and cast pin tear testing are provided in Table 3.

**Table 3** Cast Pin Tear Test parameters

| Parameter | Value |
|---|---|
| Shielding gas | Ar |
| Arc length (in./mm) | 0.6/12.7 |
| Arc current (A) | 125 |
| Arc time (s) | 5–0 |
| Mold material | C630000 |
| Set casting temperature (°C) | 1450 |
| Casting current (A) | 385 |
| Ramp down (A) | 275 |
| Ramp down (s) | 0.85 |
| Gas flow rate (CFH/L/min) | 2.5/1.18 |
| Pressure (psi/kPa) | 0.3/2.07 |
| Mold lengths (in./mm) | 0.5–2.5/12.7–63.5 |

## Results and Discussion

### *Reproducibility Study*

A reproducibility study of the new generation CPTT was conducted using one heat of filler metal 52M (heat: NX7206TK). Pins were cast in lengths ranging from 0.75 to 2.375 in. (19.0–60.3 mm). The results of this study are provided in Table 4.

Six or more pins were cast at lengths of 0.75 in. (19.1 mm), 0.875 in. (22.2 mm), and 1.75 in. (44.5 mm) and the standard deviation of their cracking response calculated. The pins cast in the 0.75 in. (19.1 mm) molds were found to have a standard deviation of 2.1 %, the 0.875 in. (22.2 mm) molds 1.60 % and 1.75 in. (44.5 mm) molds 8.99 %. These results indicate that the percent cracking is a reproducible criterion by which alloys can be ranked using the Cast Pin Tear Test. Fewer tests were performed at the intermediate pin lengths due to a large amount of scatter in the cracking response of the pins. Figure 5 shows a graphical representation of Table 4.

The circumferential cracking response of each individual pin (red diamonds) can be seen in Fig. 5 as well as the averages of all the pins (blue circles) at each length. The $R^2$ values for all the pins and their averages at each length are calculated. The results: 90.33 % for all pins and 97.52 % for the averages indicate that the trend line fits the cracking response data very well.

**Table 4** CPTT results for Alloy 52M heat: NX7206TK

| Length (in./mm) | # of pins | Min. cracking (%) | Max. cracking (%) | Avg. cracking (%) | Standard deviation (%) |
|---|---|---|---|---|---|
| 0.75/19.1 | 8 | 0 | 5 | 1 | 2.10 |
| 0.875/22.2 | 6 | 0 | 4 | 1 | 1.60 |
| 1/25.4 | 4 | 3 | 17 | 10 | |
| 1.125/28.6 | 4 | 12 | 50 | 24 | |
| 1.25/31.8 | 4 | 0 | 100 | 34 | |
| 1.375/34.9 | 3 | 25 | 49 | 37 | |
| 1.5/38.1 | 4 | 51 | 100 | 73 | |
| 1.625/41.3 | 3 | 28 | 83 | 60 | |
| 1.75/44.5 | 9 | 78 | 100 | 91 | 8.99 |
| 1.875/47.6 | 3 | 65 | 100 | 88 | |
| 2/50.8 | 3 | 100 | 100 | 100 | |
| 2.125/54.0 | 1 | 100 | 100 | 100 | |
| 2.25/57.2 | 2 | 100 | 100 | 100 | |
| 2.375/60.3 | 3 | 100 | 100 | 100 | |

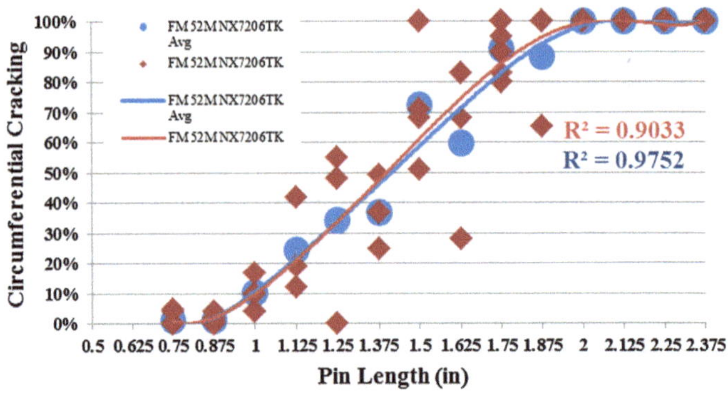


**Fig. 5** Scatter plot of CPTT results for Alloy 52M heat: NX7206TK

## *Tested Alloys*

Five alloys were tested following the reproducibility study: filler metal $82_{(A)}$, 52M [NX7206TK], TG-SN690Nb [FBH2280], 52M [NX0T85TK], and 52MSS [NX77W3UK]. A minimum of three successful casts was produced at each length for accurate comparison and a minimum of four for each length used as an alloy ranking criteria. The results of these tests are provided in Table 5 and Fig. 6.

Each tested alloy has been ranked using the criteria described above; these rankings are provided in the bar chart.

From Fig. 7 the ranking from least susceptible to most susceptible is: Filler metal $\mathbf{82_{(A)} < 52M_{[NX7206TK]} = TG\text{-}SN690Nb_{[FBH2280]} < 52M_{[NX0T85TK]} < 52MSS\text{-}C}$. These same heats of filler metals 52M and 52MSS were tested using a previous generation of the cast pin tear tester in addition to a different heat of filler metal 82. The results obtained during the previous study match the ranking obtained using the new generation CPTT using the same ranking criteria: Filler metal **82 < 52M < 52MSS** [21].

## *Dilution Study*

In this study a heat of filler metal 82 ($82_{(B)}$) was tested as well as 10 and 25 % dilutions of filler metal $82_{(B)}$ into $82_{(A)}$. The results of this testing are provided in Table 6 and Fig. 8.

The main criterion for ranking susceptibility to solidification cracking is the maximum pin length with 0 % cracking. It is evident that as more of the worse preforming $82_{(B)}$ is added to $82_{(A)}$ the cracking susceptibility of the alloy increases or the maximum pin length produced with 0 % cracking decreases.

Figure 9 indicates that when a less susceptible material is diluted with a more susceptible material, the maximum pin length at which 0 % cracking occurs is

**Table 5** CPTT results for tested alloys

| Alloy | FM $82_{(A)}$ | | | | $52M_{(NX7206TK)}$ | | | | TG-$SN690Nb_{(FBH2280)}$ | | | | $52M_{(NX0T85TK)}$ | | | | 52MSS-$C_{(NX77W3UK)}$ | | | |
|---|---|---|---|---|---|---|---|---|---|---|---|---|---|---|---|---|---|---|---|---|
| Pin length (in./mm) | # | Cracking | | | # | Cracking | | | # | Cracking | | | # | Cracking | | | # | Cracking | | |
| | | Avg (%) | Max (%) | Min (%) | | Avg (%) | Max (%) | Min (%) | | Avg (%) | Max (%) | Min (%) | | Avg (%) | Max (%) | Min (%) | | Avg (%) | Max (%) | Min (%) |
| 0.625/15.875 | | | | | | | | | | | | | 1 | 0 | 0 | 0 | | | | |
| 0.75/19.1 | | | | | 8 | 1 | 5 | 0 | 5 | 0 | 0 | 0 | 4 | 0 | 0 | 0 | 5 | 0 | 0 | 0 |
| 0.875/22.2 | 1 | 0 | 0 | 0 | 6 | 1 | 4 | 0 | 6 | 3 | 8 | 0 | 5 | 1 | 4 | 0 | 6 | 19 | 52 | 0 |
| 1/25.4 | 1 | 0 | 0 | 0 | 3 | 10 | 17 | 4 | 6 | 9 | 23 | 0 | 6 | 13 | 26 | 0 | 5 | 29 | 60 | 0 |
| 1.125/28.6 | 3 | 0 | 0 | 0 | 3 | 24 | 42 | 12 | 6 | 13 | 19 | 0 | 5 | 18 | 34 | 0 | 7 | 28 | 61 | 15 |
| 1.25/31.8 | 6 | 4 | 15 | 0 | 3 | 34 | 55 | 0 | 5 | 26 | 54 | 3 | 8 | 13 | 32 | 0 | 6 | 57 | 83 | 23 |
| 1.375/34.9 | 5 | 33 | 65 | 0 | 3 | 37 | 49 | 25 | 7 | 21 | 39 | 5 | 7 | 63 | 100 | 21 | 5 | 52 | 100 | 8 |
| 1.5/38.1 | 4 | 43 | 100 | 0 | 4 | 73 | 100 | 51 | 5 | 55 | 100 | 12 | 5 | 85 | 100 | 50 | 5 | 71 | 100 | 30 |
| 1.625/41.3 | 5 | 36 | 93 | 10 | 3 | 60 | 83 | 28 | 2 | 51 | 56 | 45 | 4 | 83 | 100 | 62 | 4 | 86 | 100 | 45 |
| 1.75/44.5 | 5 | 51 | 97 | 0 | 8 | 91 | 100 | 80 | 2 | 45 | 45 | 45 | 4 | 86 | 100 | 70 | 4 | 96 | 100 | 85 |
| 1.875/47.6 | 7 | 68 | 100 | 15 | 3 | 88 | 100 | 65 | 5 | 93 | 100 | 65 | 4 | 95 | 100 | 80 | 3 | 83 | 85 | 80 |
| 2/50.8 | 4 | 100 | 100 | 100 | 3 | 100 | 100 | 100 | 2 | 69 | 100 | 37 | 4 | 76 | 100 | 55 | 3 | 100 | 100 | 100 |
| 2.125/54 | 2 | 100 | 100 | 100 | 1 | 100 | 100 | 100 | 2 | 100 | 100 | 100 | | | | | 3 | 100 | 100 | 100 |

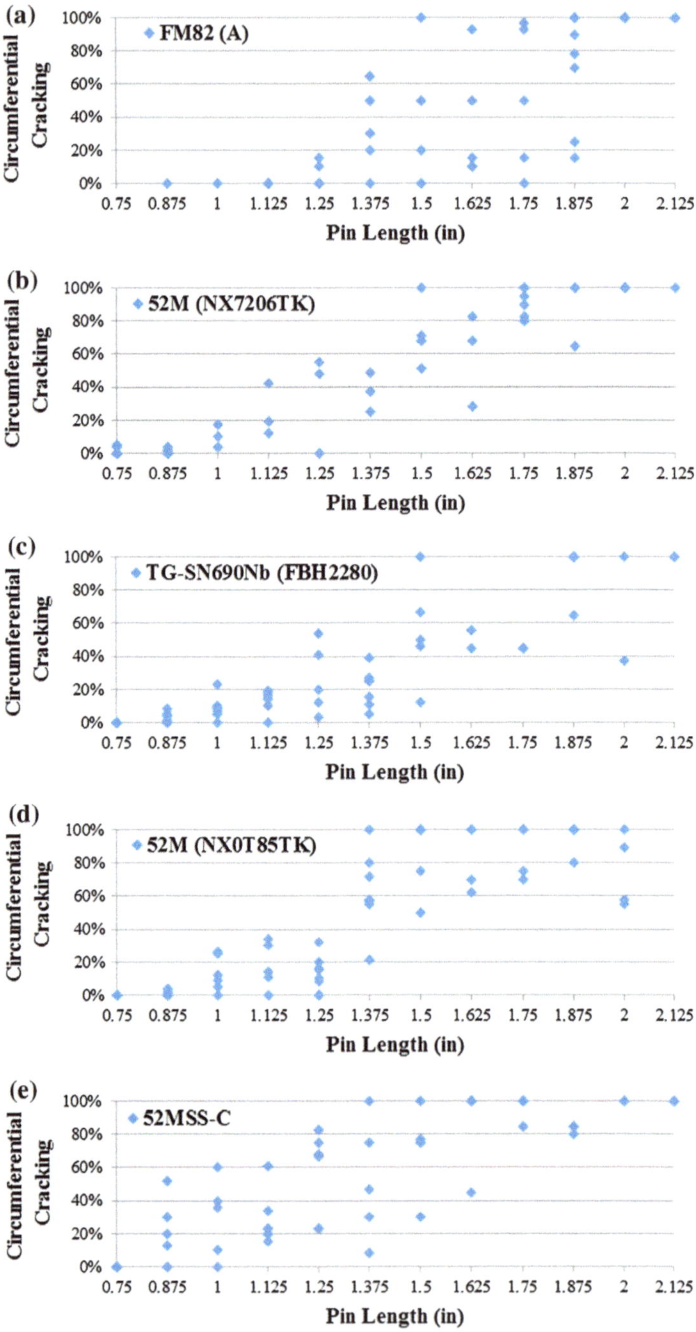


**Fig. 6** Scatter plot CPTT results for alloys: **a** 82(A), **b** 52M [NX7206TK], **c** TG-SN690Nb [FBH2280], if **d** 52M [NX0T85TK, **e** 52MSS-C [NX77W3UK]

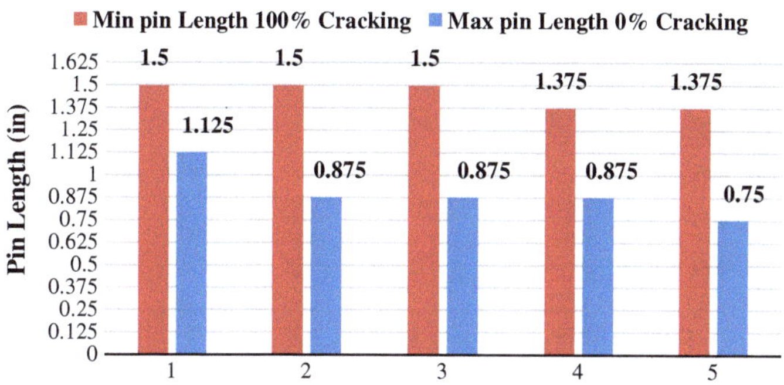


**Fig. 7** Bar Chart ranking tested alloys, The *red bars* represent the minimum pin length at which 100 % circumferential cracking occurred, the *blue bars* represent the maximum pin length at which 0 % cracking occurred (less than 10 % maximum cracking) (color figure online)

**Table 6** CPTT results for alloys $82_{(A)}$, $82_{(B)}$, 10 % dilution, and 25 % dilution

| Alloy | FMB2 (A) | | | FMB2 (A) + 10 % FMB2 (B) | | | FMB2 (A) + 25 % FMB2 (B) | | |
|---|---|---|---|---|---|---|---|---|---|
| Pin length (in./mm) | # | Cracking | | # | Cracking | | # | Cracking | |
| | | Avg (%) | Max (%) | | Avg (%) | Avg (%) | | Avg (%) | Max (%) |
| 0.75/19.05 | – | – | – | 1 | 0 | 0 | 2 | 0 | 0 |
| 0.875/22.2 | 1 | 0 | 0 | 5 | 0 | 0 | 5 | 8 | 20 |
| 1/25.4 | 1 | 0 | 0 | 5 | 3 | 10 | 2 | 28 | 30 |
| 1.125/28.6 | 3 | 0 | 0 | 3 | 13 | 20 | 3 | 8 | 20 |
| 1.25/31.8 | 6 | 4 | 15 | 4 | 21 | 30 | 3 | 16 | 30 |
| 1.375/34.9 | 5 | 33 | 65 | 2 | 33 | 50 | 4 | 14 | 29 |
| 1.5/38.1 | 4 | 43 | 100 | 4 | 60 | 100 | 3 | 67 | 100 |
| 1.625/41.3 | 5 | 36 | 93 | 2 | 55 | 100 | 4 | 56 | 100 |
| 1.75/44.5 | 5 | 51 | 97 | 5 | 74 | 100 | 3 | 73 | 80 |
| 1.875/47.6 | 7 | 68 | 100 | 5 | 74 | 100 | 4 | 100 | 100 |
| 2/50.8 | 4 | 100 | 100 | 4 | 100 | 100 | 3 | 100 | 100 |
| 2.125/54.0 | 2 | 100 | 100 | 3 | 100 | 100 | 1 | 100 | 100 |

shifted down indicating an increased susceptibility to solidification cracking. This shows that the new generation CPTT may have the capability to determine the maximum amount of dilution which does not affect the cracking susceptibility of a material.

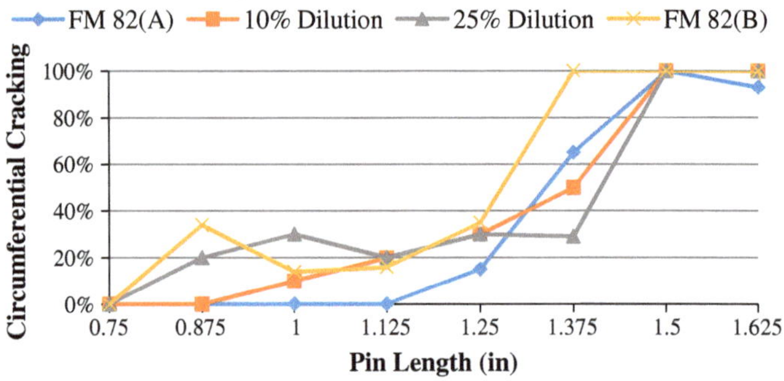


**Fig. 8** Maximum CPTT cracking response in alloy $82_{(A)}$, $82_{(B)}$, 10 and 25 % dilution of $82_{(A)}$ by $82_{(B)}$

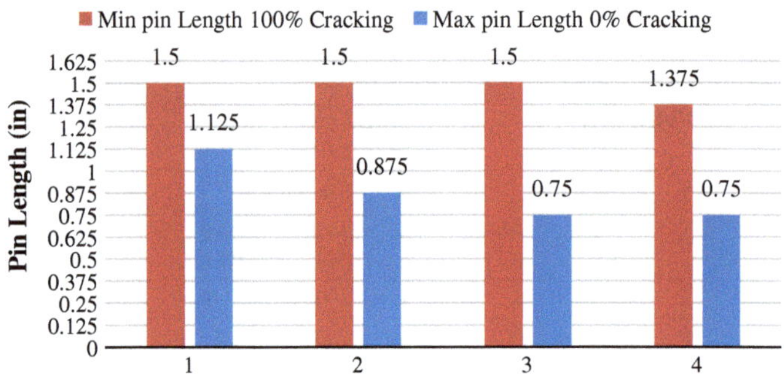


**Fig. 9** Bar Chart of alloy weldability ranking criteria for filler metal $82_{(A)}$, $82_{(B)}$, 10 and 25 % dilutions

## *Characterization*

Fracture surface analysis was performed on each sample in order to confirm the mode of failure as solidification cracking. The crack surfaces of 100 % cracked pins were analyzed and the results of this work are provided in Fig. 10. The crack surface of each material exhibits a dendritic "egg-crate" morphology which is indicative of solidification cracking. This confirms the ability of the new generation cast pin tear tester to replicate the mechanism resulting in solidification cracking.

EDS analysis preformed on cast pins of alloys 52M, 52MSS, $FM82_{(A)}$, and TG-SN690Nb revealed that the measured composition of the pins matches the nominal heat composition closely (Table 7), with the exception of Titanium. This discrepancy can be attributed to a large margin of error in identifying Titanium using EDS. The results confirm that throughout each procedural step the original material composition is maintained and that both the button melter and Cast Pin Tear Tester have been adequately purged of oxidizing elements and other contaminates.

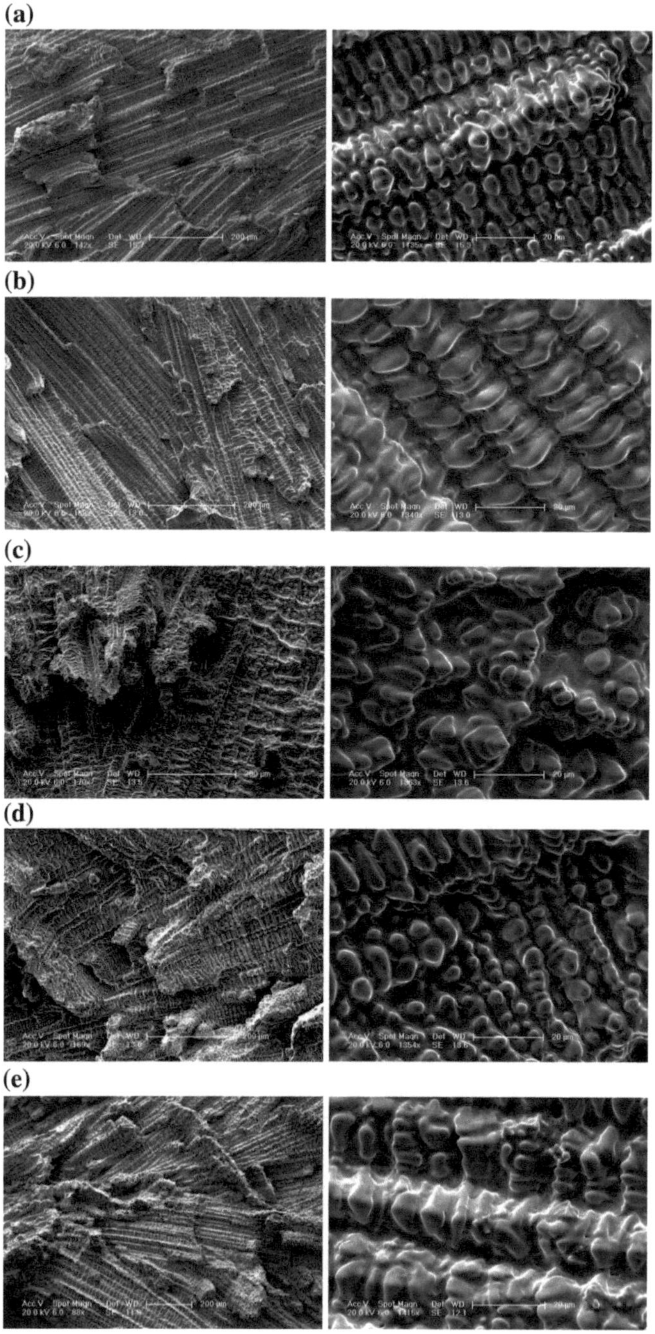

**Fig. 10** Fracture surface of 100 % cracked pins for each tested alloy: **a** 52M [NX7206TK], **b** 52M [NX0T85TK], **c** 52MSS-C [NX77W3UK], **d** TG-SN690Nb [FBH2280], **e** Filler Metal $82_{(A)}$

**Table 7** Nominal composition compared to average of 2–3 EDS scans (wt%)

| *Base Metal EDS Comparison* | | | | | | | | |
|---|---|---|---|---|---|---|---|---|
| [wt%] | 52M [NX7206TK] | | 52M [NX0T85TK] | | 52MSS-C [NX77W3UK] | | TG-SN690Nb | |
| Element | Composition (%) | Avg EDS (3) (%) | Composition | Avg EDS (3) (%) | Composition | Avg EDS (2) (%) | Composition (%) | Avg EDS (3) (%) |
| Nb | 0.81 | 0.75 | 0.87 | 0.84 | 2.51 | 1.64 | 0.60 | 0.51 |
| Ti | 0.21 | 0.71 | 0.16 | 0.63 | 0.18 | 0.49 | 0.70 | 0.93 |
| Cr | 30.01 | 30.34 | 30.11 | 29.95 | 29.24 | 30.00 | 29.70 | 29.93 |
| Fe | 8.51 | 9.23 | 8.87 | 8.84 | 8.79 | 8.32 | 8.40 | 8.63 |
| Ni | 59.41 | 58.61 | 58.95 | 59.43 | 55.12 | 52.09 | 59.84 | 59.73 |

Additional EDS analysis of 52MSS and 52M revealed Niobium-rich eutectic at crack tips of the fractured samples. The locations and results of the scans are provided in Fig. 11.

The solidification sequence of nickel-base niobium-bearing alloys can be described as such: (1) Primary solidification L → γ during this stage of solidification interdendritic liquid becomes enriched in niobium and carbon (2) Eutectic reaction L → (γ + NbC) which depletes the interdendritic regions of Carbon and (3) Secondary eutectic reaction L → (γ + Laves) which occurs during the final stage of solidification [12]. This solidification sequence describes the formation of the microstructures (Fig. 11).

From Fig. 11 it is evident that the cracks propagate and are healed interdendritically through a Nb-rich and in some cases Mo-rich eutectic constituent. This is a result of the segregation of these elements which occurs during solidification.

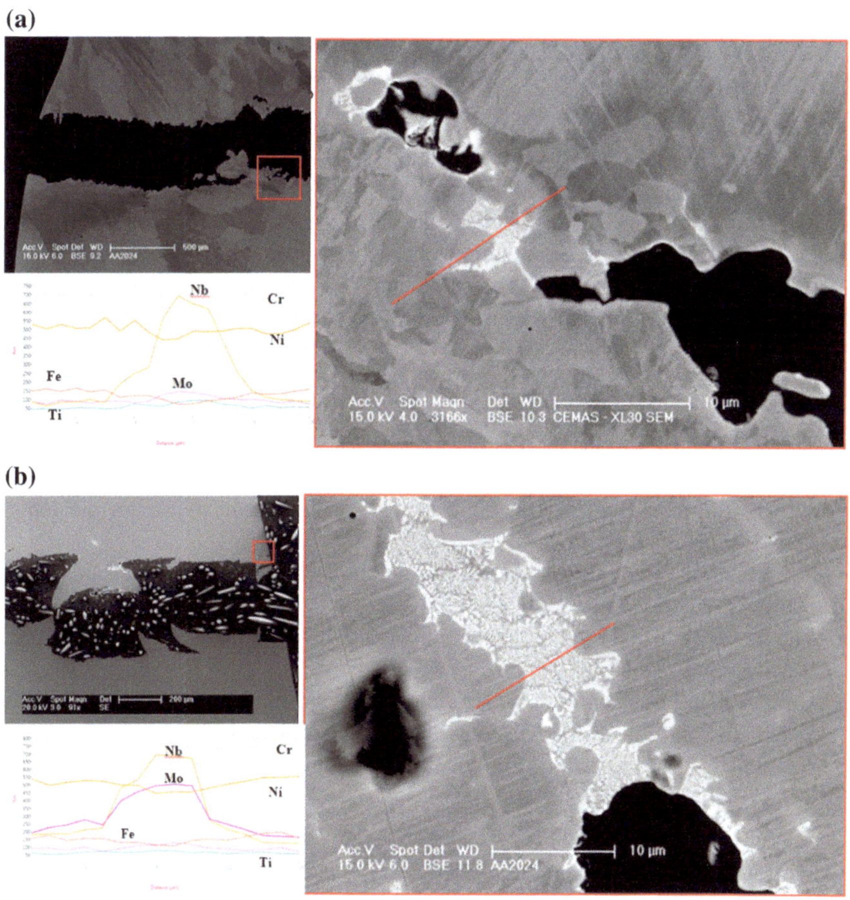


**Fig. 11** EDS analysis at crack tip of **a** 52M [NX0T85TK], **b** 52MSS [NX77W3UK]

The amount of eutectic constituent formed during the solidification process varies between these alloys based on their compositions. It is evident when comparing 52MSS and 52M that a larger fraction of eutectic exists in the microstructure of 52MSS, particularly at crack tips (Fig. 11). Nb and Mo are responsible for this difference in the quantity of eutectic formation. Alloy 52M contains 0.8–0.9 wt% Niobium and less than 0.05 wt% Molybdenum; these are fairly low alloying additions so little segregation is expected during solidification. Alloy 52MSS contains 2.51 wt% Niobium and 3.51 wt% Molybdenum which is a relatively large amount compared to alloy 52M. These elements segregate during solidification; this increased segregation expands the solidification temperature range for this alloy compared to 52M, allowing liquid films to penetrate and wet grain boundaries. The presence of these liquid films can promote cracking, if the strength of the liquid film is exceeded during solidification, or heal existing cracks, if sufficient liquid is present [10, 11].

## Conclusions

- The New Generation CPTT developed at OSU is capable of successfully casting 0.197 in. (5 mm) diameter pins from 0.5 to 2.5 in. (12.7–63.5 mm) in length (0.125 in./3.18 mm increment) and 9.5–17.5 g in mass (0.5 g increment).
- Low standard deviation at pin lengths by which alloys are ranked indicates good test reproducibility (0 and 100 % cracking).
- Rankings of the tested alloys susceptibility to solidification cracking determined using the cast pin tear test are as follows from least to most susceptible: Filler metal 82, 52M [NX7206TK], TG-SN690Nb [FBH2280], 52M [NX0T85TK], 52MSS [NZ77W3UK].
- The addition of a more susceptible material ($82_B$) to a less susceptible material ($82_A$) results in increased susceptibility to solidification cracking (a downward shift of the maximum pin length with 0 % cracking and minimum pin length with 100 % cracking).
- Solidification cracking susceptibility rankings match previously determined rankings: 82 < 52M < 52MSS [21].
- The mode of pin failure has been confirmed to be solidification cracking in each tested alloy.
- Crack healing by backfilling of Nb-rich eutectic type constituent is present in the tested alloys, with 52MSS having the most significant backfilling due to an elevated Niobium and Molybdenum content compared to alloy 52M, TG-SN690Nb, and filler metal 82.

**Acknowledgements** The authors would like to acknowledge EPRI for sponsoring this work.

## References


1. S. L. McCracken and R. E. Smith, Behavior and Hot Crack Susceptibility of FIller Metal 52M (ERNiCrFe-7A) on Cast Austenitic Stainless Steel, Springer Berlin Heidelberg, 2011.
2. J. Hickling, "Materials Reliability Program Crack Growth Rates for Evaluating Primary Water Stress Corrosion Cracking (PWSCC) of Alloy 82, 182, and 132 Welds (WRP-115NP)," EPRI, Palo Alto, California, 2004.
3. J. Ramirez, "Understanding Stress Corrosion Cracking of Welds in Nuclear Reactors," *Welding Journal,* vol. 90, pp. 38-42, 2011.
4. S. Lozano-Perez, J. Dohr, M. Meisnar and K. Kruska, "SCC in PWRs: Learning from a Bottom-Up Approach," *Metallurgical and Materials Transactions E,* vol. 1, no. 2, pp. 194-210, 2014.
5. H. Hanninen, P. Aaltonen, A. Brederholm, U. Ehrnsten, H. Gripenberg, A. Toivonen, J. Pitkanen and L. Virkkunen, "Effects of Composition and Microstructure on SCC Susceptibility," in *Dissimilar Metal Weld Joints and their Preformance in Nuclear Power Plant and Oil Refinery Conditions*, Helsinki, JULKAISIJA UTGIVARE, 2006, pp. 123-125.
6. T. M. Angeliu and G. S. Was, "The Effect of Cr, C, Yt on the Oxidation of Ni-base Alloys in High Temperature Water," *Journal of the Electrochemical Society,* vol. 140, pp. 1877-1883, 1993.
7. J. Hickling and C. King, "Materials Reliability Program Resistance to Primary Water Stress Corrosion Cracking of Alloys 690, 52, and 152 in Pressurized Water Reactors (MRP-111," EPRI, Palo Alto, California, 2004.
8. A. J. Sedricks, J. W. Schultz and M. A. Cordovi, "Inconel Alloy 690 - A New Corrosion Resistant Material," *Corrosion Engineering,* vol. 28, pp. 82-95, 1979.
9. J. C. Lippold and N. E. Nissley, "Ductility-Dip Cracking in High-Cr, Ni-Base Filler Metals," in *Hot Cracking Phenomena in Welds II*, Springer, 2008, pp. 409-426.
10. S. L. McCracken, B. T. Alexandrov, J. C. Lippold, J. W. Sowards and A. T. Hope, "Hot Cracking Study of High Chromium Nickel-Base Weld Filler Metal 52MSS (ERNiCrFe-13) for Nuclear Applications," in *ASME Pressure Vessel & Piping Conference*, Bellevue, Washington, 2010.
11. B. T. Alexandrov, A. T. Hope, J. W. Sowards and J. C. Lippold, "Weldability Studies of High-Cr, Ni-base Filler Metals for Power Generation Applications," EPRI, 2009.
12. J. N. Dupont, C. V. Robino, J. R. Micheal, M. R. Notis and A. R. Marder, "Solidification of Nb-Bearing Superalloys: Part I. Reaction Sequences," *Metallurgical and Materials Transactions A,* vol. 29, no. 11, pp. 2785-2797, 1998.
13. T. Kannengiesser and T. Boellinghaus, "Hot Cracking Tests- An Overview of Present Technologies and Applications," BAM Federal Institute for Materials Research and Testing, Berlin, Germany, 2013.
14. K. Wilken and H. Kleistner, "The Classification and Evaluation of Hot Cracking Tests for Weldments," *Welding Journal,* vol. 28, no. 7/8, 1990.
15. F. C. Hull, "Cast-Pin Tear Test for Susceptibility to Hot Cracking," *Welding Journal,* vol. 38, no. 4, pp. 176-181, 1959.
16. J. A. Talento, "Laboratory Test To Determine the Weldability of alloy 600/EN82 Ni-Cr-Fe Base-Filler Alloy Combinations," in *Effect of Minor Elements on the Weldability of High-Nickel Alloys*, New York, NY, 1967.
17. F. G. Armao and W. Yeniscavich, "Grooved Copper Mold Testing of Inconel Alloy 600," *Welding Journal,* vol. 53, pp. 170-174, 1974.
18. B. T. Alexandrov and J. C. Lippold, "Further development of the Cast Pin Tear Test for Evaluating Solidification Cracking in Welds," in *Hot Cracking Phenomena in Welds*, Columbus, OH, Springer, 2010, pp. 317-331.
19. B. T. Alexandrov, J. C. Lippold and N. E. Nissley, "Evaluation of Weld Solidification Cracking in Ni-Base Superalloys Using the Cast Pin Tear Test," in *Hot Cracking Phenomena in Welds*, Columbus, OH, Springer, 2007, pp. 193-213.

20. T. C. Luskin, "Investigation of Wledability in High-Cr Ni-Base Filler Metals," The Ohio State University, Columbus, OH, 2013.
21. B. T. Alexandrov, J. C. Lippold, A. T. Hope, B. J. Sutton, J. W. Sowards and S. L. McCracken, "Susceptibility to Solidification Cracking in High Chromium Nickel-Base Filler Metals For Nuclear Power Applications," in *Trends in Welding Research*, Chicago, IL, 2012.

# Multi-scale Modeling of the Stress-Strain State During Welding of Ni-Based Alloys

**K.A. Yushchenko, V.S. Savchenko, N.O. Chervyakov, A.V. Zvyagintseva and E.A. Velikoivanenko**

**Abstract** Numerical modeling of changes of stress-strain state in welding of alloys of Ni-Cr–Fe alloying system at macro- and mesoscales was carried out. The kinetics of changes of stresses and deformations in the weld and heat-affected zone at sites of probable formation of hot cracks was considered. Calculations data of the stress-strain state at the macroscale were used in modeling of thermal-deformation processes at the mesoscale. During modeling the experimental data on anisotropy of physical properties were used depending on crystallographic orientation of grains in the heat affected zone of a real welded joint. The modeling was performed considering the changes of properties of the material depending on temperature. It was shown that depending on anisotropy of physical and mechanical properties of the metal in the limits of neighboring grains, the non-uniform distribution of plastic deformation becomes apparent. The change of deformation exhibits a gradient with localization of deformation near the grain boundaries.

## Introduction

High-chromium, nickel based filler materials of the Alloy 690 type are used in manufacture of units of nuclear power plants. These alloys have good resistance against stress corrosion cracking. At the same time they showed a tendency for formation of ductility dip cracks, especially in welding with multi-pass welds [1–4]. Formation of cracks in welded joints in the ductility-dip temperature range is a limiting stage of degree of degradation of the weld metal.

The crack formation in welds is connected with formation of a coarse-grain macrostructure. Such increased sensitivity to crack formation is due to change in properties of the metal of weld and heat affected zone. One hypothesis of crack

K.A. Yushchenko (✉) · V.S. Savchenko · N.O. Chervyakov · A.V. Zvyagintseva · E.A. Velikoivanenko
The E.O. Paton Electric Welding Institute, Kiev, Ukraine
e-mail: yu.kon@paton.kiev.ua



Th. Boellinghaus et al. (eds.), *Cracking Phenomena in Welds IV*,
DOI 10.1007/978-3-319-28434-7_14

formation is connected with the effect of thermal-deformation loading and significant local deformation in the range of temperatures of the ductility dip and also segregation of impurity elements to the grain boundaries [5–7].

The mechanism of crack formation remains still to be a matter of discussions [8–10]. The aim of this work was to evaluate the influence of welding conditions and high-temperature deformation on the processes which predetermine the intergranular fracture. Numerical modeling was used during investigation of the stress-strain state. Modeling of thermal-deformation processes at macro- and mesoscales was performed applying the computer program "Weldpredictions" developed at the E. O. Paton Electric Welding Institute of the NASU.

## Experimental Procedures

Thermo-deformation processes in the heat-affected zone were studied in the weld produced using the wire ERNiCrFe-7 (IN52) under the conditions, typical for GTA welding. The chemical composition of investigated material is presented in Table 1.

The conditions of welding were modeled as-applied to the case of deposition on the plate of sizes $B_x$, $B_y$ and $B_z$ thickness (Fig. 1). In the process of welding heating the temperature field T(x, y, z) is determined with time and the problem on kinetics of distribution of stresses and deformations is solved by average values of temperature T(x, y, t). As the basis the method of successive tracking in time with the pitch $\Delta t$ and method of finite elements in space was used, i.e. the considered region was presented by the totality of elementary volumes of sizes $h_x$, $h_y$, $h_z$. During calculations at macroscale the size of element amounted to 0.5 × 0.5 × 0.5 mm, at mesoscale—0.025 × 0.025 × 0.025 mm.

For calculations the physical and mechanical properties of the material including the elastic modulus, yield strength and coefficient of linear expansion are necessary. All these characteristics were determined experimentally at temperatures up to 1200 °C. The modulus of elasticity and yield strength were determined in a tensile machine (MTS-810) and the coefficient of linear expansion was determined using a contactless laser dilatometer.

Appearance of the specimens after welding and the macrostructure of a plan view section of the multilayer weld are shown in Figs. 2 and 3, respectively.

Specimens for evaluation of mechanical properties were cut transverse to the joints. Appearance of the specimens cut from a plate is shown in Fig. 4.

The specimens for evaluation of coefficients of linear expansion, $\alpha(T)$ were manufactured of deposits, made by GTA welding with filler wire IN52 into the copper chill mould. Schematically the specimen is shown in Fig. 5.

The data used in calculations of thermo-physical properties are shown in Fig. 6. Size of the plate was 70 × 40 × 3 mm. Welding heat input was 2385 J/mm.

**Table 1** Chemical composition of welding wire and base metal

| | Elements, wt% | | | | | | | | | |
|---|---|---|---|---|---|---|---|---|---|---|
| | C | Mn | Ni | Cr | Fe | Nb | Mo | Ti | Al | Si |
| Inconel 690 | 0.025 | 0.24 | Base | 29.72 | 10.3 | – | – | 0.28 | 0.87 | 0.32 |
| IN52 | 0.026 | 0.31 | Base | 28.8 | 8.5 | 0.03 | 0.03 | 0.51 | 0.72 | 0.12 |

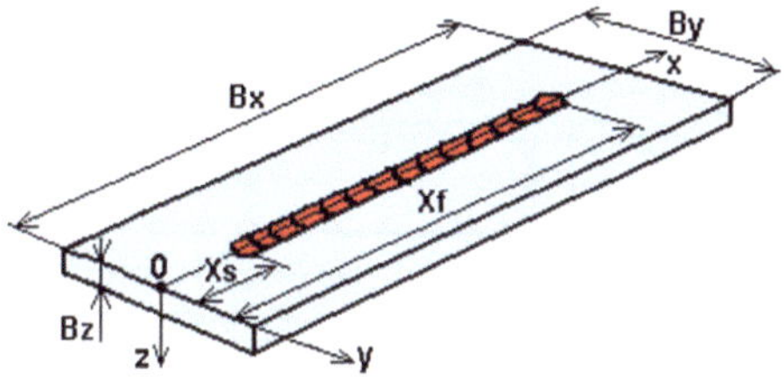


**Fig. 1** Geometry of investigated specimen ($B_x$ is the length of plate, $B_y$ is the width of plate, $B_z$ is the thickness of plate, $x_s$ and $x_f$ are the points of beginning and end of welding, respectively)

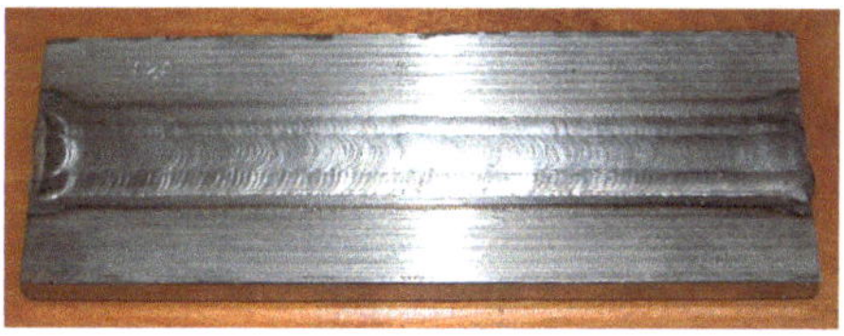

**Fig. 2** Appearance of the surface of an Inconel 690 specimen after multi-pass welding using wire IN52 (ERNiCrFe-7)

**Fig. 3** Macrostructure of specimen of nickel alloy Inconel 690 with preliminary weld deposition using welding wires IN52

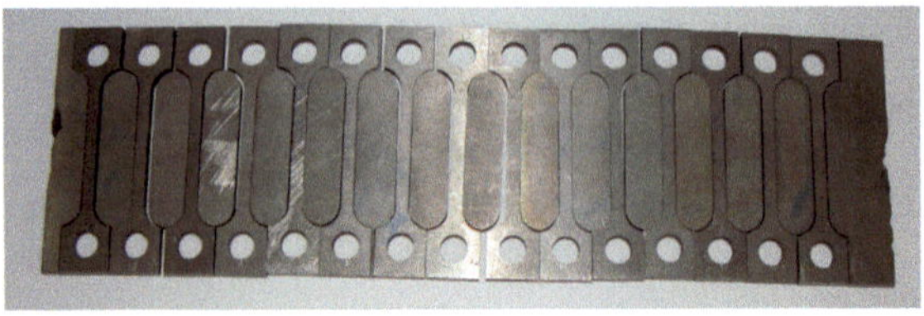

**Fig. 4** Mechanical test specimens cut transverse to deposited welds

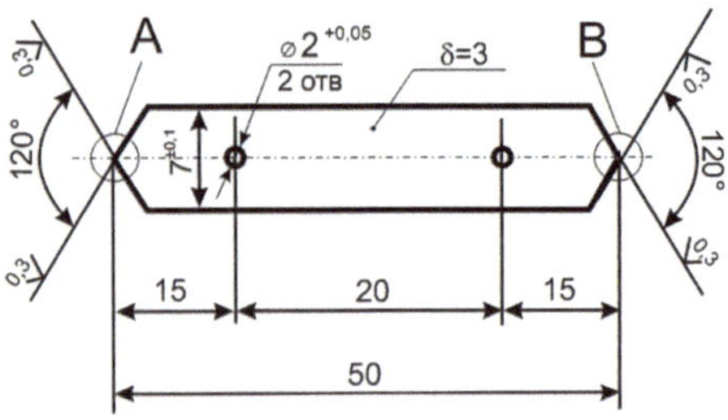


**Fig. 5** Specimen for evaluation of coefficients of linear expansion

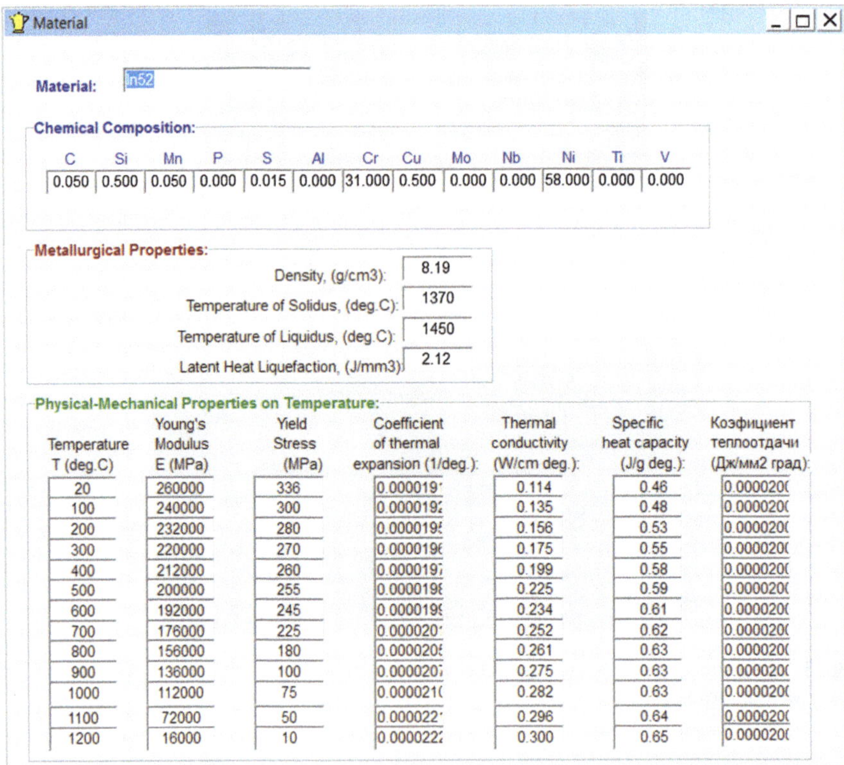

Material

Material: In52

Chemical Composition:

| C | Si | Mn | P | S | Al | Cr | Cu | Mo | Nb | Ni | Ti | V |
|---|---|---|---|---|---|---|---|---|---|---|---|---|
| 0.050 | 0.500 | 0.050 | 0.000 | 0.015 | 0.000 | 31.000 | 0.500 | 0.000 | 0.000 | 58.000 | 0.000 | 0.000 |

Metallurgical Properties:

Density, (g/cm3): 8.19
Temperature of Solidus, (deg.C): 1370
Temperature of Liquidus, (deg.C): 1450
Latent Heat Liquefaction, (J/mm3): 2.12

Physical-Mechanical Properties on Temperature:

| Temperature T (deg.C) | Young's Modulus E (MPa) | Yield Stress (MPa) | Coefficient of thermal expansion (1/deg.): | Thermal conductivity (W/cm deg.): | Specific heat capacity (J/g deg.): | Коэфициент теплоотдачи (Дж/мм2 град): |
|---|---|---|---|---|---|---|
| 20 | 260000 | 336 | 0.0000191 | 0.114 | 0.46 | 0.0000200 |
| 100 | 240000 | 300 | 0.0000192 | 0.135 | 0.48 | 0.0000200 |
| 200 | 232000 | 280 | 0.0000195 | 0.156 | 0.53 | 0.0000200 |
| 300 | 220000 | 270 | 0.0000196 | 0.175 | 0.55 | 0.0000200 |
| 400 | 212000 | 260 | 0.0000197 | 0.199 | 0.58 | 0.0000200 |
| 500 | 200000 | 255 | 0.0000198 | 0.225 | 0.59 | 0.0000200 |
| 600 | 192000 | 245 | 0.0000199 | 0.234 | 0.61 | 0.0000200 |
| 700 | 176000 | 225 | 0.0000201 | 0.252 | 0.62 | 0.0000200 |
| 800 | 156000 | 180 | 0.0000205 | 0.261 | 0.63 | 0.0000200 |
| 900 | 136000 | 100 | 0.0000207 | 0.275 | 0.63 | 0.0000200 |
| 1000 | 112000 | 75 | 0.0000210 | 0.282 | 0.63 | 0.0000200 |
| 1100 | 72000 | 50 | 0.0000221 | 0.296 | 0.64 | 0.0000200 |
| 1200 | 16000 | 10 | 0.0000222 | 0.300 | 0.65 | 0.0000200 |

**Fig. 6** Mechanical and thermo-physical properties of IN52 weld for numeric modeling

## Modeling at the Macrolevel

Distribution of temperature fields in the plane and across the thickness of a plate in welding with energy input of 238 J/mm is shown in Fig. 7. Due to the symmetry of the specimen only half of the plate is shown.

Consider the influence of given temperature distributions on the development of thermal-deformation processes in welding. The main type of cracks in welding is transverse cracks relative to the direction of welding. Moreover, they are arranged both in the weld as well as in the heat-affected zone at the distance of 1.5 mm from the fusion line. With account for these peculiarities the kinetics of stress-strain state in heat-affected zone near the fusion line was considered.

Typical distribution of time and residual longitudinal and transverse stresses in welding are shown in Figs. 8 and 9, respectively.

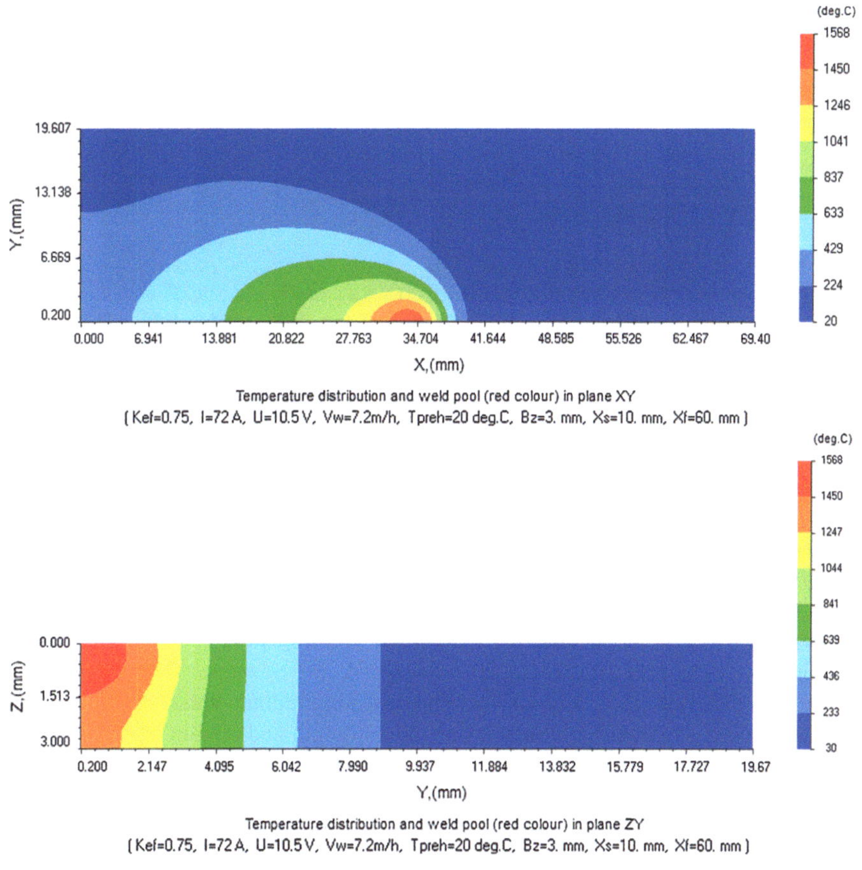


**Fig. 7** Distribution of temperatures in plane and across the thickness of plate

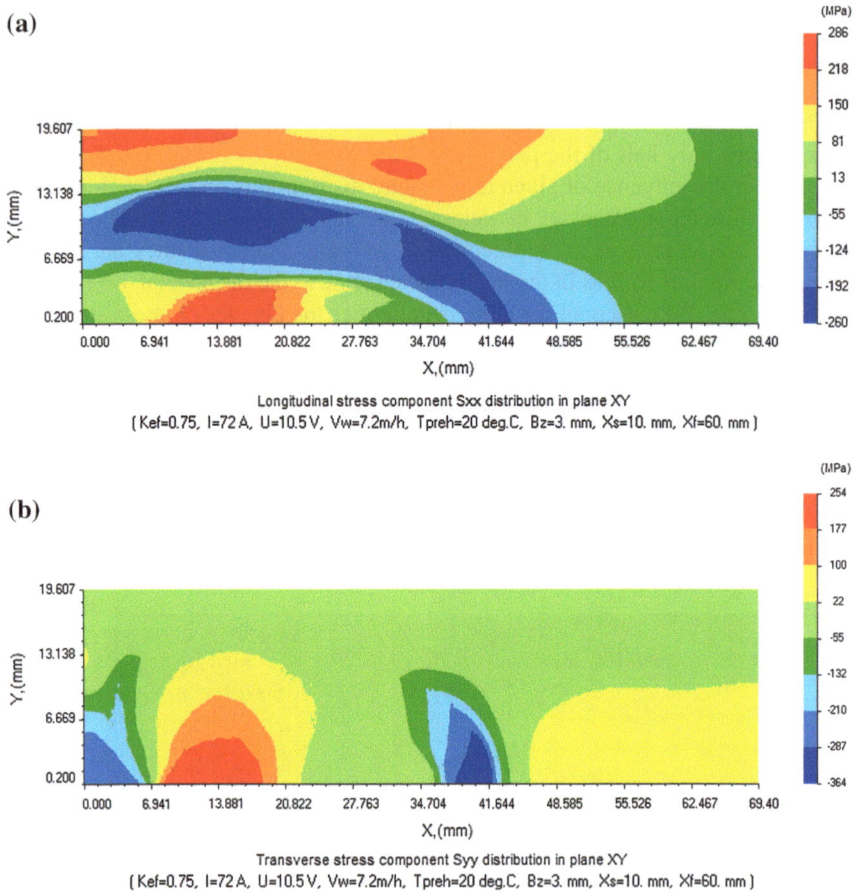


**Fig. 8** Distribution of longitudinal (**a**) and transverse (**b**) temporary stresses in plane of plate

Analysis of data shows that in welding at the mentioned energy input the temporary stresses relatively to the direction of welding dominate which exist along the entire length of the weld and in the heat-affected zone.

Ahead of the weld pool the longitudinal and transverse stresses are compressive, however already at some distance behind the welding pool the stresses become tensile.

Behind the weld pool, a quick increment of longitudinal stresses occurs which during cooling of welded joint reach the yield strength and remain along the entire length of a weld.

As compared to the longitudinal welds, the transverse stresses are more localized. The tensile transverse stresses exist not along the entire length of a weld, they are concentrated at some distance from trailing edge of the weld pool and maximums of stresses move together with the movement of welding pool and are fixed in the zone of welding heating completion.

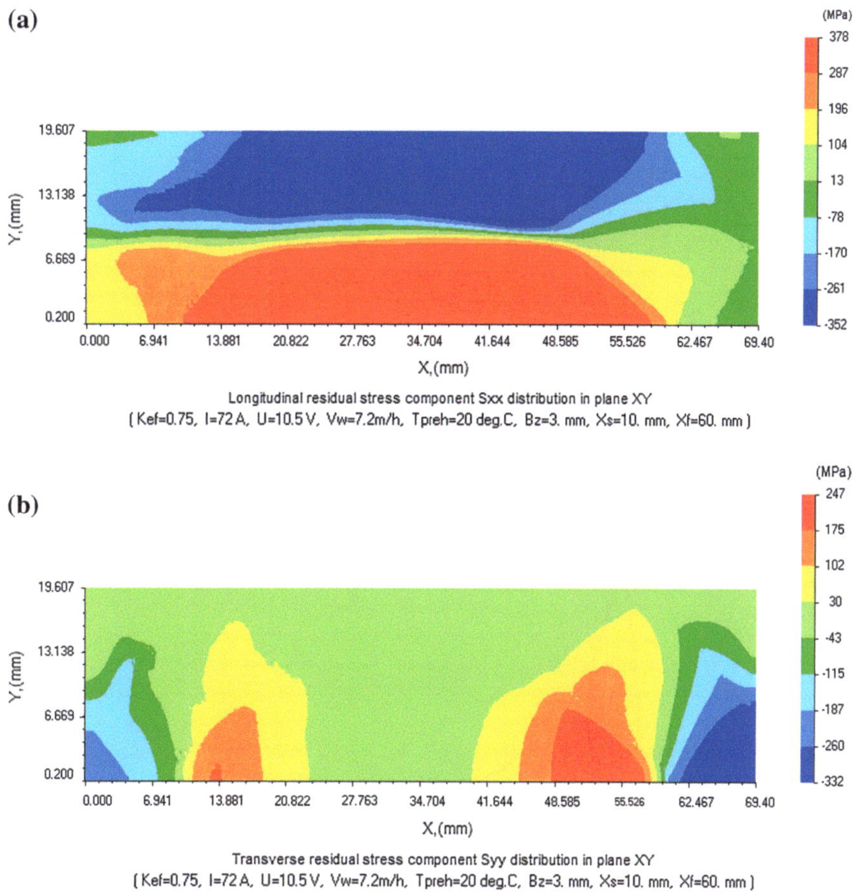


**Fig. 9** Distribution of residual longitudinal (**a**) and transverse (**b**) stresses in plane of plate

Longitudinal plastic strains have the similar character of distribution (Fig. 10). The strains start growing also at some distance behind the weld pool, amounting to a maximum value in cooling at a temperature of 700 °C.

Figure 11 shows the change of temperature, longitudinal and transverse stresses and also longitudinal plastic strains at a point 0.5 mm from the fusion line where crack formation is the most possible.

With the start of the heating cycle the metal in the point at the distance of 0.5 mm from the fusion line is subjected to compressive longitudinal ($\varepsilon_{xx}$) and transverse($\sigma_{yy}$) stresses. Longitudinal plastic strains ($\varepsilon_{xx}$) are also compressive, their value reaches $\sim -1.6$ %. During cooling the tensile stresses arise.

Longitudinal stresses reach the values of 250 MPa (which is somewhat lower than yield strength) and significantly exceeds the transverse stress which is equal to −50 MPa. Longitudinal stresses reach maximum values in cooling down to the temperature 650–700 °C. Upon further cooling the longitudinal stresses remain at

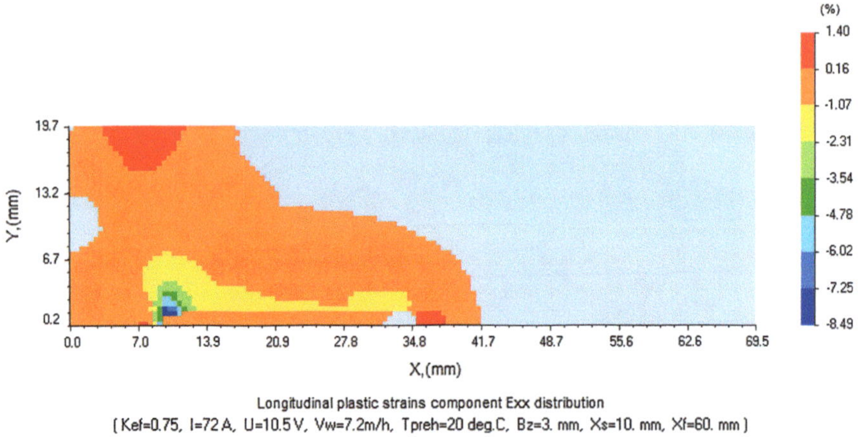


**Fig. 10** Distribution of temporary longitudinal plastic strains in plane of plate

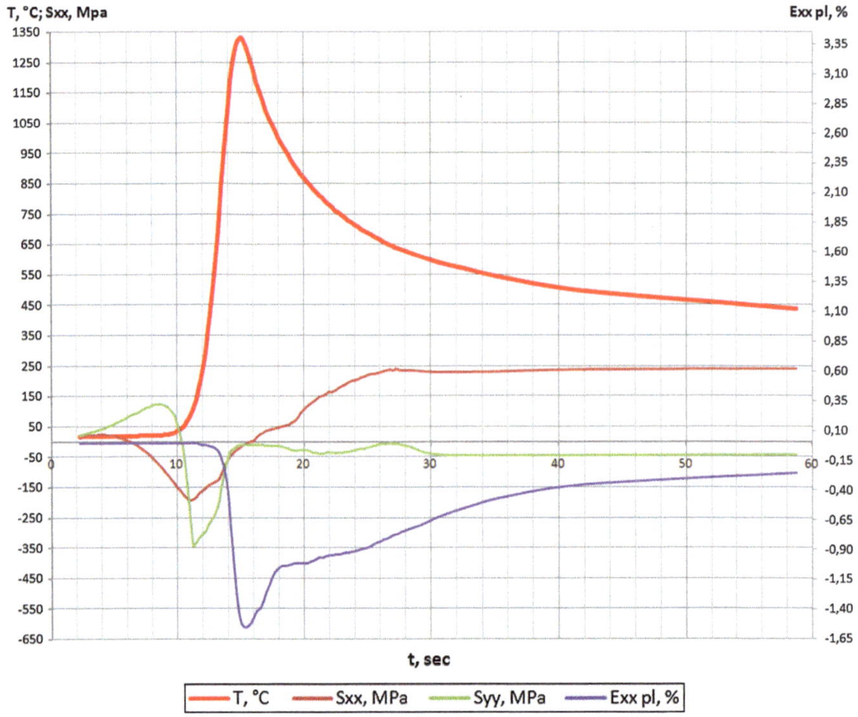


**Fig. 11** Kinetics of variations in temperature (T), longitudinal ($\sigma_{xx}$) and transverse ($\sigma_{yy}$) stresses, and longitudinal plastic strains ($\varepsilon_{xx}$) in HAZ at the distance of 0.5 mm from fusion line during welding

the constant level and the transverse gradually decrease to zero. The longitudinal plastic strains in cooling have a positive increment up to 0.7 % and coincide with the temperature interval of brittleness of the alloy (DTR ~700–1000 °C) and are developing simultaneously with the growth of longitudinal tensile stresses.

Thus, the calculations performed at the macrolevel showed that for welds produced using wire IN52 in the interval of temperatures 1100–700 °C the longitudinal plastic strains have positive increments simultaneously with the growth of longitudinal tensile stresses, that will facilitate the probability of crack initiation in that temperature interval.

## Modeling at Mesoscale

In polycrystalline materials each of the grains has a crystallographic orientation with its characteristic properties. For modeling of the stress-strain state at mesoscale it is necessary to consider crystallographic orientations of separate grains of the welded joint.

To plot the model distribution of grains, the specimen of weld joint produced by the wire IN52 after forced deformation (PVR-test), at which ductility dip cracks in the deposited weld were formed, was selected (Fig. 12).

Using methods of EBSD the crystallographic orientations of grains in the zone of deposited metal (Fig. 13) were determined. According to the microstructure of IN52, a finite element model was constructed, as shown in Fig. 14.

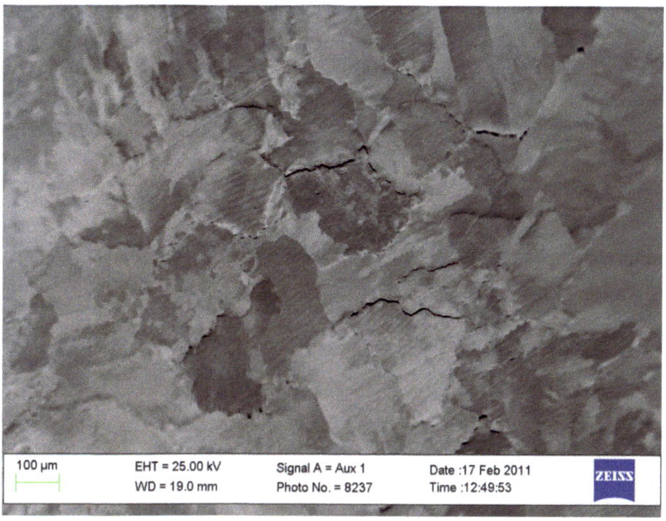


**Fig. 12** Cracks in the heat-affected zone of welds, made by IN52 after the PVR-test

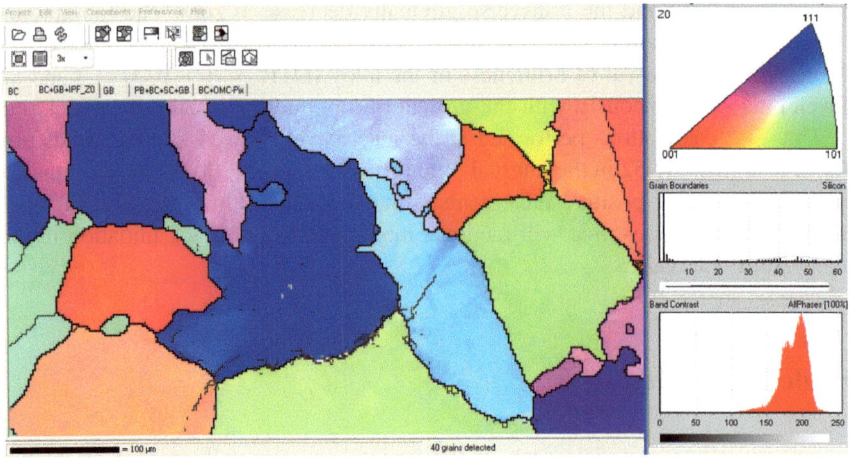


**Fig. 13** Crystallographic orientations of grains obtained using EBSD analysis

**Fig. 14** Schematic illustration of finite element model (*1* crystallographic orientation similar to <111>, *2* crystallographic orientation similar to <011>, *3* crystallographic orientation similar to <001>)

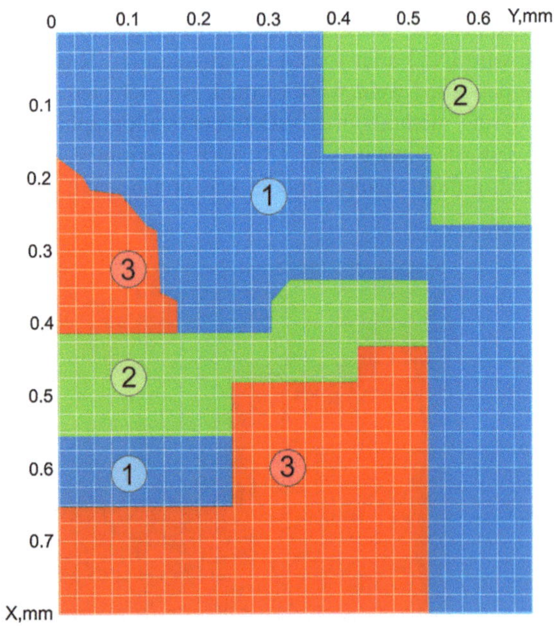


For the grains of crystallographic orientation <001>, <011> and <111> the corresponding temperature dependencies of elasticity modulus and yield strength were preset. It is known [11], that the anisotropy of properties for single-crystal alloys, including elastic properties and strength depends on crystallographic orientation relative to direction of applied external load. Thus, the modulus of

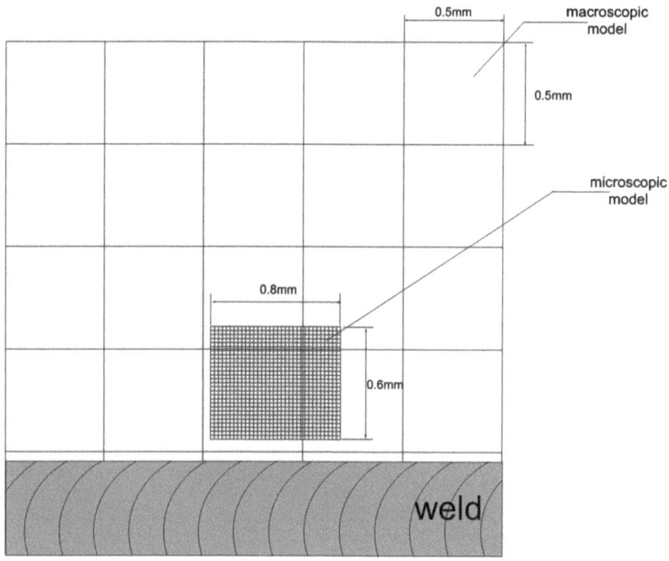


**Fig. 15** Fragment of modeling region

elasticity for nickel-based single-crystal alloy of is changed depending on orientation by almost 1.5–2.5 times, and yield strength value is differed by 30 %. There is a procedure [12] of elasticity modulus determination by measurement of microhardness in nickel-based single crystal.

The measurement of microhardness of grains of weld, made by wire IN52, was performed using laboratory nanoindenter Nano Indenter—II type. Berkovich indenter was used. The maximum load on the indenter was 200 mN. The difference in values of elasticity modulus of separate grains was found. These data were used during numerical modeling of stress-strain state within the limits of neighboring grains.

Data on distribution of temperatures and displacements in modeling at macrolevel were transferred to the mesoscale. Region of 0.6 × 0.8 mm size of HAZ was analyzed, for which the distribution of grains according to Fig. 14 was used.

A fragment of modeling of the HAZ region is shown in Fig. 15.

It is shown from the results of modeling that in calculations with account for anisotropy of physical-mechanical properties within the limits of neighboring grains the non-uniform distribution of plastic deformations and stresses in the grain body occurred (Figs. 16 and 17). White lines identify the grain boundaries. Change in deformations has a gradient nature with localizing of deformation near the grain boundaries.

The highest gradients of deformations occur at the junction of grains with orientations 011 and 111, moreover, the deformations are maximum in grain with orientation 011. Intensity of plastic deformations is maximum in HAZ in sites of

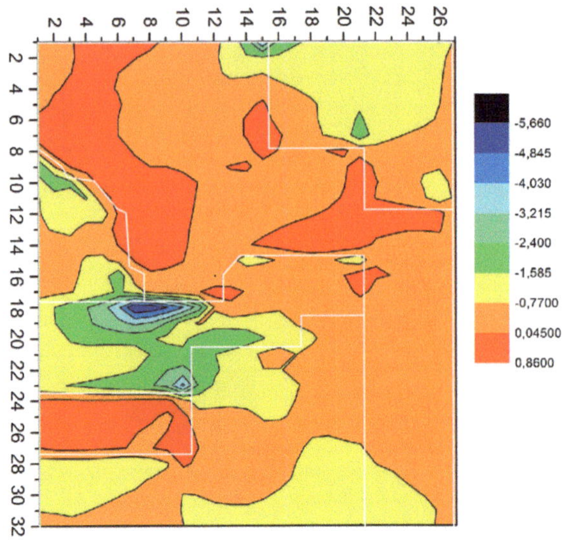


**Fig. 16** Distribution of longitudinal plastic deformations $\varepsilon^{\rho}_{xx}$, % in grains (*white lines* identify the grain boundaries)

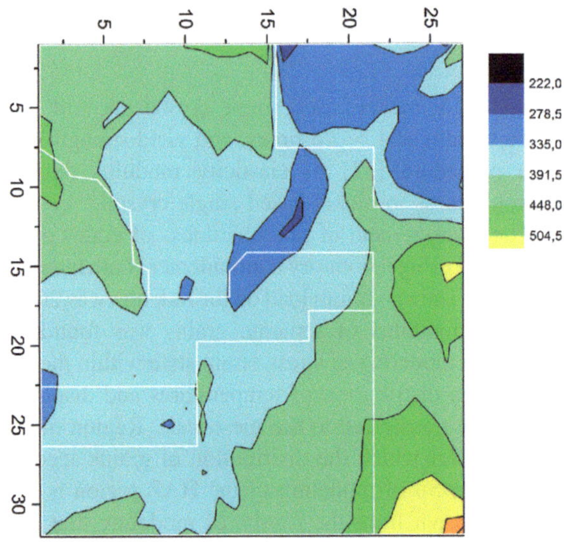


**Fig. 17** Distribution of longitudinal stresses $\varepsilon_{xx}$, MPa in grains

triple junction of grains and reaches 6 % (Fig. 18). This can serve as one of causes of crack initiation in this zone in the temperature interval of ductility dip, becoming apparent in real welded joints.

Kinetics of change of intensity of plastic deformation of sites in the near-boundary zone of grains with orientations 111(1) and 011(2) is shown in Fig. 19.

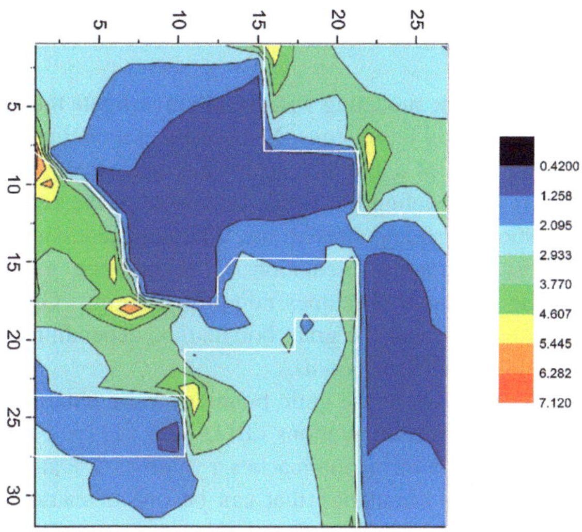


**Fig. 18** Field of intensity of plastic deformations ε, %

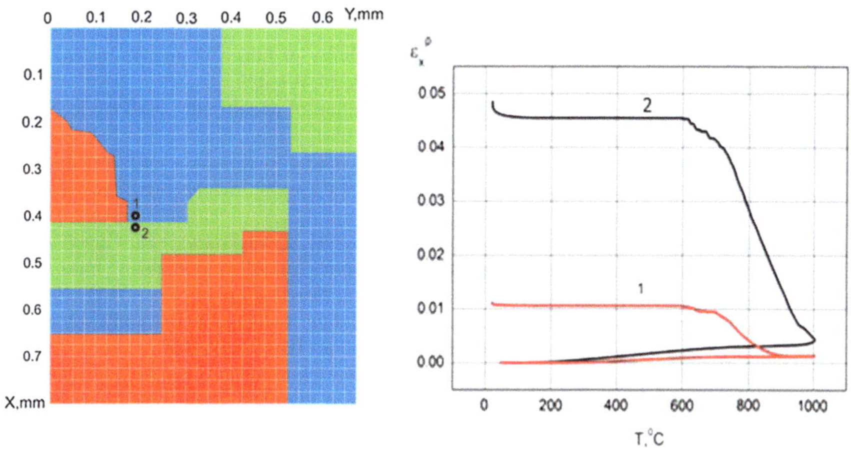


**Fig. 19** Kinetics of change of intensity of plastic deformation in points (*1*) and (*2*)

It is seen that intensity of plastic deformation in grain with orientation <111> (curve 1) is much lower than in grain with orientation <011> (curve 2). Intensity of growth and value of plastic deformation occurs in the temperature interval of 1000–600 °C and preserved down to room temperatures.

## Conclusions

1. The method of multiscale modeling was used to investigate the stress-strain state of welded joints of nickel base alloys. Mathematical modeling at the mesoscale is possible on the basis of data obtained for the macroscale.
2. Under conditions of modeling the single-pass weld at the macroscale, the distribution of plastic strains and stresses has a homogeneous nature. Accounting for different crystallographic orientation of neighboring grains, the thermal stress state at the mesoscale becomes non-homogeneous. Gradients of plastic deformation are present near the grain boundaries, depending on the crystallographic orientation of adjacent grains.
3. Maximal gradients relative to the grain boundary were obtained at the junction of neighboring grains with orientations <011> and <111>.
4. The intensity of plastic deformation reaches a value of 6 % at the triple junction of grains with different orientation that can be one of causes of initiation of ductility dip cracks at this site in real welded joints.

## References


1. Weite W., Tsai C.H. (1999) Hot cracking susceptibility of fillers 52 and 82 in alloys 690 welding. Metallurgical and Materials Transactions A 30:417–425.
2. M.G. Collins and J.C. Lippold. (2003). An investigation of ductility-dip cracking in Ni-base filler metals-Part 1, Welding Journal, 82(10):288s–295s.
3. Nishimoto K., Saida K., Okauchi H. (2006) Microcracking in multipass weld metal of alloy 690. Part 1: Microcracking susceptibility in reheat weld metal. Science and Technology of Welding and Joining 11(4):455–461.
4. K. Yushchenko, V. Savchenko, N. Chervyakov, A. Zvyagintseva, E. Guyot (2011) Comparative hot cracking evaluation of welded joints of Alloy 690 using filler metals Inconel 52 and 52 MSS, Welding in the World, v.55, №9-10, pp. 28-35.
5. Collins M.G., Ramirez A.J., Lippold J.C. (2004) Welding Journal 83 (2), An investigation of ductility-dip cracking in Nickel-based weld metals - Part III: 39s–49s.
6. Ramirez A. J., Lippold J. C. (2004) . High temperature cracking in nickel - base weld metal, Part 2 — Insight into the mechanism, Materials Science and Engineering A, 380: pp. 245 – 258.
7. E. A. Torres, F. G. Peternella, R. Caram and A. J. Ramírez (2010) In Situ Scanning Electron Microscopy High Temperature Deformation Experiments to Study Ductility Dip Cracking of Ni–Cr–Fe Alloys. In-situ Studies with Photons, Neutrons and Electrons Scattering pp. 27-39.
8. Ramirez A.J. and Lippold J.C. (2005) New insight into the mechanism of ductility-dip cracking in Ni-base weld metals. Hot Cracking Phenomena in welds, Part I: Phenomena and mechanism, pp. 19–41.
9. Lippold J.C. and Nissley N.E. (2008) Ductility-dip cracking in high chromium Ni-base filler metals. Hot Cracking Phenomena in welds II, Part V: Ductility-dip cracking, pp. 409–426.
10. J.S. Unfried, E.A. Torres, and A.J. Ramirez In Situ Observations of Ductility-Dip Cracking Mechanism in Ni-Cr-Fe Alloys. Hot Cracking Phenomena in welds III, Part III: Nickel-Base Alloys, pp. 295-314.

11. D. Goldschmidt (1994) Einkristalline Gasturbinenschaufeln aus Nickelbasis-Legierungen. Teil II: Wärmebehandlung und Eigenschaften, Materialwissenschaft und Werkstofftechnik, V.25, Issue 9, pp. 373-382.
12. Hidenari Takagi, Masami Fujiwara, Koji Kakehi Measuring Young ~ s modulus of Ni-based superalloy single crystals at elevated temperatures through microindentation/Materials science and Engineering A 387-389 (2004) pp. 348-351.

# Weldability of Cast and Wrought Nickel Base Alloys 59, 617 and 625

C. Fink, M. Zinke and S. Jüttner

**Abstract** In this study, the hot cracking susceptibility of centrifugal and sand cast nickel base alloys was evaluated compared to the wrought products. Three solid-solution strengthened nickel base alloys—alloy 59, alloy 625 and alloy 617—were studied with respect to the formation and propagation of hot cracking in the cast and wrought microstructure. Hot cracking tests were performed by PVR (programmable deformation crack) test to rank the cracking susceptibility of the cast base metals against the wrought equivalents. Further investigation on the crack susceptible region in the base metal heat-affected zone (HAZ) of cast and wrought alloy 625 was conducted by hot ductility testing using a Gleeble® system. PVR testing indicates a considerably higher hot cracking susceptibility of the cast nickel base alloys compared to the wrought forms of the same composition. The much more severe HAZ liquation cracking of the cast alloys is attributed to the presence of low melting constituents or constitutional liquation of constituents in the large grained and very segregated cast microstructure. Results of hot ductility testing showed that the brittle temperature range (BTR) of the cast alloy 625 is almost three times as wide as for the wrought equivalent. Indicating a much higher hot cracking susceptibility of the cast alloy, this is in good correlation to the PVR test results. Metallurgical and fractographic evaluation of hot ductility samples revealed that in the cast microstructure severe liquation initiates in the vicinity of MC (NbC) carbides. This carbide constitutional liquation reaction leads to the formation of grain boundary liquid films providing a strong driving force for HAZ liquation cracking in cast alloy 625.

C. Fink (✉) · M. Zinke · S. Jüttner
Institute of Materials and Joining Technology, Otto-von-Guericke-University Magdeburg, Universitätsplatz 2, 39106 Magdeburg, Germany
e-mail: manuela.zinke@ovgu.de



Th. Boellinghaus et al. (eds.), *Cracking Phenomena in Welds IV*,
DOI 10.1007/978-3-319-28434-7_15

## Introduction

A wide variety of nickel base alloys is used in chemical and petrochemical applications as well as in industrial furnaces and in the power generation industry. Efficiency aspects require an increasing number of large-sized or complex shaped components made of cast nickel alloys which operate at highly corrosive or high pressure, high temperature conditions, e.g. pumps, fittings, fan blades, tubes. Welding operations on structural castings require maintenance and repair welding as well as joint welding. However, welding of nickel base alloy castings is often a far greater challenge than the welding of their wrought equivalents, which are much more homogenous and finer grained than the cast counterparts. Most of the welding literature on nickel base alloys is based on the weldability of wrought product, but little has been published about the welding metallurgy of cast nickel base alloys.

In most instances successful nickel base wrought alloy chemistries have been turned into cast alloys. Therefore, welding recommendations for nickel base alloy castings are often simply based upon the wrought product of the same composition in terms of heat input limitation, joint preparation or interpass temperature control. However, welds on cast nickel base products are particularly affected by hot cracking when conducted under equal conditions as for the wrought equivalents, see Fig. 1; the heat-affected zone (HAZ) of the cast base metal being the most crack susceptible region.

A review of the literature reveals limited investigations of the HAZ cracking susceptibility in cast nickel base alloys and a lack of understanding of its metallurgical nature as compared with that previously reported for the wrought materials. The purpose of the present work was to develop this understanding by studying the relationship between the heat-affected cast microstructure and the tendency for HAZ hot cracking for different nickel base alloys.

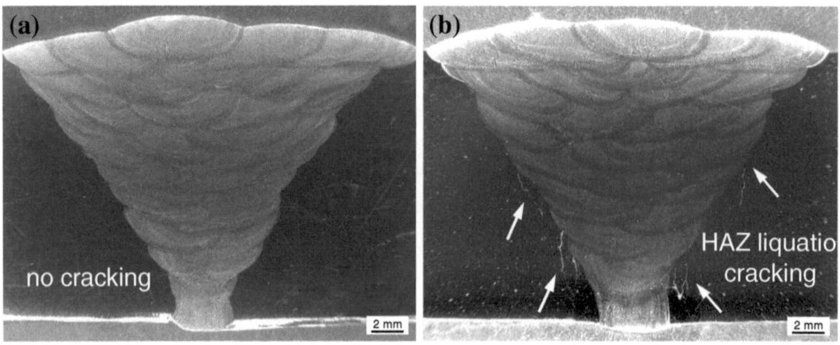


**Fig. 1** Gas tungsten arc (GTA) welded joints of wrought (**a**) and cast nickel base alloy 59 (**b**) under constant welding parameters (heat input 6–8 kJ/cm)

# Experimental Procedure

## *Materials*

Nickel base cast materials produced by two different methods, centrifugal and sand casting, were investigated in this study, see Table 1. Nickel-chromium-molybdenum alloy 59, nickel-chromium alloy 625 and nickel-chromium-cobalt alloy 617 are industrially relevant solid-solution strengthened nickel base alloys for highly corrosive or high pressure, high temperature conditions. Wrought forms of these alloys were used as a reference. The composition and heat treatment of all test materials are listed in Table 2; note the similarity of the compositions of the cast and wrought base metals.

PVR testing for quantifying material-specific cracking susceptibility was performed for all cast and wrought test materials listed in Table 2. Studies on HAZ liquation cracking susceptibility and on the metallurgical causes for cracking using hot ductility testing were only performed for centrifugal cast alloy 625 and the equivalent wrought product, also given in Table 2.

The microstructure of the alloy 625 hot ductility test materials in as-delivered condition is shown in Fig. 2. The mill-annealed wrought base metal consists of an austenitic matrix with some intergranular and intragranular precipitates in the form of blocky Nb-rich and Ti-rich carbides. Note that the wrought microstructure is fine grained (ASTM No. 9.8) and homogenous compared to the cast microstructure of the same composition, which is characterized by coarse, both equiaxed and columnar as-cast grains, which may range from 0.2 to 4 mm or larger in the major dimension (ASTM No. 00). The cast alloy 625 was tested in as-cast condition; there was no additional heat treatment applied. Alloy 625 is a nickel-chromium alloy with additions of molybdenum and niobium for solid-solution hardening. The strong segregation of niobium in the dendritic structure makes the metallurgy of the cast alloy 625 material considerably different than that of its wrought counterpart of the same nominal composition. For the interdendritic sites in the centrifugal cast base metal niobium levels as high as 13 wt% were determined by energy dispersive X-ray (EDX) analysis even though the bulk composition is only 2.9 wt%, see Table 2. Due to niobium segregation during solidification the interdendritic sites of

**Table 1** Test materials of nickel base cast and wrought base metal

| Alloy | DIN No. | UNS No. | Centrifugal casting | Sand casting | Wrought grade |
|---|---|---|---|---|---|
| 59 | 2.4605 | N06059/N26059 | X | X | X |
| 625 | 2.4856 | N06625/N26625 | X | X | X |
| 617 | 2.4663 | N06617/N26617 | X | – | X |

**Table 2** Chemical composition (Element, wt%) and heat treatment of nickel base cast and wrought materials

| Alloy | 59 | | | 625 | | | 617 | |
|---|---|---|---|---|---|---|---|---|
| Material | Centrifugal casting | Sand casting | Wrought grade | Centrifugal casting | Sand casting | Wrought grade | Centrifugal casting | Wrought grade |
| Ni | Bal. | Bal. | Bal. | Bal. | Bal. | Bal. | Bal. | Bal. |
| Cr | 23.6 | 24.0 | 22.9 | 21.5 | 21.3 | 21.3 | 22.1 | 22.2 |
| Fe | 0.03 | 0.15 | 0.60 | 2.30 | 2.13 | 4.80 | 0.24 | 1.36 |
| C | 0.010 | 0.012 | 0.004 | 0.030 | 0.055 | 0.024 | 0.070 | 0.080 |
| Mo | 15.9 | 15.6 | 15.5 | 9.0 | 9.0 | 8.6 | 9.0 | 8.6 |
| P | 0.004 | 0.002 | 0.006 | 0.012 | 0.012 | 0.007 | 0.003 | 0.002 |
| S | 0.004 | 0.004 | 0.002 | 0.005 | 0.001 | 0.001 | 0.003 | < 0.002 |
| Other | – | – | – | Nb 2.94 | Nb 3.52 | Nb 3.34 | Co 11.4<br>B 0.0045<br>Al 1.56<br>Ti 0.28 | Co 11.5<br>B 0.001<br>Al 1.06<br>Ti 0.43 |
| Heat treatment | Solution annealed | Solution annealed | Solution annealed | As-cast condition | Solution annealed | Mill annealed | Solution annealed | Solution annealed |

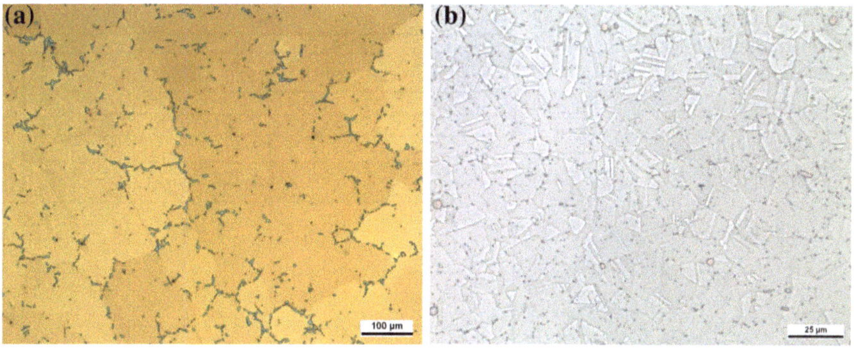


**Fig. 2** Microstructure of centrifugal cast alloy 625 (**a**) and wrought microstructure (**b**) of the same composition used for hot ductility testing. Note the difference in magnification

the cast microstructure contain a large amount of MC (NbC) carbides with blocky or Chinese script [1, 2] morphology, see Fig. 2a. No Laves phase ($Ni_2Nb$) was found in the cast base metal microstructure.

## PVR Testing

PVR (programmable deformation crack) testing was used to quantify the hot cracking susceptibility of the nickel base cast materials against their wrought equivalents (listed in Table 2). The PVR test is an externally loaded test method, which is described in ISO/TR 17641-3:2004. A gas tungsten arc (GTA) bead-on-plate weld on a flat sample is superimposed by a linearly increasing tension speed in welding direction, see Fig. 3. The initiation of hot cracking in the sample corresponds to a critical tension speed ($v_{cr}$), which directly correlates to the critical deformation rate [%/°C] described in the hot cracking theory of Prokhorov [3]. The critical tension speed can therefore be used as a criterion for assessing hot cracking resistance.

PVR samples with nominal dimensions displayed in Fig. 3 were machined from all test materials in a uniform thickness of 10 mm. Prior to testing, the surface of the samples was prepared by grinding and cleaning with acetone. Welding was performed using a fully-mechanized GTAW torch for bead-on-plate welds without filler metal. The conditions for PVR testing, listed in Table 3, have been specifically developed and were used for nickel base materials before [4]. Welding and testing parameters remained constant for all test materials. A minimum of three PVR samples was examined for each cast and wrought material. The initial macro- and micro-cracking on the sample was determined using dye penetrant testing (PT) and stereo microscopy at 25× magnification.

**Fig. 3** PVR testing—schematic illustration. Dimensions in mm

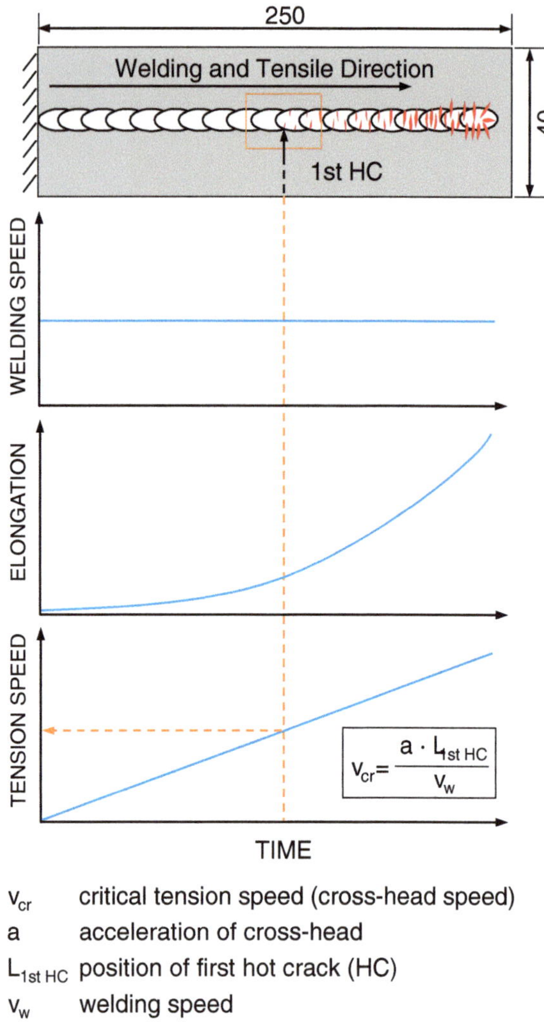


**Table 3** Conditions for PVR testing

| Parameter | |
|---|---|
| Current | 180 A |
| Voltage | 12 V |
| Welding speed | 20 cm/min |
| Maximum cross-head speed | 45 mm/min |
| Heat input | 6.5 kJ/cm |
| Sample thickness | 10 mm |
| Shielding gas | Argon |

## Hot Ductility Testing

Hot ductility testing was performed on cast and wrought alloy 625 base metal using a Gleeble® 3500 thermo-mechanical simulator. Wrought samples were machined from 16 mm thick plate. Cast samples were created by sectioning a 250 mm inner-diameter pipe of 42 mm wall thickness. For each section three samples were machined from across the wall thickness, see Fig. 4. The nominal dimensions of the hot ductility samples are displayed in Fig. 4. Temperature was controlled by a K-type thermocouple (NiCr-Ni, 0.25 mm in diameter) percussion welded on the surface at the sample center.

The conditions for Gleeble® hot ductility testing are listed in Table 4. Although the hot ductility test is roughly described in ISO/TR 17641-3:2004, there is no standardized testing procedure. The testing parameters used in this study are geared to follow the procedure of Lin et al. [5], intending to evaluate the reproducibility of hot ductility testing by comparing the results for the wrought alloy 625 to previous work. The nil-strength temperature (NST) of the cast and wrought test material was determined by subjecting the sample to a small static load of approximately 10.2 kg and heating at a linear rate of 111 K/s until sample failure. The NST was determined as mean value of three tested samples each at the point where the recorded temperature begins to deviate from the programmed temperature. On-heating tests

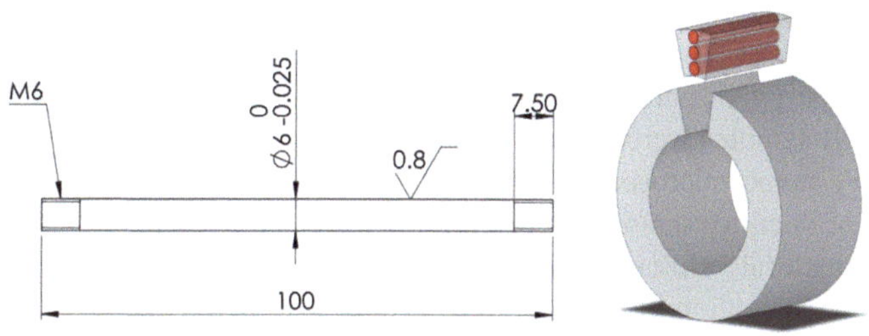


**Fig. 4** Hot ductility sample schematic (*left*) and sampling from centrifugal cast alloy 625 pipe (*right*). Dimensions in mm

**Table 4** Conditions for Gleeble® hot ductility testing

| Parameter | On-heating | On-cooling |
|---|---|---|
| Heating rate (K/s) | 111 | 111 |
| Peak temperature | $T_{test}$ | 1235 °C (NDT < $T_{peak}$ < NST) |
| Holding time at peak temperature | N/A | 0.2 s |
| Cooling rate | N/A | free-cool (35 K/s) |
| Holding time at test temperature (s) | 0.3 | 0.3 |
| Stroke rate (mm/s) | 50 | 50 |
| Sample freespan (mm) | 25 | 25 |
| Atmosphere | Argon | Argon |

were conducted by heating samples at a rate of 111 K/s to the testing temperature and pulling them to failure at a rate of 50 mm/s after a short holding time of 0.3 s. On-cooling tests were heated at a rate of 111 K/s to a temperature of 1235 °C. This peak temperature is between NDT and NST for both the cast and wrought test material and had to be chosen because of extensive liquation near the NST especially for the cast samples. The samples were held at 1235 °C for 0.2 s and then free cooled to the testing temperature, where they were pulled to failure after a short holding time of 0.3 s. Free-cooling was through conduction with the Gleeble® jaws at a rate of 35 K/s, which was the maximum achievable without externally assisted cooling.

After testing, the cross-sectional area of both fractured surfaces was measured by image analysis using a stereo microscope. Sample ductility, in terms of reduction-in-area, was calculated as mean value by comparing to the original cross-sectional area of the sample. Plotting the reduction-in-area versus testing temperature allowed the development of a ductility "signature" [5, 6] for the wrought and cast alloy 625 test material.

## Metallurgical Characterization

Microstructures of as-received cast and wrought base metals were characterized in the etched condition using an optical microscope at magnifications up to 500×. Nickel base cast metallurgy was further examined by scanning electron microscope (SEM) and energy dispersive X-ray (EDX) analysis. Surface sections of representative PVR samples were metallographically prepared to confirm the test results conducted by stereo microscope with respect to the type and position of initial hot cracking. Hot ductility samples were ground to the center preparing cross-sections for microstructural characterization by optical microscopy up to 1000×. Fractographic evaluation of the fractured surfaces was performed using a scanning electron microscope.

## Results

### PVR Tests

The PVR test results are presented in Fig. 5. An increased critical tension speed indicates a greater resistance of the examined test material to the occurrence of hot cracking. As shown in Fig. 5, for all tested nickel base alloys the wrought forms exhibit a considerably greater resistance to hot cracking compared to the centrifugal and sand cast materials of the same composition. All cast nickel base alloys are characterized by a high tendency towards hot cracking, since cracking initiates at a

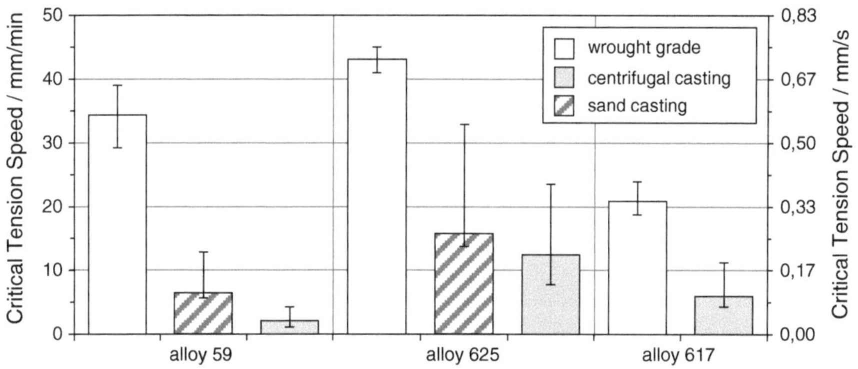


**Fig. 5** PVR test results for nickel base cast and wrought materials

very low critical tension speed during PVR testing. Cast materials of alloy 59 exhibit the highest susceptibility.

The first type of hot cracking to occur in all cast and wrought nickel base test materials was liquation cracking (LC) in the base metal HAZ adjacent to the fusion line of the GTA bead-on-plate weld. Cracking formed along the grain boundaries of the heat-affected cast and wrought microstructure, see Fig. 6. Metallurgical evaluation revealed evidence of local liquation at the grain boundaries adjacent to the fusion line. For all wrought nickel base alloys the first liquation cracks in the surface sections of the PVR samples were quite short (approx. 25 μm), located clearly in the partially melted zone (PMZ) near to the fusion line, as is illustrated in Fig. 6 (bottom). In the cast nickel base materials, the cracks also initiated at liquated grain boundaries near the fusion line, but propagated further along the grain boundaries into the base metal HAZ microstructure, see Fig. 6 (top). Therefore, all cast PVR samples were characterized by a significantly higher crack length of the initial cracks (up to one millimeter) compared to the wrought samples.

## Hot Ductility Tests

The hot ductility test results for the wrought alloy 625 are presented in Fig. 7 together with previous results from Lin et al. [5]. The on-heating ductility decreases rapidly above 1100 °C and approaches zero at 1220 °C, see Fig. 7 (left). The temperature at which the ductility drops to zero measured on-heating is termed the nil-ductility temperature (NDT). For the on-heating hot ductility curve the results obtained in this study are in good correlation with the work of Lin et al. [5], since the fairly rapid ductility drop at temperatures above 1100 °C is nearly the same and the NDT is equivalent, see Fig. 7 (right). The same goes for the point at which the strength is zero, termed the nil-strength temperature (NST), which is approximately 1308 °C and was determined by Lin et al. [5] at 1313 °C.

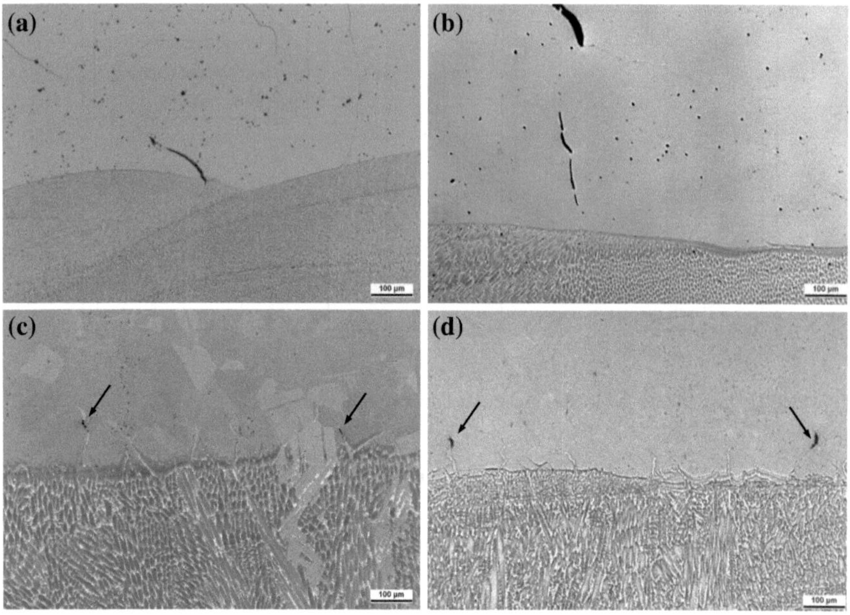


**Fig. 6** Initial HAZ liquation cracking in PVR samples of centrifugal cast (*top*) and wrought (*bottom*) nickel base alloy 617 (**a**, **c**) and alloy 59 (**b**, **d**)

Significant differences compared to the previous results were obtained in terms of the on-cooling ductility curve. This is due to the different peak temperature. Lin et al. [5] heated to the NST before cooling to the testing temperature and pulling the sample to failure. The ductility of their test material did not recover significantly until it had been cooled below 1150 °C, see Fig. 7 (right). This point at which measurable ductility is observed is termed the ductility recovery temperature (DRT), normally considered at 5 % reduction-in-area on-cooling. In this study

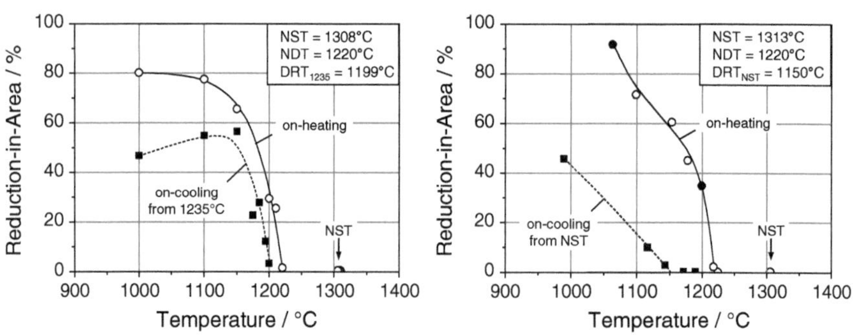


**Fig. 7** Comparison of hot ductility test results for wrought alloy 625 (*left*) with results from Lin et al. [5] (*right*)

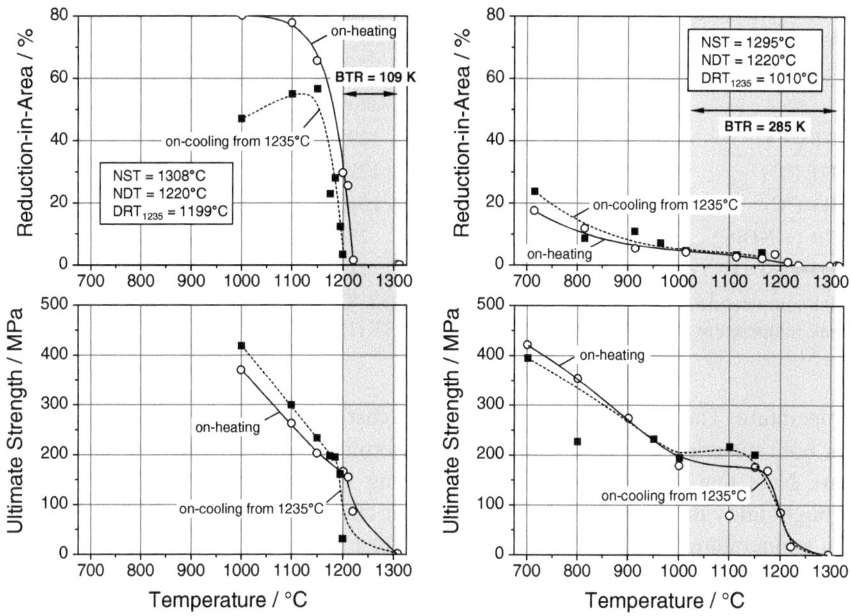


**Fig. 8** Hot ductility test results for wrought (*left*) and cast alloy 625 (*right*)

on-cooling tests were performed using a constant peak temperature of 1235 °C for the cast and wrought test material, since there was extensive liquation at the NST for the cast samples. The lower peak temperature (between NDT and NST) resulted in a fairly abrupt recovery of ductility at temperatures below 1200 °C, see Fig. 7 (left), hence the DRT is higher than the one determined by Lin et al. [5]. This is in agreement with the literature, since the selection of the peak temperature for the on-cooling tests is reported to be critical to the hot ductility response [7].

Figure 8 compares the hot ductility test results for the wrought and cast alloy 625. Note that the on-heating and on-cooling ductility curves are quite different for the cast alloy. Since the on-heating ductility was fairly low, testing of the cast samples was expanded to temperatures as low as 700 °C, but the ductility did not significantly increase at decreased temperatures. The NDT is equivalent to the wrought alloy with ductility approaching zero at 1220 °C, while the NST is slightly lower at 1295 °C. On-cooling from the peak temperature of 1235 °C as for the wrought samples, the ductility of the cast alloy was as low as on-heating and did not recover significantly in the testing temperature range. Therefore, the $DRT_{1235}$ (1010 °C) for the cast alloy is well below the one for the wrought material. The difference in the DRT temperatures indicates that the brittle temperature range (BTR), which is the temperature range between NST and DRT, differs widely for the cast and wrought alloy. The BTR for the cast alloy (285 K) is more than twice as large as for the tested wrought equivalent (109 K).

Besides the reduction-in-area of the fractured samples, the ultimate tensile strength was determined after fracturing and plotted as a function of the test

**Table 5** Hot ductility results of tested materials of nickel base alloy 625 and previous results from Lin et al. [5]

| | Wrought grade[1] alloy 625 | Centrifugal casting[1] alloy 625 | Wrought grade[2] Lin et al. [5] |
|---|---|---|---|
| NST (°C) | 1308 | 1295 | 1313 |
| NDT (°C) | 1220 | 1220 | 1220 |
| DRT (°C) | 1199 | 1010 | 1150 |
| BTR (= $NDR_{on\text{-}cooling}$) (K) | **109** | **285** | **163** |
| NDT-DRT (K) | 21 | 210 | 70 |

[1]Peak temperature for on-cooling ductility curve: 1235 °C
[2]Peak temperature for on-cooling ductility curve: NST (1313 °C)

temperature. The results for the wrought and cast alloy are also presented in Fig. 8. For both materials the NST temperatures determine the point where the strength is zero. Note that the on-heating and on-cooling strength curves are higher for the wrought alloy than for its cast counterpart for temperatures below 1200 °C. Above that temperature the strength curves for both test materials are nearly identical. The hot ductility test results are summarized in Table 5.

The microstructures near the fracture surface of cast and wrought hot ductility samples tested at NST and NDT are shown in Fig. 9. For the wrought alloy there is no grain boundary liquation apparent at the NDT. However, there is evidence of liquation around grain boundary carbides for the cast alloy, which resulted in the formation of thin films as well as cavities along grain boundaries in the cast microstructure, see Fig. 9b. Extensive liquation occurs at the NST temperatures for both tested materials. Note the complete coverage of the grain boundaries by thick liquid films, which formed a dense network of intergranular cavities adjacent to the fractured surface especially for the wrought alloy. The cast sample is likewise characterized by massive intergranular liquation. The fine grained structure indicates a refinement of the coarse as-cast grains probably due to melting and resolidification in the near-fracture region of the sample during the NST testing.

The fractured surfaces of representative wrought and cast hot ductility samples are shown in Figs. 10 and 11, respectively. The wrought sample tested at NST in Fig. 10a is characterized by a fine structure of rounded grain faces due to thick liquid films, which solidified at the fracture surface. On the on-heating hot ductility curve the wrought samples exhibit a dimpled ductile fracture morphology for temperatures up to 1100 °C. As the temperature increases to 1150 °C, there is an intergranular fracture with a spongy morphology at the grain faces, which eventually disappears near the NDT, where the fracture surface is characterized by an intergranular fracture mode with more rounded grain faces, see Fig. 10b. When cooling from the peak temperature of 1235 °C to the DRT of 1199 °C the fracture surface in Fig. 10c exhibits a featureless intergranular morphology with already slightly distinct grain faces. With decreasing temperature up to 1150 °C there is again an intergranular spongy morphology before the fracture mode changes to ductile at temperature below 1000 °C, see Fig. 10d.

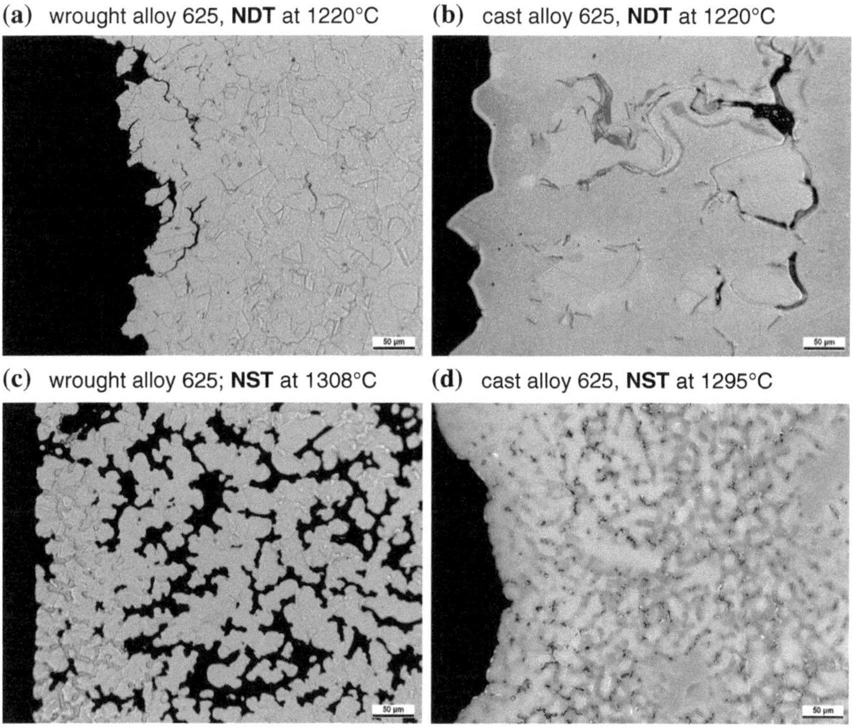


**Fig. 9** Microstructure of hot ductility samples of wrought and cast alloy 625 tested at NDT and NST

The fracture surface of the cast hot ductility sample tested at NST is presented in Fig. 11a. It is nearly identical to that of the wrought sample in Fig. 10a, exhibiting a fine structure with rounded grain faces covered in a thick layer of solidified liquid. On the on-heating curve the cast samples are characterized by a brittle fracture morphology with a low-ductile portion below temperatures of 1000 °C. The coarse interdendritic carbide particles in the cast microstructure undergo little plastic deformation and act as a crack initiation site for brittle fracture. Up until the NDT signs of plastic deformation fully decrease; the fracture surface exhibiting a nearly complete intergranular morphology with an increasing portion of smooth rounded grain faces, see Fig. 11c. In the temperature range between 1000 °C and NDT, the fracture surface is interrupted by coarse brittle carbide particles surrounded by a spongy morphology, which is evidence of partial liquation, see Fig. 11b.

During cooling from the peak temperature in the temperature range between 1235 °C and DRT (1010 °C), the fracture surface exhibits a primarily intergranular fracture with large-scale smooth grain faces and partial spongy morphology around carbides as the temperature decreases to 1100 °C. The spongy morphology diminishes and eventually disappears near the DRT. Further decrease in on-cooling test temperature up to 700 °C results in a primarily brittle fracture mode with an increasing ductile portion, see Fig. 11d.

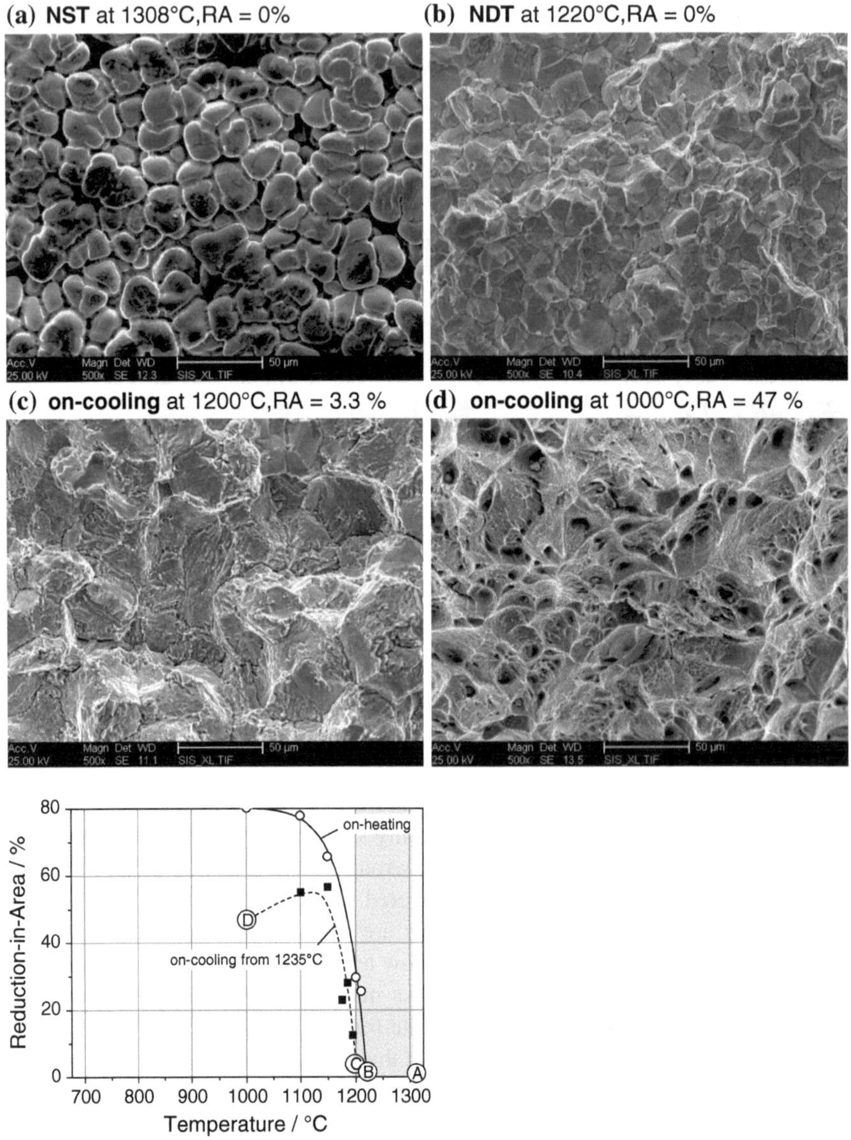


**Fig. 10** Fracture surfaces of representative hot ductility samples of wrought alloy 625 and location of fractographs with respect to the hot ductility curve

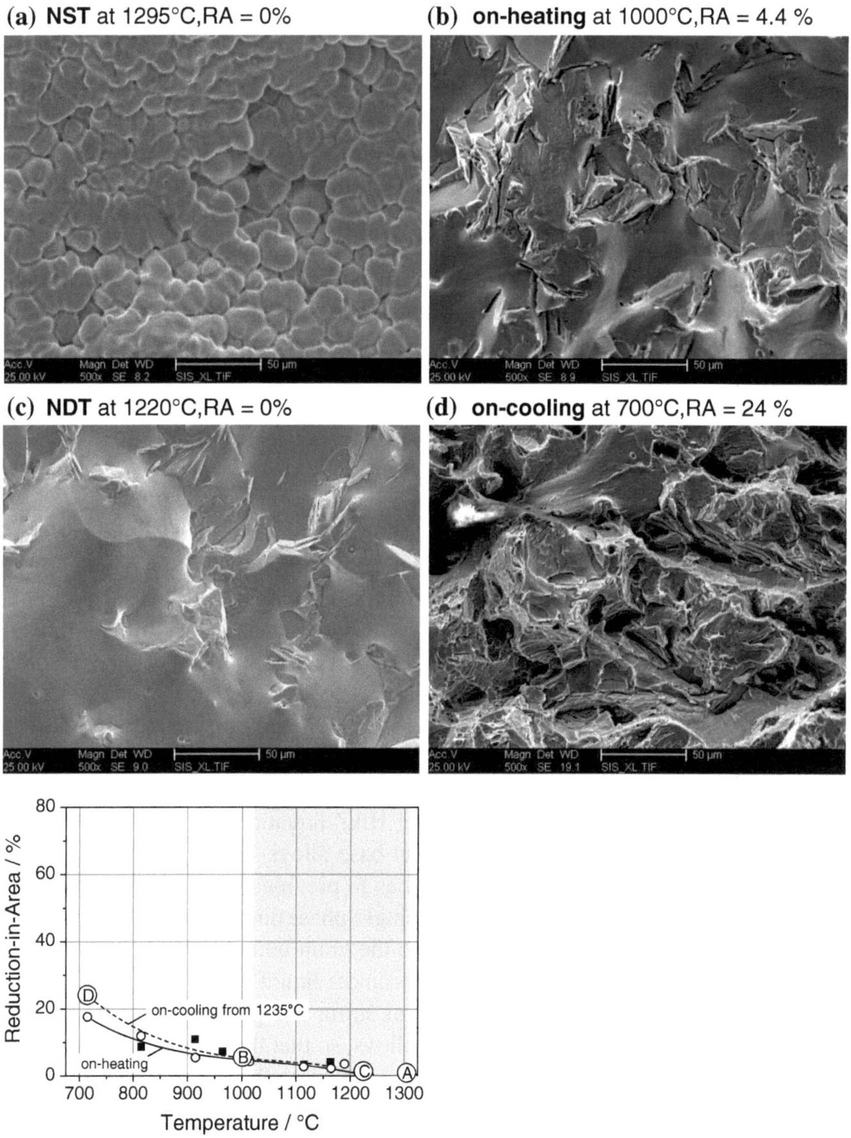


**Fig. 11** Fracture surfaces of representative hot ductility samples of centrifugal cast alloy 625 and location of fractographs with respect to the hot ductility curve

## Discussion

Evaluation of hot cracking susceptibility by means of the PVR test has shown that in accordance to actual welds on cast nickel base alloys the tested cast materials are characterized by a high tendency towards liquation cracking in the base metal heat-affected zone (HAZ). Their cracking susceptibility is much higher than that of the wrought products of the same composition.

Crack initiation in the cast PVR samples occurred at much lower critical tension speed compared to the wrought samples. However, for both the cast and wrought alloys the first cracking to occur in the samples was HAZ liquation cracking; the crack susceptible region being the partially melted zone (PMZ) just adjacent to the fusion line of the GTA bead-on-plate weld, which was characterized by localized grain boundary melting. The liquid within this locally melted region could not accommodate the externally induced stresses experienced during PVR testing and formed cracks along HAZ grain boundaries.

For the cast samples the first cracks propagated much further into the base metal after crack initiation than for the wrought samples, resulting in much higher crack lengths. This is due to the coarse microstructure and pronounced segregation of the cast alloys. Cracks propagated at the long grain boundaries and formed up to one millimeter in length, as they were less blocked by grain boundary triple points. In contrast, the first hot cracks in the wrought samples were quite short and occurred only at locally liquated grain boundaries just adjacent to the fusion line.

Besides this structural effect on crack propagation and crack length in the PVR samples, for the explanation of the much higher liquation cracking susceptibility of the cast materials the mechanism of grain boundary liquation in the PMZ has to be considered. The phenomenon of base metal HAZ liquation has been observed for the tested solid-solution strengthened nickel base alloys—alloy 59, alloy 617 and alloy 625—by several investigators as well as in previous work of the authors [4]. For nickel base alloys which are primarily single-phase austenite, the segregation of alloy and impurity elements (S, P, Pb, B) to the grain boundaries can cause a local depression of the melting temperature and promote liquid grain boundary films [6]. The segregation of impurity elements such as sulfur and phosphorus is expected to have a detrimental effect in all nickel base alloys, so that those impurities should be held to levels as low as possible. In conjunction with boron, the segregation mechanism has been proposed for alloy 617, where boron is added to improve creep properties, but may promote liquation cracking [8]. The second mechanism promoting HAZ liquation cracking is known as penetration mechanism and is due to constitutional liquation [9] or eutectic melting around secondary constituents, such as intermetallics, carbides or TCP phases [6]. This type of liquation has been observed for nickel base alloy 625, where constitutional liquation is initiated by the interaction of the austenitic matrix and NbC carbides [5]. The liquid then penetrates the grain boundaries and can promote cracking in the HAZ.

For the tested wrought and cast nickel base alloys, impurity elements such as sulfur and phosphorus are controlled to very low levels. However, the coarse microstructure of the cast alloys results in significant less grain boundary area and therefore, a much higher concentration of detrimental i.e. liquation-inducing material at the grain boundaries. Although, this might be the reason for the higher cracking susceptibility of the tested cast materials of alloy 59, hot cracking promoting elements such as sulfur and phosphorus could not be detected at the liquated grain boundaries near the fusion zone. For alloy 617 it is obvious though, that the tested cast alloy contains four times as much boron (0.0045 wt%) as the wrought product (0.001 wt%). Based on the discussion above and the even higher concentration at the grain boundaries due to the coarse microstructure, this may cause the higher cracking susceptibility of the cast alloy 617 in the PVR test. However, it was not possible to detect boron in the liquid grain boundary films.

The centrifugal and sand cast structure of alloy 625 contains a large amount of interdendritic MC (NbC) carbides due to niobium segregation during solidification. The crack susceptible region of the cast PVR samples of alloy 625 showed extensive liquation adjacent to the fusion line of the GTA bead-on-plate weld. As illustrated in Fig. 12b, evidence of severe liquation in the vicinity of the NbC carbides in the high temperature zone of the HAZ and subsequent wetting of base metal grain boundaries was apparent for the cast alloy. Since constitutional

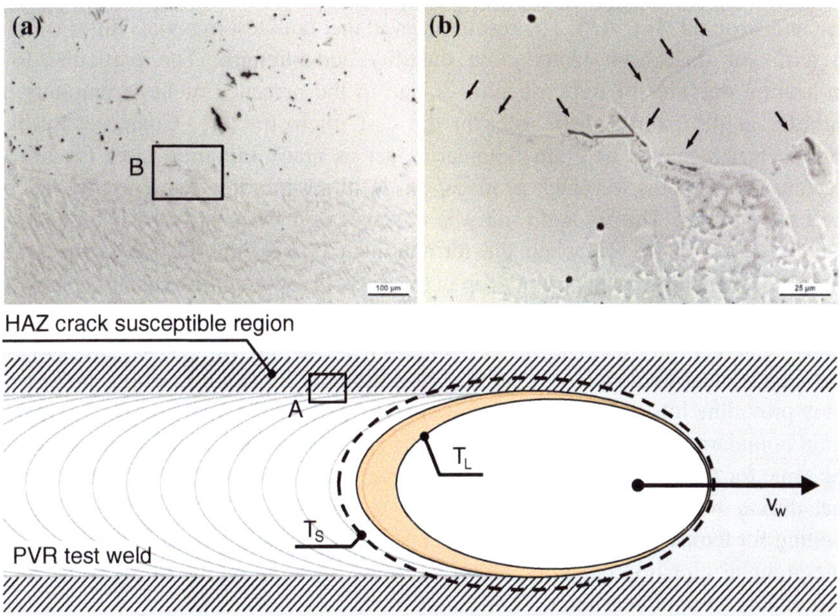


**Fig. 12** Crack susceptible region in the heat affected zone (HAZ) of cast alloy 625 near the fusion line of the PVR test weld; note HAZ liquation around MC (NbC) carbides and liquid films at grain boundaries (*arrows*)

liquation of NbC carbides promotes HAZ liquation cracking in the wrought condition [5], this indicates an intense carbide constitutional liquation reaction in the cast structure. If the degree of liquation is a function of carbide distribution and grain size, there is more residual liquid in the PMZ of the cast PVR samples as for the wrought equivalent. The liquid penetrates along the coarse as-cast microstructure and wets the base metal grain boundary to a greater extent, promoting subsequent HAZ liquation cracking even at some distance from the fusion line, see Fig. 12a.

Based on the results of the PVR testing and the discussion above, there are apparent differences between the amount of liquation and the associated degree of HAZ liquation susceptibility between the cast and the wrought nickel base alloys. Metallurgically, HAZ liquation cracking is associated with an exhaustion of available high temperature ductility due to the presence of liquid grain boundary films in the partially melted zone. During PVR testing, the HAZ cracking susceptible region forms around the advancing weld pool, as is illustrated in Fig. 12. Gleeble hot ductility testing is used to simulate the HAZ before (on-heating) and behind (on-cooling) the weld pool and to measure its mechanical properties. By this, the magnitude of the HAZ crack susceptible region can be determined and compared for the cast and wrought nickel base products.

Hot ductility testing allowed the development of a ductility "signature" for the cast and wrought alloy 625. The results showed that both test materials differ widely in terms of their high temperature ductility and strength. The particular low on-heating ductility of the cast alloy is due to the presence of large amounts of carbides at the interdendritic sites in the cast microstructure. Coarse chain-like carbide arrangements at grain boundaries act as crack initiation sites for brittle fracture and as paths for crack propagation, resulting in a low reduction-in-area of the cast samples. The wrought alloy is characterized by a fairly high on-heating ductility up to 1100 °C, when the formation of liquid films begins at the grain boundaries resulting in an abrupt drop in ductility. Fractographic evaluation showed that on-heating grain boundary liquation in the cast alloy begins at lower temperatures (above 1000 °C). The spongy morphology around the coarse NbC carbide particles is clear evidence of a carbide constitutional liquation reaction in the cast alloy providing lots of liquid for grain boundary penetration and wetting. Complete grain boundary liquation is not reached until the NDT, which was determined to be the same for the cast and wrought alloy. The on-heating hot ductility results indicate that during weld heating, the cast alloy is susceptible to partial grain boundary melting for temperatures above 1000 °C and sufficient imposed strain may result in intergranular liquation cracking. However, the on-heating region of the weld HAZ is probably in compression.

On-cooling hot ductility testing had to be performed for both test materials by cooling from a peak temperature of 1235 °C instead from the NST. The lower peak temperature had a significant influence of the on-cooling hot ductility behavior, as

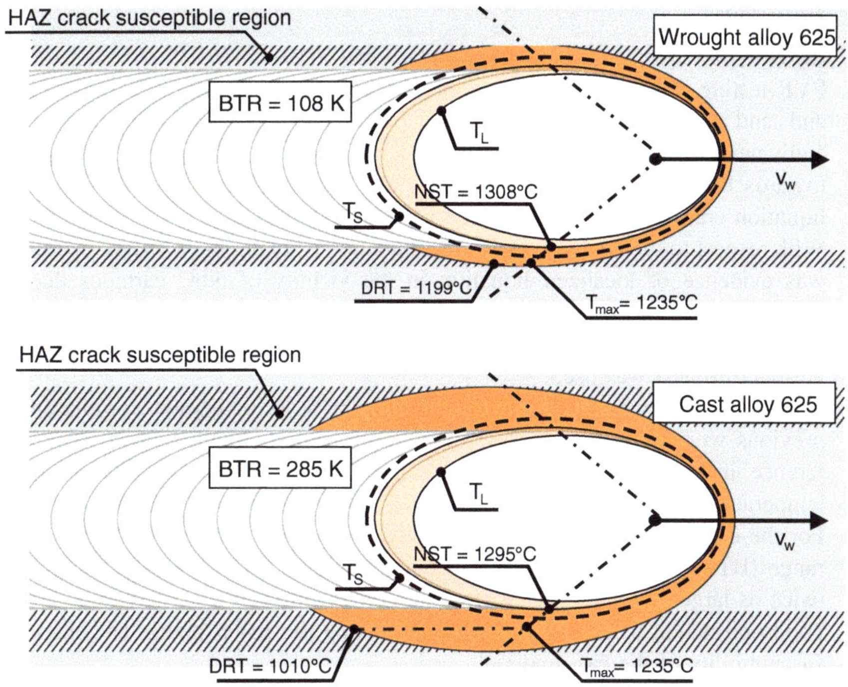


**Fig. 13** Schematic illustration of the HAZ crack susceptible region of wrought (*top*) and cast (*bottom*) alloy 625 as determined by hot ductility testing

the determined DRT temperatures were shifted to higher temperatures for both test materials. At the DRT the alloy begins to recover ductility upon cooling from the peak temperature, since intermittent solid grain contact re-develops at the grain boundaries. For weld cooling, the temperature range between the DRT and the NST therefore represents a critical temperature range within which the material is susceptible to liquation cracking due to localized loss of grain boundary ductility. This so called brittle temperature range (BTR) is material- specific and is typically used as an indicator for HAZ cracking susceptibility. Alloys with a wider BTR value exhibit more extensive liquation and associated cracking susceptibility. Based on this approach, the extent of the BTR as obtained from the hot ductility testing differs widely between the tested cast and wrought alloy 625. The BTR of the tested cast alloy (285 K) is almost three times as wide as for the wrought material (108 K). It can be concluded, that the magnitude of the HAZ crack susceptible region around an advancing weld pool is much higher during welding of cast alloy 625, see Fig. 13; resulting in a higher HAZ cracking susceptibility as compared to the wrought product. This is in good correlation to the PVR test results.

## Conclusions

1. PVR testing was used to evaluate the hot cracking susceptibility of centrifugal and sand cast nickel base alloy 59, alloy 617 and alloy 625 against their wrought equivalents. The cast alloys were characterized by a significant higher tendency towards hot cracking in the PVR test. The first cracking to occur was HAZ liquation cracking for all cast nickel base alloys.
2. With respect to the occurrence of HAZ liquation cracking in cast alloy 625 there was evidence of localized liquation in the vicinity of NbC carbides due to constitutional liquation. To further study the HAZ crack susceptible region (partially melted zone) of cast and wrought nickel base alloy 625 the Gleeble® hot ductility test was used.
3. Hot ductility test results for wrought alloy 625 showed good correlation to previous work from Lin et al. [5] for the on-heating ductility curve. The difference in the on-cooling behavior can be explained by the different peak temperature used in this study.
4. For the cast alloy 625 tested in this study, the extent of the brittle temperature range (BTR) determined by the hot ductility test is 285 K. This is more than twice as large as for the tested wrought equivalent (109 K). The results are in good correlation to the PVR testing, indicating the much higher hot cracking susceptibility of the cast material.

**Acknowledgements** These tests were funded by the Federal Ministry of Economics and Energy (BMWi) assisted by the Arbeitsgemeinschaft industrieller Forschungs- vereinigungen (AiF—industrial research consortium) "Otto von Guericke" e.V. (AiF-No. 17.403 B/DVS-No. 01.069) and supported by the DVS Forschungsvereinigung Schweißen und verwandte Verfahren e.V. (DVS—German Research Association on Welding and Allied Processes).

## References

1. Sabol, G. P.; Stickler, R.: Microstructure of Nickel-Based Superalloys. physica status solidi (b) 35(1), 1969, p. 11–52.
2. Floreen, S.; Fuchs, G. E.; Yang, W. J.: The Metallurgy of Alloy 625. Superalloys 718, 625, 706 and various derivatives, 1994, p. 13–37.
3. Prokhorov, N.: The technological strength of metals while crystallizing during welding. IIW-Doc. IX-479-65, 1965.
4. Fink, C.; Zinke, M.; Keil, D.: Evaluation of Hot Cracking Susceptibility of Nickel- Based Alloys by The PVR Test. Welding in the World 56(7-8), 2012, p. 37–43.
5. Lin, W.; Nelson, T. W.; Lippold, J. C., et al.: A study of the HAZ crack susceptible region in Alloy 625. International trends in welding science and technology, Tagungsband, 1993, p. 695–702.
6. DuPont, J. N.; Lippold, J. C.; Kiser, S. D.: Welding metallurgy and weldability of nickel-base alloys, Wiley Hoboken, NJ, 2009.

7. Lundin, C. D.; Qiao, C. Y. P.; Lee, C. H.: Standardization of Gleeble Hot Ductility Testing: Part II: Experimental Evaluation. In: Patterson, R. A. (Hrsg.) Weldability of materials. ASM International Materials Park, Ohio, 1990, p. 9–22.
8. Lundin, C. D.; Qiao, C. Y. P.; Swindeman, R.: HAZ hot cracking behavior of HD 556 and Inconel 617. International trends in welding science and technology, Tagungsband, 1993, p. 801–06.
9. Pepe, J. J.; Savage, W. F.: Effects of constitutional liquation in 18-Ni maraging steel weldments. Welding Journal, Welding Research Supplement 46(9), 1967, p. 411s-422s.

# Part V
# Cold Cracking Phenomena

# Numerical Investigations on Hydrogen-Assisted Cracking in Duplex Stainless Steel Microstructures

**T. Mente and Th. Boellinghaus**

**Abstract** Duplex stainless steels (DSS) are used in various industrial applications, e.g. in offshore constructions as well as in chemical industry. DSS reach higher strength than commercial austenitic stainless steels at still acceptable ductility. Additionally, they exhibit an improved corrosion resistance against pitting corrosion and corrosion cracking in harsh environments. Nevertheless, at specific conditions, as for instance arc welding, cathodic protection or exposure to sour service environments, such materials can take up hydrogen which may cause significant property degradation particularly in terms of ductility losses which, in turn, may entail hydrogen-assisted cracking (HAC). The cracking mechanism in DSS is different from steels having only a single phase, because hydrogen diffusion, stress-strain distribution and crack propagation are different in the austenite or ferrite phase. Therefore, the mechanism of HAC initiation and propagation as well as hydrogen trapping in DSS have not been fully clarified up to the present, as for most of the two-phase microstructures. At this point the numerical simulation can bridge the gap to a better insight in the cracking mechanism regarding the stress-strain distribution as well as hydrogen distribution between the phases, both austenite and ferrite, of the DSS. For that purpose, a two dimensional numerical mesoscale model was created representing the microstructure of the duplex stainless steel 1.4462, consisting of approximately equal portions of austenite and ferrite. Hydrogen assisted cracking was simulated considering stresses and strains as well as hydrogen concentration in both phases. Regarding the mechanical properties of austenite and ferrite different statements can be found in the literature, dependent on chemical composition and thermal treatment. Thus, various stress-strain curves were applied for austenite and ferrite simulating the HAC process in the DSS microstructure. By using the element elimination technique crack critical areas can be identified in both phases of the DSS regarding the local hydrogen concentration and the local mechanical load. The results clearly show different cracking behavior

T. Mente (✉) · Th. Boellinghaus
BAM Federal Institute for Materials Research and Testing, Berlin, Germany
e-mail: Tobias.Mente@bam.de

Th. Boellinghaus
e-mail: Thomas.Boellinghaus@bam.de



Th. Boellinghaus et al. (eds.), *Cracking Phenomena in Welds IV*,
DOI 10.1007/978-3-319-28434-7_16

with varying mechanical properties of austenite and ferrite. Comparison of the results of the numerical simulation to those of experimental investigations on DSS will improve understanding of the HAC process in two phase microstructures.

## Introduction

During the last decades there is an increasing demand on steels applied to aggressive environments. Economical and design aspects lead to increasing requirements for steels concerning the resistance to corrosion as well as the toughness of used steels. Thus, duplex stainless steels (DSS) are used replacing commercial austenitic stainless steels [1, 2], i.e. in offshore structures as well as in the chemical and petrochemical industry. Duplex stainless steels have higher strength than conventional austenitic stainless steels and they exhibit a better ductility as well as an improved corrosion resistance in harsh environments compared to ferritic stainless steels. However, despite these good properties the literature shows some failure cases of DSS in which hydrogen plays a crucial role for the cause of the damage.

As for example, cracking of a subsea manifold hub was reported by Taylor et al. [3]. The crack appeared in a forged manifold of DSS in a short distance to the weld due to hydrogen uptake by cathodic protection, Fig. 1. Additionally, the DSS had a coarse microstructure and the geometry of the hub results in different locations of maximum load during pressure test and during service. This leads to the absence of beneficial effects from local stress relaxation (shakedown) [3]. Thus, during service local plastic deformation occurs while hydrogen is present in the microstructure which may lead to cracking. For investigating the reason of cracking Taylor et al. used simple laboratory tests to full scale tests supported by numerical simulations.

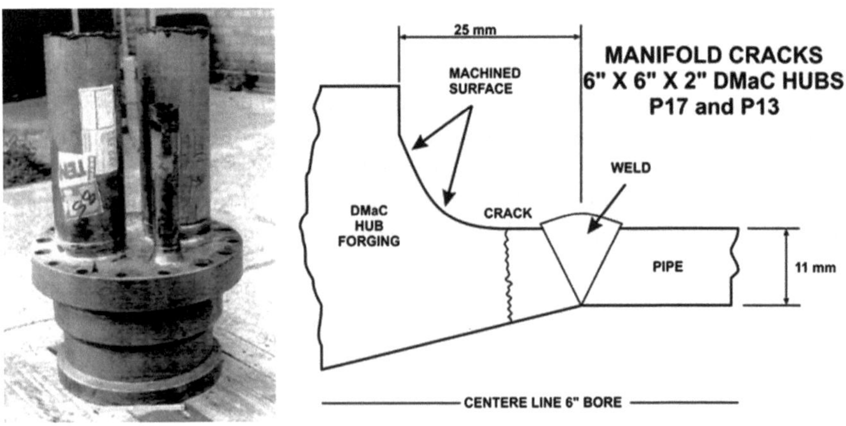


Fig. 1 Crack location in the manifold hub made of forged DSS [3]

Numerical simulations can give a significant contribution in clarifying the damage mechanisms. They help to interpret experimental results as well as help to transfer results from laboratory tests to component tests and vice versa. So far most numerical simulations of hydrogen-assisted material damage were performed at the macroscopic scale [4–8]. However, DSS consist of approximately equal portions of austenite and δ-ferrite. Both phases have different mechanical properties [9–15] as well as hydrogen transport properties [16, 17]. Thus, the sensitivity for hydrogen-assisted damage is different in both phases, too. By comparing both phases of the DSS the austenitic phase should be less sensitive to a degradation of its mechanical properties due to hydrogen and the ferritic phase should have a higher sensitivity [18–20]. The different cracking behavior was shown by investigations of hydrogen-assisted cracking (HAC) in DSS [21, 22], Fig. 2. The cracks mainly initiate in the ferritic phase and often the cracks are arrested by austenite.

As seen from the reported failure of a DSS [3] hydrogen-assisted material damage rarely occur by single hydrogen uptake into the material. Rather the local microstructure with its specific properties and last but not least the magnitude of local mechanical load influences the HAC process most recently. Thus, in order to investigate HAC in DSS all influencing factors have to be taken into account, in particular local hydrogen concentration, local mechanical load and local microstructure, Fig. 3. Additionally, all three main factors may be influenced by manufacturing/fabrication process or by service conditions.

Therefore, the objective of this research was to develop a numerical mesoscale model of a duplex stainless steel microstructure enabling simulation of hydrogen transport, mechanical stresses and strains as well as crack initiation and propagation in both phases. Thus, the critical locations for cracking in the DSS microstructure as well as crack critical combinations of hydrogen concentration and local mechanical load can be identified. Hence, it helps clarifying the damage mechanisms and improves the understanding of hydrogen-assisted material damage in two phase microstructures.

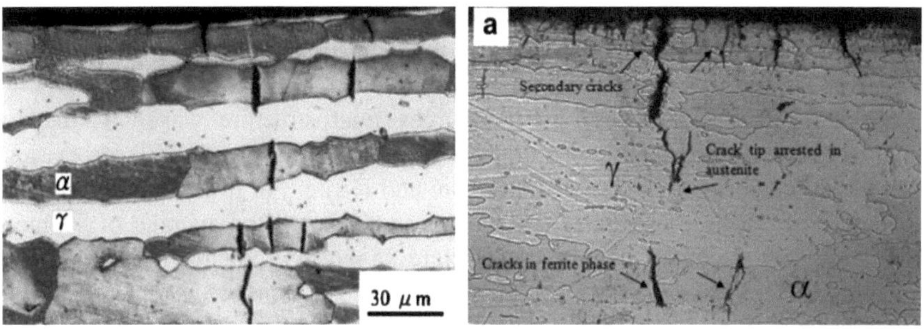


**Fig. 2** Microscopic images of hydrogen-assisted cracks in DSS, cracking in the ferritic phase and crack arrest in the austenitic phase [21, 22]

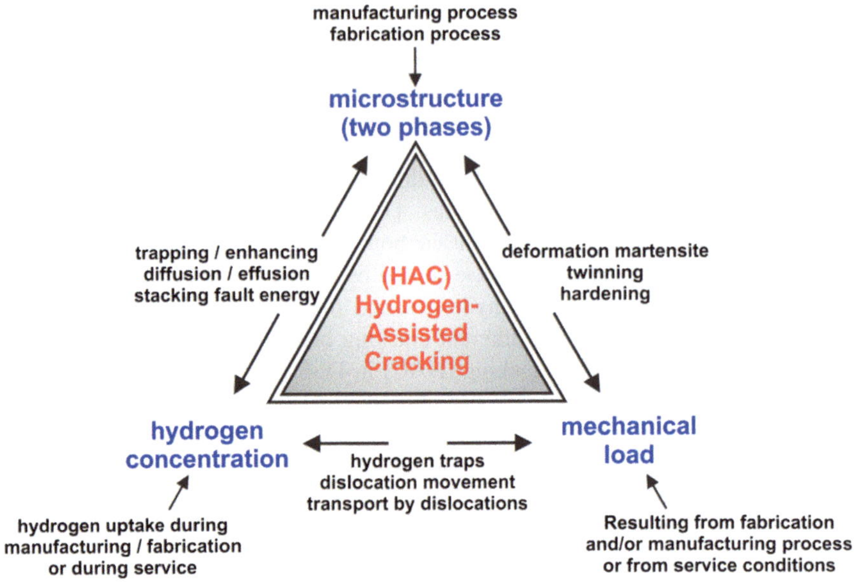


**Fig. 3** Influences on hydrogen-assisted cracking

## Numerical Procedure for Simulating Hydrogen-Assisted Cracking

### *Material Used—Chemical Composition and Mechanical Properties*

Realizing the numerical simulation of HAC the material properties of the DSS X2CrNiMoN 22-5-3 (Mat.-No.: 1.4462 according to DIN EN 10088-2:2012) were used. The chemical composition is listed in Table 1. The mechanical properties of the DSS were evaluated by tensile tests, with tensile specimens cut from a wrought DSS sheet of 6 mm thickness transverse to the rolling direction. The results are shown in Table 1.

Metallographic examination of the duplex microstructure was performed in order to check if intermetallic precipitates ($\sigma$-phase, $\chi$-phase, etc. [20]) are existent as well as to determine the ratio of austenite and ferrite. For evaluating the ratio of austenite and ferrite the etchant Beraha II was used and the average of 47 measurements was calculated. The microstructure of the DSS is shown in Fig. 4. No precipitates were visible in the DSS microstructure. The ratio of austenite and ferrite was 52:48 vol.% and is near the optimum ratio of 50:50 vol.% [9].

**Table 1** Chemical composition and mechanical properties of the duplex stainless steel X2CrNiMoN 22-5-3

| *Chemical composition according to DIN EN 10088-2 in wt.%* | | | | | | | | |
|---|---|---|---|---|---|---|---|---|
| | Fe | C | Cr | Ni | Mo | Mn | N | Si |
| Min. | Bal. | – | 21.00 | 4.50 | 2.50 | – | 0.100 | – |
| Max. | Bal. | 0.030 | 23.00 | 6.50 | 3.50 | 2.00 | 0.220 | 1.00 |
| *Measured chemical composition in wt.%* | | | | | | | | |
| **Ist** | Rest | 0.027 | 21.70 | 5.80 | 2.80 | 1.92 | 0.138 | 0.32 |
| *Mechanical properties according to DIN EN 10088-2* | | | | | | | | |
| $R_{p0.2}$ in MPa | | $R_m$ in MPa | | $A_{gt}$ in % | | | $A_5$ in % | |
| ≥460 | | 640–840 | | – | | | ≥25 | |
| *Measured mechanical properties (gauge length: 35 mm)* | | | | | | | | |
| 570 | | 743 | | 20 | | | 36 | |

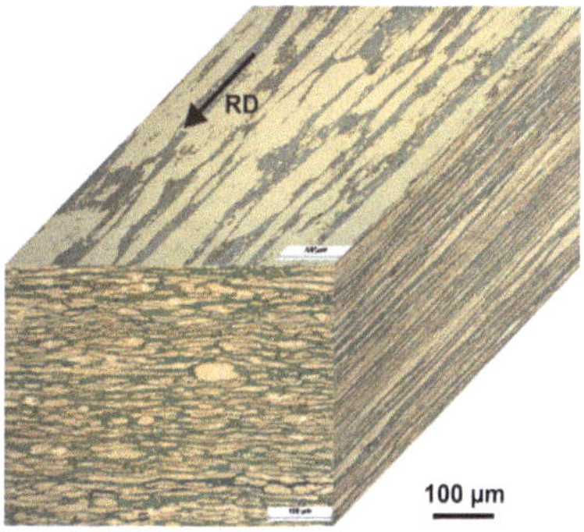


**Fig. 4** Microstructural orientation of the ferritic phase (*dark*) and austenitic phase (*bright*) in the DSS, RD-rolling direction, etched using Beraha II

## *Numerical Simulation of HAC*

Following the approach, the required material properties as well as the input parameters for the simulation of HAC are described in detail. First of all a numerical model of the duplex microstructure was created. Afterward, the simulation of each decoupled part of the procedure for simulating HAC, i.e. analysis of hydrogen transport, stress and strain distribution as well as crack initiation and crack propagation is explained. Additionally, input parameters and assumptions have to be defined in order to perform the numerical simulation.

## Numerical Mesoscale Model of the DSS Microstructure

In order to perform the numerical simulation of HAC in DSS a numerical model of the real duplex microstructure is needed, because the mechanical behavior as well as cracking behavior of DSS strongly depends on the microstructural alignment of austenite and ferrite [14, 15, 23–25]. Thus, a microscopic image of the DSS microstructure transverse to the rolling direction was used and converted to a black and white image, Fig. 5.

A representative section of 100 μm x 100 μm out of the cross section was used for the generation of the model. The chosen section then was vectorised and exported to readable format for the commercial finite-element-program ANSYS® version 12.1. By connecting the resulting key points by splines the according areas of austenite and ferrite can be generated. Finally, the numerical model was meshed with approx.

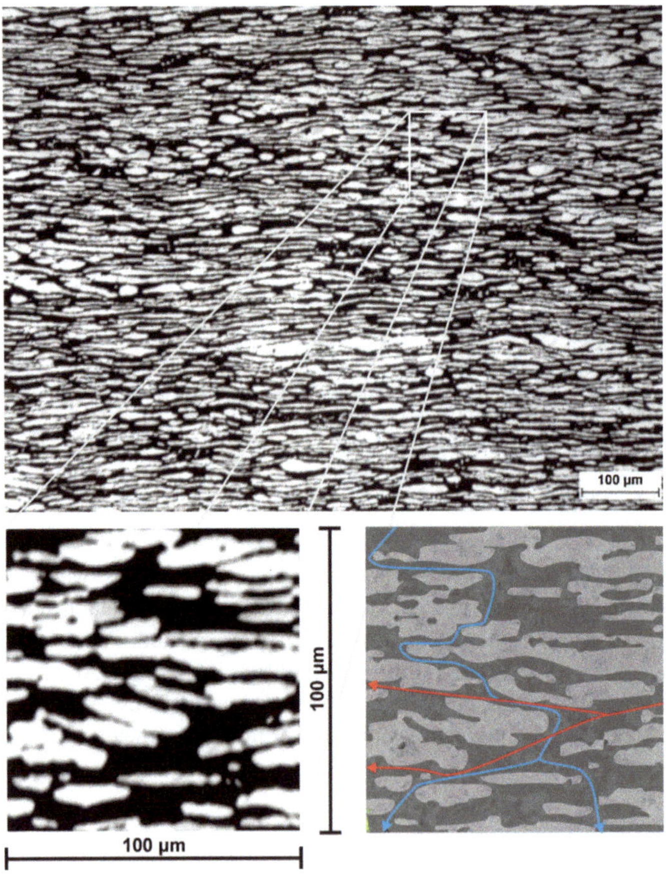


**Fig. 5** Cross section of the DSS microstructure, chosen area of 100 μm x 100 μm and numerical model, ferrite = *dark* and austenite = *bright* (Color figure online)

145,400 elements. As can be seen in Fig. 5 by comparing the mesoscale model and the chosen cross section, some small microstructural features are not included in the model. This is due to the simplifications made in order to accurately mesh the model. The ratio of austenite and ferrite in the mesoscale model is approx. 46:54 vol.% and slightly different from the averaged measured values of 52:48 vol.%. However, the value is within the typical range of 40:60–60:40 vol.% for DSS [9]. Nevertheless, the model appropriately represents the DSS microstructure.

## Numerical Simulation of Hydrogen Diffusion

For the simulation of hydrogen diffusion the thermal module of ANSYS® was used according to the analogy of Fick's second law and Fourier's heat equation law. This is due to the fact that ANSYS® version 12.1 does not provide a separate module for the simulation of hydrogen diffusion. Moreover, other authors used this approach for simulating hydrogen diffusion and achieved reliable results [7, 8, 26]. Assuming that no irreversible trapping occurs, hydrogen diffusion considering reversible trapping and enhancing of hydrogen diffusion can be simulated by using an effective diffusion coefficient $D_{eff}$ [7], (Eq. 1):

$$D_{eff} = D_0 \cdot e^{\frac{E_A}{k_B \cdot T}}, \qquad (1)$$

with $D_0$-lattice diffusion coefficient, $E_A$-activation energy, $k_B$-Boltzmann constant and T-absolute temperature.

The simulation of hydrogen diffusion and hydrogen distribution was performed using the effective diffusion coefficients for austenite and ferrite determined by Owczarek and Zakroczymski via permeation experiments [16], i.e. for the ferritic phase of the DSS a value of $1.5 \times 10^{-5}$ mm$^2$ s$^{-2}$ and for the austenitic phase of the DSS a value of $1.4 \times 10^{-10}$ mm$^2$ s$^{-2}$. Owczarek and Zakroczymski mentioned that with increasing thickness of the permeation sample the macroscopic effective diffusion coefficient decreases due to an increasing path for hydrogen transport in the ferritic phase. With increasing sample thickness hydrogen has to be transported around the austenitic islands which results in longer diffusion paths. They refer to the investigations of Turnbull and Hutchings [17] who introduced a tortuosity factor describing the tortuous diffusion path within the ferritic phase of a duplex stainless steel. Thus it can be seen, that hydrogen diffusion depends on the microstructural alignment of austenite and ferrite. In order to investigate the effect of a tortuous diffusion path in ferrite, hydrogen diffusion was simulated longitudinal and transversal to the microstructural alignment. Therefore, simulation of hydrogen diffusion was performed charging the model from the right side (red arrow, longitudinally) and from the top side (blue arrow, transversally) while the other sides were adiabatic, as indicated by the red and blue arrows in Fig. 5. As pictured in Fig. 5 the path for hydrogen transport in ferrite has more or less straight connections from the right side to the left side of the model (red arrow) and is more tortuous from the top side to the bottom side of the model (blue arrow). For the investigation of hydrogen effusion a homogeneous hydrogen concentration in the whole model was initially assumed. All boundaries of the model were set to a hydrogen concentration of zero in order to simulate unhindered effusion of hydrogen. Accumulation of hydrogen due to plastic deformation as well as hydrostatic stresses was neglected for all simulations in order to reduce the effort of the numerical simulation. Considering hydrogen accumulation additional terms must be introduced in order to calculate the hydrogen concentration. Thus, homogeneous constant diffusion coefficients were applied to the phases of DSS microstructure.

## Numerical Simulation of Phase Specific Stresses and Strains

Performing the analysis of phase specific stresses and strains mechanical properties of both phases are needed. For the DSS corresponding stress-strain curves are available in literature and were also measured. For the respective phases of the DSS no stress-strain curves are available. Additionally, there are no steels available having the same chemical composition as well as the same grain size of the respective phases austenite and ferrite in order to measure corresponding stress-strain curves. Furthermore, the stress-strain behavior of both phases is discussed differently in the literature [9–15]. Sometimes the austenitic phase has a higher strength than the ferritic phase and sometimes vice versa. As for example Johansson et al. [14] found higher strength values for the austenitic phase in their x-ray experiments. Moverare and Odén [15] confirmed the higher strength for the austenitic phase of DSS and argue it by a finer grain size of austenite and a higher nitrogen content. Otherwise, Dakhlaoui et al. [12, 13] found in their experiments using tensile tests as well as compression tests that austenite deforms plastically at first in both tensile and compression test. The ferritic phase has a higher strength compared to the austenitic phase. Additionally, they found an increasing strength for both phases of DSS with increasing alloy content of chromium, molybdenum and nitrogen. Other authors [27, 28] used tie-line alloys in order to determine the strength properties of DSS. They found a decreasing yield and tensile strength for DSS with decreasing amounts of δ-ferrite. Also, Nilsson and Chai [10] as well as Tsuchida et al. [11] reported different stress-strain curves for both phases of DSS depending on thermal treatment or chemical composition of the DSS.

Thus, two different stress-strain behaviors were assumed for the numerical simulation of phase specific stresses and strains according to the findings in literature. On the one hand it was assumed that the ferritic phase has a higher yield strength than DSS and austenite has a lower yield strength than DSS (case one) and on the other hand vice versa (case two). For the DSS the stress-strain curve was fitted according to the measured values, see Table 1. The corresponding stress-strain curves for both cases are shown in Figs. 6 and 7.

All curves were fitted using the law of Ramberg-Osgood [29] in the slightly modified version as it can be found in the work of Rasmussen [30], (Eq. 2). The values used for (Eq. 2) are listed in Table 2. In order to calculate the key values for

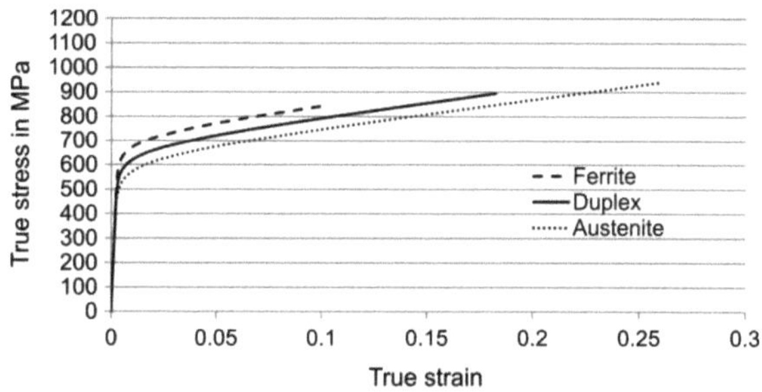


**Fig. 6** True stress-strain curves for the DSS (*Duplex*) as well as for the ferritic phase (*Ferrite*) and austenitic phase (*Austenite*) in case one

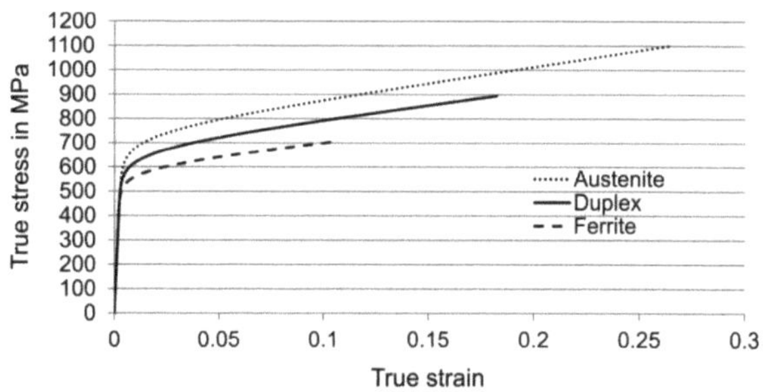


**Fig. 7** True stress–strain curves for the DSS (*Duplex*) as well as for the ferritic phase (*Ferrite*) and austenitic phase (*Austenite*) in case one

**Table 2** Parameters used for the fit of stress-strain curve for DSS as well as both phases for both cases

| | Case one | | | Case two | |
|---|---|---|---|---|---|
| | Austenite | δ-Ferrite | Duplex | Austenite | δ-Ferrite |
| Phase fraction vol.% | 52 | 48 | – | 52 | 48 |
| Yield strength $R_{p0.2}$ in MPa | 520 | 626 | 570 | 626 | 520 |
| Tensile strength $R_m$ in MPa | 725 | 760 | 745 | 845 | 635 |
| Uniform strain $A_{gt}$ in % | 30 | 11 | 20 | 30 | 11 |
| Elongation at fracture $A_5$ in % | 45 | 20 | 36 | 45 | 20 |
| Density in kg/dm$^3$ | 8 | 7.7 | 7.8 | 8 | 7.7 |
| Poisson ratio | 0.310 | 0.288 | 0.3 | 0.310 | 0.288 |
| Young's modulus in MPa | 195,000 | 206,000 | 200,000 | 195,000 | 206,000 |
| Exponent n | 15 | 20 | 17.13 | 16.7 | 20 |

the stress-strain curves in both cases, i.e. the phase specific yield strength and tensile strength, the "modified law of mixture" according to Tamura et al. [28] was adopted. Thus, the phase fractions of austenite and ferrite as well as the macroscopic mechanical properties of the DSS were taken into account. For the reason that ANSYS® needs true stress-strain values as input parameter, true stress-strain curves were calculated from technical values according to Dieter [31].

$$\varepsilon = \frac{\sigma}{E} + 0.002\left(\frac{\sigma}{\sigma_{0.2}}\right)^{n} \tag{2}$$

The boundary conditions used for analyzing the phase specific stresses and strains are in accordance with those of Weber [32] and are shown in Fig. 8. Symmetric boundary conditions were applied to the left side and bottom side of the model. While at the top side and at the right side of the model the border was even. The load in the form of displacements was applied to the right side of the model in x-direction. The maximum displacement applied to the mesoscale model was evaluated from a real tensile test. Therefore, a two dimensional model of the tensile specimen was created and was fine meshed with an element size of 0.1 mm × 0.1 mm within the measuring area. The measured displacement at maximum force from the tensile test of 10 mm then was applied to the two dimensional model of the tensile specimen. Additionally, for every time step the displacements of one element in the middle of the tensile specimen were recorded in order to obtain the maximum displacements for the mesoscale model dependent on the macroscopic loads. The global displacement of 10 mm results in approx. 23 % elongation in x-direction of one element in the middle of the tensile specimen. Thus, the displacement which has to be applied to the nodes located at the right side of the mesoscale model is approx. 0.023 mm.

El Bartali et al. [33] performed tensile tests as well as compression tests using a DSS and found that kinematic hardening describes best the behavior of the material. Also, Johansson et al. [14] obtained reliable results for their two dimensional numerical simulations by applying kinematic hardening as well as plane stress. Thus, for the numerical simulations performed here kinematic hardening and plane stress was applied. For all numerical simulations thermal micro strains resulting

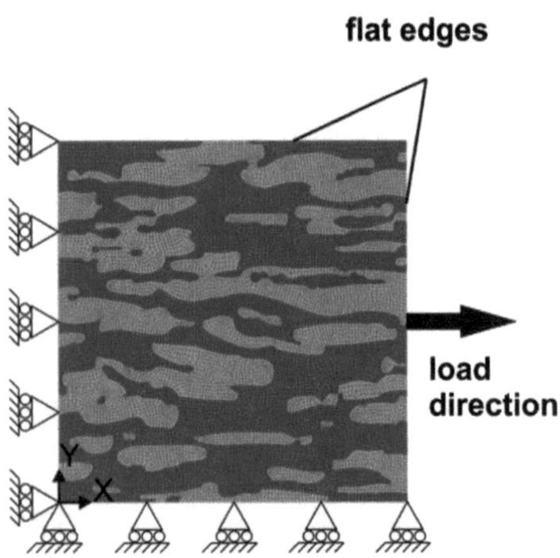


**Fig. 8** Boundary conditions used for phase specific stress-strain analysis, according to [32]

from the manufacturing process due to the different thermal expansion of austenite and ferrite were neglected.

In order to obtain the phase specific mechanical behavior, the stresses and strains of each element in the mesoscale model were recorded for every time step and multiplied by the corresponding area of the element. All values of one phase in the numerical model were summed and divided by the whole area of the phase. Thus, the averaged stress and strain can be calculated for each phase and each time step. Additionally, the macroscopic stress-strain values for the DSS were calculated from the phase specific values according to Tamura et al. [28] regarding the volume fractions of austenite and ferrite in the numerical model.

## Numerical Simulation of Material Damage

The numerical simulation of material damage was performed using the Element-Elimination-Technique (EET) similar to the approach described by Boellinghaus and Hoffmeister [8], but independent from predefined crack paths. In ANSYS® this is realized by the "birth and death" technique. The EET allows taking into account all main influences on HAC, i.e. the local hydrogen concentration and the local mechanical load in the corresponding phases of DSS.

The crack criteria used is represented by a value of true elongation at fracture depending on the local hydrogen concentration. By comparing the local strain to the crack criteria, crack initiation and crack propagation is modeled independent from a predefined crack path by failure of the respective elements. The cracking process can be understood as failure of a sequence of fictive tensile specimens which were represented by the elements of the model, Fig. 9. This procedure is repeated for

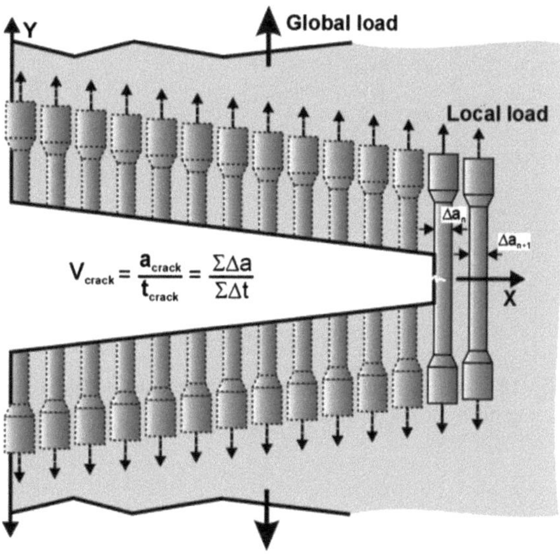


**Fig. 9** Schematic illustration of the concept of fictive tensile specimens to simulate cracking [8]

every element at every time step. After each time step the phase specific stresses and strains are recalculated using the new geometry.

In order to perform the numerical simulation the true elongation at fracture for the DSS as well as for the respective phases are needed. For example, such criteria can be evaluated by tensile tests using hydrogen charged tensile specimens. For high-strength low alloyed (HSLA) [34, 35] and for supermartensitic stainless steels (SMSS) [7, 36] crack criteria are available in literature. Also, for DSS values of true elongation at fracture depending on hydrogen concentration can be found in the literature [18, 25, 37, 38]. However, for simulating material damage in the respective phases of DSS for the austenitic phase as well as for the ferritic phase corresponding values are needed. Due to the fact that there are no steels available having the same microstructure as well as chemical composition compared to the phases of DSS, those values have to be adopted from steels having a similar chemical composition and microstructure. Thus, for the austenitic phase values of the austenitic stainless steels AISI 316 and AISI 304 were adopted. In case of the ferritic phase values of the SMSS were used, because for ferritic stainless steels there are nearly no or only less values available in literature. However, it is known that ferritic stainless steels are more sensitive to a degradation of its mechanical properties due to hydrogen compared to austenitic stainless steels [18–20]. This is also seen in Fig. 10 comparing the HSLA (S690QL and S1100QL) and the SMSS to the austenitic stainless steels AISI 316 [39–42] and AISI 304 [39, 43, 44]. The averaged curve from the austenitic stainless steel AISI 316 and the SMSS is nearly similar compared to the curve of the DSS up to a hydrogen concentration approx. 10 wt.-ppm. For further numerical investigations on hydrogen-assisted cracking the shown values of true elongation at fracture depending on hydrogen concentration are used as clue in order to assess the effect of hydrogen on the phase specific mechanical properties.

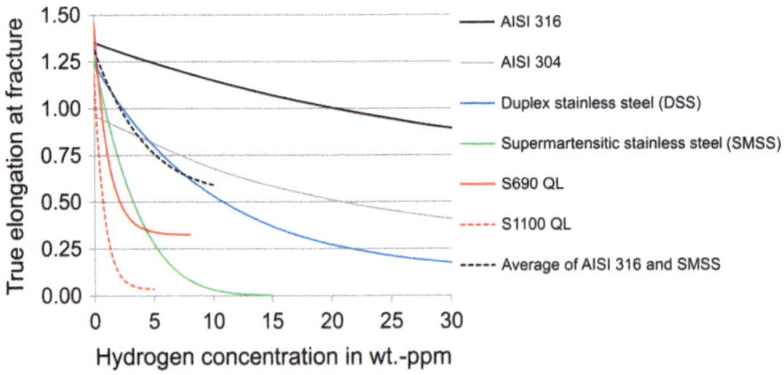


**Fig. 10** True elongation at fracture dependent on hydrogen concentration for different steels having different microstructures

## Results and Discussion

Due to the numerical effort simulating material damage, the numerical analysis of hydrogen diffusion was decoupled from the mechanical analysis as well as from the cracking analysis. First of all, the results of the diffusion analysis are shown followed by the analysis of the mechanical behavior and cracking analysis.

### *Phase Specific Hydrogen Diffusion and Hydrogen Distribution*

Two simulations using different boundary conditions were performed to analyze the diffusion behavior of DSS. Assuming hydrogen diffusion longitudinal to the microstructural orientation the model was charged from the right side and assuming transverse hydrogen diffusion the model was charged from the top side. The results of the corresponding phase specific hydrogen distribution is shown in Fig. 11 for 100 s, 1 and 12 h after the beginning of hydrogen charging.

As can be seen in Fig. 11, hydrogen diffusion in the ferritic phase is much faster compared to the austenitic phase, due to the different hydrogen transport behavior of both phases. The diffusion coefficient of ferrite is higher by five orders of magnitude than the diffusion coefficient of austenite. Thus, hydrogen diffusion in DSS is mainly realized by hydrogen transport in ferrite while austenite acts as diffusion barrier for global hydrogen transport in DSS. By comparing the longitudinal and transverse diffusion behavior in Fig. 11 it is clearly seen that longitudinal hydrogen diffusion (left column) is much faster than in the transverse direction (right column). This is due to the hindered diffusion path in ferrite in case of transverse diffusion. Here the diffusion path in ferrite is interrupted by austenitic islands and hydrogen has to be transported around these islands. In case of longitudinal diffusion hydrogen can be transported on more or less straight paths within the ferritic phase, as already indicated in Fig. 5 by the red arrow. One hour of charging (3600 s) in longitudinal direction results in a nearly homogeneous hydrogen distribution within the ferritic phase as high as the applied hydrogen concentration of HD = 5 ml/100 g Fe, while at the same time of transverse charging hydrogen is not homogeneously distributed in ferrite.

For elucidation of the fast hydrogen transport in ferrite and the strong dependency on the microstructural alignment, the phase specific normalized hydrogen concentration for twelve hours of charging is shown in Fig. 12. Therefore, the hydrogen concentration of each phase in the DSS is normalized to the applied hydrogen concentration for each time step, with respect to the area of each phase in the mesoscale model. Remember the ratio of austenite to ferrite in the mesoscale model is 46:54.

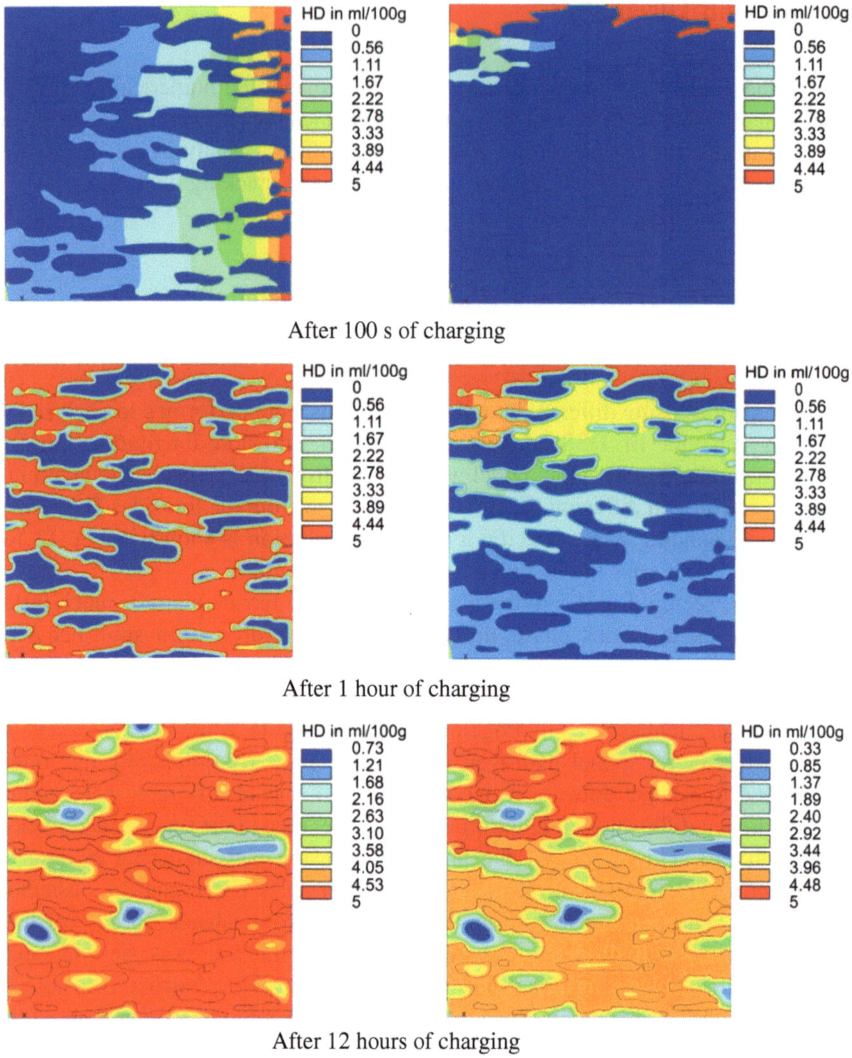


**Fig. 11** Simulated hydrogen distribution depending on microstructural orientation (longitudinal-*left column*, transversal-*right column*). After 100 s of charging, After 1 h of charging, After 12 h of charging

As can be seen in Fig. 12, for longitudinal hydrogen diffusion (red coarse dashed line) hydrogen diffuses very fast in ferrite and after one hour of charging (3600 s) the whole ferrite reaches a normalized hydrogen concentration of nearly 100 %, resulting in approx. 70 % for the whole DSS (red solid line). With increasing charging time more and more hydrogen diffuses from ferrite to austenite and the

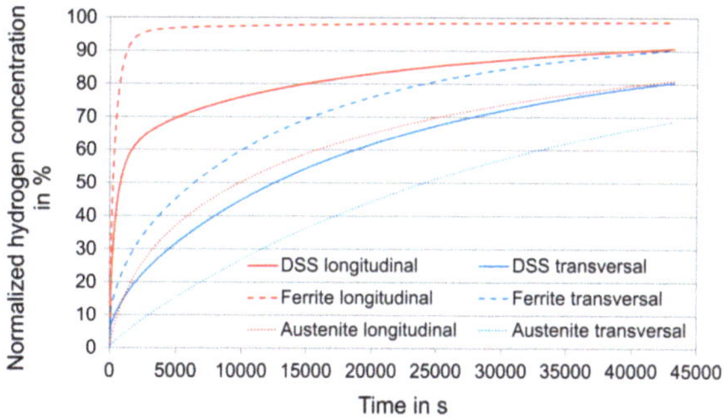


**Fig. 12** Comparison of phase specific hydrogen diffusion depending on microstructural orientation (Color figure online)

overall hydrogen concentration increases. The complete saturation of ferrite, by the meaning of reaching 100 % of the applied hydrogen concentration in the whole ferrite, is reached very slowly above 90 %, because some ferritic islands are surrounded by austenite. Thus, hydrogen has to be transported through the austenite at lower diffusion coefficients. The austenitic phase does not reach as high hydrogen concentrations in the whole phase as the applied concentration within the simulated charging time of 12 h.

By comparing the transverse hydrogen diffusion to longitudinal charging, the hydrogen diffusion in the ferrite phase as well as for the whole model is much slower due to the tortuous diffusion path in ferrite. The whole ferrite phase is supplied with hydrogen only by a small path of ferrite in the upper left part of the model. Due to longer diffusion times in order to reach a homogeneous hydrogen concentration in ferrite, also hydrogen supply from ferrite to austenite is lower. Thus, the whole charging process needs more time in order to reach high hydrogen concentrations in the whole microstructure. Therefore, only a normalized hydrogen concentration of 80 % is reached within twelve hours of transverse hydrogen charging compared to 90 % for longitudinal hydrogen charging. Those dependencies of hydrogen diffusion have to be taken into account for charging experiments on DSS.

The literature shows a similar behavior for hydrogen diffusion in DSS. As for example some authors performed charging experiments and used the microprint technique in order to make hydrogen visible in the DSS microstructure [45–47]. The experimental results showed that hydrogen was only visible in ferrite after short charging times. Also, the dependencies of hydrogen transport on the microstructural orientation was shown [16, 17].

As a result from the numerical simulation of phase specific hydrogen diffusion it can be confirmed that hydrogen transport in DSS mainly occurs in ferrite and strongly depends on the microstructural alignment of austenite and ferrite as well as on the shape and size of austenitic islands. In case of longitudinal hydrogen diffusion HAC may occur earlier due to the fast hydrogen transport resulting in higher hydrogen concentrations in ferrite within shorter charging times.

## *Phase Specific Hydrogen Effusion*

The diffusion behavior of hydrogen can be transferred to the effusion of hydrogen. Effusion of hydrogen to all boundaries of the mesoscale model was modeled applying a hydrogen concentration of zero to all surface nodes, assuming an initially homogeneous hydrogen concentration of HD = 5 ml/100 g Fe in the whole model, Fig. 13. Hydrogen first effuses from the ferrite within short times and high hydrogen concentrations remain at austenite.

The hydrogen remaining in austenite also is called trapped hydrogen due to the very slow effusion or diffusion of hydrogen out of austenite [16]. One hour after the beginning of effusion nearly no hydrogen remains in ferrite, despite some small ferritic islands surrounded by austenite. The hydrogen concentration in austenite remains as high as the initial concentration at the same time. Even after twelve hours of effusion there remain high hydrogen concentrations of about 85 % from the initial concentration.

Similar results were achieved by Straub et al. [48] in experiments using time-of-flight secondary ion mass spectroscopy (TOF-SIMS). They investigated a deuterium charged DSS specimen one hour after charging. The results of Straub et al. [48] are compared to the numerical simulation in Fig. 14. As can be seen from

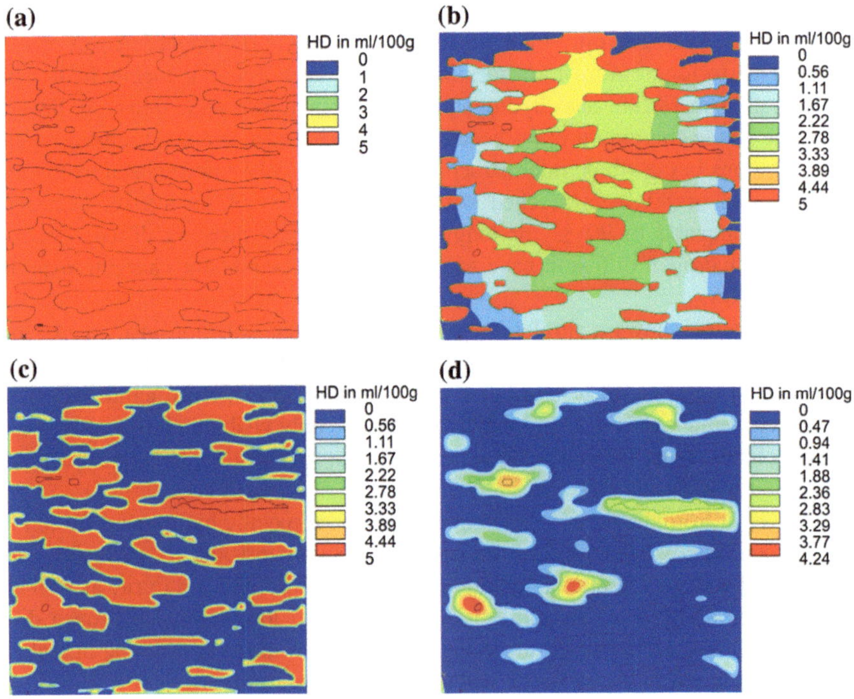


**Fig. 13** Simulated hydrogen distribution for 100 s, 1 and 12 h of effusion to all sides of the model. **a** Initial condition **b** after 100 s of effusion **c** after 1 h of effusion **d** after 12 h of effusion

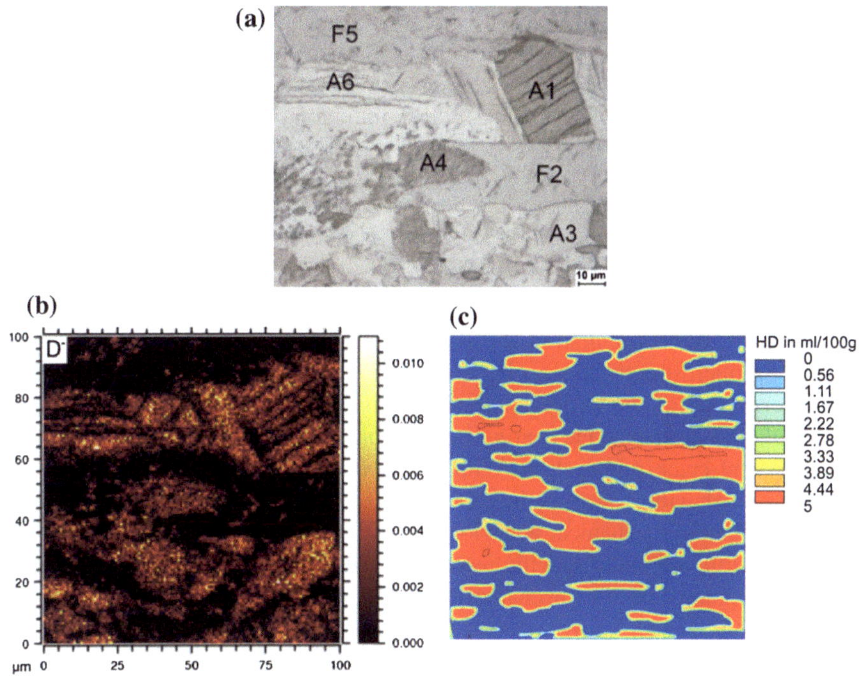


**Fig. 14** Comparison of the hydrogen/deuterium distribution in a DSS microstructure determined using (**b**) TOF-SIMS [48] (scale: normalized ion counts; counts of $D^-$ ion divided by total ion counts) and (**c**) numerical simulation **a** analyzed section from the DSS (*A* = austenite, *F* = ferrite) **b** TOF-SIMS image showing the deuterium distribution **c** simulated hydrogen distribution for 1 h of effusion

Fig. 14b, high amounts of deuterium can be detected in austenite of the DSS, e.g. areas A1, A3, A4 and A6 in Fig. 14a. Thus, the deuterium initially located in ferrite was totally effused from the DSS specimen.

However, the reversibly trapped hydrogen may be released due to thermal load as well as mechanical load [44, 49–52]. Thus, hydrogen from austenite then maybe diffuses into highly stressed and strained areas in ferrite which may initiate HAC. Moreover, high plastic deformation in addition with high hydrogen concentrations may enhance martensitic transformation of austenite, which may also enhance crack initiation and crack propagation [20, 53–56].

Additionally, Straub et al. [48] found some microstructural features in the DSS microstructure after charging the sample with deuterium, as can be seen in Fig. 14a in marked area A1. Parallel cracking was observed in the austenite. The authors assume that plastic deformation in austenite occurred due to high hydrogen concentrations resulting in twinning and parallel cracking of austenite. Also, Głowacka et al. [57] observed parallel cracking in the austenitic phase of DSS resulting solely from hydrogen charging. Additionally, they also found fractions of martensite within the former austenite. Barnoush et al. [58] observed in situ formation of slip lines at the surface of austenite grains in DSS with increasing charging times.

Thus, the mechanical behavior of DSS is strongly affected by hydrogen. If high hydrogen concentrations are present in the DSS microstructure the risk for HAC is increased for a long time. Even during service trapped hydrogen possibly is released from austenite which enhances crack initiation or crack propagation in both phases of the DSS.

## *Phase Specific Stress–Strain Analysis*

Besides hydrogen concentration, the amount and distribution of stresses and strains in the DSS microstructure have a big impact on hydrogen assisted material damage. The interactions of the hydrogen concentration and mechanical load within the phases of DSS determine the bearing capacity of the microstructure as well as crack initiation and crack propagation. With regard to the literature, the phase specific mechanical behavior as well as the effects of load sharing is not fully clarified up to date and different mechanical behavior can be found for both phases of the DSS. Thus, two different phase specific mechanical behavior were assumed for investigating the phase specific mechanical interaction. Therefore, a displacement of approx. 0.009 mm (9 %) was applied to the nodes located at the right side of the mesoscale model. The resulting phase specific stresses and strains are shown in Fig. 15 for both cases of the mechanical behavior.

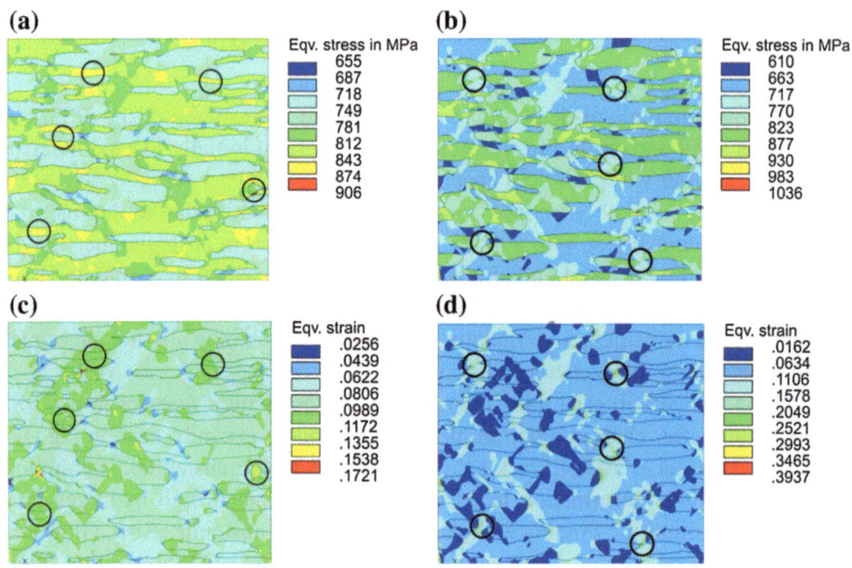


**Fig. 15** Comparison of the phase specific mechanical behavior for assumed material properties of case one (ferrite is stronger, *left column*) and case two (austenite is stronger, *right column*) **a** Eqv. stress case one **b** Eqv. stress case two **c** Eqv. strain case one **d** Eqv. strain case two (Color figure online)

As can be seen in Fig. 15 the phase specific strains as well as the stresses are locally higher than the applied global strain of approx. 9 %. This was also found by Johansson et al. [14] for their numerical investigations. Additionally, the authors performed x-ray experiments on DSS in order to measure phase specific stresses, but those measurements can only give averaged values of the stresses in each phase. However, those localized high stresses and strains are crucial regarding HAC in DSS. Thus, a crack critical region within the DSS microstructure can only be localized by numerical simulations of the phase specific mechanical behavior taking into account the real DSS microstructure.

By comparing the mechanical behavior of both cases it is seen that for case two (right column in Fig. 15, austenite has higher strength) the interval between the highest and lowest value of stresses and strains as well as the absolute values are much larger than for case one (ferrite has a higher strength). For case one the highest stresses are located in ferrite and the highest strains in austenite. The opposite behavior is seen for case two of the mechanical behavior. The locations where the highest stresses and strains in austenite and ferrite can be found for case one are now the locations where the lowest stress and strain values are located for case two and vice versa. Thus it can be confirmed that the phase which exhibits lower strength values will preferably experience higher strains and the phase which has higher strength values will bear higher loads (stresses), as it was proposed by Fischmeister and Karlsson [59] as well as Cho and Gurland [60] and numerically determined by Johansson et al. [14].

Nevertheless, the locations of the highest strains within the microstructure, especially for ferrite, are more important regarding HAC. As can be seen from Fig. 15, for case one there are much lower strains than for case two and they are more homogeneously distributed in the microstructure. For case two of the phase specific behavior, there is a preferred orientation of the maximum strains arranged at an angle of 45°. This was also observed for the numerical simulation performed by Sun et al. [61] for a martensitic-ferritic steel and they found that the ductile failure later on will follow those shear bands. However, in case one of the simulated phase specific mechanical behavior the highest strains can be found within small elongated (tapered) areas of ferrite parallel to the loading direction which are surrounded by large austenitic islands, as indicated by the black circles in Fig. 15c. In case two of the simulated phase specific material behavior, the highest strains in ferrite can be found in small (tapered) areas between austenitic islands with an orientation perpendicular to the load direction and they are more localized ("hot spots"). Hence, the behavior of crack initiation as well as crack propagation will strongly differ between case one and case two, too. Quite similar numerical results for the phase specific stresses and strains in a DSS microstructure as well as the locations of maximum stresses and strains were achieved by Johansson et al. [14].

Regarding the susceptibility for HAC both cases are critical, but the simulated phase specific material behavior of case two seems to be more critical than case one. Here large ferritic areas are located on both sides of the determined crack critical locations oriented perpendicular to the applied load. However, case one of the phase specific mechanical behavior makes more sense, due to the fact that the higher

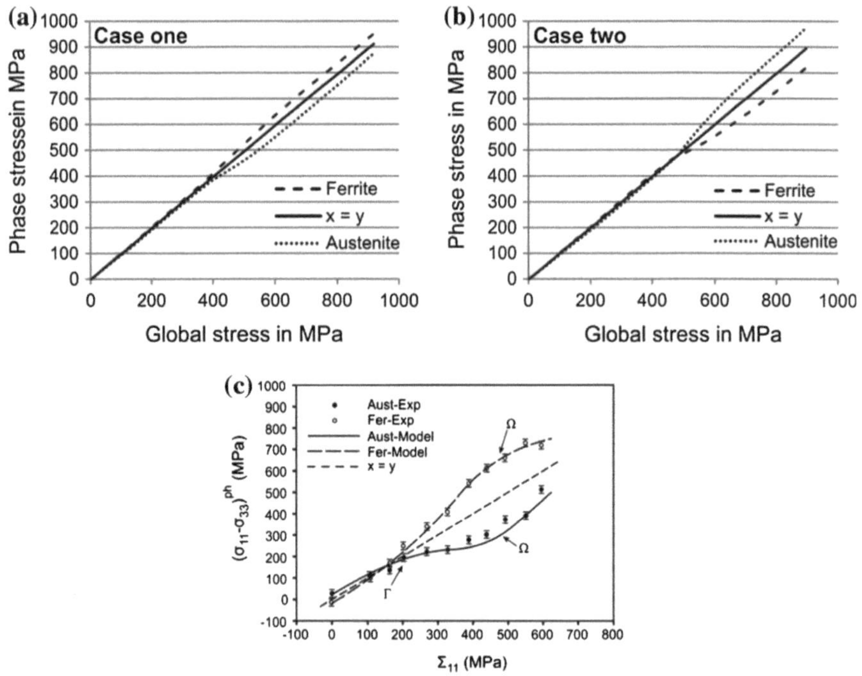


**Fig. 16** Comparison of numerically and experimentally determined phase specific stresses depending on global or macroscopic stresses **a** phase specific stresses for case one **b** phase specific stresses for case two **c** experimentally determined phase specific stresses [12, 13]

strains can be found in austenite as expected for the austenitic phase of DSS. This becomes clearer by comparing the numerically determined phase specific stresses to the experimental results of Dakhlaoui et al. [12, 13] determined by neutron and x-ray experiments for the duplex stainless steel UR45 N (X2CrNiMoN 22-5-3), Fig. 16. It is clearly seen that the phase specific stresses of simulated case one (Fig. 16a) correspond qualitatively with the measured phase specific stresses. For both, simulated case one and experimentally determined results, the ferritic phase of the DSS has higher phase specific stress values than austenite. The only difference is the distance from the straight line (x = y). This is due to the short distance of the applied phase specific stress-strain curves to the stress–strain curve of the DSS. With increasing distance of the phase specific stress-strain curves to the stress–strain curve of the DSS, as shown by the results of additional numerical simulations not published yet. Moreover, the slope of the curves become more similar to those from the x-ray experiments and the yield points of the austenitic phase (Γ in Fig. 16c) as well as of the ferritic phase (Ω in Fig. 16c) is clearly seen, too.

Regarding the results of the numerical phase specific stress-strain analysis it can be said that there are overall good agreements to numerical and experimental results from the literature.

## *Numerical Simulation of Material Damage*

The numerical simulation of hydrogen-assisted material damage was decoupled from the hydrogen diffusion analysis, due to the numerical effort. The respective crack criteria for austenite and ferrite of the DSS were adopted with respect to the results of the hydrogen diffusion analysis as well as the represented hydrogen dependent true elongation at fracture for steels having an austenitic or ferritic and martensitic microstructures (see Fig. 10). That means that the crack criteria was not calculated for every time step as was done for the numerical simulations by Boellinghaus and Hoffmeister [8] or Viyanit [7]. For the numerical simulation of hydrogen-assisted material damage performed in this work it was assumed that high hydrogen concentrations will be present in the ferritic phase of the DSS within short times. The austenite at the same time has very low hydrogen concentrations, as was shown for the numerical hydrogen diffusion analysis as well as for experiments [45–47]. Furthermore, ferritic microstructures are more prone for a degradation of its mechanical properties due to hydrogen compared to austenitic microstructures [18–20]. Thus, two main failure cases were assumed for simulating hydrogen-assisted material damage.

On the one hand, high hydrogen concentrations are present in the DSS microstructure and homogeneously distributed in the ferritic phase resulting in totally brittle ferrite, while austenite has nearly no degradation of its mechanical properties and remains ductile. This behavior is also being seen for the SMSS and AISI 316 in Fig. 10, i.e. for a hydrogen concentration about 10 wt.-ppm. Thus, for the first assumed failure case referred to as “high hydrogen” the true elongation at fracture for ferrite was 0.5 % and austenite reaches his true elongation corresponding to the tensile strength. On the other hand, low hydrogen concentrations were assumed to be present in the DSS microstructure and both phases reach their elongation corresponding to their tensile strength, i.e. 11 % for ferrite and 30 % for austenite, referred to as “low hydrogen”.

The respective crack criteria were limited to values corresponding to the phase specific tensile strength in order to avoid convergence problems as they occurred for preliminary investigations simulating HAC independent from predefined crack paths. Hence, after reaching the tensile strength no higher loads are needed for further elongation of a material, e.g. as can be seen for tensile tests after necking. Additionally, no construction is designed for stresses higher than the yield strength and with regard to HAC failure of a component loaded near the yield strength or having only local and very small amounts of plastic deformation would be more interesting. The results of the simulated material damage is shown in Figs. 17 and 18 for case one and case two of the assumed phase specific material behavior.

In Fig. 17 the true stresses and true strains are shown for case one of the phase specific material properties (ferrite has a higher strength). As can be seen in Fig. 17a and b assuming high hydrogen concentrations, the crack initiates at a low global strain of 0.35 % in a small elongated area of ferrite which is orientated within the load direction and surrounded by large austenitic islands, marked by the black circle

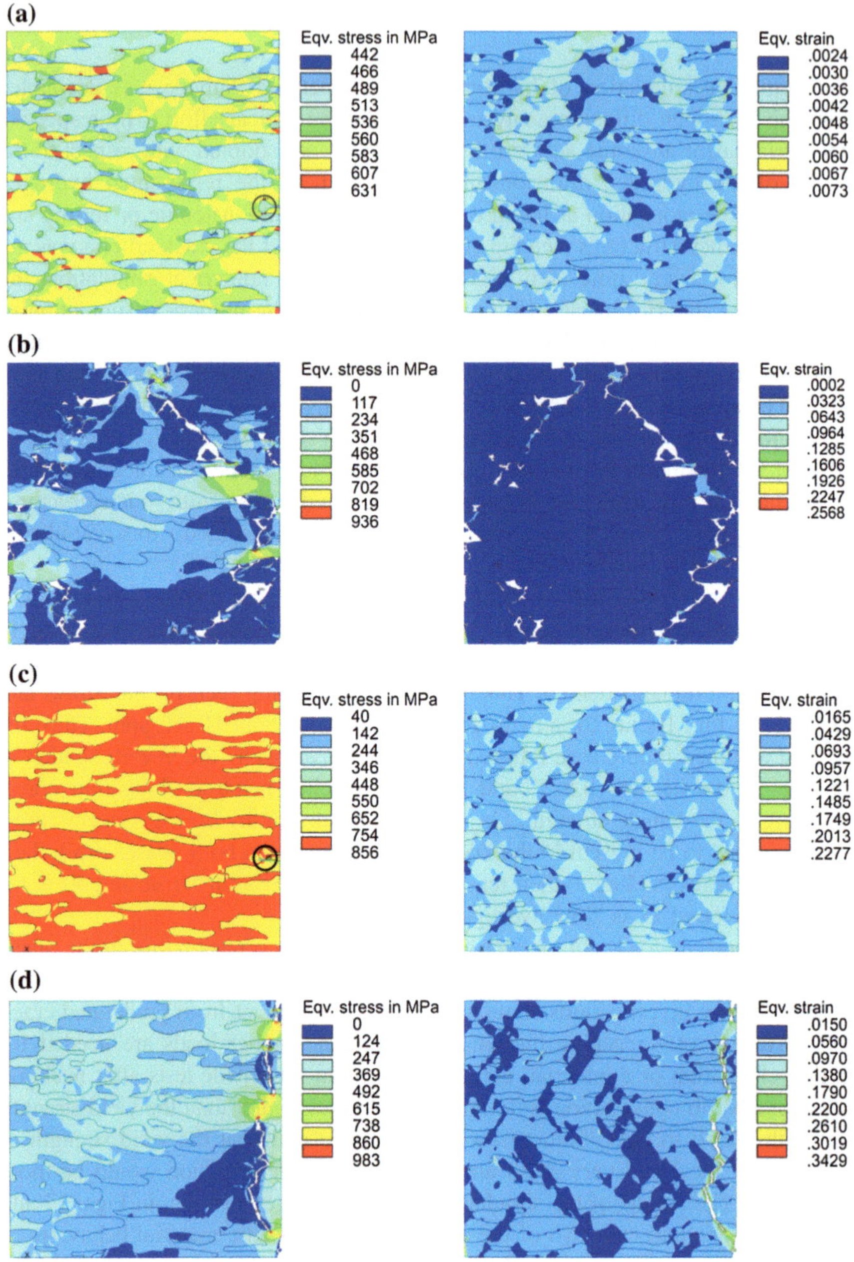


**Fig. 17** True stress (*left*) and true strain (*right*) of simulated cracking for "high hydrogen" (**a**, **b**) and "low hydrogen" (**c**, **d**) in case one of the phase specific material behavior **a** crack initiation in ferrite at 0.35 % global strain (high hydrogen) **b** crack propagation at 0.79 % global strain (high hydrogen) **c** crack initiation in ferrite at 6.2 % global strain (low hydrogen) **d** crack propagation at 7.3 % global strain (low hydrogen) (Color figure online)

in Fig. 17a. With increasing load the cracks start to propagate through the ferritic phase until they reach the austenitic phase where they were arrested. Several additional cracks initiate at the same time in small elongated areas of ferrite and start to propagate. Only with increasing load the cracks also start to propagate through the austenite linking the cracks in ferrite. Due to failure of large ferritic areas causing convergence problems, the numerical simulation was interrupted at a global load of 0.79 %. When regarding the plastic deformation in the microstructure at the end of the simulation (Fig. 17b, right) it is seen that there are low overall plastic strains within the microstructure. The highest plastic strains can be found on the crack's surface within the austenite.

Assuming low hydrogen concentrations, the cracks initiate at the same location in the microstructure as for high hydrogen concentrations, but at a higher global strain of 6.2 %, Fig. 17c. With increasing load several new cracks initiate in ferrite and start to grow until they reach austenite and start to grow again with rising load. The numerical simulation was interrupted at a global strain of 7.3 % due to convergence problems. The difference to the assumption of high hydrogen concentrations is that only one main crack develops and no secondary cracks are visible, Fig. 17d. Additionally, higher overall plastic deformation occurred in the DSS microstructure with higher deformations alongside the crack. Also, the highest plastic deformations can be found in austenite.

For case two of the phase specific material properties the results of the cracking analysis are shown in Fig. 18 for high hydrogen concentrations (a and b) and for low hydrogen concentrations (c and d). As can be seen in Fig. 18a assuming high hydrogen concentrations, several cracks initiate in ferrite at the same global strain of 0.35 % as for case one of the phase specific material properties, but at different locations. Here the cracks initiate at small elongated areas between austenitic islands which have an orientation perpendicular to the loading direction ("hot spots"). Also, these cracks start to propagate with increasing load until they reach austenitic islands. With further increasing load some of the cracks start to propagate trough the austenite linking the cracks in ferrite, but at a global load of 0.4 % the numerical simulation was aborted by convergence problems, Fig. 18b. Nonetheless, several initiation sides as well as several separate cracks can be seen in the microstructure and only small paths have to be linked between the cracks in order to get a fully cracked model. Also, less plastic deformation occurred in the microstructure and the highest plastic strain can be found in austenite along the crack surface.

In Fig. 18c and d the results of the cracking analysis for case two of the phase specific material properties assuming low hydrogen concentrations are shown. As can be seen, only one crack initiates at a "hot spot" within the ferritic phase at a global strain of 3 %. With increasing load several other cracks at "hot spots" initiate within ferrite and then start to propagate following the previously mentioned shear bands until they reach austenitic islands. With further increasing load one main crack through the whole model developed due to cracking of austenite which links the existing cracks in ferrite, Fig. 18d. Additionally, high plastic strains can be found in the microstructure whilst the highest plastic deformation occurs along the surface of the crack.

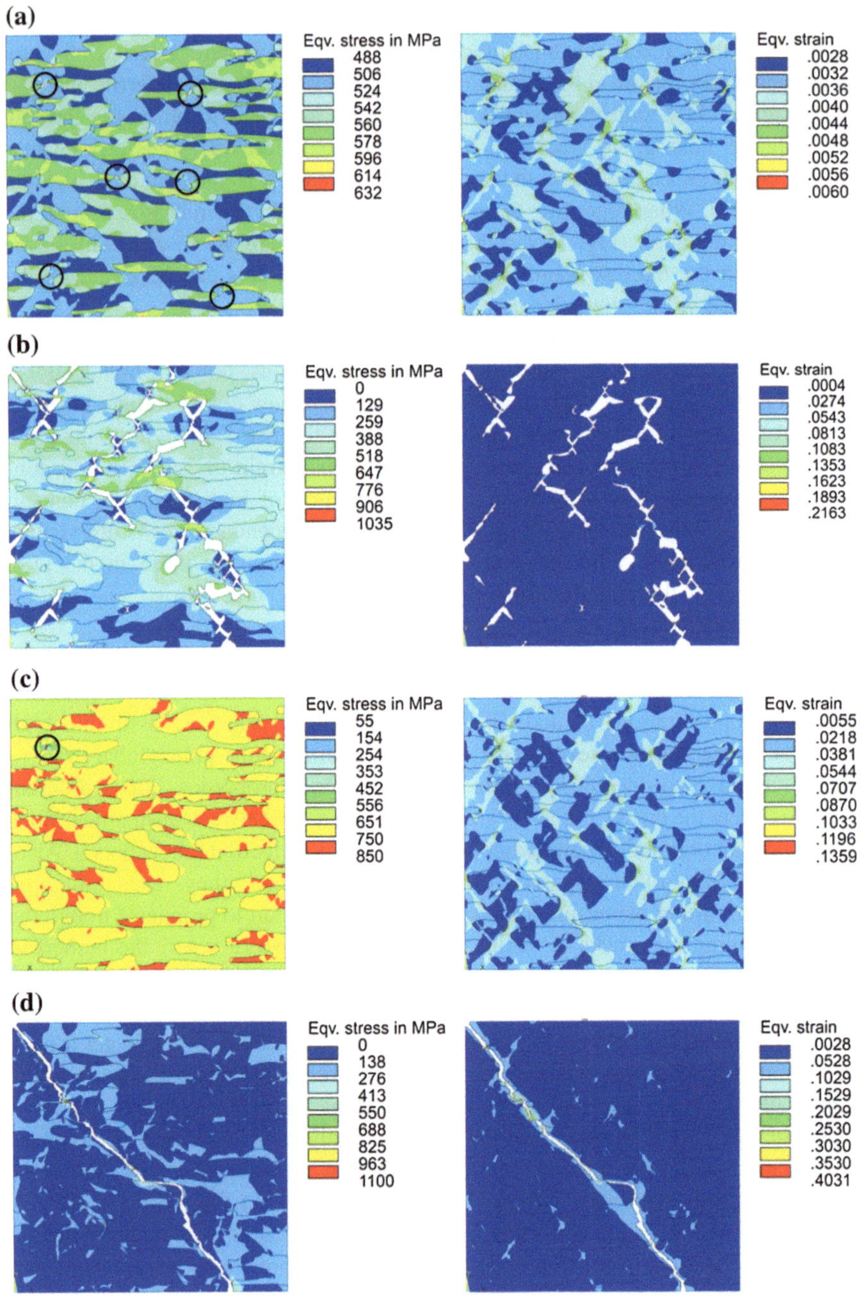


**Fig. 18** True stress (*left*) and true strain (*right*) of simulated cracking for "high hydrogen" (**a**, **b**) and "low hydrogen" (**c**, **d**) in case two of the phase specific material behavior **a** crack initiation in ferrite at 0.35 % global strain (high hydrogen) **b** crack propagation at 0.4 % global strain (high hydrogen) **c** crack initiation in ferrite at 3 % global strain (low hydrogen) **d** crack propagation at 3.8 % global strain (low hydrogen)

By comparing both cases of the phase specific material properties it is seen that crack initiation always occurred in ferrite, independent from the material behavior as well as independent from the hydrogen concentration present in the DSS microstructure. The cracks always propagated until they were arrested in austenitic islands and higher loads were needed for further crack propagation. Similar results for hydrogen-assisted cracking can be found in the literature [10, 21, 22, 25, 62]. Additionally, some similarities can be found comparing the results of "high hydrogen" and "low hydrogen" for both cases of the phase specific material properties. Assuming high hydrogen concentrations several cracks initiate in the DSS microstructure resulting in one main crack and several secondary cracks and less overall plastic deformation occurs. Assuming low hydrogen concentrations, several cracks initiate in the DSS microstructure, but they were all linked together and form only one main crack.

The results of the numerical cracking analysis are in good accordance to other investigations on HAC. As for example, comparing the results of the simulated crack propagation to the theoretical model for hydrogen-assisted crack propagation in DSS proposed by San Marchi et al. [18], Fig. 19. San Marchi et al. describe hydrogen-assisted crack propagation by cleavage micro-cracks forming in ferrite and induce high stresses in austenite which results in fracture of austenite, Fig. 19a, b point A. Also, the fracture can proceed alongside the austenitic-ferritic interface, Fig. 19a, b point B. The cracks in austenite or along interfaces link the cleavage cracks already propagated in ferrite to form the nominal fracture plane.

Another comparison can be made regarding the roughness of the fracture plane, Fig. 19b, c and d. Michalska et al. [62] investigated fracture profile lines of hydrogen charged and uncharged specimens. They found that hydrogen charged specimens show higher values of fracture linear roughness. Furthermore, they saw large numbers of secondary cracks and micro-cracks which formed mainly in ferrite parallel to the nominal fracture surface.

Thus it can be said that the numerical simulation of hydrogen-assisted material damage performed here as well as the hydrogen analysis and phase specific stress-strain analysis qualitatively reflects the behavior of duplex stainless steels with and without the influence of hydrogen. However, more investigations are needed in order to improve the numerical simulation quantitatively and qualitatively.

With regard to welding of duplex stainless steels, the microstructure and resulting hydrogen transport behavior leads to higher risk for hydrogen-assisted cracking in the weld metal. Compared to the base metal, the weld metal has higher amounts of δ-ferrite, i.e. up to 70 vol.% [63–65]. The large δ-ferrite grains are enclosed by allotriomorphic austenite having Widmanstätten shaped austenite within the grain. Thus, high amounts of hydrogen introduced by welding will remain in δ-ferrite for a long time, due to the austenite acting as diffusion barrier for hydrogen [66]. Therefore, high amounts of δ-ferrite in addition with high amounts of hydrogen will result in a high risk for crack initiation and fast crack propagation. However, further numerical investigations are needed applying the numerical model to a real weld metal microstructure of a duplex stainless steel, in order to evaluate and verify the cracking behavior.

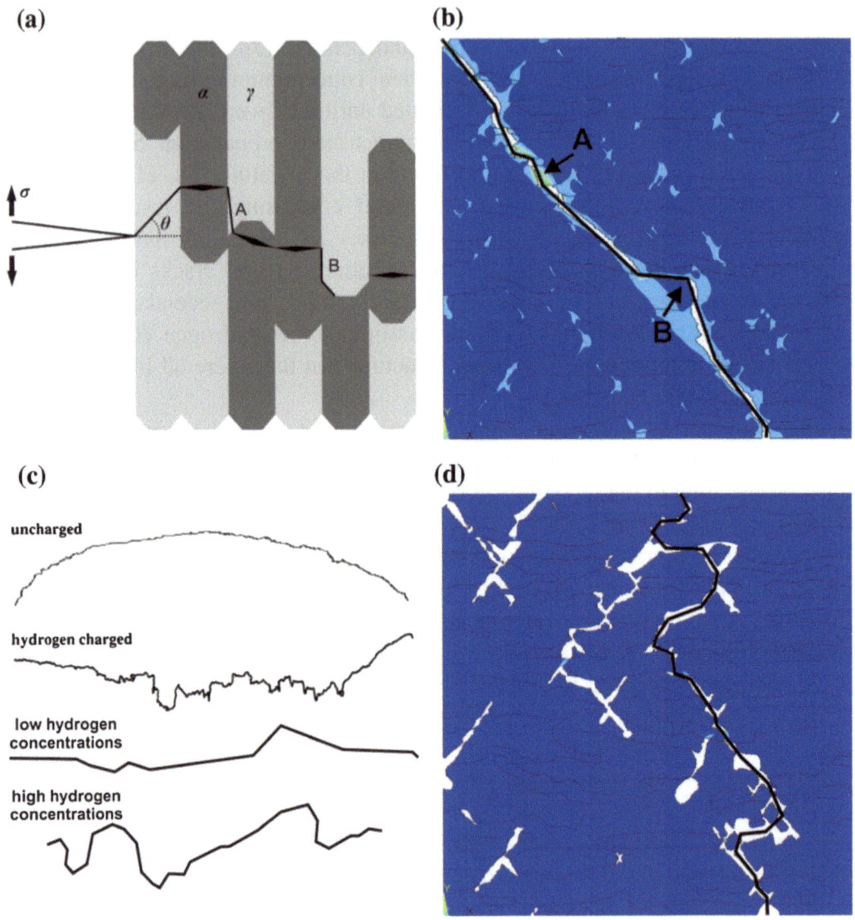


**Fig. 19** Comparison of the numerical results of the cracking analysis to the model for HAC in DSS of San Marchi et al. [18] and fracture profile lines measured by Michalska et al. [62] **a** model for HAC in DSS [18] **b** simulated fracture of case two, low hydrogen concentrations **c** comparison of fracture profile lines determined experimentally [62] and numerically **d** simulated fracture of case two, high hydrogen concentrations

## Conclusions

In this study a numerical mesoscale model of a duplex stainless steel microstructure was created in order to simulate hydrogen-assisted material damage independent from predefined crack paths and taking into account hydrogen distribution as well as stress-strain distribution within the DSS microstructure. Therefore, hydrogen diffusion analysis and stress strain analysis was performed first followed by numerical simulation of HAC assuming different phase specific mechanical properties and different hydrogen concentrations. Regarding results of the numerical simulation of hydrogen assisted cracking, the following conclusions can be drawn:

1. The hydrogen transport in DSS at room temperature mainly occurs in ferrite, due to the high diffusion rate compared to austenite. Hydrogen diffusion in austenite needs more time for total saturation of austenite. With regard to the lower deformation capability of ferrite compared to austenite this will result in a higher risk for HAC in ferrite. For the effusion of hydrogen a similar behavior can be observed. Hydrogen first effuses from ferrite and high amounts of hydrogen remain in austenite. Hydrogen in austenite possibly is released due to thermal or mechanical load and may diffuse in highly strained areas in ferrite and possibly initiate cracking. Also, high hydrogen concentrations in austenite may enhance twinning or martensitic transformation which may also enhance crack propagation in austenite.
2. Different phase specific mechanical properties result in a shift of the crack critical locations in the DSS microstructure and thus a shift of crack initiation sites. If ferrite has a higher yield and tensile strength compared to austenite, small elongated areas of ferrite orientated in line with the mechanical load are crack critical locations. If austenite has a higher yield and tensile strength, the crack critical areas shift to small areas in ferrite ("hot spots") orientated perpendicular to the direction of loading. Additionally, higher plastic deformation occurs in ferrite resulting in lower global loads needed for crack initiation as well as crack propagation.
3. Crack initiation always occurred in ferrite and the cracks propagated in ferrite until they reached an austenitic island where they were arrested. Only with increasing load, cracking of austenite occurs linking the preceding cleavage cracks in ferrite. If high hydrogen concentrations are present in the DSS microstructure several cracks initiate in ferrite to form one main crack and several secondary cracks. If a low hydrogen concentration or nearly no hydrogen is present in the microstructure, several cracks will initiate in ferrite to form one main crack and no secondary cracks are visible.
4. The mesoscale model enables simulation of phase specific crack initiation and crack propagation independent from predefined crack paths. The achieved results of the mesoscale model are in good qualitative accordance to the literature on hydrogen assisted cracking in duplex stainless steels. Thus, the mesoscale model allows evaluation of crack critical conditions for hydrogen assisted cracking in DSS. Additionally, interaction of hydrogen and local mechanical load within the phases of the DSS can be investigated. Transferring the macroscopic material behavior to the mesoscale model and vice versa helps to improve the understanding of hydrogen assisted cracking phenomena in two phase microstructures on the one hand. And on the other hand those numerical simulations improve the understanding of the micromechanical behavior of DSS and thus help to optimize and explain experiments as well as allowing development or improvement of theoretical models for failure mechanisms.

Further investigations are needed in order to improve the numerical model with regard to a quantification of phase specific mechanical properties as well as phase specific crack criteria. Additionally the model has to be improved in order to overcome the convergence problems.

## References

1. M. Liljas: 80 years with duplex steels, a historic review and prospects for the future. In: *6th European Stainless Steel Conference*, Helsinki, 2008, pp. 535–540
2. I. Alvarez-Armas: Duplex Stainless Steels: Brief History and Some Recent Alloys. *Recent Patents on Mechanical Engineering*, Vol. 1 (2008), No. 1, pp. 51–57
3. T.S. Taylor; T. Pendlington; R. Bird: Foinaven Super Duplex Materials Cracking Investigation. In: *Offshore Technology Conference*, Houston, Texas, Society of Petroleum Engineers, 1999, pp. 467–480
4. V. Olden; C. Thaulow; R. Johnsen: Modelling of hydrogen diffusion and hydrogen induced cracking in supermartensitic and duplex stainless steels. *Materials & Design*, Vol. 29 (2008), No. 10, pp. 1934–1948
5. W. Brocks; R. Falkenberg; I. Scheider: Coupling aspects in the simulation of hydrogen-induced stress-corrosion cracking. *Procedia IUTAM*, Vol. 3 (2012),, pp. 11–24
6. P. Wongpanya; T. Böllinghaus; G. Lothongkum; H. Hoffmeister: Numerical Modelling of Cold Crack Initiation and Propagation in S 1100 QL Steel Root Welds. *Welding in the World*, Vol. 53 (2009), No. 3/4, pp. R34–R43
7. E. Viyanit: Numerical Simulation of Hydrogen Assisted Cracking in Supermartensitic Stainless Steel Welds. 2005, p. 229
8. T. Boellinghaus; H. Hoffmeister: Numerical Model for Hydrogen-Assisted Cracking. *Corrosion Science*, Vol. 56 (2000), No. 6, pp. 611–622
9. M. Knyazeva; M. Pohl: Duplex Steels: Part I: Genesis, Formation, Structure. *Metallography, Microstructure, and Analysis*, Vol. 2 (2013), No. 2, pp. 113–121
10. J. Nilsson; G. Chai: The physical metallurgy of duplex stainless steels. In: *International Conference & Expo Duplex 2007*, Gardo, Italien, Associazione Italiana di Metallurgia, 2007, pp. 73–82
11. N. Tsuchida; T. Kawahata; E. Ishimaru; A. Takahashi; H. Suzuki; T. Shobu: Static Tensile Deformation Behavior of a Lean Duplex Stainless Steel Studied by In Situ Neutron Diffraction and Synchrotron Radiation White X-rays. *ISIJ International*, Vol. 53 (2013), No. 7, pp. 1260–1267
12. R. Dakhlaoui; A. Baczmański; C. Braham; S. Wroński; K. Wierzbanowski; E.C. Oliver: Effect of residual stresses on individual phase mechanical properties of austeno-ferritic duplex stainless steel. *Acta Materialia*, Vol. 54 (2006), No. 19, pp. 5027–5039
13. R. Dakhlaoui; C. Braham; A. Baczmanski: Mechanical properties of phases in austeno-ferritic duplex stainless steel—Surface stresses studied by X-ray diffraction. *Materials Science and Engineering: A*, Vol. 444 (2007), No. 1-2, pp. 6–17
14. J. Johansson; M. Odén; X.-H. Zeng: Evolution of the residual stress state in a duplex stainless steel during loading. *Acta Materialia*, Vol. 47 (1999), No. 9, pp. 2669–2684
15. J.J. Moverare; M. Odén: Influence of elastic and plastic anisotropy on the flow behavior in a duplex stainless steel. *Metallurgical and Materials Transactions A*, Vol. 33 (2002), No. 1, pp. 57–71
16. E. Owczarek; T. Zakroczymski: Hydrogen transport in a duplex stainless steel. *Acta Materialia*, Vol. 48 (2000), No. 12, pp. 3059–3070
17. A. Turnbull; R.B. Hutchings: Analysis of hydrogen atom transport in a two-phase alloy. *Materials Science and Engineering: A*, Vol. 177 (1994), No. 1-2, pp. 161–171

18. C. San Marchi; B.P. Somerday; J. Zelinski; X. Tang; G.H. Schiroky: Mechanical Properties of Super Duplex Stainless Steel 2507 after Gas Phase Thermal Precharging with Hydrogen. *Metallurgical and Materials Transactions A*, Vol. 38 (2007), No. 11, pp. 2763–2775
19. T. Michler; J. Naumann: Microstructural aspects upon hydrogen environment embrittlement of various bcc steels. *International Journal of Hydrogen Energy*, Vol. 35 (2010), No. 2, pp. 821–832
20. K.H. Lo; C.H. Shek; J.K.L. Lai: Recent developments in stainless steels. *Materials Science and Engineering: R: Reports*, Vol. 65 (2009), No. 4-6, pp. 39–104
21. S.-L. Chou; W.-T. Tsai: Hydrogen embrittlement of duplex stainless steel in concentrated sodium chloride solution. *Materials Chemistry and Physics*, Vol. 60 (1999), No. 2, pp. 137–142
22. A.M. Elhoud; N.C. Renton; W.F. Deans: Hydrogen embrittlement of super duplex stainless steel in acid solution. *International Journal of Hydrogen Energy*, Vol. 35 (2010), No. 12, pp. 6455–6464
23. W.B. Hutchinson; K. Ushioda; G. Runnsjö: Anisotropy of tensile behaviour in a duplex stainless steel sheet. *Materials Science and Technology*, Vol. 1 (1985), No. 9, pp. 728–736
24. A. Sharrfeddin; S.M. Musa; F.M. Elshawesh: Role of structural orientation on the susceptibility of 2205 duplex stainless steel to hydrogen embrittlement. In: *2nd International Advances in Applied Physics and Materials Science Congress*, M.H. Aslan, A.Y. Oral (Ed.), Antalya, Turkey, American Institute of Physics, 2012, pp. 199–203
25. W. Zheng; D. Hardie: Effect of Structural Orientation on the Susceptibility of Commercial Duplex Stainless Steels to Hydrogen Embrittlement. *Corrosion*, Vol. 47 (1991), No. 10, pp. 792–799
26. T. Böllinghaus; H. Hoffmeister: FINITE ELEMENTE CALCULATIONS OF PRE- AND POST-HEATING PROCEDURES FOR SUFFICIENT HYDROGEN REMOVAL IN BUTT JOINTS. In: *Mathematical Modelling of Weld Phenomena 3*, H. Cerjak (Ed.), Maney Materials Science, 1997, pp. 727–756
27. S. Floreen; H.W. Hayden: The Influence of Austenite and Ferrite on the Mechanical Properties of Two-Phase Stainless Steels Having Microduplex Structures. *Transactions of ASM*, Vol. 61 (1968),, pp. 489–499
28. I. Tamura; Y. Tomoto; A. Akao; Y. Yamaoka; M. Ozawa; S. Kanatani: On the Strength and Ductility of Two-phase Iron Alloys. *Transactions of the Iron and Steel Institute of Japan*, Vol. 13 (1973),, pp. 283–292
29. W. Ramberg; W.R. Osgood: Description of Stress–Strain Curves by three Parameters. Washington, Technical Note No. 902, NASA Scientific and Technical Information Facility, 1943, p. 32
30. K.J.R. Rasmussen: Full-range stress–strain curves for stainless steel alloys. *Journal of Constructional Steel Research*, Vol. 59 (2003), No. 1, pp. 47–61
31. G.E. Dieter: Elements of the Theory of Plasticity. In: *Mechanical Metallurgy*, New York, McGraw Hill Higher Education, 1988, pp. 69–102
32. D.U. Weber: Modellierung von Verformung und Schädigung in Werkstoffgefügen mit unterschiedlich großen Teilchen und unter Wasserstoffeinfluss Vorwort. (2006), No. April
33. a. El Bartali; P. Evrard; V. Aubin; S. Herenú; I. Alvarez-Armas; a. F. Armas; S. Degallaix-Moreuil: Strain heterogeneities between phases in a duplex stainless steel. Comparison between measures and simulation. *Procedia Engineering*, Vol. 2 (2010), No. 1, pp. 2229–2237
34. P. Zimmer; D.M. Seeger; T. Böllinghaus: Hydrogen Permeation and Related Material Properties of High Strength Structural Steels. In: *High Strength Steel for Hydropower Plants*, H.H. Cerjak, N. Enzinger (Ed.), Graz, Verlag der Technischen Universität Graz, 2005, pp. 17–1 ff
35. P. Zimmer: Zur Bewertung der Kaltrisssicherheit von Schweißverbindungen aus hochfesten Feinkornbaustählen. 2007, p. 169

36. T. Bollinghaus; H. Hoffmeister; L. Reuter: Material properties of as delivered and quenched modified martensitic stainless steels dependent on hydrogen concentration. In: *Proceedings of the 1st International Conference Supermartensitic Stainless Steels*, Brüssel, Belgian Welding Institute, 1999, pp. 264–271
37. T. Zakroczymski; a. Glowacka; W. Swiatnicki: Effect of hydrogen concentration on the embrittlement of a duplex stainless steel. *Corrosion Science*, Vol. 47 (2005), No. 6, pp. 1403–1414
38. A.A. El-yazgi; D. Hardie: The embrittlement of a duplex stainless steel by hydrogen in a variety of environments. *Corrosion Science*, Vol. 38 (1996), No. 5, pp. 735–744
39. C. San Marchi; T. Michler; K. a. Nibur; B.P. Somerday: On the physical differences between tensile testing of type 304 and 316 austenitic stainless steels with internal hydrogen and in external hydrogen. *International Journal of Hydrogen Energy*, Vol. 35 (2010), No. 18, pp. 9736–9745
40. T. Matsuo; J. Yamabe; S. Matsuoka; Y. Murakami: Influence of Hydrogen and Prestrain on Tensile Properties of Type 316 l Austenitic Stainless Steel. In: *Proceedings of the 2008 International Hydrogen, Effects of Hydrogen on Materials*, B. Somerday, P. Sofronis, R. Jones (Ed.), ASM International, 2009, pp. 105–112
41. C.S. Marchia; B.P. Somerdaya; X. Tangb; G.H. Schiroky: Effects of alloy composition and strain hardening on tensile fracture of hydrogen-precharged type 316 stainless steels. *International Journal of Hydrogen Energy*, Vol. 33 (2008), No. 2, pp. 889–904
42. C. Skipper; G. Leisk; A. Saigal; D. Matson; C.S. Marchi: Effect of Internal Hydrogen on Fatigue Strength of Type 316 Stainless Steel. In: *Proceedings of the 2008 International Hydrogen Conference, Effects of Hydrogen on Materials*, B. Somerday, P. Sofronis, R. Jones (Ed.), ASM International, 2009, pp. 139–146
43. C.M. Younes; a. M. Steele; J. a. Nicholson; C.J. Barnett: Influence of hydrogen content on the tensile properties and fracture of austenitic stainless steel welds. *International Journal of Hydrogen Energy*, Vol. 38 (2013), No. 11, pp. 4864–4876
44. M.R. Louthan; G.R. Caskey; J.A. Donovan; D.E. Rawl: Hydrogen embrittlement of metals. *Materials Science and Engineering*, Vol. 10 (1972),, pp. 357–368
45. D. Yang; K. Chasse; P.M. Singh: Stress and Environmental Factors Affecting Stress Corrosion Cracking of Duplex Stainless Steels in White Liquor Environments. In: *TAPPI Engineering, Pulping, Environmental Conference*, Memphis, Tennessee, 2009, p. 19
46. S.S. Chen; T.I. Wu; J.K. Wu: Effects of deformation on hydrogen degradation in a duplex stainless steel. *Journal of Materials Science*, Vol. 39 (2004), No. 1, pp. 67–71
47. W.C. Luu; P.W. Liu; J.K. Wu: Hydrogen transport and degradation of a commercial duplex stainless steel. *Corrosion Science*, Vol. 44 (2002), No. 8, pp. 1783–1791
48. F. Straub; T. Böllinghaus; W. Unger; T. Mente: In-situ detection of Deuterium in duplex stainless steels by time-of-flight secondary ion mass spectrometry (TOF-SIMS). In: *Hydrogen-materials interactions - Proceedings of the International Hydrogen Conference (IHC 2012)*, P. Sofronis, B.P. Somerday (Ed.), Moran, Wyoming, USA, ASME, 2014, pp. 505–512
49. T. Böllinghaus; H. Hoffmeister; C. Middel: Scatterbands for hydrogen diffusion coefficients in steel having a ferritic or martensitic microstructure and steels having an austenitic microstructure at room temperature. *Welding in the World*, Vol. 37 (1996), No. 1, pp. 16–23
50. T. Böllinghaus; H. Hoffmeister; A. Dangeleit: A scatterband for hydrogen diffusion coefficients in micro-alloyed and low carbon structural steels. *Welding in the World*, Vol. 35 (1995), No. 2, pp. 83–96
51. J. Tien; A.W. Thompson; I.M. Bernstein; R.J. Richards: Hydrogen transport by dislocations. *Metallurgical Transactions A*, Vol. 7 (1976), No. 6, pp. 821–829
52. J.A. Donovan: Accelerated evolution of hydrogen from metals during plastic deformation. *Metallurgical Transactions A*, Vol. 7 (1976), No. 11, pp. 1677–1683
53. T.-P. Perng; C.J. Altstetter: Cracking kinetics of two-phase stainless steel alloys in hydrogen gas. *Metallurgical Transactions A*, Vol. 19 (1988), No. 1, pp. 145–152

54. E. Dabah; V. Lisitsyn; D. Eliezer: Performance of hydrogen trapping and phase transformation in hydrogenated duplex stainless steels. *Materials Science and Engineering: A*, Vol. 527 (2010), No. 18-19, pp. 4851–4857
55. M. Dadfarnia; P. Novak; D.C. Ahn; J.B. Liu; P. Sofronis; D.D. Johnson; I.M. Robertson: Recent advances in the study of structural materials compatibility with hydrogen. *Advanced materials (Deerfield Beach, Fla.)*, Vol. 22 (2010), No. 10, pp. 1128–35
56. B. Gołebiowski; W.A. Swiatnicki; M. Gaspérini: Microstructural changes induced near crack tip during corrosion fatigue tests in austenitic-ferritic steel. *Journal of microscopy*, Vol. 237 (2010), No. 3, pp. 352–8
57. A. Głowacka; M.J. Woźniak; G. Nolze; W.A. Świątnicki: Hydrogen induced phase transformations in austenitic-ferritic steel. *Solid State Phenomena*, Vol. 112 (2006), pp. 133–140
58. A. Barnoush; M. Zamanzade; H. Vehoff: Direct observation of hydrogen-enhanced plasticity in super duplex stainless steel by means of in situ electrochemical methods. *Scripta Materialia*, Vol. 62 (2010), No. 5, pp. 242–245
59. H. Fischmeister; B. Karlsson: Plastizitätseigenschaften grob-zweiphasiger Werkstoffe. *Zeitschrift für Metallkunde*, Vol. 68 (1977), No. 5, pp. 311–327
60. K. Cho; J. Gurland: The law of mixtures applied to the plastic deformation of two- phase alloys of coarse microstructures. *Metallurgical Transactions A*, Vol. 19 (1988), No. 8, pp. 2027–2040
61. X. Sun; K.S. Choi; W.N. Liu; M. a. Khaleel: Predicting failure modes and ductility of dual phase steels using plastic strain localization. *International Journal of Plasticity*, Vol. 25 (2009), No. 10, pp. 1888–1909
62. J. Michalska; M. Sozańska; M. Hetmańczyk: Application of quantitative fractography in the assessment of hydrogen damage of duplex stainless steel. *Materials Characterization*, Vol. 60 (2009), No. 10, pp. 1100–1106
63. E.M. Westin: Microstructure and properties of welds in the lean duplex stainless steel LDX 2101 ®. 2010, p. 72
64. I. Varol; J.C. Lippold; W.A. Baeslack: Welding of Duplex Stainless Steels. *Key Engineering Materials*, Vol. 69-70 (1992),, pp. 217–252
65. R.A. Walker; T.G. Gooch: Hydrogen cracking of welds in duplex stainless steel. In: *International Conference on Duplex Stainless Steel, Les editions de physique*, J. Charles (Ed.), Beaun, Frankreich, 1991, pp. 1053–1063
66. T. Mente; T. Bollinghaus: Modeling Of Hydrogen Distributionin A Duplex Stainless Steel. *Welding in the World*, Vol. 56 (2013), No. 11-12, pp. 66–78

# Hydrogen Assisted Cracking of a Subsea-Flowline

Th. Boellinghaus, E. Steppan and T. Mente

**Abstract** Since the mid-nineties, supermartensitic stainless steels (SMSS) have increasingly been applied to welded subsea-pipeline systems in the North Sea oil and gas fields, especially to flowlines at mild sour service conditions. However, in 2001 cracking and leaks occurred during installation and service start-up of two SMSS flowlines in the Norwegian Tune gas condensate field, welded with a new developed matching filler wire. Brittle transgranular cracking started especially at inter-run lack of fusion and propagated brittle, predominantly through the weld metal. The present paper provides a brief overview of the original failure case and respective sequence of events leading to complete replacement of the SMSS by carbon steel flowlines in 2002. Then, detailed investigations of a circumferential weld sample of the failed Tune flowline are highlighted, targeted at comparison of the failure appearance to previous investigations of this filler material type and to search for possible explanations for the brittle fracture at the crack initiation area. SEM investigations of the fracture surface revealed brittle areas only in the direction towards the top side of the weld while the major part of the investigated surface exhibited ductile fracture. As an approach to clarify, if the fracture was a consequence of hydrogen assisted cracking, five small sized specimens have been cut out of the original sample. Cracking has been introduced parallel to the original fracture surface in these specimens at respective saw cuts and bending. The results show that brittle transgranular cracking appeared only in the specimen cooled down to very low temperatures by liquid nitrogen and in the sample charged with hydrogen to an average concentration of about 15 ml/100 g. However, a fracture similar to the original surface was observed only in the hydrogenized specimen. As a further result, very similar fracture surfaces of supermartensitic stainless steel weld metals had been observed on specimens subjected to hydrogen assisted cold cracking (HACC) as well as to hydrogen assisted stress corrosion cracking (HASCC).

Th. Boellinghaus (✉) · E. Steppan · T. Mente
Federal Institute for Materials Research and Testing (BAM),
Unter den Eichen 87, 12205 Berlin, Germany
e-mail: Thomas.Boellinghaus@bam.de

 
Th. Boellinghaus et al. (eds.), *Cracking Phenomena in Welds IV*,
DOI 10.1007/978-3-319-28434-7_17

In total, the results indicate that brittle fracture starting at the inter-run lack of fusion were not initiated by high notch tip deformation rates, but rather influenced by hydrogen, probably taken up during welding.

## Introduction

Supermartensitic stainless steels, as a new generation of high-alloyed martensitic chromium steels, have been developed since the mid-eighties in several variants, i.e. as low carbon (LC) steels with Ni ≤ 4.5 and Mo ≤ 1.5 wt% and a couple of years later as extra low carbon (ELC) type with Ni ≤ 6.5 and Mo ≤ 2.5 wt%. At considerable material market shortages and cost saving issues in the oil and gas industry in the mid-nineties, such materials were especially intended to be used as pipeline materials, replacing much more expensive nickel alloys as well as much higher alloyed steels, like superaustenitic or duplex stainless steels, SASS or DSS, respectively. Particularly at pipelines connecting the well heads and the risers to the offshore oil and gas production units, so-called flowlines, such materials are subjected to an increased risk for hydrogen assisted stress corrosion cracking, due to sour service applications. However, the susceptibility of such materials to hydrogen assisted cracking was increased as compared to DSS or SASS anyway, due to the martensitic microstructure, in particular the occurrence of hard quenched martensite in the HAZ after welding of the soft annealed martensitic base materials, containing a considerable amount of retained and annealing austenite. SMSS found their first applications as flowline material in the Gullfaks South and Åsgard North Sea oil and gas fields. The materials have been welded so far using DSS as filler materials, providing a triplex microstructure, in particular near the fusion line. The SMSS pipeline segments have been welded using a new SMSS filler wire more or less for the first time, matching much better the microstructure and the strength of the base materials, than up to that time usually applied DSS.

The Tune flowlines were carried out as twin lines with an OD of 323.9 mm (12″) and a thickness of d = 15.9 mm in Zone 1 (Pipeline 1—PL1) and of d = 17.5 mm in Zone 2 (Pipeline 2—PL2), respectively. The assembly of the pipes was performed by semi-automatic hot wire TIG-welding up to 600 m long segments on-shore which were respectively reeled and fixed. Afterwards such segments were welded together to a complete pipeline on a lay-barge. The more than 11 km long twin line was spooled and laid in the typical J-form, connecting the Oseberg D platform and the Tune subsea gas production unit. It should be noted that the maximum allowed strain during reeling and installation operations up to 3 % are approved. Tables 1 and 2 show the chemical composition, mechanical properties and service conditions.

**Table 1** Chemical composition (wt%) and mechanical properties of martensitic Cr-steel [1]

| Element | C | Cr | Mo | Ni | P | S | Si | Mn | Ti | Fe |
|---|---|---|---|---|---|---|---|---|---|---|
| Base material | 0.011 | 12.3 | 2.2 | 5.9 | 0.01 | 0.001 | 0.2 | 0.2 | 0.01 | Bal. |
| Filler material | 0.004 | 12.4 | 2.7 | 6.4 | 0.02 | 0.0003 | 0.4 | 0.6 | 0.002 | Bal. |
| YS ($R_{p\ 0.2}$) | | Tensile test results | | | NKK tensile test results | | | | | |
| | | 742 MPa | | | 678 MPa | 712 MPa | | | 699 MPa | |
| UTS ($R_m$) | | 848 MPa | | | 810 MPa | 811 MPa | | | 805 MPa | |
| YS/UTS | | 088.8 | | | 0.84 | 0.88 | | | 0.87 | |
| Elong. $A_5$ (%) | | 19.8 | | | 24 | 23 | | | 23 | |

**Table 2** Environmental characteristics inside the flowline (gas condensate) [1]

| Property | Dimension | Value |
|---|---|---|
| Pressure | bar | 250–420 |
| Temperature, inlet | °C | 115 |
| $CO_2$ | mol% | 3.1 |
| $H_2S$ | ppm | 10 |
| *Formation water* | | |
| pH in reservoir | | 5.6 |
| $Na^+$ | mg/l | 14,800 |
| $Cl^-$ | mg/l | 23,600 |
| $HCO_3-$ | mg/l | 1072 |
| Organic acid | mg/l | 363 |

# Failure Sequence and Initial Situation

## *Sequence of Events*

The first failure occurred during pipeline laying operations in the Tune field in June 2001. A fracture occurred in a weld of Pipeline 2 (PL2) due to the laying in the straightening machine. This represented the last section of the fabrication process on the laying vessel, i.e. the failure occurred after 2 % straining of the pipe, due to the reeling and spooling operations. The weld was cut out and the rest of the installation was completed. During the final pressure test, a leak occurred in a straight section of each flowline. Both leaks were repaired by hyperbaric welding and, after completion of the second repair, a third leak was found and repaired. During testing of the last repair, at pressure in both flowlines, a sudden large leak occurred in the first hyperbaric repair weld of PL1.

In addition, the riser sections and the expansion loop near the Oseberg D Platform were inspected, because these locations experience higher stresses during production, as a consequence of thermal expansion, than during a normal pressure

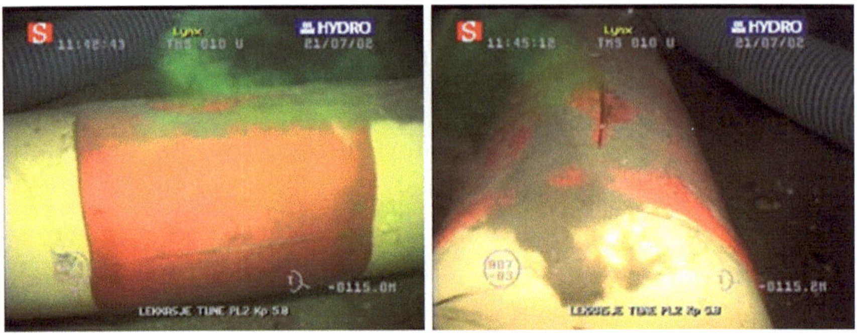


**Fig. 1** Leak during pressure test in pipeline 2, visible due to colored water [1]

test. To prove that there were no cracks and weld defects in these sections that would survive pressure testing, but would not survive 25 years of cyclic service loading, ultrasonic testing was carried out from inside the riser up to 130 m into the flowline. After finalizing such activity, the Tune flowlines were considered as fit for purpose in April 2002, if a complete system pressure test would be successful.

Due to the fact that the hyperbaric repair welds passed the first leak test, but one more weld failed during the following leak tests, hyperbaric welding was not considered as a suitable repair procedure any more, due to oxygen and hydrogen contamination. Instead, it was decided to repair the three leaks by installation of straight spools with compact flanges which was successfully completed—one year after occurrence of the first failure.

The final pressure test exhibited three new leakages in June/July 2002. A first small leak occurred at PL1 with 3 bar per day pressure loss during searching for leakage at constant pressure. During search of this leak and after passing the 24 h pressure test, a large leakage occurred at 6 bar per day pressure loss in PL2 and the same occurred in PL1 24 h later (Fig. 1). Subsequently, all repairs were stopped, the SMSS flowlines were scrapped and replaced with inhibited carbon steel by autumn 2002. The design life was 12 years and this corresponds to the production time of the field.

## *Preliminary Investigations*

All cracks observed in the pipeline fabrication welds were associated with inter-run lack of fusions (IRLOF). Even the crack growing in the straightener, before the pipeline was laid, was initiated at an approximately 200 mm long IRLOF, having an effective length of 157 mm at a height of 2.7 mm perpendicular to the axial direction of the pipe. The crack located at the 6 o'clock position during welding started brittle in the weld metal and propagated ductile, as can be seen from the dimple surface of the remainder. This pipeline was strained twice by about 2 %, in particular during the reeling and de-reeling as well as by a considerable restraint due to welding, being sufficient to cause such fracture.

Also the following three cracks which appeared during pressure testing of PL1 and PL2 between July and December 2001 were of the same type, starting brittle at an IRLOF and exhibited ductile final fracture. These cracks were smaller than the above mentioned one occurring in the straightener, probably due to less deformation. Also, the dimensions of the IRLOFs were 50–220 mm long, but had a lower height of 1.0–1.5 mm, Fig. 2. All cracks were located in the 12–2 o′clock position after the installation.

Also the fractures occurring in fabrication welds during the final pressure testing showed the same characteristics. The fracture always spread from both sides of a lack of fusion towards the top and root side of the weld [2]. In the first part of the

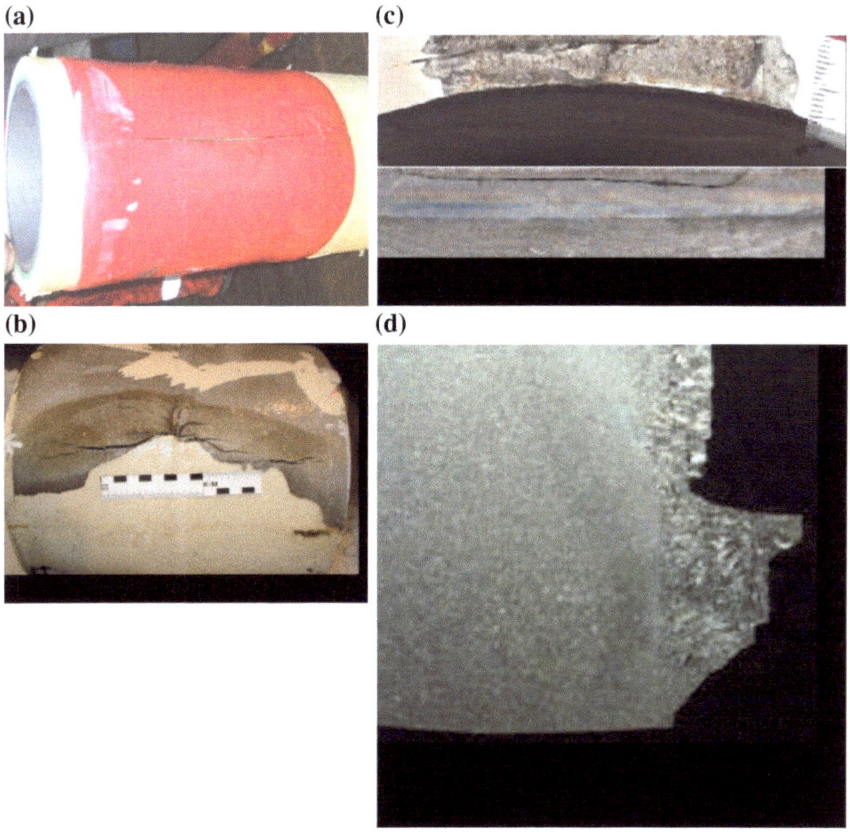


**Fig. 2** Preliminary investigations of recovered pipeline sections. **a** Leaking section of PL2 after retrieval, **b** crack propagating from the weld metal into the base material on both sides in the above section of PL2, **c** crack appearing on the outer weld surface, **d** fracture surface after opening up, showing the IRLOF in the center, **c** cross section showing the IRLOF with low height in the middle of the pipeline wall and way of crack propagation [1]

fracture surface the crack propagated into the weld metal and then shifted more towards the heat affected zone (HAZ). However, at the crack initiation site the fracture surface appeared brittle as transgranular cleavage or cleavage-like mode [2]. The rest and the major part of the fracture surface in the weld metal and in the heat affected zone, respectively, was identified as ductile microvoid coalescence. Up to the present, the fracture behaviour of the weld metal generally appeared to be ductile, as documented, for instance, in the welding procedure specification and further prefabrication tests [3, 4]. In the preliminary failure investigation report [2], hot cracking and cold cracking had been excluded as origins of the brittle part on the fracture surface.

## Investigations

Analyses of an original failure part was targeted to elucidate the brittle start of the cracks originated at the IRLOF of the pipeline welds, since such undesired brittle behavior of the matching filler material had been claimed by the end user as well as by the pipeline welding manufacturer. For this, fractographic SEM inspections of the fracture surface as well as respective simulation experiments have been carried out on a part of an originally cracked weld.

### *Original Fracture Surfaces*

As illustrated in Fig. 3, the provided sample had a length of about 40 mm and was taken from a region of a failed weld joint including parts of the IRLOF as well as brittle and ductile fracture surface parts. Already from this specimen, it becomes

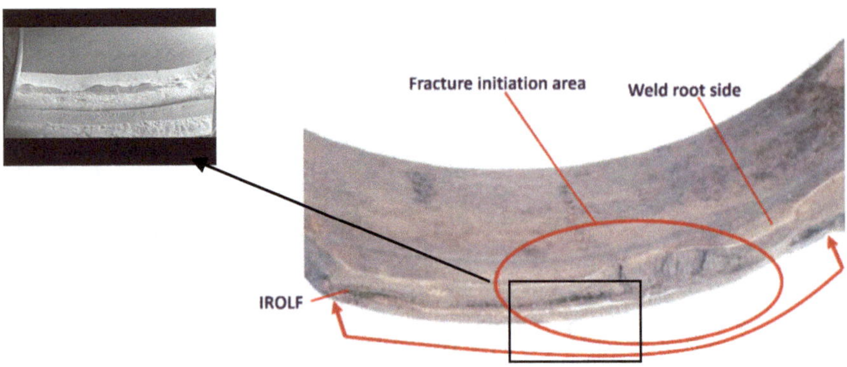


**Fig. 3** Location of the investigated 40 mm long sample taken from a segment of a failed supermartensitic stainless steel weld

obvious that the crack initiated at the IRLOF and then spread first to the top side as well as the root side of the weld and then propagated into further regions, here circumferentially. The specimen was cut out wide enough into the direction transverse to the weld by more than 20 mm, and thus included weld metal, the HAZ and also base material. The specimen was covered with a dirt oil film when it was delivered and was prepared for SEM inspection by cleaning with citric acid as well as by ultrasonic baths in ethanol and acetone.

Fractography of the original fracture surface was performed at nine locations at various magnification levels, predominantly at a magnification of 150 and of 750 (Fig. 4).

The SEM photograph in Fig. 4 demonstrates that at least at the location of the delivered sample, the inter-run lack of fusion had an extension of about 5 mm into the transverse direction of the weld and is located towards the top side of the weld. The various fracture modes of the different regions on the original fracture surface are shown in Fig. 5.

Generally, most of the fracture surface was also covered with additional very small pits far below one μm in diameter, independent of a brittle or ductile fracture mode. The appearance of such locations is most obviously shown in the Regions 1 and 5, at a magnification of 750× (Fig. 5) and might be attributed to corrosion effects on the fracture surfaces during crack opening and exposure to seawater until retrieval of the respective pipeline segment or etching effects caused by the SEM preparation and cleaning procedure.

From Fig. 5 and the characterization in Table 3 it becomes obvious that only the part of the surface above the lack of fusion towards the top side of the weld exhibited a brittle transgranular cleavage-like fracture mode. The part of the fracture surface below the lack of fusion, i.e. towards the root side, was identified as completely ductile including also shearing in Region 9 and below Region 3. Since Region 4 appeared also as ductile, the brittle failure of the weld below the lack of

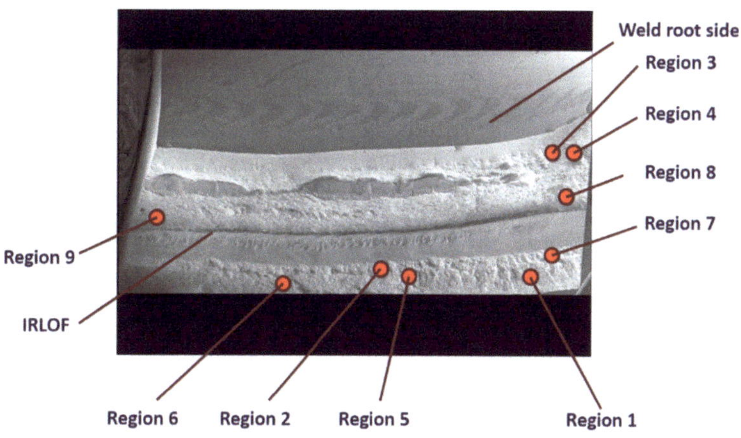


**Fig. 4** Locations at the original fracture surface investigated by SEM fractography

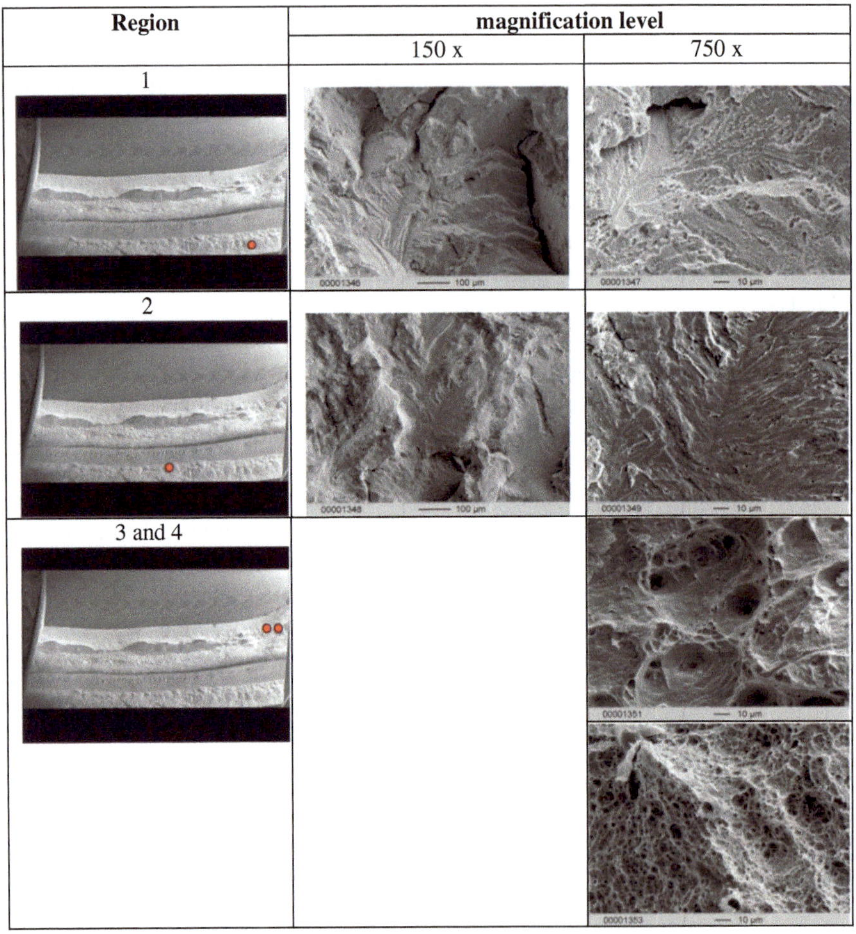


**Fig. 5** Original fracture surface of different regions in Fig. 4

fusion must have a lower extension in the circumferential direction than it has it in the upper parts of the welds towards the top side. The more brittle behaviour of the upper part of the weld might be attributed to less annealing of the last layers by subsequent passes [4]. In addition to the SEM investigation also visual inspection showed that the brittle transgranular cleavage-like part of the upper fracture surface disappeared towards the side located away from the brittle crack initiation area. This means that the brittle failure at both sides of the lack of fusion stopped already in the weld metal. This also indicates that the weld metal of this type of filler materials shows normally ductile fracture behaviour. Table 3 summarizes the fracture modes at the various different locations.

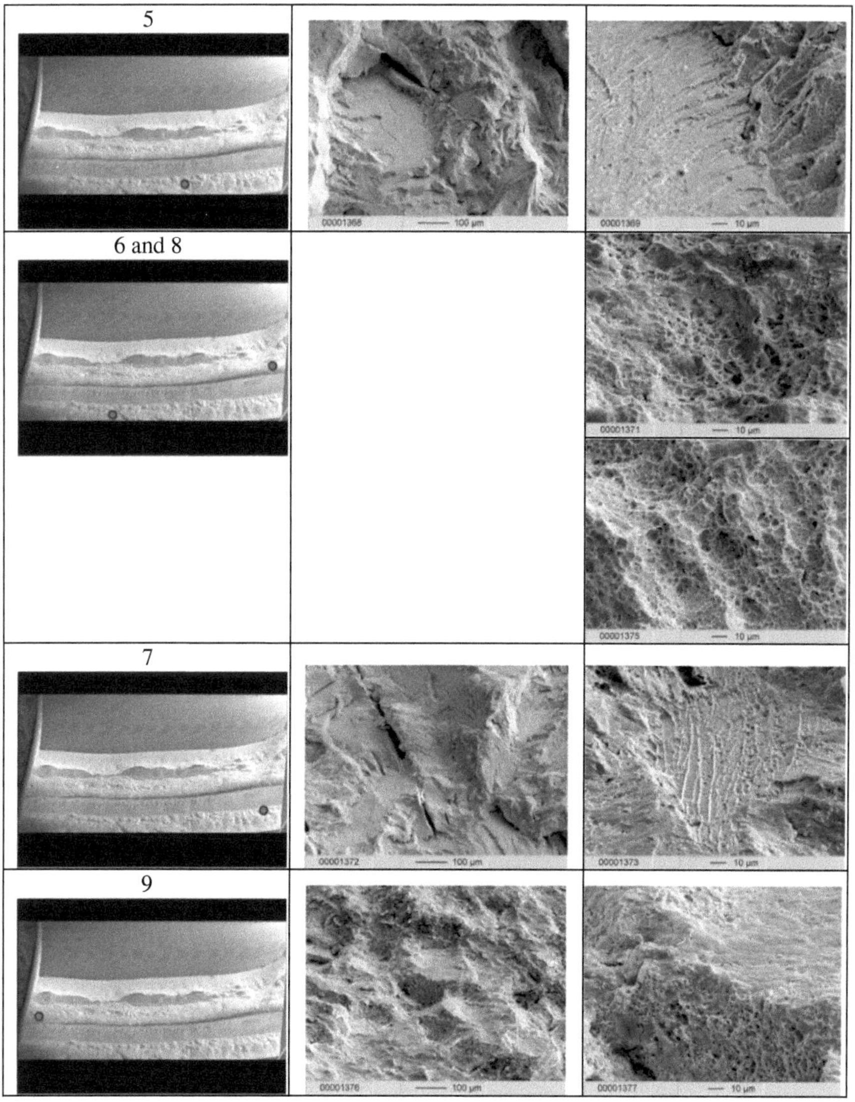


**Fig. 5** (continued)

## *Simulation Experiments*

Five additional simulation experiments using the original failure part have subsequently been carried out. For this, small specimens have been cut out of the original part and saw cuts with a width of 0.2 mm similar to the IRLOF have been introduced into the weld metal parallel to the original fracture plane. The specimens

**Table 3** Overview of the fracture types on the original fracture surface

| Location | Fracture mode | Remarks |
|---|---|---|
| 1 | Brittle—cleavage or cleavage-like fracture | Small pits on the cleavage facets |
| 2 | Brittle—cleavage or cleavage-like fracture | No pits on the cleavage facets |
| 3 | Ductile—microvoid coalescence | Deep and large dimples slightly deformed and indicating shear loading |
| 4 | Ductile—microvoid coalescence | Typical dimples |
| 5 | Brittle—cleavage or cleavage-like fracture | Small pits on the cleavage facets |
| 6 | Ductile—microvoid coalescence | Typical dimples with additional small pits on the facets |
| 7 | Brittle—cleavage or cleavage-like fracture | Small pits on the cleavage facets |
| 8 | Ductile—microvoid coalescence | Typical dimples |
| 9 | Ductile—shearing | Shear lips and dimples |

have subsequently been bent in a way to initiate fracture at such saw cuts, respectively, to characterize the topography. The investigation was targeted to identify conditions that cause a brittle quasi-cleavage fracture of the weld metal near the IRLOF. Furthermore, they were aimed to confirm that no brittle fracture might be attributed to application of the matching filler material to the specified welding procedure.

These simulation experiments were named Laboratory Fracture 1–5. The first one was conducted by breaking such a specimen at about −30 °C (Fig. 6a). The fracture surface exhibited ductile fracture with dimples oriented towards the loading direction. The second experiment was carried out at very low temperature of about −100 °C. As shown in the Fig. 7a, b, the fracture topography appears now predominantly as cleavage. But, still, small regions of this laboratory fracture showed microvoid coalescence, indicating ductile portions, which are shown in Fig. 7c.

In order to elucidate the reasons for the small pits detected particularly on the cleavage facets of the brittle parts of the original fracture surface (Fig. 5, Regions 1 and 5 at a magnification of 750×), the Laboratory fractures 1 and 2 have been subjected to the same cleaning procedure as the original specimen. SEM micrographs of such prepared surfaces are shown in the Figs. 6b and 7d. Comparing Fig. 6a with b reveals that the edges of the dimples might slightly be rounded by such procedure. Also the cleavage facets of the brittle transgranular fracture seem at least partly be affected by the cleaning procedure which becomes evident by comparison of Fig. 7a–c with d. Thus, the cleaning procedure might be one reason for the occurrence of the small pits on the cleavage facets on the original fracture surface. Cleavage planes and dimples might occur adjacent to each other, showing a mixture of ductile and brittle regions of fracture. However, there is no

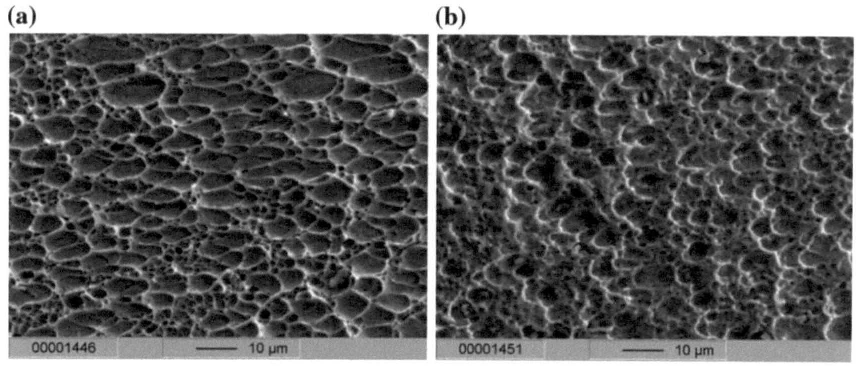


**Fig. 6** Fracture topography of laboratory Fracture 1 (ca. −30 °C without hydrogen charging). **a** 1200×, without cleaning **b** 1200×, with cleaning

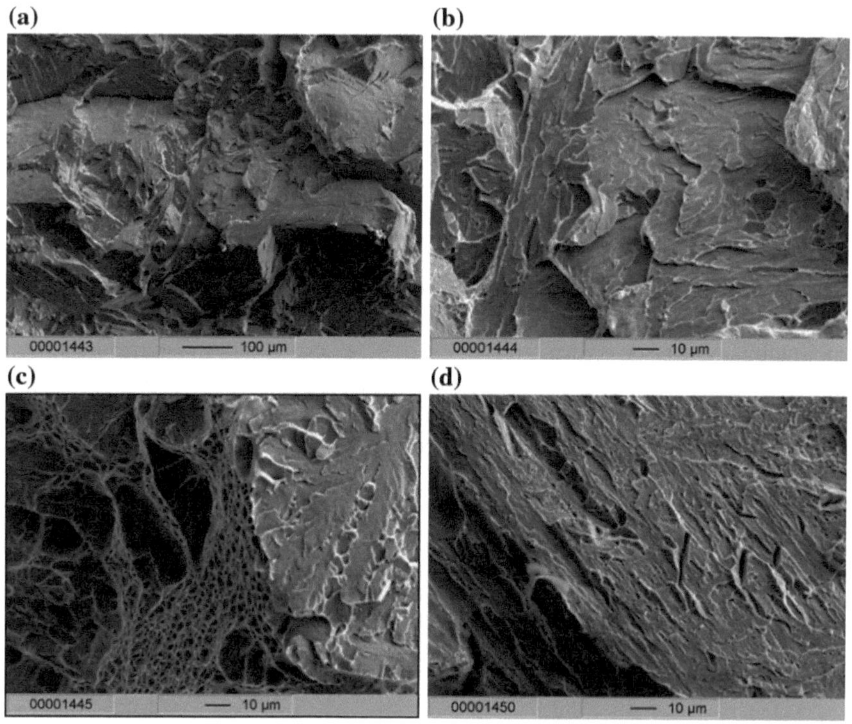


**Fig. 7** Fracture topography of laboratory Fracture 2 (ca. −100 °C without hydrogen charging). **a** 150× **b** 750× **c** 750× **d** 750×, with cleaning

micromechanical reason for the occurrence of equally distributed pits of the same size on cleavage fracture planes. It can thus be assumed that such pits have been introduced into the already fractured surface topography.

Since the effects are less pronounced on the cleaned laboratory fracture, it can anticipated also that corrosion of the original fracture surface after opening of the crack and exposure to the seawater before retrieval of the respective pipe segment had a major influence. This assumption is also supported by the fact that cold deformed regions represent generally sites for a preferential and also quite rapid pitting attack on supermartensitic stainless steels exposed to chloride environments. It can only be emphasized that such pits on the cleavage facets are not relevant for brittle fracture initiation and, in particular, do not indicate ductile failure at such locations.

Laboratory Fracture 3 was initiated at room temperature to test if any brittle fracture might be initiated in the weld metal at a notch providing stress concentrations similar to the edges of an IRLOF. As shown by the SEM micrographs in Fig. 8, this fracture showed a completely ductile topography.

This fracture behaviour of the filler material is consistent to previous findings and indicates that the filler material has a ductility that is at least high enough that any brittle fracture initiation at the edges of the IRLOF without other influences is quite unlikely. Some of the flatter dimples might tentatively be attributed to δ-ferrite that occurs in the weld microstructures of this SMSS type, especially in the weld metal and along the fusion line.

For reproducibility, Laboratory Fracture 4 was also initiated at room temperature in the center of the weld metal and parallel to the original surface, as shown in Fig. 9a. Two different regions of the fracture topography have been investigated in detail (Fig. 9b). Region 1 represents a representative topography in the center of the fracture. Region 2 was selected in a region where the fracture has been initiated from the root side of the weld directly at the saw cut to investigate if any strain or stress accumulation at the tip of the saw cut might have an influence on the fracture topography. As can be drawn from the Fig. 9c–e at different magnifications, both regions exhibited completely ductile fracture characterized by microvoid coalescence. This can be taken as another evidence that the matching weld metal will normally not fail brittle by transgranular cracking at notches having a width similar to an IRLOF.

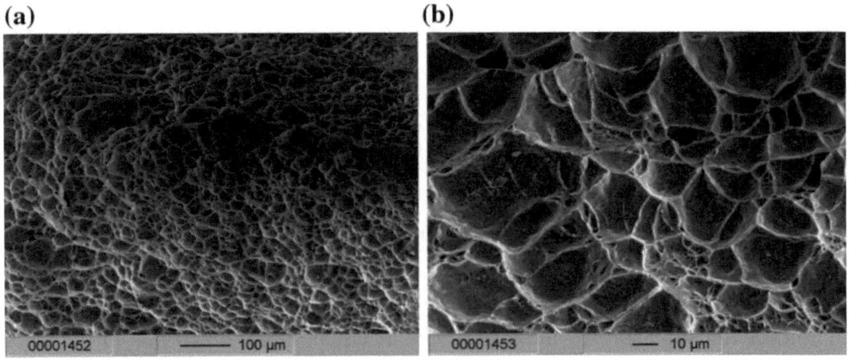


**Fig. 8** Topography of laboratory Fracture 3 (at RT without hydrogen charging). **a** 150× **b** 750×

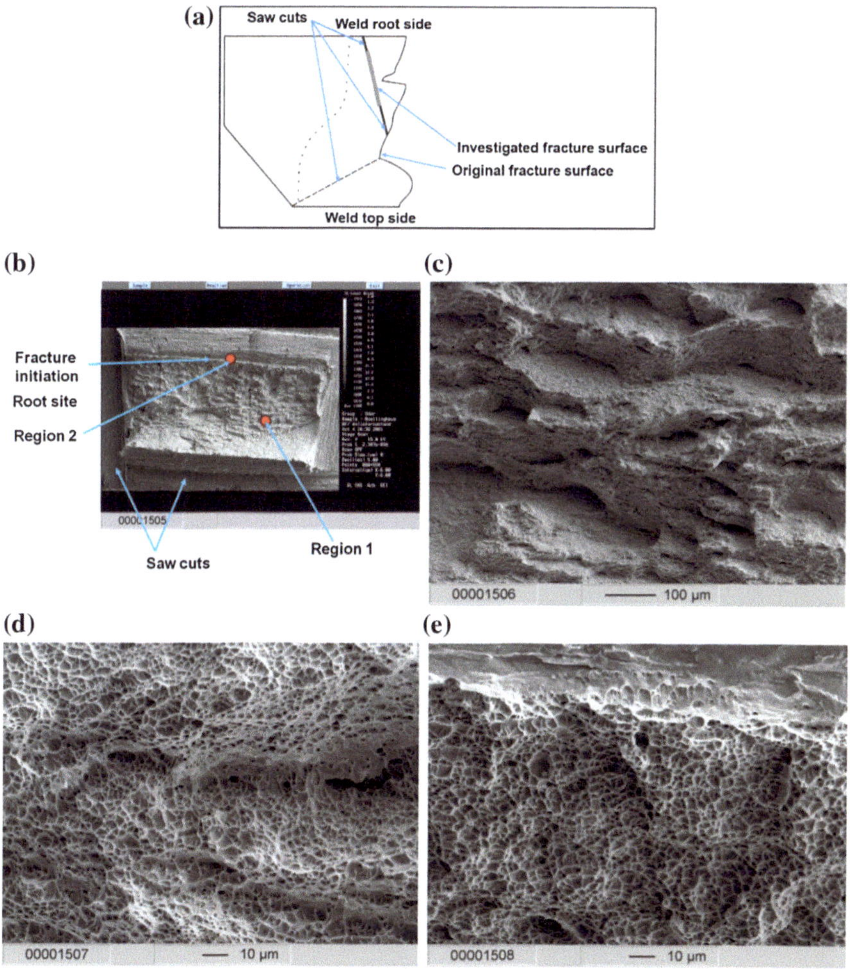


**Fig. 9** Topography of laboratory Fracture 4 (at RT without hydrogen charging). **a** Location of laboratory fractures with respect to the original fracture surface **b** investigated regions **c** region 1–150× **d** region 1–750× **e** region 2–750×

As shown in Fig. 10a, Laboratory Fracture 5 has been initiated similar to the previous Fracture 4 in the weld metal and parallel to the original fracture surface. However, this specimen has been hydrogen charged for two days in a 50 % $H_2S$ saturated buffered NACE solution (50 g/l NaCl + 5 g/l $CH_3COOH$) at a potential of −1050 mV (against the Standard Calomel Electrode—SCE). As controlled by respective carrier gas hot extraction, the hydrogen concentration in this completely saturated specimen was about 15.9 ml/100 g corresponding to app. 14 ppm. Three different regions of this fracture topography have been investigated in detail. Region 1 was located quite in the center of the fracture. Region 2 was selected to

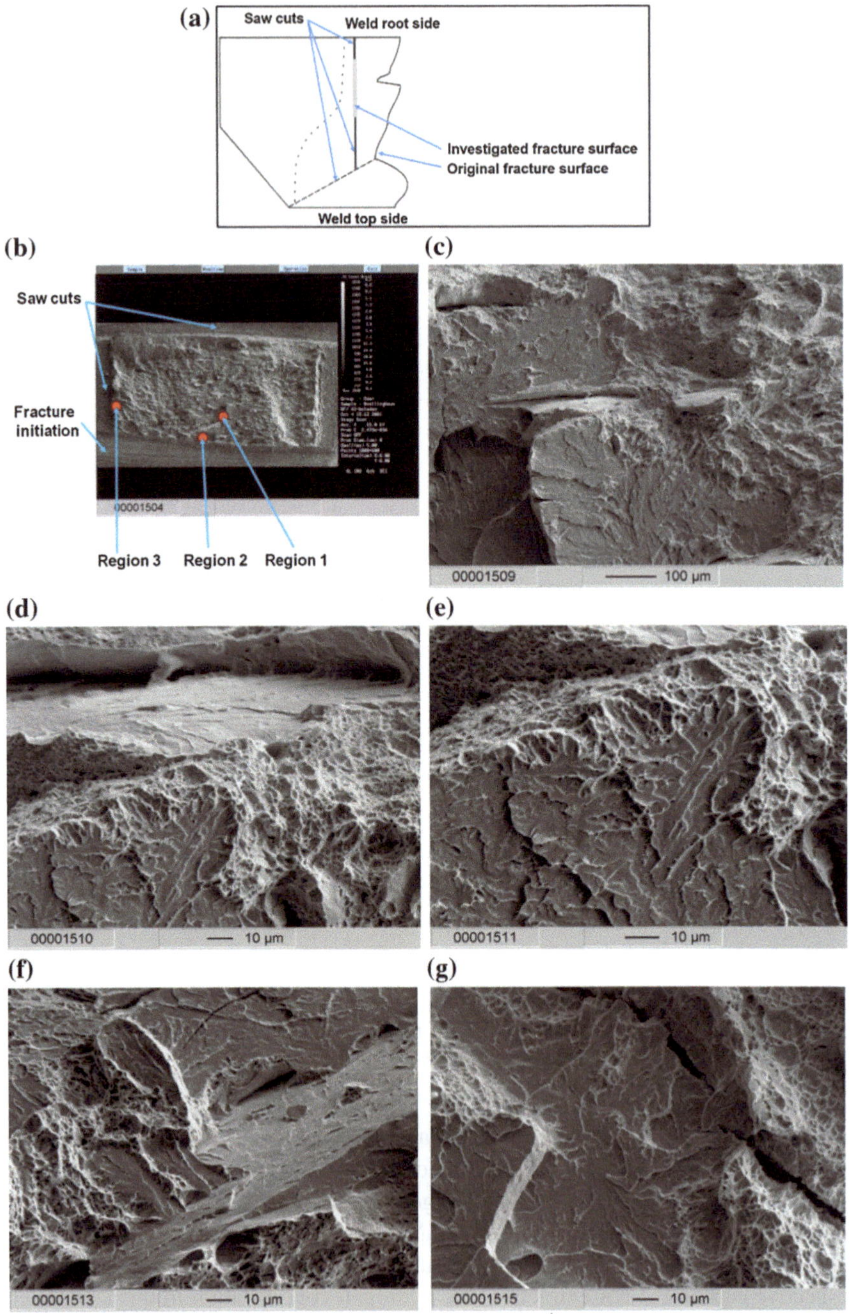


**Fig. 10** Topography of Laboratory Fracture 5 (specimen hydrogen charged at RT for 2d in 50 % $H_2S$ saturated NACE solution at −1050 mV-SCE). **a** Location of laboratory fracture parallel to the original fracture surface **b** investigated regions **c** region 1—150× **d** region 1—750× **e** region 1—1000× **f** region 2—750× **g** region 3—750×

investigate the fracture topography directly at the fracture initiation site near the saw cut to find out whether the notch tip might have any influence on the fracture type. Region 3 represents a location also close to the saw cut for similar reasons. From the Fig. 10c–g viewing the three regions at different magnifications it becomes obvious that all three regions exhibit brittle quasi-cleavage transgranular cracking, mixed with microvoid coalescence at the margins of the cleavage planes. In comparison to the other laboratory fractures, the topography is completely changing for the hydrogen charged weld metal microstructure.

## *Comparison of Fracture Topographies*

To confirm that topography of Laboratory Fracture 5 can be attributed to hydrogen assisted cracking and to relate these findings to the original fracture surface, the various topographies of cleavage cracking observed in SMSS weld metals have been compared. Figure 11 provides an overview of the respective SEM micrographs. Figure 11a, b allow a comparison of the real fracture topography with that one initiated at −100 °C in the weld metal at the same magnification of 750×. Despite the small pits caused by corrosion of the original fracture topography or etching during cleaning it can be seen that the original surface exhibits also some ductile microvoid coalescence at the margins of the cleavage planes. Additionally, feather-like markings on the cleavages planes as well as fan-like structures at their edges can be observed. Such features are usually associated with hydrogen assisted cracks in SMSS microstructures. In contrast to this, the fracture surface of the cold fractured specimen exhibits pure transgranular cleavage planes. A comparison of the Fig. 11a, c having also the same magnification reveals that the laboratory fracture initiated parallel to the original fracture surface in electrochemically hydrogen charged specimen at a concentration of 15.9 ml/100 g shows very similar markings of quasi-cleavage cracking as the original topography. In comparison with the respective literature [5–8], this can be taken as an evidence that also the original fracture was initiated in a hydrogenized microstructure.

For further evidence, Fig. 11d shows the topography of a fracture initiated in a SMSS weld metal by Hydrogen Assisted Stress Corrosion Cracking (HASCC) at the three dimensional stress–strain state after cooling of the restrained orbital welded pipeline in a typical formation water composition saturated with 1 vol.% $H_2S$ in $CO_2$ [9, 10]. Such cleavage-like cracking of hydrogen containing weld metals has also been found in previous investigations (Fig. 11e and f). Those experiments were carried out by application of the Instrumented Restraint Cracking test to investigate the resistance of supermartenstic stainless steel TIG welds without filler wire against hydrogen assisted cold cracking and hydrogen assisted stress corrosion cracking [5, 6]. It has to be mentioned that in contrast to the failed weld metal investigated here, a medium alloy was used for those investigations. But, the welds were also not heat treated, and thus, the weld metal exhibited a similar quenched martensite. From the Fig. 11e and f in comparison to Fig. 11a,

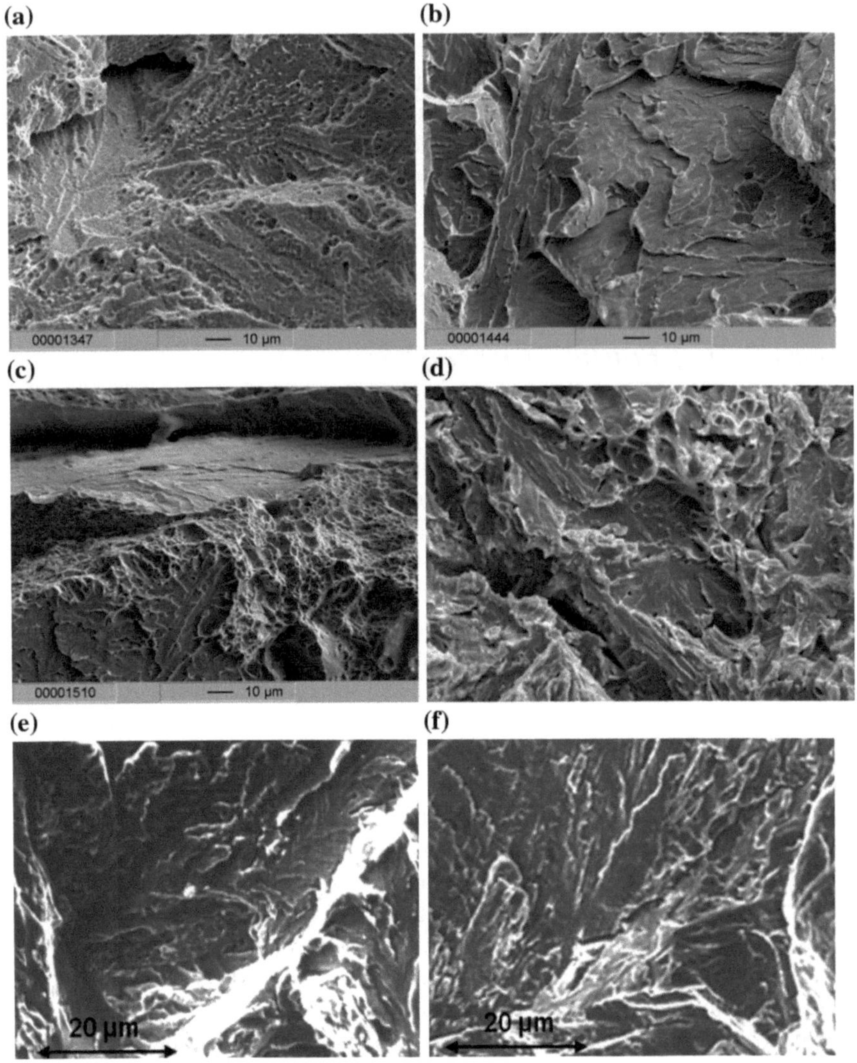


**Fig. 11** Fracture topographies related to HAC in supermartensitic stainless steel (SMSS) weld metal **a** original fracture surface—750× **b** fracture surface at low temperature (ca. −100 °C)—750× **c** fracture surface hydrogen charged (ca. 14 ppm)—750× **d** HASCC (1 vol.% $H_2S/CO_2$ saturated formation water) **e** fracture surface of a SMSS weld metal, HACC (10 vol.% hydrogen to the shielding gas) **f** SMSS HASCC (pH = 2.7, 15 % $H_2S$ saturated NACE solution)

it can be drawn that the medium alloy also shows cleavage-like cracking features in the hydrogenized state which are very similar to that of the high alloy investigated for this failure analysis.

Also already pointed out previously, the same quasi-cleavage transgranular fracture appears by Hydrogen Assisted Cold Cracking (HACC) or HASCC of the weld metal. This means, the appearance of the quasi-cleavage fracture topography in SMSS weld metals is independent of the way of hydrogen uptake, i.e. via a welding arc at HACC or electrochemically at HASCC under sour service conditions.

Finally, it also has to be mentioned in this context that small scaled Charpy tests have been carried out at 30 °C of such weld metals which were also not post weld heat treated. Fractographic SEM investigations of those specimens exhibited also complete ductile failure and, in particular, did not show any similarities to the brittle transgranular cleavage-like mode observed on the original fracture surface and also on those produced at low-temperature or after hydrogen uptake.

Furthermore, the hardness of the weld metal was not specified up to the present time and, with respect to the measured values between 246 and 352 HV, ranges are within the tolerances given by the DNV Specifications (2000) for carbon steel subsea pipelines. Consequently, the hardness values of the SMSS weld metal show the common range and, based on the present experiences, no brittle fractures have to be expected under the loading rate during the pipeline laying operations, even not in the presence of sharp notches.

Summarizing these findings, transgranular quasi-cleavage brittle fracture topographies similar to that on the original fracture surface can only be identified in weld metal that had taken up hydrogen at significant concentrations.

## Conclusions

The objective of the present investigations was to investigate the brittle cracking occurring during pipeline laying and first pressure tests, initiated at IRLOFs in the circumferential direction. From the present results of investigations carried out using a sample of the failed pipeline, the following conclusions can be drawn:

1. The IRLOF exhibited dimensions that had to be tolerated according to the existing regulations and standards at the time of failure occurrence. This, however, is only related to linear welds. Regarding orbital and circumferential welds which are experiencing bending during pipeline laying, there were no limitations at that time [8]. For future installation of welded offshore pipeline it should be considered that the circumferential welds are generally subjected to severe straining of up to several percent and that such LOF might represent sharp notches when being loaded perpendicular to the edges during pipeline laying operations. Respective weld specifications and standards providing tolerances for such IRLOFs should thus be reviewed in this regard.
2. Since additional Charpy tests, having a loading speed much higher than the mechanical loading rate during the pipeline laying operations and which normally exhibit no hydrogen assisted cracking, showed different fracture topography than that of the original specimen and those initiated during the simulation experiments. The brittle fracture initiation at the weld IRLOF ends can thus not be attributed to a low weld metal toughness.

3. From the simulation experiment at min. −30 °C, the original weld metal exhibits ductile fracture at a mechanical load comparable to the original failure sequence with respect to location and rate. At −100 °C, the weld metal exhibits typical brittle transgranular fracture, but the topography is significantly different from that of the original fracture surface.
4. A topography similar to the original fracture exhibiting typical features of transgranular quasi-cleavage cracking, i.e. feather-like markings on the cleavage planes, fan-like structures at the edges of the cleavage planes and a partial mixture with ductile microvoid coalescence at the cleavage edges, has only been found in the simulation experiments on fracture surfaces of hydrogen charged weld metal specimens. The quasi-cleavage transgranular fracture topographies of SMSS weld metals resulting from HACC and from HASCC are very similar.

Summarizing, it has been shown that the original fracture has to be related to hydrogen assisted cracking. The SMSS twin flowline has never been in operation in the Tune field and was not long enough exposed to cathodic protection to allow electrochemical hydrogen uptake in crack relevant concentrations. Thus, HASCC cannot be associated with the failure, also, because the hydrogen diffusion time from the surface to the IRLOFs located deep inside the weld metal would not have been sufficient. As root cause of failure, it only comes into consideration that hydrogen has been introduced into the welds during the pipeline manufacturing process. It has thus to be anticipated that the welds had been contaminated with hydrogen via leakages in the TIG torch cooling systems during onshore orbital welding of respective sections. It has to be emphasized that it took more than two years in the present case to combat this frequent experience in failure analysis of welded components and until it finally was confirmed that hydrogen had been traced in during the manufacturing of the pipelines by leakages of the cooling systems—and thus the failure origin falls under the category how to turn a good filler material into a bad weld [1].

## References

1. R. Mollan: Experience with 13Cr Supermartensitic Stainless Steel in the Tune Submarine Flowlines, NACE-05092, Corrosion 2005, NACE International
2. DNV-Report No. 2001-3142: Tune Project – Failure Analysis and Testing of Weld Connections, Oslo, July, 2001.
3. Information provided by Böhler Thyssen Welding, August-October, 2001
4. Information provided by Norsk Hydro, October, 2001
5. Th. Boellinghaus, H. Hoffmeister, M. Littich: Online SCC Monitoring of LC 13 % Cr Welds at Realistic Weld Restraint Conditions in the IRC Test, Advances in Corrosion Control and Materials in Oil and Gas Production, P. S. Jackman, L. M. Smith (Ed.), EFC Publication No. 26, pp. 286-303
6. G. Lange, M. Pohl (Eds.): Systematische Beurteilung technischer Schadensfälle, Wiley-VCH, Weinheim, Germany (2014)
7. VDI Guideline 3822 – Part 7: Failures at Welded Components, Draft, to be published 2015

8. Th. Boellinghaus, H. Hoffmeister: A Numerical Model for Hydrogen Assisted Cracking, Corrosion 56 (2000), No. 6, pp. 611 – 622
9. Th. Boellinghaus: Hydrogen Assisted Cracking of Supermartensitic Stainless Steels, N. R. Moody et al. (Eds.): Proc. of Hydrogen Effects on Materials Behaviour and Corrosion Deformation Interactions, TMS 2003, pp. 1009-1018
10. Th. Boellinghaus, Th. Kannengiesser: Effect of Filler Material Selection on Stress Strain Build Up and Stress Corrosion Cracking Resistance of Welded Supermartensitic Stainless Steel Pipes, Intern. Conf. Corrosion, Denver 2002, Paper 02061

# Part VI
# Stress Corrosion and Cold Cracking Phenomena

# Numerical Modelling of Hydrogen Assisted Cracking in Steel Welds

**Th. Boellinghaus, T. Mente, P. Wongpanya, E. Viyanit and E. Steppan**

**Abstract** Hydrogen assisted stress corrosion and cold cracking represent still a major topic regarding the safety of welded steel components against failure in many industrial branches. Hydrogen might be introduced during fabrication welding or might be taken up from an environment during sour service or at cathodic protection. Additionally, understanding and avoidance of hydrogen entry into weld microstructures from gaseous pressurized environments becomes increasingly important for renewable energy components. There are two types of metallurgical mechanisms associated with hydrogen assisted cracking, i.e. the cracking as well as hydrogen transport and trapping mechanisms. For numerical modelling, it has to be considered that both types are not independent of each other, that the mechanisms are not yet completely clarified and that validation of such models strongly depends on implementation of the correct hydrogen related materials properties. However, quite significant achievements have been made in modelling of hydrogen assisted cracking by indirect coupling of thermal, stress-strain as well as hydrogen uptake and diffusion analyses. After a brief introduction into the subject and by revisiting various proposed cracking mechanisms, the present contribution focuses on recent developments of a numerical model based on a comparison of actual hydrogen concentrations and mechanical loads with respective hydrogen dependent material properties as crack initiation and propagation criteria. The basic procedure for numerical simulation of crack initiation and propagation is outlined and it is shown how such numerical simulations can be validated experimentally. Furthermore, it is highlighted how such a procedure has been extended to a comprehensive model for life time prediction of welded steel pipeline components and experimentally verified. Finally, it is outlined how the model can be extended to simulate cracking in heterogeneous steel microstructures on the different scales.

Th. Boellinghaus (✉) · T. Mente · P. Wongpanya · E. Viyanit · E. Steppan
BAM Federal Institute for Materials Research and Testing, Berlin, Germany
e-mail: Thomas.Boellinghaus@bam.de



Th. Boellinghaus et al. (eds.), *Cracking Phenomena in Welds IV*,
DOI 10.1007/978-3-319-28434-7_18

## Introduction

Economic design of modern lightweight components requires steels having increased strength aiming at significant reduction in wall thickness. Advanced steels are thus precisely designed today for specific applications. Steels are thus representing still the most innovative structural materials, due to the excellent capability to combine load sustainability, microstructural diversity and improved joinability in one material. However, due to increasing strength and increasing fitness-for-purpose design, the parameter windows for welding manufacturing and service application generally become more and more narrow. Consequently, service life prediction has to be an inherent part in designing constructions using modern high strength steels and increasingly, numerical simulations using Finite Element Analyses (FEA) are integrated into such concepts, as they can contribute to failure avoidance. Particularly to this respect, industry and R&D are working today much more closely together to establish such concepts for safety relevant large scale components than it has been in the past.

It is generally accepted that with increasing strength of the applied steels, welded components become more susceptible to Hydrogen Assisted Cracking (HAC). A lot of research efforts have been made to clarify the metallurgical mechanisms of such failure phenomena which can be divided into cracking and hydrogen transport mechanisms. Research approximately between the seventies and the nineties has mainly been focused on clarification of such mechanisms, which limited success. Since about 25 years, research is concentrating more on the phenomenology of this failure type. The weld manufacturing of modern steel components is usually carried out at much more narrow parameter windows. Consequently, many procedures have been developed for successful prevention of cracking up to the present. For instance, post weld heat treatment procedures for hydrogen effusion have been identified which are sufficient to avoid cold cracking in steel welds. However, such procedures are normally restricted to specific materials applied to a specific component. The same applies to analytical calculation procedures and fracture mechanics based models introduced to prevent HAC during service, frequently neglecting the different cracking resistance of weld microstructures. In contrast, numerical modelling represents a much more comprehensive tool to cover HAC phenomena during fabrication welding and component service life time [1], as will briefly be outlined by the respective advantages in the following passages.

After an outline of the respective advantages of numerical HAC modelling, the present contribution focuses on a quantitative, partly descriptive numerical model [2] based on FEA of the temperature, stress-strain and hydrogen distribution which has been developed further in the recent years with respect to transferability to different welded components from various industrial branches, to different high strength steel types and towards HAC modelling in heterogeneous microstructures.

## Hydrogen Assisted Cracking (HAC)—Types and Mechanisms

As visualized in Fig. 1, the most important life cycle stages with respect to failure avoidance and safety of welded components are the production and service period. Both can be evaluated by the interaction of the influencing factors material, load and design. In other words: If the local load of a specific material microstructure at a specifically designed part of the components is exceeding the loading capacities, the probability of failure is significantly increasing. In contrast to considerations elsewhere, where the weld manufacturing process is regarded as a fourth influencing factor in the life time evaluation of welded components, this approach allows a much more decent investigation of the weldability and the service behavior of components. It has to be emphasized that both, in particular the weldability, represent component properties and not material properties. This also allows evaluating influences of the manufacturing process on the service life behavior of welded components via the axis material and design. Such approach gains particular importance for the evaluation of HAC at welded components. For instance, the weld residual stresses and strains can significantly influence initiation and propagation of Hydrogen Assisted Stress Corrosion Cracking (HASCC) during service or nests of small Hydrogen Assisted Cold Cracks (HACC) can be widened or be accumulated to larger cracks by respective loads introduced during the first in-service implementation of a component.

Regarding the service behavior it has to be distinguished between stress corrosion cracking (SCC, usually pure anodic processes without the influence of hydrogen) and Hydrogen Assisted Stress Corrosion Cracking (HASCC, with influence of hydrogen). Both occur in corroding environments, whereas HASCC frequently results from preceding pitting or crevice corrosion due to hydrogen uptake in the

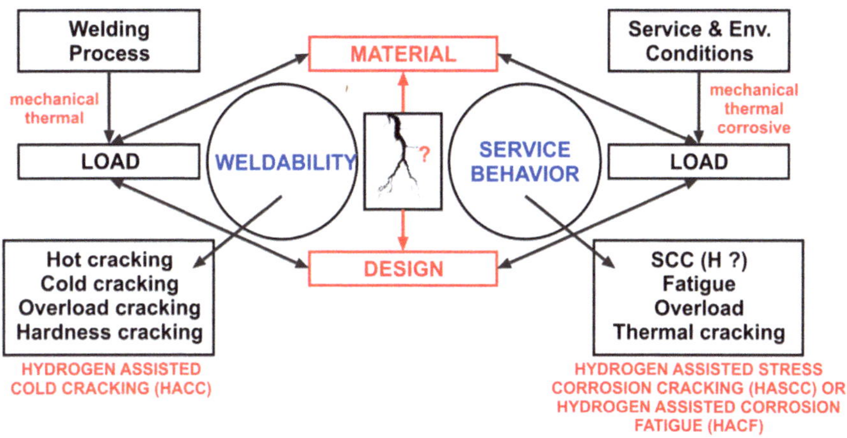


**Fig. 1** Weldability and service behavior of welded components

material. Possible sources for hydrogen are sour service environments, especially $H_2S$, or acidification inside pits and crevice which are decoupled from the environment. Thus, regions having low pH values and low chemical potentials as well as due to oxygen depletion or hydrolysis of metals hydrogen uptake occur, this may develop HASCC. As a special form, Hydrogen Assisted Corrosion Fatigue (HACF) might have to be considered during service life of welded components, especially, if those are subjected to Low Cycle fatigue (LCF). Recently, with emphasis on welded components in the areas of renewable energy and hydrogen storage/transport, HAC avoidance for component service life in highly pressurized environments has gained increased importance.

All types of HAC itself depend on three influencing factors, as illustrated in Fig. 2, i.e. HAC generally occurs, if a critical combination of local hydrogen concentration and local mechanical load in a local microstructure is exceeded. With a special focus on welded components, it has to be considered that the various microstructures behave differently with respect to their HAC resistance.

Also, two phase microstructures like Duplex Stainless Steels (DSS) will influence HAC differently, due to different mechanical and hydrogen transport properties of the constituent phases as well as by generating different residual stresses and strains during manufacturing due to different thermal expansion of the phases. It can only be emphasized that cracking is not initiated by just the presence of hydrogen. Thus, a consistent numerical model should provide the capabilities to take such wide variation of all three influencing factors into account for a precise

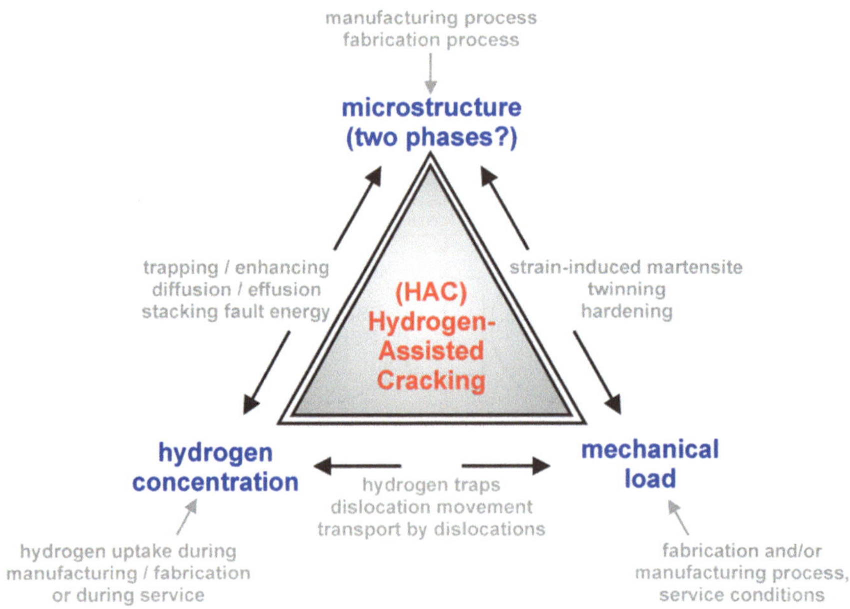


**Fig. 2** Influences on hydrogen assisted cracking

weldability and life time evaluation of welded components, but also, to separate their significance on HAC initiation and propagation.

## Mechanisms

It has been documented various times that the fracture topographies for both HACC and HASCC are quite similar, at least in more or less homogeneous martensitic microstructures. This implies at least some evidence that the metallurgical cracking mechanisms are quite similar for a specific microstructure [3, 4]. Following, a brief overview of HAC mechanisms is given.

Various metallurgical mechanisms have been suggested in the past seventy years, in order to explain and to describe the HAC phenomenon in metals. Only a few overviews have been published, focusing predominantly on corrosion problems [3, 5]. Hirth [6] published an overview of HAC mechanisms by which it also became clear that the metallurgical mechanisms of hydrogen assisted cracking are independent from the process by which hydrogen is introduced into the metal. After this, also a couple of further theories have been published and reviewed in detail [1, 7] which are summarized in the following passages.

### Pressure Theory

The pressure theory, presented by Zapffe and Sims [8], is one of the oldest theories regarding the effect of hydrogen in metallic materials. The authors describe that hydrogen atoms are combining with molecules at internal defects, i.e. interfaces at inclusions, microvoids, microcracks. Recombination of hydrogen with molecules is associated with a volume expansion resulting in high internal pressures. Thus, blistering may occur near the surfaces or initiation and coalescence of micro voids in the bulk material. Regarding this mechanism, high hydrogen concentrations are needed. Moreover, those failures are associated to steels having lower strength and/or toughness values. It is generally anticipated that cracking, as a consequence of high internal pressure, takes place by formation of microvoids or microcracks which are then connected to more macroscopic cracks [5, 8].

### Adsorption Theory

The idea of the adsorption theory proposed by Petch and Sables [9] is that adsorbed hydrogen at the crack surface lowers the surface energy needed for propagating a crack. Brittle fracture is then promoted, if cohesion strength at the crack tip is reduced to a level below the crack critical shear stress.

Application of this theory to welding is difficult, because rupture mostly occurs in the bulk in a short distance ahead of the crack tip. Following the arguments by

Oriani the adsorption theory does not explain why cracking occurs at that surface at which the material absorbs hydrogen [10]. Additionally Robertson [11] emphasized that the effect of a reduced surface energy becomes inconsequential, if it is considered in combination with the plastic work of separation. Thus, the adsorption theory is particularly not applicable to the period of crack incubation, i.e. the time before crack surfaces of sufficient size exist. Additionally, the theory does only consider environmental hydrogen uptake and, for instance, does not explain cracking caused by hydrogen introduced by arc welding.

## Inner and Outer Hydride Formation

Hydrogen can combine with metals of subgroups IV and V from the periodic table to form brittle hydrides, i.e. Titanium, Zirconium, Niobium, Tantalum [4, 12, 13]. Whereby, inner and outer hydride formation can be distinguished. Outer hydride formation results in a brittle surface layer, which plays a major role regarding stress corrosion cracking of nonferrous metals, i.e. Titanium [5, 13, 14]. At the inner hydride formation embrittlement of the material occurs due to the volume expansion of hydrides inducing stresses and strains as well as from the very low deformability of hydrides [5, 12, 13]. Thus, brittle crack propagation occurs due to continuous formation and brittle cracking of hydrides. Those hydrides preferentially form at regions with high plastic deformations. Regarding welding of iron based alloys, unstable hydrides only form at high hydrogen partial pressure and thus, such cracking mechanisms are negligible for steel welds.

## Hydrogen Enhanced Decohesion Theory (HEDE)

Contrary to the adsorption theory, crack nucleation in the material ahead of a crack tip is considered. This theory originally was developed by Trojano [15] and further improved by Oriani [10, 16]. It is presumed that hydrogen diffusion into regions of high hydrostatic stresses, as for instance ahead of notches and crack tips, is accelerated. As shown in Fig. 3, hydrogen interacts with the metal atoms by reducing the bonding energy. Thus, hydrogen together with high tensile stresses leads to purely elastic failure of the material. Following Trojano [15], the effect of hydrogen on the cohesive energy of the metal lattice can be explained by the metal electron hull structure: Atomic hydrogen shares its electron to the 3d-shell of the iron atoms and is present as a proton. Presence of hydrogen in the metal as an ion has been experimentally confirmed several times, by application of strong polarization currents to metal bars leading to respective accumulation of originally homogeneously distributed hydrogen at the negative electrode. Anyway, it is presumed that due to the energy increase by such hydrogen metal lattice interactions, the metal lattice cohesion strength is decreased. In particular, the HEDE theory is not related to a specific hydrogen uptake process, because hydrogen diffusion into a

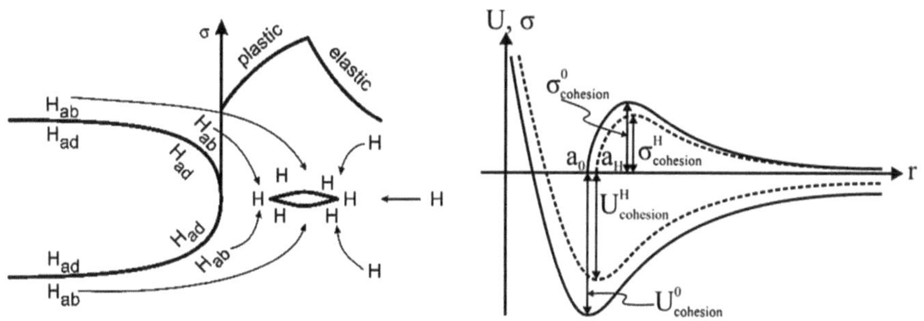


**Fig. 3** Schematic illustration of crack nucleation by HEDE (*left*, [5]) and the lowered cohesive energy (U) and cohesive strength (σ) with regard to the critical displacement of the interface (r), $a_0$ = lattice parameter (*right*, [17])

crack critical region has to take place before cracking. Thus, the HEDE theory is also independent of the location of a specific crack critical region.

## Plasticity Theory

This expression summarizes all metallurgical mechanisms assuming hydrogen affecting the local plasticity of a material. According to mechanical metallurgy basics [18] such effects are commonly attributed to hydrogen dislocation interactions [4]. For instance, diffusion enhancing can take place by transport of hydrogen with moving dislocations [19]. According to Nelson [4] those processes definitely are not degradation mechanisms themselves, but might assist the embrittlement of materials. Also, the author describes an increased crack propagation rate due to decreased critical stresses needed for propagation of the crack. In general, two different theories have to be distinguished with respect to hydrogen dislocation interactions:

On the one hand it is assumed that hydrogen forms Cottrell clouds in the dilation zone of dislocations hindering dislocation movement which entails a hardening effect. Generally, the plasticity of the material is reduced ahead of the crack tip which favors brittle fracture.

On the other hand it is assumed that hydrogen might locally increase the dislocation activity. Beachem [20] observed in experimental investigations with high strength steels, that crack propagation might be increased by an increased dislocation velocity, resulting in a locally enhanced deformation capability (Fig. 4).

From those studies, Sofronis et al. [17, 22–25] derived the hydrogen enhanced localized plasticity (HELP) mechanism from in situ experiments of hydrogen related dislocation activities. Under the term HELP several hydrogen dislocation interactions are summarized:

Regarding the HELP mechanism it is assumed that hydrogen locally increases the plasticity of a metallic material by the interaction of hydrogen with the stress

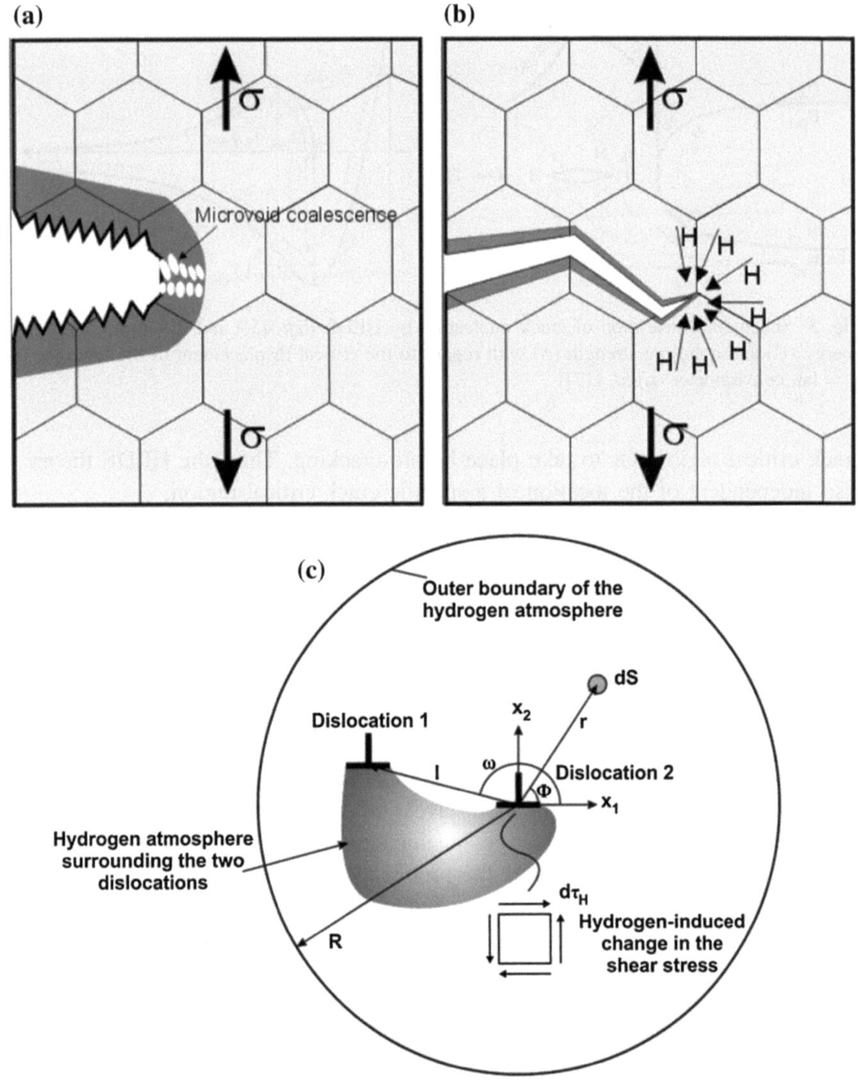


**Fig. 4** Schematic illustration of crack propagation based on Beachem's assumption: **a** In the absence of hydrogen crack growth occurs by coalescence of micro void, **b** In the presence of hydrogen deformation becomes easier and crack growth occurs as result of localized deformation at the crack tip [20, 21], **c** The interaction of hydrogen with dislocations according to Sofronis and Birnbaum [22, 23]

field of the dislocation. Hydrogen segregates to the stress field of a dislocation and modifies it such that it is increased in one direction and decreased in the other direction [11]. Thus, in direction where the stress field is decreased also the interaction of the dislocation with obstacles is reduced resulting in enhanced dislocation motion at lower stress levels. By this shielding effect, cross-slip is limited and slip planarity is increased by hydrogen.

This means that the transport of hydrogen with a dislocation explains the limited cross slip and the preferential planar slip of dislocations. In other words, hydrogen would have to be relocated inside the stress field to allow again cross slip of a specific dislocation and respective screw movement. But, there is no effect supporting this, since there is no driving force for such constellation and thus, such dislocations will be kept in the planar slip mode.

Additionally, hydrogen modifies dislocation-dislocation interactions by decreasing the equilibrium separation distance between dislocations in a dislocation pile-up [11]. The hydrogen-hydrogen interactions in those regions leads to formation of hydride-like areas resulting in lower hydrogen concentrations needed for the above mentioned shielding effect and thus favoring the HELP mechanism.

To sum this up, it has to be considered on the one hand that hydrogen can move alongside a respective dislocation, dependent on the temperature range and the local strain rate. On the other hand it has generally to be considered that hydrogen acts as an interstitial and thus, retards dislocation movement.

Due to the enhanced slip processes, the yield strength of the material might be lowered locally accompanied by a reduction in strength. Thus, high tensile stresses in front of the crack tip might result in a type of ductile fracture of the material.

In a recent paper, Robertson [11] summarizes various experimental verifications of the hydrogen enhanced dislocation movement in a number of materials and also, how such accelerated dislocation movement acts on crack propagation inside a grain or alongside grain boundaries to understand the transfer of dislocations into another grain. The mechanism for dislocation transfer over grain boundaries is discussed generally with respect to plastic failure of a microstructure without interaction with hydrogen. The microstructure underneath the crack surface shows quite a similar behavior in the hydrogen free and charged condition. In both cases, the microstructure shows significant plastic deformation. However the hydrogen affected dislocation cells are somewhat smaller, due to the interaction with hydrogen. Roberston [11] also discussed the respectively different fracture topography.

## Adsorption Induced Dislocation Emission Theory (AIDE)

The theory of AIDE was proposed by Lynch [26] and has some similarities to the mechanisms summarized under the HELP, as it is also proposed that hydrogen influences the dislocation activity. The author describes that hydrogen which is adsorbed in the first few atomic layers reduces cohesion of the metal atoms in the lattice (HEDE). Thus, nucleation of dislocations is facilitated (Fig. 5). Due to an enhanced localized plasticity (HELP) in the crack tip zone micro voids will be nucleated. Additionally, if hydrogen is transported in the plastic zone of the crack tip by dislocations, micro voids also will nucleate at inclusions or precipitations in the bulk material. Thus, cracking occurs by coalescence of cracks and micro voids

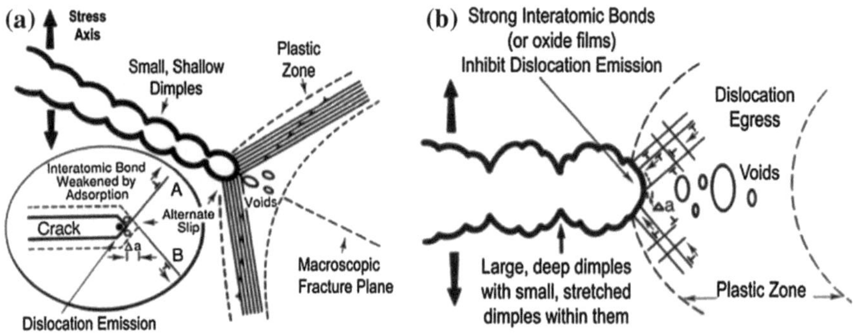


**Fig. 5** Schematic illustration of AIDE (**a**) and ductile cracking due to coalescence of cracks with micro voids (**b**) [26]

producing shallower dimples at the fracture surface [26]. It is proposed that both ductile as well as brittle transgranular and intergranular fracture can occur by AIDE, depending on where dislocation and void nucleation emerged most easily.

Regarding the AIDE versus the HEDE mechanism, they are meanwhile referenced in literature as a chicken or egg type of process as to whether plasticity nucleates voids at interfaces which then nucleate cleavage or intergranular fracture or whether local cleavage occurs followed by ligament tearing [7, 11, 27].

As worked out elsewhere [1, 7, 11], most of the mechanisms presented above only represent aspects of the hydrogen-materials-interaction, but do not provide a holistic solution. Additionally, the pressure theory, adsorption theory as well as hydride formation are nearly not applicable to explain HAC in welded steel components.

Detailed information regarding the mechanisms of HAC can be found in the referenced literature. Some publications provide summaries of the various mechanisms [1, 7] and, additionally, experimental and theoretical milestones with regard to HEDE and HELP have ben pointed out [7].

Following the discussion at conferences and the literature of the last decade, a combination of both mechanisms HEDE and HELP has been proposed several times. In summary, hydrogen-assisted intergranular fracture is attributed to a hydrogen-enhanced plasticity-mediated decohesion process [11]. The role of the hydrogen-enhanced plasticity is to impact the structure of the grain boundary through the process of slip transfer, to provide the grain boundary with the critical hydrogen concentration such that it is now the weak link in the system responding to the local stress state that is imposed by the dislocations that interact with the grain boundary and the surrounding microstructure.

As for example Novak et al. [28] investigated the effects of hydrogen on dislocations experimentally and numerically. They argue that HAC cannot be characterized only by HELP mechanism, but rather can be explained by a synergism of HEDE and HELP. Barnoush and Vehoff [29] found in their nano-indentation experiments on nickel single crystals, that the pop-in load is reduced by hydrogen.

The pop-in load is directly connected to plastic deformation and reflects the required shear stress for nucleation and propagation of dislocations. Also, here the initiation of dislocations is explained by HEDE, while the enhanced propagation of dislocations is associated to HELP. Thus, both mechanisms obviously occur at the same time. Also, Gerberich [7] as well as Robertson and coauthors [11, 30] pointed out that HEDE or HELP may occur parallel.

## Numerical Simulation of HAC

All above mentioned mechanisms can only be incorporated into numerical modelling of HAC, if their effects have been quantified experimentally. Models for HAC can generally be divided into more statistical-empirical and into more mathematical approaches [31]. Typically, mathematical approaches have a quantitative character and are predominantly based on the calculation of crack initiation and propagation. Thus, in such models the typical three stages of cracking have to be considered (Fig. 6). The first stage is the crack incubation, characterized by the development of a very small crack usually below 100 μm. Following, stable slow crack growth occurs in stage two. This progress of cracking takes place, if an initiated microcrack achieved a larger size and is characterized by the continuous process that the hydrogen concentration is increasing by diffusion towards the crack tip and reducing the respective mechanical load the material can withstand. If the material does not crack by overload this process of crack advance can last for the major period of the life time of a component until finally unstable rapid fracture takes place. This stage of cracking will be reached if the crack length is increased to

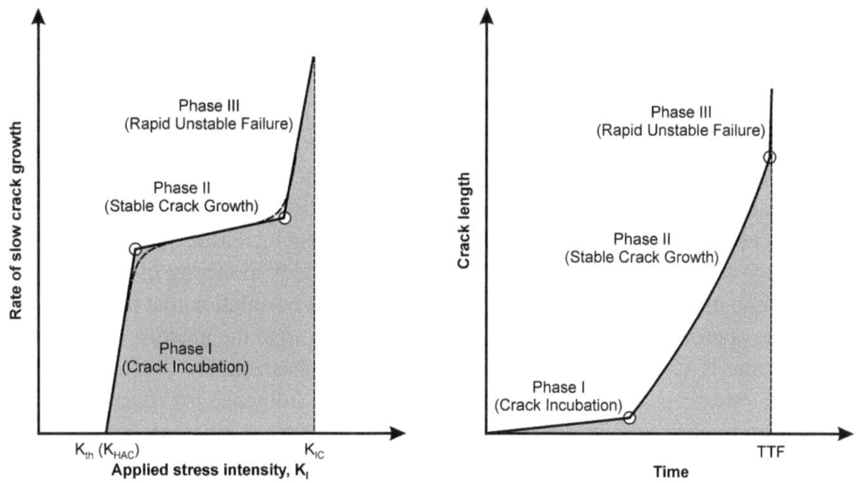


**Fig. 6** Illustration of the three stages of crack propagation [21]

a size that the rest of the material is overloaded. Usually, this stage of crack propagation is not dependent on the interaction between hydrogen permeation into the strain field ahead of the propagating crack and respective materials degradation, although hydrogen absorbed in the ligament might enhance rapid failure.

During the last two decades, research more intensively concentrates on numerical modelling of HAC. The most important studies will be reflected here.

The majority of numerical models deal with the period of crack propagation, i.e. take into consideration the influence of highly elastic-plastic fields on hydrogen transport and accumulation of hydrogen in front of crack tips, notches or around inclusions [28, 32–44]. Most of such models take into consideration HEDE and/or HELP, as defined in earlier publications, from Oriani [45] as well as Sofonis and McMeeking [24].

Such models also account for the influence of lattice imperfections like dislocations or vacancies on the hydrogen permeation in the material. Both, consideration of cracking as well as hydrogen transport mechanisms, are predominantly the implemented by respective formulas in commercial Finite-Element Software codes, as discussed for example by Leo and Anand [44]. In many cases, such models are applied to classical fracture mechanic experiments like crack tip opening displacement (CTOD) tests using compact tensile (CT) specimens to study the influence of hydrogen on ductile and brittle crack propagation in metallic materials as well as in order to validate the models. With regard to modelling and simulation of hydrogen transport and distribution at crack tips as well as the interaction of hydrogen and dislocations, significant achievements have been made by Sofronis and coauthors [24, 28, 32–37, 46] as well as by Turnbull et al. [47], Krom et al. [40, 48], Serebrinsky et al. [49] and Brocks et al. [42, 50], as already summarized elsewhere [1, 7, 51–56].

Within the past five years, modelling of hydrogen assisted cracking has been extended towards the atomistic level using molecular dynamic simulations and first principle calculation [57–62]. Considering above mechanisms, the interaction of hydrogen with the metal lattice and lattice defects has been investigated [59, 61, 62]. For instance, Song and Curtin [62] investigated the influence of hydrogen on dislocation motion and found that hardening occurs most likely according to the original plasticity theory by Beachem [20]. Itakura et al. [61] found that softening occurs due to the interaction of hydrogen and dislocations, i.e. according to the HELP mechanism. However, such developments make obvious that in the course of further developments of computational hardware having increased computing power and storage capacities, a more detailed and more complex modeling of the effects of hydrogen on metallic materials becomes possible. The postulation that prospectively multiscale approaches reaching from the atomic scale up to the macroscale will help to clarify the HAC phenomenon [7] can only be emphasized.

Beside stress assisted hydrogen diffusion, the influence of microstructural effects (grain size, grain boundaries and microstructural phases) on hydrogen transport [58, 63–65] as well as hydride formation or blistering (internal hydrogen cracking due to high internal pressure) [66–69] has been drawn into the focus of recent research. With regard to hydrogen assisted stress corrosion cracking

(Hydrogen Assisted Stress Corrosion Cracking—HASCC), hydrogen uptake due to electrochemical reactions in crevices, pits and inside cracks has been considered [42, 65, 70–80].

For the simulation of crack initiation and propagation, different techniques were used, i.e. cohesive zone modelling, node release technique, element elimination technique and Gurson-Tvergaard-Needleman model and have been outlined in detail with respect to FEM applications elsewhere [81, 82]. However, most of these models involve classical or advanced fracture or failure mechanics. Some of them only take account of stationary cracks and can only be applied to isotropic materials. As an additional disadvantage, such models often consider only stress based values as crack criterion. But, it is well-known that hydrogen reduces the material properties much more with respect to ductility than in terms of strength. Considering above mentioned hydrogen dislocations interactions as well plasticity effects, quite often a strength increase is observed while introducing hydrogen in metallic microstructures at lower concentrations, typically between 2 and 4 ml/100 g Fe in structural steels, which cannot be considered in many of these models. With respect to such modelling techniques it has also to be considered that increasing hydrogen concentrations continuously change the materials stress-strain behavior in very different ways, i.e. cause fracture type alterations from a plane stress state at lower concentrations to plane strain state at high concentrations. To model this accurately would require a continuous adaption of the flow curve and the respective material properties to the hydrogen concentration in each element dependent on time and location. On top of this, such models normally consider only very homogeneous microstructures and cannot account for multiphase materials, like Duplex Stainless Steels (DSS), for instance, or the various weld microstructures which evolve different sensitivities against HAC.

It has generally to be kept in mind that numerical simulations fitted to experimental results in order to evaluate missing parameters for the model cannot be transferred to other materials at all. Additionally, experiments for validating the numerical simulations are complex or require high effort in preparing the samples, i.e. CTOD experiments using pre-cracked CT-specimens.

When it comes to consideration of hydrogen transport mechanisms like stress or strain assisted diffusion or accumulation of hydrogen in the plastic zones of a crack tips, complex mathematical formulations have to be integrated into the FEM code via user written subroutines. Frequently additional parameters are needed for solving those equations, i.e. trap density or occupation of trap sides and lattice sides. Furthermore, quantifying the effects of hydrogen in metallic materials or the amount of hydrogen which was accumulated ahead of a crack tip is quite difficult or requires complex experimental investigations.

To sum this up, a quantitative numerical model for evaluating HAC in welded components should be kept simple in the first approach, in order to avoid over-modelling of non-problem orientated aspects of hydrogen transport and cracking [1, 2]. For welded components this definitely means that a quantitative procedure for modelling HAC should focus on the three local influences, microstructure, hydrogen concentration and mechanical load and, in particular, should contain a comparison between the actual mechanical load at a specific local hydrogen concentration and the critical hydrogen dependent value. In other words, a quantitative model for hydrogen assisted cracking is only useful for life time prediction of components if it contains a continuous comparison of the actual to the crack relevant values. Hydrogen distribution at a crack tip can only calculated numerically and should consider geometrical effects of crack growth on the hydrogen distribution. Additionally, it should be possible to investigate easily hydrogen dependent material properties as basis for HAC modeling. In this context it has to mentioned again that strain based crack criteria rather than stress based values will be more effective due to the fact that hydrogen degrades more the ductility rather than strength [1, 2, 27].

## Modelling Approach and Application to Welded Components

### *Numerical Model*

Considering above developments, a numerical model has been developed that generally accounts for the three main local influences on hydrogen assisted cracking, microstructure, mechanical load and hydrogen concentration [1, 2]. The procedure can best be understood, if a sequence of tensile specimens is considered alongside a specific crack path, as shown in Fig. 7a. The approach is based on the finite element method and each tensile specimen is represented by a respective finite element, as illustrated in Fig. 7b.

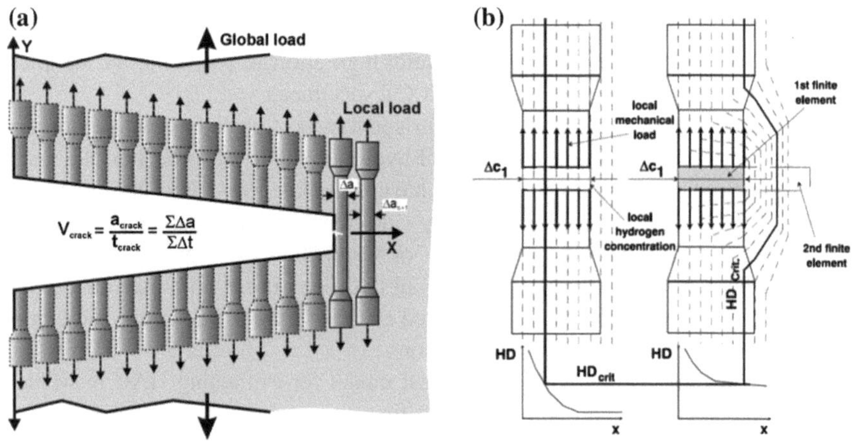


**Fig. 7** Schematic illustration of the concept of the fictive tensile specimen to simulate cracking: **a** crack propagation as sequence of cracked tensile specimens; **b** hydrogen distribution caused by crack propagation demonstrated for the first specimen [2]

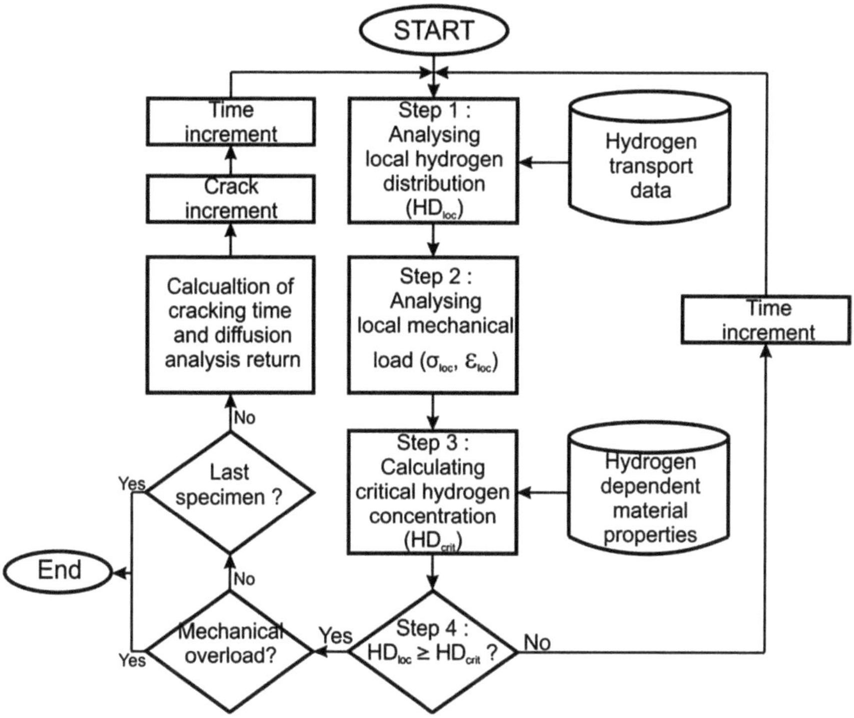


**Fig. 8** Flow chart for numerical solutions for hydrogen cracking [2]

Figure 8 illustrates the modelling procedure by a flow chart. The main part consists of the numerical calculation of the hydrogen concentration distribution within a specific time interval based on Fick's second law, as shown previously [2]. In the second step, the local mechanical load is calculated by finite element analysis which has been coupled indirectly to the hydrogen diffusion analysis. For very simple geometries such calculations have previously also been carried out analytically. Calculation of the crack critical hydrogen concentration from the hydrogen dependent material properties from a respective materials and microstructure related database represents the third step. In the fourth step the calculated hydrogen concentration in the respective finite element, i.e. tensile specimen, adjacent to the crack tip is compared to the critical hydrogen, i.e. to that hydrogen concentration that the respective materials microstructure can tolerate dependent on the respective mechanical load. During the period in which the respective critical hydrogen concentration is not reached the calculations of the hydrogen distribution and of the mechanical load ahead of the crack tip are continued for the next time intervals. In case of reaching the hydrogen concentration the exact time $t_{\Delta cn}$ of cracking of the respective tensile specimen $\Delta c_n$, i.e. finite element, within the time interval is calculated, if the total specimen is still not completely cracked and if no rupture by mechanical overload without hydrogen takes place. Due to the used program, a respective rerun of the transient diffusion analysis up to that time has to be performed. Crack propagation is simulated by deleting the respective element $\Delta c_n$ at the tip (Figs. 7 and 8). Subsequently, the total procedure is repeated for the next time interval with the such modified finite element model, focusing on the subsequent finite element $\Delta c_{n+1}$, i.e. tensile specimen, at the proceeded crack tip.

Thus, such numerical simulations of crack propagation reveal the advantage that the geometrical effects of crack propagation on the hydrogen diffusion profile are considered correctly. Additionally, the extent of crack propagation is simulated correctly, because it is directly dependent on the number of elements at the crack tip for which the critical hydrogen concentration is reached.

## *HASCC in a SSRT Specimen of Supermartensitic Stainless Steel*

For validation, the numerical model has been applied to simulations of hydrogen assisted cracking of a supermartensitic stainless steel (SMSS) with 0.025 C, 13.3 Cr, 3.85 Ni and 0.96 Mo (wt%) in the slow strain rate test (SSRT) [1, 83–87]. In order to reduce computation time and storage obtainment, the gage length of the standard specimen has been modelled two-dimensionally [85, 86]. Also for simplicity, hydrogen cracking has been simulated in the as delivered base material. The diffusion coefficient and the subsurface concentration required as input data for the finite element calculations of the hydrogen distribution have been determined by respective permeation experiments. It turned out that subsurface concentrations of up to 70 ml/100 g Fe can be absorbed in the as delivered steel [87]. In general, the hydrogen concentration profiles have been calculated numerically by modification of the thermal module of a commercially available finite element program [1, 84]. Similar to a previous study of a four point bend specimen [2], the local strains have been calculated analytically by multiplication of the global strain rate with time and with a constant factor $k$ as already suggested by Lidbury [88].

Figure 9 shows a sequence of diffusion profiles at different stages of the cracking progress in the gage length at a constant global strain rate of $1 \times 10^{-6}$ 1/s and at a constant relation of k = 10. It was assumed that hydrogen was taken up by a constant subsurface concentration of 25 ml/100 g Fe corresponding to an assumed exposition to the 1 % $H_2S$ saturated NACE solution at room temperature [87]. Also for simplicity, an incubation time for cracking of $t_{inc}$ = 10,000 s has been assumed. Such predefinition of the incubation time is allowed, because it has been proven ahead of further numerical simulations of hydrogen assisted stress corrosion cracking of slow strain rate specimens that the time to failure differs only marginally at increasing incubation times. This also becomes obvious by Fig. 10 which demonstrates that different incubation times only influence the crack propagation during the next step of cracking at a constant subsurface concentration of 25 ml/100 g Fe and a constant diffusion coefficient of $4.5 \times 10^{-6}$ mm$^2$/s. The mistake produced by too short or too long incubation times in computing crack

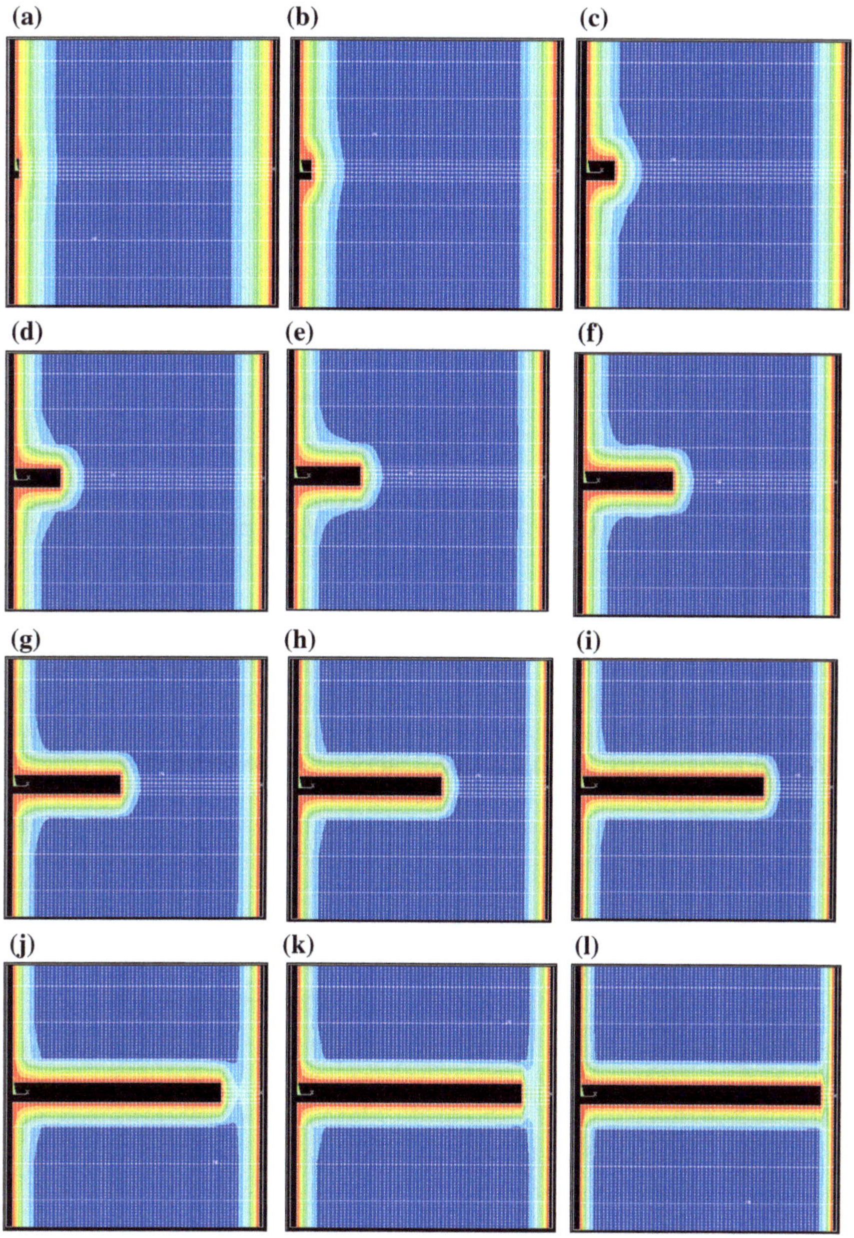


**Fig. 9** Simulated crack propagation in the gage length of a slow strain rate specimen of a supermartensitic stainless steel tested in the state as delivered at a strain rate of $\varepsilon = 1 \times 10^{-5}$ 1/s (assumptions: $k$ = 10, *HDss* = 25 ml/100 g Fe, $D = 4.5 \times 10^{-6}$ mm²/s, $t_{inc}$ = 10,000 s) [1]. **a** t = 10,000 s, **b** t = 18,273 s, **c** t = 28,236 s, **d** t = 35,094 s, **e** t = 41,088 s, **f** t = 47,487 s, **g** t = 54,181 s, **h** t = 59,278 s, **i** t = 62,406 s, **j** t = 67,461 s, **k** t = 67,465 s, **l** t = 67,466 s

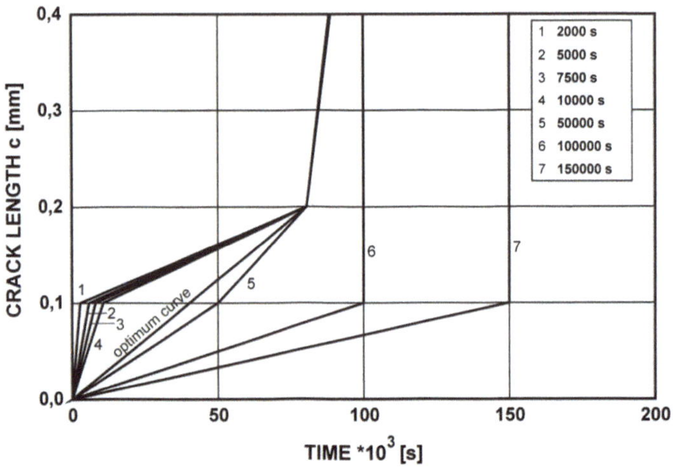


**Fig. 10** Crack length versus time for different incubation times for the first four finite elements of crack propagation (assumptions: $E = 1 \times 10^{-6}$ 1/s, $k = 10$, $HDss$ = ml/100 g Fe, $D = 4.5 \times 10^{-5}$ $mm^2/s$) [1]

propagation is recovered within calculating cracking of the adjacent finite element. In other words, the incubation time can be predefined within reasonable limits.

In order to investigate the influence of different constant ratios *k* between the local and the global strain rate recommended by Lidbury [88], cracking has been calculated assuming the above mentioned subsurface concentration and diffusion coefficient. The result was that for different relations *k* between 1 and 20 time stable crack propagation can be simulated by the numerical model over a length of 4.5 mm in the slow strain rate specimen. Figure 11 shows the time to failure dependent on

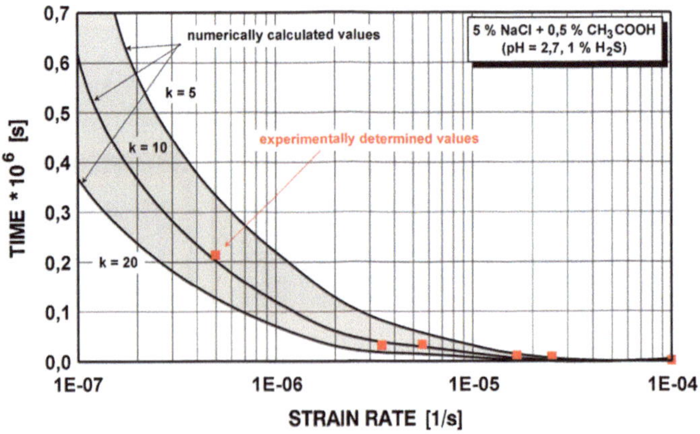


**Fig. 11** Calculated time to failure dependent on strain rate for various ratios *k* in comparison to experimentally determined data in the 1 % $H_2S$ saturate NACE solution [1, 84]

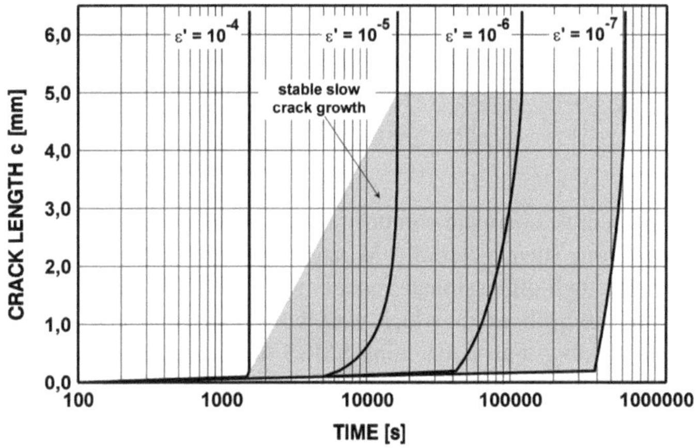


**Fig. 12** Crack length versus time for different strain rates at various ratios $k$ between local and global strain rate ($HDss$ = ml/100 g Fe, $D = 4.5 \times 10^{-6}$ mm$^2$/s, $t_{inc}$ = 10,000 s) [1, 84]

global strain rate for different constant ratios $k$ between 5 and 20 at the same environmental conditions. This diagram demonstrates very clearly that the curve for a relation of $k = 10$ agrees very well to the experimentally determined time to failure in the same environment. The numerical simulations of hydrogen assisted cracking can thus be regarded as sufficiently verified, considering particularly that a ratio between the local and the global strain rate of $k = 10$ was also recommended by Lidbury [88]. For such a ratio of $k = 10$ the crack propagation curves are shown in Fig. 12 at four different global strain rates between $1 \times 10^{-4}$ and $1 \times 10^{-7}$ 1/s, but for the same environmental conditions. It can be seen that stable crack growth has only been calculated for strain rates of and below $1 \times 10^{-5}$ 1/s. The reason for such behavior is that the mechanical load in terms of strain increases very rapidly at a higher strain rate of $1 \times 10^{-4}$ 1/s and the exposition time is too low for hydrogen uptake and diffusion. In other words, the hydrogen concentration at the crack tip does not reach a critical concentration to allow a stable hydrogen assisted crack growth. It has also been confirmed by the corresponding time-strain-fracture (TFS) diagram for the same material in the same environment that such high strain rates are not recommended to show any hydrogen related material degradation [89, 90] which emphasize the agreement between numerical simulations and experimental results.

## *HASCC in a Welded Supermartensitic Stainless Steel*

The model then was applied to component tests of orbitally welded supermartensitic stainless steel (SMSS) tubulars for offshore applications [21, 91–102], in order to investigate the HASCC behavior of a welded pipeline dependent on the load

**Table 1** Chemical composition of the base material and the matching filler wire [91]

| [wt%] | C | Cr | Ni | Mo | Mn | Si | N | S | P |
|---|---|---|---|---|---|---|---|---|---|
| BM | 0.015 | 12.35 | 4.48 | 1.41 | 1.11 | 0.29 | 0.013 | 0.001 | 0.025 |
| FM | 0.013 | 11.77 | 6.16 | 2.78 | 0.61 | 0.47 | n. d. | n. d. | 0.002 |

history. The respective chemical compositions of the SMSS base material (BM) as well as the matching filler material (FM) are shown in Table 1.

For this, the two-dimensional model was modified and applied to the cross-section of the girth welds which were built up in four layers by pulsed and oscillated TIG welding with a matching SMSS filler wire, as shown in Fig. 13.

In order to identify the region where cracking is most likely, three different crack paths in the weld metal, the heat affected zone and the base material have been predefined in the finite element model (Fig. 13). As a consequence, all three crack tips have to be controlled for crack propagation within the respective time intervals and thus, the total modelling procedure illustrated by the flow chart in Fig. 14 becomes considerably more complex as compared to small scale specimens referring only to a single crack propagating into the material [2].

For simplicity, a homogeneous microstructure has been assumed in the weld metal and the heat affected zone as well as in the base material, respectively.

Accordingly, different hydrogen diffusion coefficients of $D = 1.8 \times 10^{-5}$ $mm^2$ $s^{-1}$ have been assigned to the quenched heat affected zone and weld metal as well as of $D = 3.7 \times 10^{-6}$ $mm^2$ $s^{-1}$ to the as delivered base material (Fig. 15a). The diffusion coefficients for the various microstructures have been determined by permeation experiments at room temperature [87] and range within the scatterband of temperature dependent hydrogen diffusion coefficients in SMSS [94, 95, 99] (Fig. 15b).

For correlation to the experiments with a typical formation water composition [94, 100], exposure of the internal pipe surface to a similarly 1 % $H_2S$ saturated NACE standard solution has been presumed in the numerical simulations. In contrast to the diffusion coefficient, the hydrogen subsurface concentration is significantly dependent on the $H_2S$ saturation of the electrolyte [87], as shown by the diagram in Fig. 15a for the as delivered as well as for the quenched material. Thus, a subsurface concentration of 20.12 ml/100 g for the as delivered base material and of 3.15 ml/100 g for the as quenched weld material and heat affected zone has been assigned to the finite element model by a first grade boundary condition at the nodes representing the internal surface of the pipe. Unhindered hydrogen effusion has been assumed at the outer surface, and thus the concentration at these nodes has been set at zero in the model [1, 2].

For simplicity, it has been assumed that hydrogen enters the material by the same subsurface concentration alongside the crack flanks and at the crack tip as at the pipe bulk surface. For better comparison and avoidance of time consuming non-linear mechanical analyses, the strain in the respective crack tip elements has been calculated as described above using a factor of k = 10 [1, 2, 88].

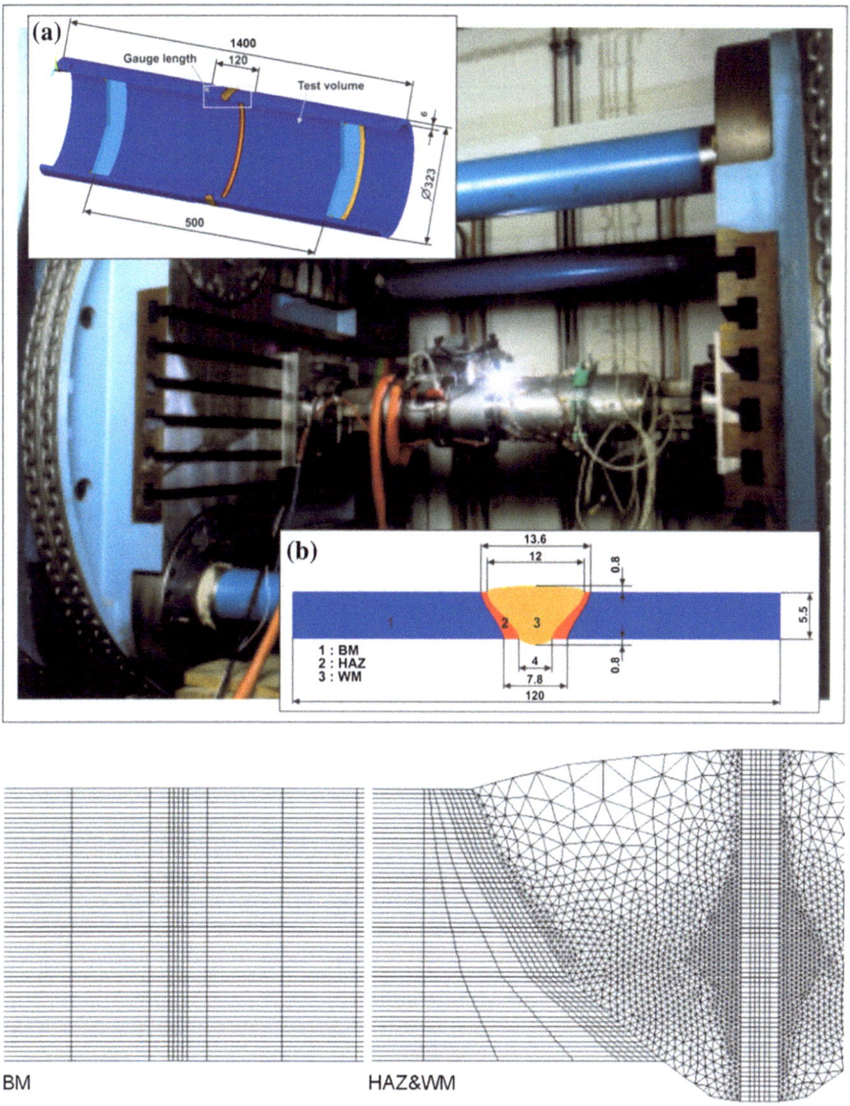


**Fig. 13** Large scale test facility for performance of complete welding tests and further tests on the orbitally welded SMSS pipeline for investigating the susceptibility to HAC: **a** Dimension of the assembled pipeline, **b** Cross-section of the gauge length of the assembled pipeline and at the bottom picture the numerical model of the cross section including the crack path in the base material (*BM*), heat affected zone (*HAZ*) and weld metal (*WM*) is shown [94, 95]

Again, the local hydrogen concentration $HD_{loc}$ exceeding the critical hydrogen concentration $HD_{crit}$ at a specific local strain level has been considered as crack criterion. Such critical hydrogen concentrations were calculated from the

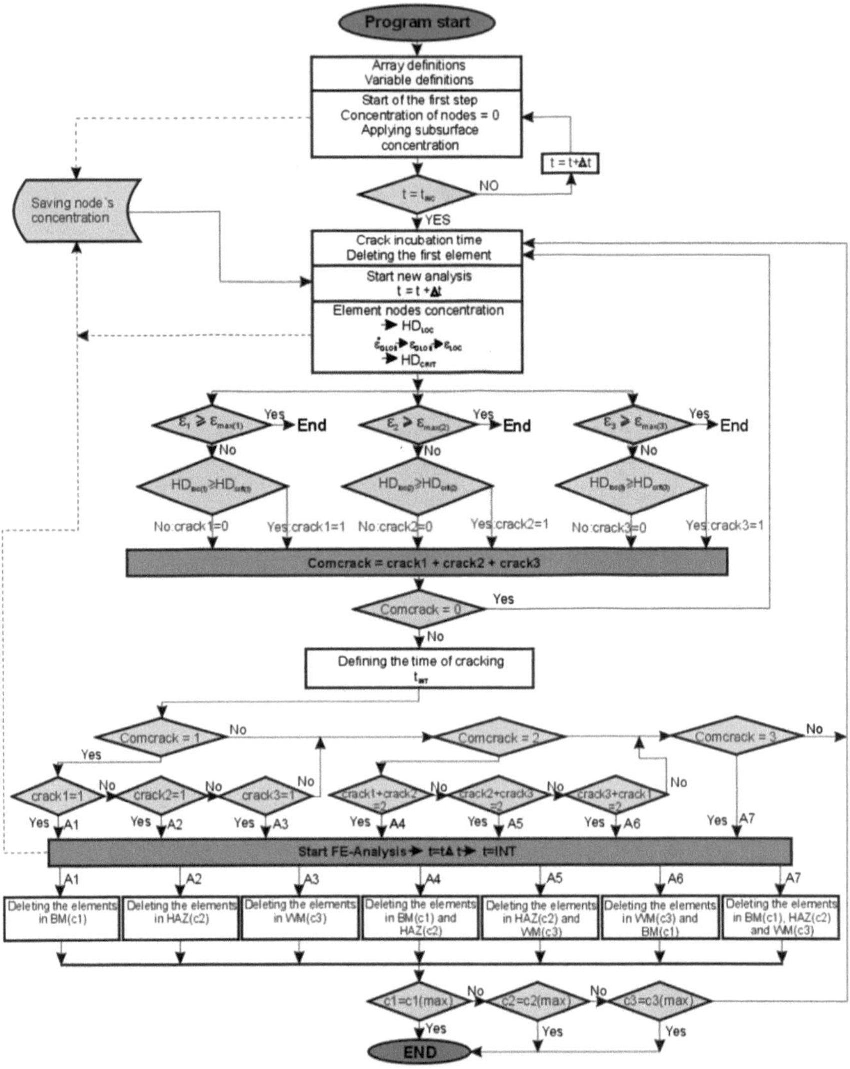


**Fig. 14** Flow chart for numerical modelling of hydrogen assisted cracking in a orbitally welded supermartensitic stainless steel pipeline by assumption of three crack paths [94]

experimentally evaluated hydrogen dependent true fracture strain curve [98] (Fig. 16). Considering that different hydrogen diffusion coefficients and, in particular, solubilities in the as delivered and quenched condition [87] entail different crack criteria, different formulas for the critical hydrogen concentration depending on the local strains have been evaluated from the experimental results for the as delivered base material (Eq. 1) and for the weld metal and the heat affected zone (Eq. 2) and considered for the various microstructural regions in the model.

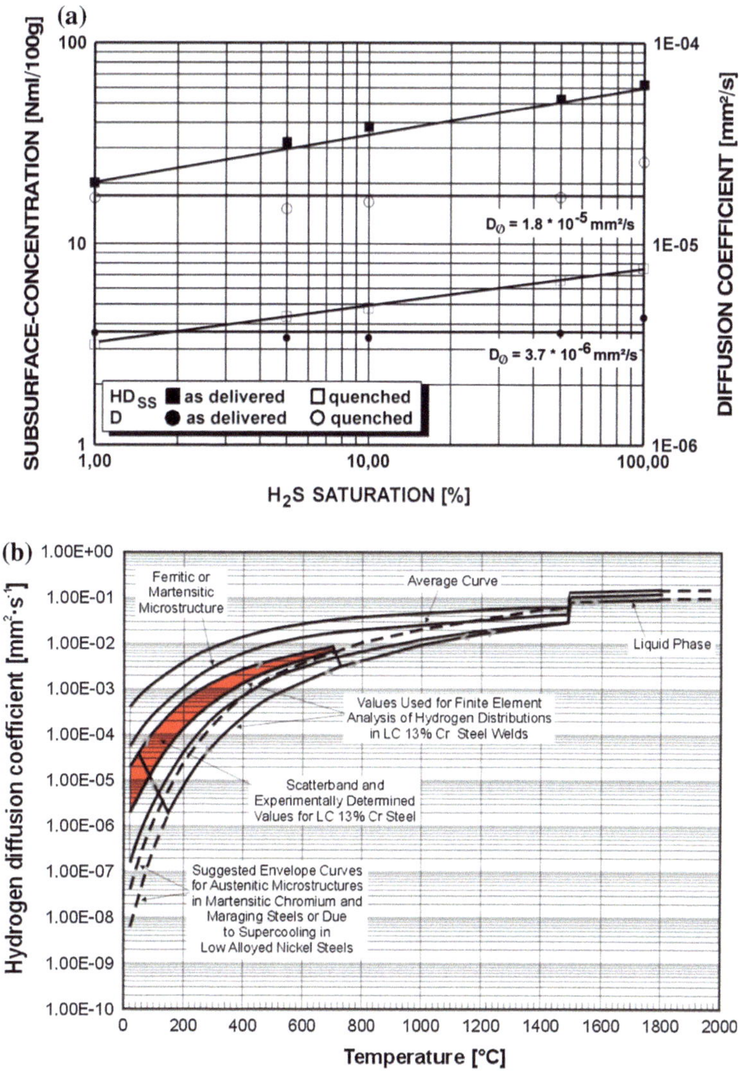


**Fig. 15** Hydrogen subsurface concentration HDss and diffusion coefficient D dependent on $H_2S$ saturation of the NACE solution in a medium alloyed SMSS (**a**) [87, 91] and (**b**) scatterband of hydrogen diffusion coefficients in steels having austenite decomposition microstructure and in SMSS as a function of temperature [94, 95, 99]

Such comparison of the hydrogen concentrations is conducted at the nodes adjacent to the crack tip, and thus crack propagation takes only place if the critical concentration is achieved in the whole element ahead of the crack tip. In other words, cracking takes place over a limited distance ahead of the crack tip, represented by the respective finite element width, and is modelled as an internal

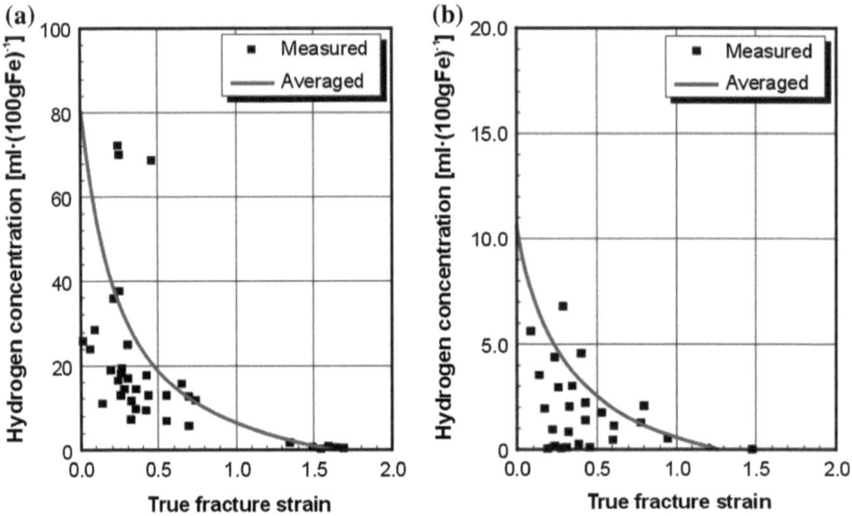


**Fig. 16** Critical hydrogen concentration dependent on the true fracture strain of supermartensitic stainless steel: **a** as delivered state, **b** as quenched state [21, 98]

phenomenon which, naturally, requires some hydrogen diffusion to reach concentration levels relevant for cracking. In the present model, this distance ranged between 0.1 and 0.2 mm.

$$HD = 64.2490\exp(-4.5 \cdot \varepsilon) - (9.9447 \cdot \varepsilon) + 16.3060 \quad (1)$$

$$HD = 6.3410\exp(-5.2 \cdot \varepsilon) - (3.5953 \cdot \varepsilon) + 4.5320 \quad (2)$$

Crack propagation itself is modelled by deletion of the number elements ahead of the crack tip for which the actual hydrogen concentration exceeds the critical level. By such procedure, the crack growth rates are modelled very accurately by subsequent rupture of that amount of elements treated as micro tensile specimens ahead of the crack tip for which the critical hydrogen concentration is exceeded. Crack propagation is thus represented in the model by an autogenous process.

The simulation procedure has been kept as consistent to the real component test as possible [97, 100]. Thus, in the model, constant global straining of the pipe at $\varepsilon = 0.001$ has been assumed, corresponding to the stress distribution of around 250 MPa introduced during welding and cooling of the pipe at realistic low shrinkage restraints. For three weeks, the pipe has internally been exposed to realistic sour service conditions at this stress strain level. After that period, the welded component was stepwise strained each day at a rate of around $1 \times 10^{-5}\ s^{-1}$ to 70 and 100 % of the weld metal yield strength, corresponding to a simulated global strain of $\varepsilon = 0.003$ and $\varepsilon = 0.005$. Subsequently, straining of the pipe in the weld metal up to $\varepsilon = 3$ and finally $\varepsilon = 5$ % has been performed in the experiments and has been numerically modelled, respectively. Similar to other experiments [87, 99] measurements during the component tests revealed a pH value of about 4.5 and an electrochemical potential of about −510 mV-SCE. In the model this has been considering by application of the same subsurface hydrogen concentrations as obtained from respective permeation experiments with the respective weld microstructures subjected to real formation water.

Figure 17 shows the hydrogen concentration profiles at different stages of crack propagation. The respective crack lengths versus time curves are illustrated in Fig. 18. As seen, cracking appears faster in the heat affected zone than in the base material and in the weld metal. As shown by the macrosection in Fig. 19a, base metal dissolution is very low in the root pass and crack initiation sites are particularly established in the adjacent heat affected zone. Thus, the low resistance of the heat affected zone against hydrogen assisted cracking is reflected in the component test as well.

Similar to respective slow strain rate tests, the fracture topography exhibited intergranular cracking alongside the former austenite grains, followed by transgranular cleavage-like cracking and finally by more ductile microvoid coalescence (Fig. 19). This indicates that the formation water composition used in the test corresponds to the hydrogen concentration data of the NACE solution simulated in the numerical analyses.

A study of the failure sequences and fracture modes of supermartensitic stainless steels in sour environments [3] showed that such intergranular cracking has to be attributed to high hydrogen concentrations under relatively low mechanical loads. By the hydrogen distributions presented in Fig. 17 it becomes obvious that such high levels are provided by hydrogen ingress before crack initiation takes place. In particular, hydrogen diffuses from the base material into the heat affected zone, due to respective concentration gradients.

The plots of the actual hydrogen concentration and of the strain at the crack tip versus crack length (Fig. 20a) and versus time (Fig. 20b) confirm that high hydrogen concentrations relevant for intergranular cracking are particularly piled up in the heat affected zone during the long period of three weeks at considerable low strains of $\varepsilon = 0.001$, in contrast to the above discussed results for immediate straining of the pipe at a constant strain rate. Due to the assumption of a standard test solution in the numerical simulations, the absolute depth of the experimentally determined fracture modes should not be directly related to the calculated crack tip positions or to the local hydrogen concentrations and strains. However, the numerical simulations supported very valuable information to identify crack critical regions and to explain the various fracture topographies.

## HACC in a Welded Supermartensitic Stainless Steel

The principal model has also been applied to evaluate the sensitivity of SMSS welds against hydrogen-assisted cold cracking (HACC), which may occur in welded components during fabrication welding, for instance during offshore pipeline assembly [95]. Such simulations were predominantly targeted at determination

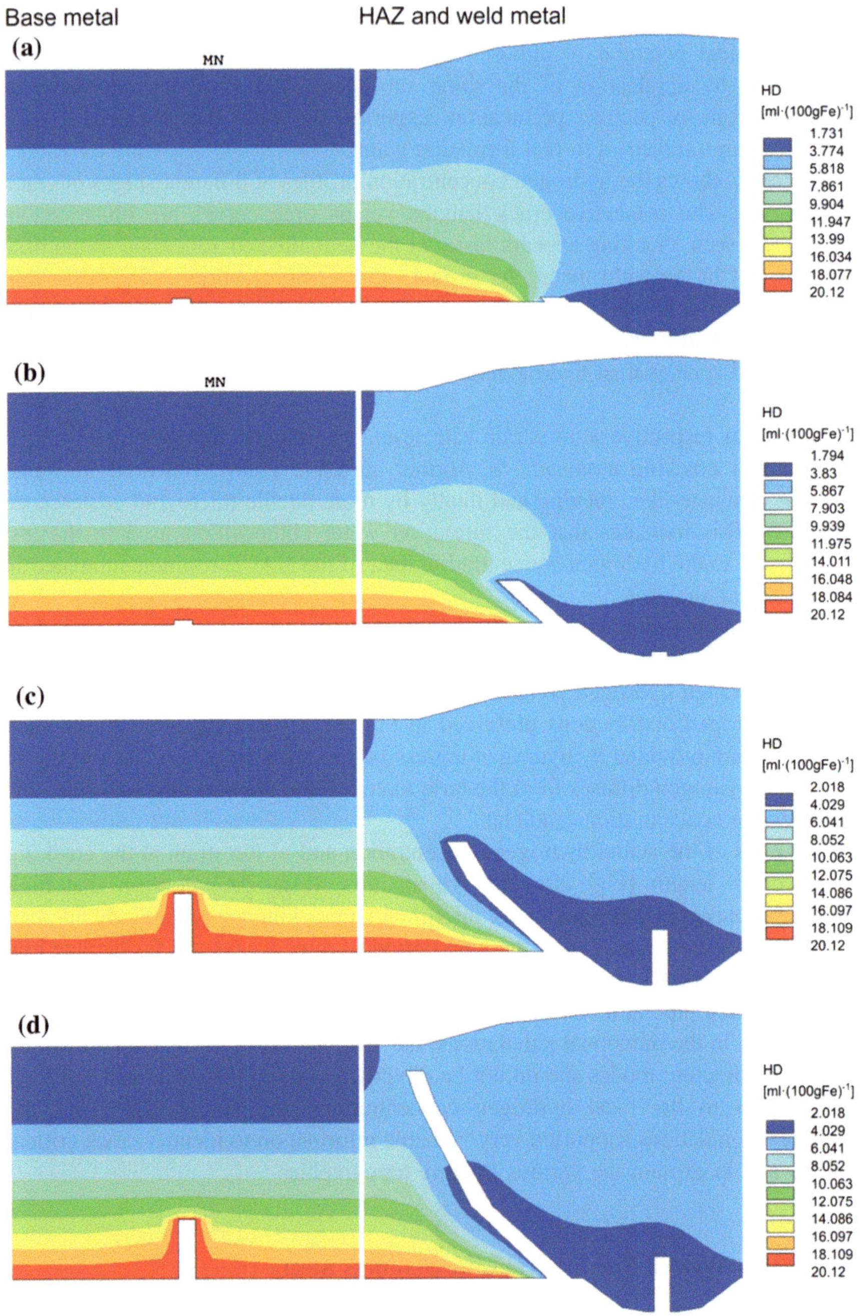


**Fig. 17** Simulated crack propagation in base metal, heat affected zone and weld metal of an orbitally welded supermartensitic stainless steel pipe exposed to the 1 % $H_2S$ saturated NACE solution at the load history applied to the component test [21, 94]. **a** t = 1,987,000 s, **b** t = 2,013,353 s, **c** t = 2,103,366 s, **d** t = 2,103,366 s

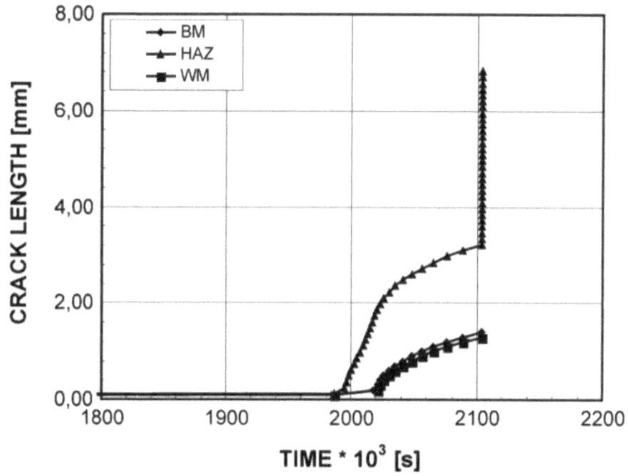


**Fig. 18** Crack length of the three different cracks in the orbitally welded supermartensitic stainless steel pipe exposed to the load history of the component test versus time [94]

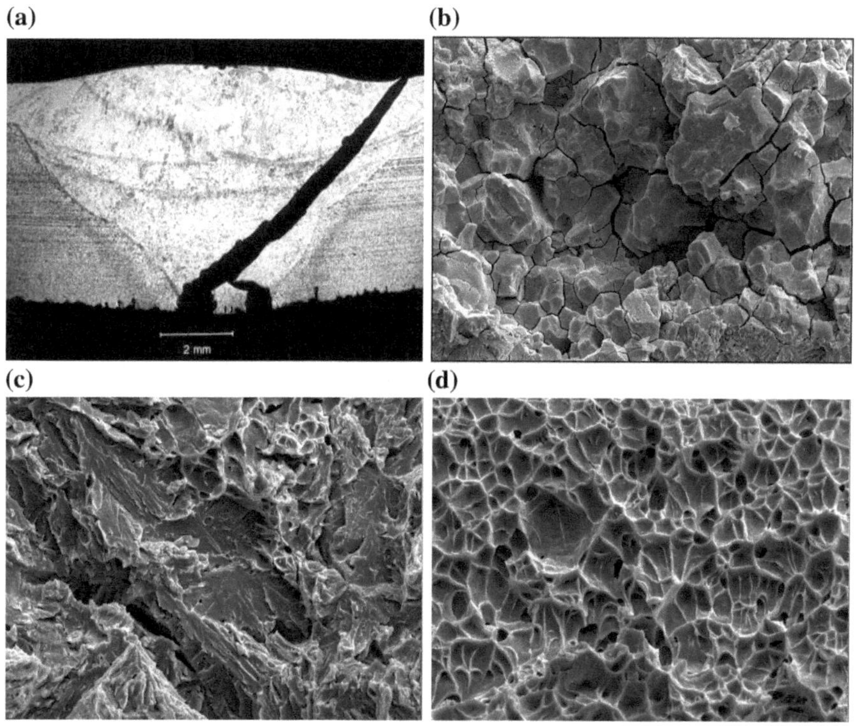


**Fig. 19** Cracking in the heat affected zone of the orbital TIG weld: **a** overview, **b** SEM micrograph of intergranular cracking (IC), **c** SEM micrograph of transgranular cracking (TC) and **d** SEM micrograph of ductile cracking (DC) by microvoid coalescence [97]

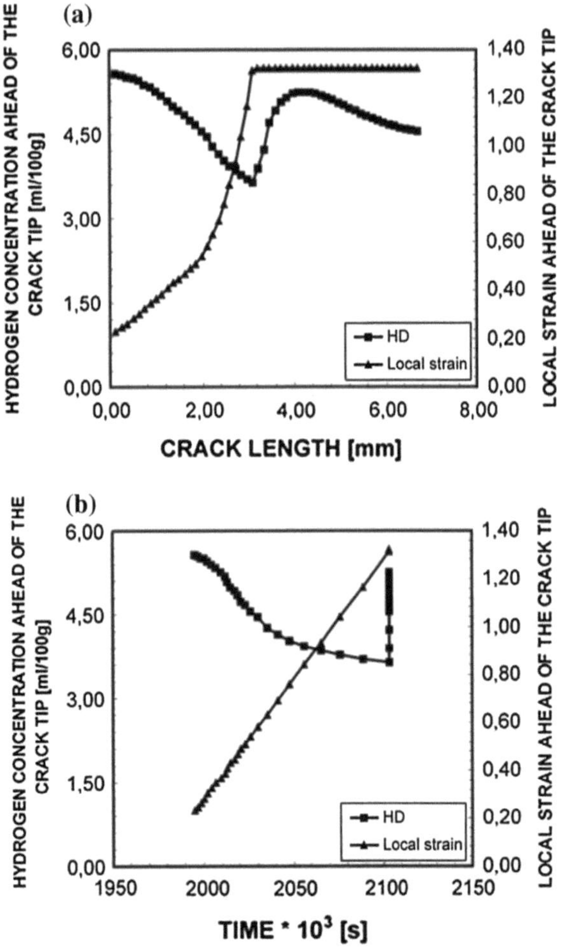


**Fig. 20** Local hydrogen concentration and strains at the crack tip in the heat affected zone of the simulated orbital component test weld **a** dependent on crack length and **b** dependent on time [94]

of sufficient post weld heat treatments (PWHT) for reduction of the hydrogen concentrations in the welds and, if possible, also of the respective stress-strain level. Thus, a slightly modified two-dimensional model of the cross section showed in Fig. 13 was created with the assumption of two predefined crack paths in the weld metal and the heat affected zone, as shown in Fig. 21.

According to the flow chart shown in Fig. 22, three indirect coupled numerical simulations have to be carried out, i.e. thermal analysis, structural analysis, hydrogen diffusion analysis. For even more accurate crack propagation modelling, the stresses and strains around the crack are also calculated numerically. The modelling procedure of each analysis step is described in detail elsewhere [21, 95]. As seen from Fig. 22, the range of input variables increased to cover the whole temperature range for the welding and, particularly, the cooling process, i.e. in contrast to previous HASCC modelling, the thermo-mechanical and thermo-physical material properties

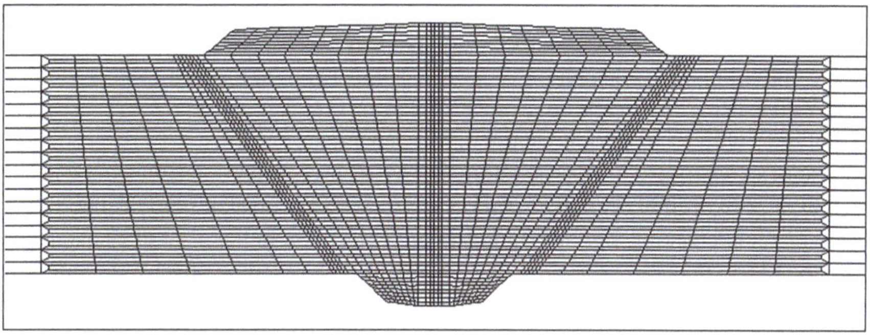

**Fig. 21** Modified two-dimensional finite element model according to the cross-section of the gauge length for calculation of the susceptibility of welded pipelines to hydrogen assisted cold cracking [21, 95]

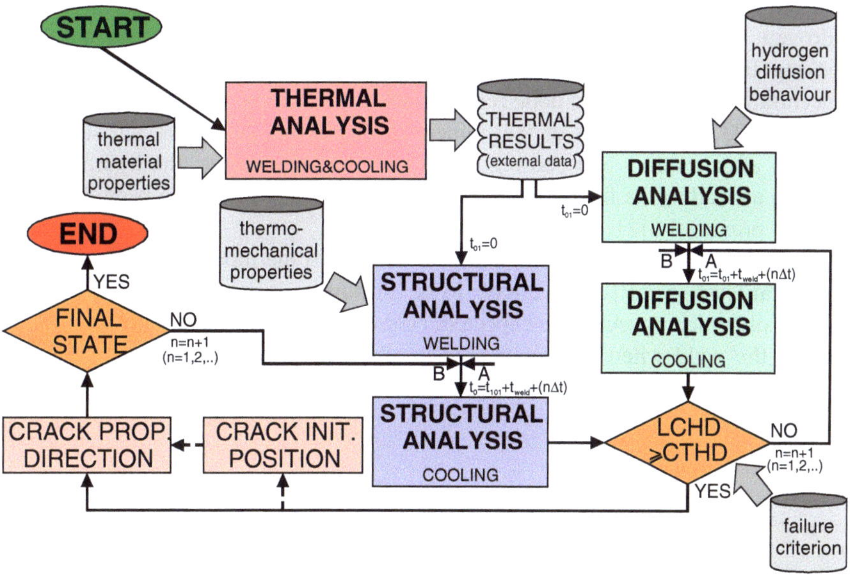


**Fig. 22** Flow chart of numerical modelling of hydrogen assisted cold cracking in orbitally welded supermartensitic stainless steel pipelines [95]

as well as temperature dependent diffusion coefficients have to be implemented for modelling of HACC. Similar to previous investigations on HASCC, identical crack criteria as well as the temperature dependent diffusion behavior were assumed for the weld metal and the heat affected zone. But, thermo-mechanical properties were different for weld metal, heat affected zone and base material and a bilinear stress-strain relationship was used. The respective material properties have been published elsewhere [21, 95].

For HACC it can be assumed that cracking usually occurs at temperatures below the $M_f$ -temperature of supermartensitic stainless steels. Numerical simulation of HACC initiation and propagation process has thus been be focused to near ambient temperature using the crack criteria illustrated in Fig. 16.

Not only with respect to the temperature-dependence of the materials properties, but also to an additional extent modelling of HACC is more complex than of HASCC: In contrast to hydrogen stress corrosion cracking which propagates only at one side of the crack, cold cracking can propagate at both ends of an initiated crack. This means that for HACC the crack can propagate either in downward or upward direction, or both, and thus, an exact definition of the crack growth direction is needed. This is accomplished by comparing the concentrations at the first node opposite to the crack front in both possible crack growth directions. If the local hydrogen concentration of the respective node reaches the critical value, crack propagation takes place in both directions concurrently. In the other case, crack propagation will appear in one favourite direction if the local hydrogen concentration at the specific node in that direction reaches the critical value. Also, crack arrest occurs more frequently during hydrogen assisted cold cracking than during hydrogen assisted stress corrosion cracking. But, this is easily recognised during the simulation when the local hydrogen concentration at the specific node remains below the critical value.

In Fig. 23, some results of the numerical simulation, i.e. stress and strain distribution after welding and cooling down to room temperature, the hydrogen distribution as well as the final state of HACC, are shown, which have been discussed in detail previously [21, 95].

By application of various initial concentrations to the model, it has been found out that at least 15 ml/100 g Fe have to be introduced by welding of each layer to initiate cold cracking. As to be expected, cold cracking is first initiated in the upper third of the weld where hydrogen was accumulated to the highest levels during welding. However, the local hydrogen concentration was first decreased to about

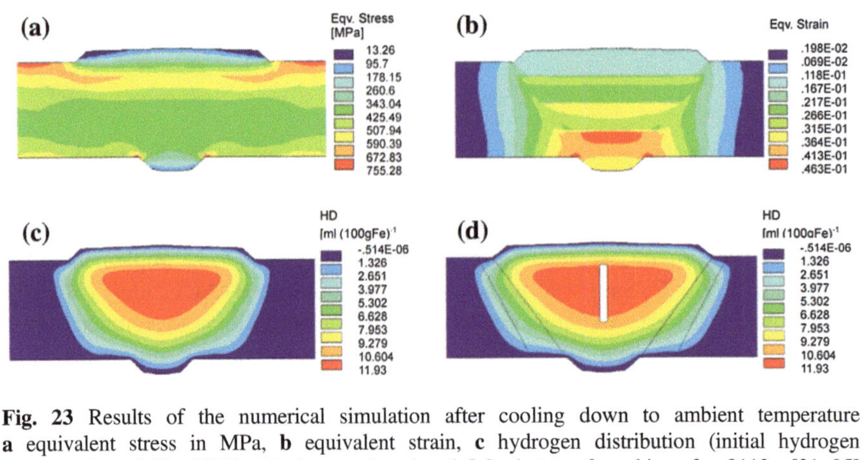


**Fig. 23** Results of the numerical simulation after cooling down to ambient temperature **a** equivalent stress in MPa, **b** equivalent strain, **c** hydrogen distribution (initial hydrogen concentration of 15 ml/100 g Fe for each layer) and **d** final state of cracking after 9112 s [21, 95]

11.92 ml/100 g Fe by effusion and diffusion before cracking started at about 9080 s (within 3 h) after finishing welding. This also makes obvious the risk of delayed failure by hydrogen assisted cold cracking.

The final state of hydrogen assisted cold cracking in the orbitally welded supermartensitic stainless steel pipeline is represented in Fig. 23d and indicates that the crack went not through the total weld metal. This also shows very obviously how dangerous HACC might be in weakening the weld cross sectional area and that it has to be detected before a component is given into service and subjected additional loads.

As a procedure to avoid HACC, different post weld heat treatments (PWHT) were investigated, which might decrease the local hydrogen concentration and also might improve the properties by reduction of hardness and stresses. In general, post weld heat treatment can be categorised into two main procedures:

1. Long-term PWHT (soaking) at temperatures between 200 and 250 °C.
2. Short-term PWHT at temperatures between 600 and 650 °C.

The first one, soaking is commonly applied up to now for hydrogen removal, particularly in high strength structural steels. The latter one is generally more used for hardness and stress reduction after welding of low alloyed steels, but also of supermartensitic stainless steels. In order to find out how post weld heat treatment can be most purposeful to be applied to avoid the risk of hydrogen assisted cold cracking in welded supermartensitic stainless steel components, both procedures have been investigated by numerical simulations.

The results of the respective post weld heat treatments for an initial hydrogen concentration of 10 ml/100 g Fe applied to each layer during welding are shown in Fig. 24. The diagram in Fig. 24 helps to get a better understanding of effects of the

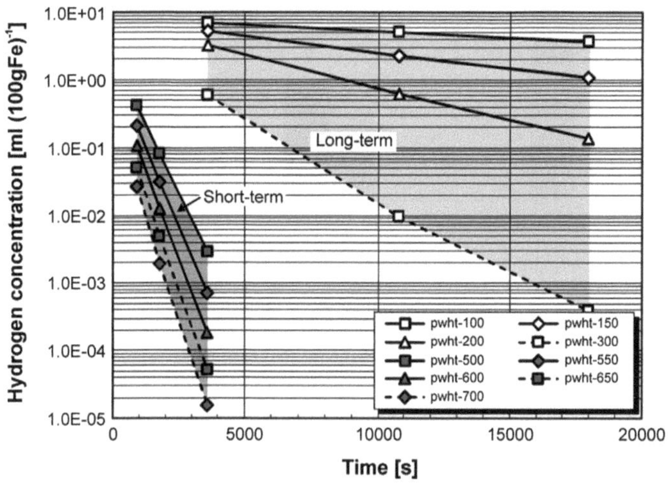


**Fig. 24** Influence of post weld heat treatment on the reduction of local hydrogen concentration by assumption of an initial concentration of HD = 10 ml/100 g Fe [21, 95]

different heat treatment approaches and clearly indicates also for a lower initial concentration that the local hydrogen concentration can be removed from the hydrogen assisted cold cracking susceptible zones quite fast by short-term PWHT.

## *HACC in S 1100 QL High Strength Structural Steel Welds*

The numerical model of HACC in SMSS welds was further modified and transferred to a welded high strength low alloyed (HSLA) steel having yield strength of 1100 MPa [104–109], in order to evaluate the cracking behavior of S 1100 QL and to determine sufficient postheating procedures for avoiding HACC. From Fig. 25 showing the rapid decrease of that critical hydrogen concentration indicating a complete ductility loss by coincident yield and tensile strength it becomes very obvious that with increasing design yield strength of such steels the various weld microstructures become very susceptible for hydrogen assisted cracking.

The chemical composition and mechanical properties of the base metal and weld metal are listed in Tables 2 and 3. In contrast to previous investigations on SMSS, here all important weld metal microstructures with regard to HAC susceptibility have been considered by applying the respective material properties to the elements

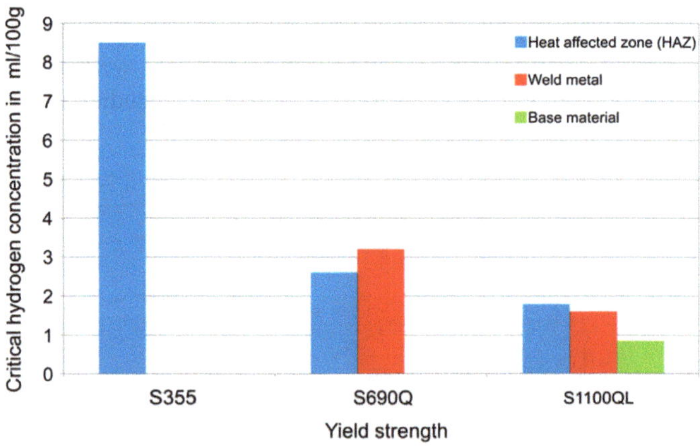


**Fig. 25** Critical hydrogen concentration indicating a ductility loss by coincidence of the tensile strength with the 0.2 % proof stress dependent on the strength level of structural steels [104–111]

**Table 2** Chemical composition of S 1100 QL and Union X 96, in wt% [104]

| Material | C | Si | Mn | Cr | Mo | Ni |
|---|---|---|---|---|---|---|
| S 1100 QL | 0.1700 | 0.271 | 0.854 | 0.459 | 0.45 | 1.88 |
| Union X 96 | 0.1206 | 0.780 | 1.860 | 0.46 | 0.53 | 2.36 |

**Table 3** Mechanical properties of hydrogen-free S 1100 QL and weld microstructure [104]

| Mechanical properties | S1100QL | | | WM Union X 96 |
|---|---|---|---|---|
| | BM | CHAZ | FHAZ | |
| Tensile strength (MPa) | 1348 | 1366 | 1399 | 1071 |
| Yield strength (MPa) | 1150 | 1000 | 1010 | 985 |
| Rupture strain (%) | 15 | 15 | 15 | 14 |
| Reduction of area (%) | 69 | 64 | 65 | 51 |

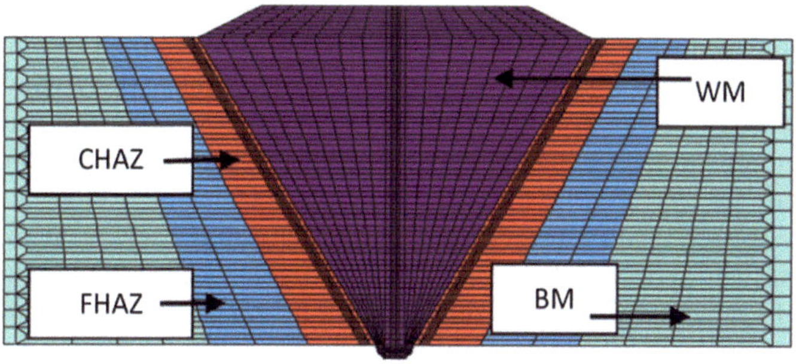


**Fig. 26** Different weld microstructures considered in the weld joint of S 1100 QL and crack paths in weld metal and CHAZ (fine meshed paths)

accordingly, i.e. base material (BM), weld metal (WM), coarse grained heat affected zone (CHAZ) and fine grained heat affected zone (FHAZ), as shown in Fig. 26.

The procedure for simulating HACC in HSLA steel was quite similar to the approach shown in the flow chart in Fig. 22. The thermo-physical material properties were taken from the diagrams of Richter [103, 106] for steel with a composition similar to the S 1100 QL, i.e. density, heat convection coefficient, enthalpy and thermal conductivity. Since the chemical composition of the four microstructures (Fig. 26) do not differ significantly, the same thermo-physical properties were applied to all four microstructures.

For structural analysis, temperature dependent true stress-strain curves were applied to the respective microstructures. Since no temperature dependent stress strain curves are available for the different microstructures of S 1100 QL the stress-strain curves determined by Brinkmann [112] for a S 460 structural steel were scaled up to S 1100 QL. The curves were scaled with respect to the ratio of yield strength to tensile strength at room temperature of S 1100 QL (Table 3). With increasing temperature the stress-strain curves become more ideal plastic and the yield strength decreases [104–109]. Additionally, phase transformation of the weld metal and HAZ has been considered by interpolation between the various temperature dependent stress-strain curves and temperature dependent thermal expansion coefficient.

Performing the diffusion analysis, the thermal module of the FEM software ANSYS was used taking into account the analogy of Fourier's heat equation and Fick's second law, as already described previously. The temperature dependent diffusion coefficients were taken from the scatterband for hydrogen diffusion coefficients in micro-alloyed and low carbon structural steels (Fig. 27) [110]. For hydrogen diffusion near room temperature the values determined by Seeger [111] for the different weld microstructures of the S 1100 QL were considered. With increasing temperature the diffusion coefficients of all microstructures converge again and the lower envelope curve in the scatterband [110] was used as worst case assessment. Thus, diffusion coefficients covering the whole temperature range of welding are available. Additionally, phase transformation is considered.

Regarding the hydrogen-assisted cracking analysis, the required crack criteria for the respective weld microstructures were taken from a similar study with hydrogen charged tensile specimens like it has been performed with the SMSS weld microstructures. Moreover, the authors determined a crack critical hydrogen concentration of 1.8 ml/100 g Fe for all weld microstructures of S 1100 QL, Fig. 25. The respective pre- and postheating procedures have thus been evaluated as to be sufficient, if they were capable to reduce the hydrogen concentration at any location in the weld below this value. Two different heat treatments have been investigated:

1. Preheating for 30 min by application of heating pads providing a constant temperature on the surface of the model. The temperature used for the heat treatment procedure ranged between 50 and 300 °C.
2. The onset postheating from 30 up to 240 min at preheating temperature has been simulated after completion of welding reaching the respective postheating temperature during cooling.

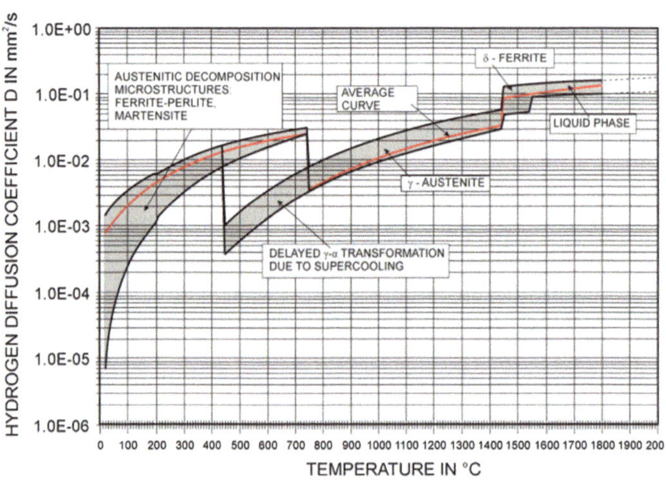


**Fig. 27** Scatterband for hydrogen diffusion coefficients in micro-alloyed and low carbon structural steels [110]

Additionally, if preheating was applied, also a controlled interpass temperature at preheating temperature has been simulated. The numerical model for HACC was applied to two different weld geometries: A single-pass butt-weld of a 20 mm steel plate (root weld) and a 12 mm multi-pass butt-weld consisting of five weld beads [104–109].

In addition to the evaluation of sufficient hydrogen removal heat treatments (HRHT), the effect of pre- and postheating on stress-strain distribution was investigated. For validation of the models, the results of the structural analysis, especially of the single-pass root weld, were compared to experimental results obtained by instrumented restraint cracking (IRC) test. The IRC test also allows investigating residual stresses and strains during and after welding at realistic shrinkage restraint levels [113–115]. Detailed information on the numerical procedure and additional results have been published previously and here, only selected results about HACC of S 1100 QL welds are summarized [104–115].

In Fig. 28, the influence of pre- and postheating on residual stresses at different restraint intensities is shown for the root weld of the 60° V butt joint. Heating was applied for 30 min to the surface of the steel plate. As seen from the stress-heating-restraint diagrams, stresses will rise more extensively due to elevated preheating temperature than due to increasing restraint intensities (Fig. 28b). In contrast, residual stresses are not increased by hydrogen removal heat treatment after welding (Fig. 28c) [104]. Inhomogeneous preheating thus increases the HACC risk of restrained welds by increased residual stresses and strains and non- sufficient reduction of the hydrogen concentration below a crack critical limit represented by 1.8 ml/100 g Fe for the steel S 1100 QL.

Similar to the experiments, the numerical analyses of the root weld showed that cracking occurs mainly in the weld metal center line (Fig. 29). Only at increased increasing initial hydrogen concentrations above 5 ml/100 g Fe cracking appears also in the coarse grained HAZ at sufficiently long diffusion times. As another reason for the predominant weld metal HACC it has to be considered that the selected filler wire has to be cooled very rapidly to achieve matching strength levels, resulting in a martensitic microstructure prone for HACC.

For a better understanding of the HACC process in these welds, the results have been compiled into a Time-Strain-Fracture (TSF) diagram, as shown in Fig. 30. In this diagram, the local strain at three different locations in the root weld and the critical strain at the same position dependent on different initial hydrogen concentrations are plotted versus time, under the presumption that no HRHT was applied to the S 1100 QL root weld. As generally implied by the fictitious tensile specimen concept, cracking will occur, if the local strain exceeds the critical strain. The critical strain is determined by the local hydrogen concentration at this location dependent from the diffusion time and is given by the respective experimentally determined strain-hydrogen relationship [114, 115].

During the welding process, hydrogen is introduced via the arc into the weld metal and at the respective location, i.e. at the middle height of the weld metal (Loc. 3), this concentration persists over quite a while when hydrogen is diffusing into the adjacent regions and is also effusing at the weld top and bottom surface.

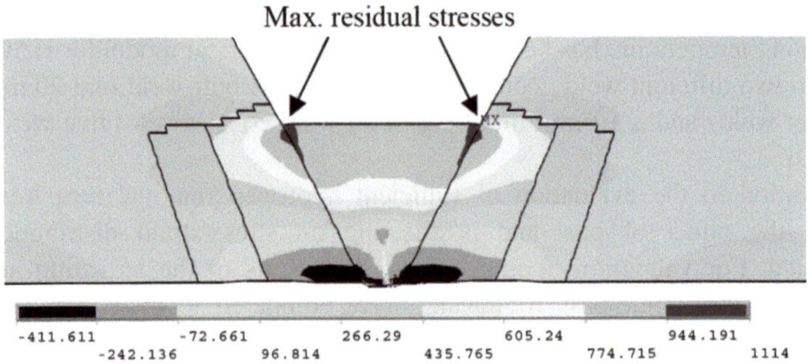


(**a**) Residual Stresses in MPa in the root weld of 20 mm S1100 QL at a restraint intensity of $R_{Fy}$ = 14.91 kN/(mm*mm)

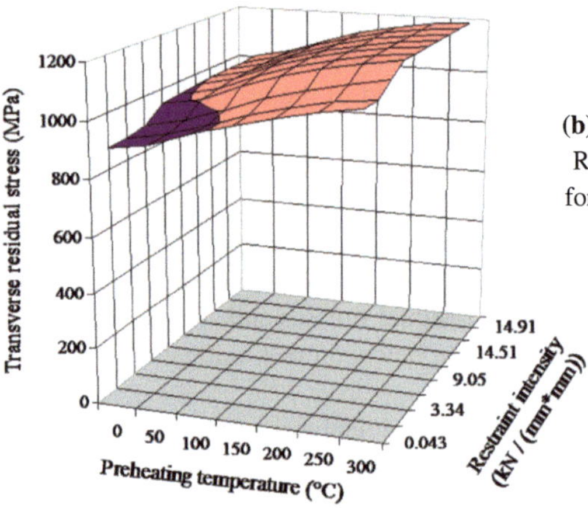


(**b**) Stress-Heating-Restraint diagram for 30 min preheating

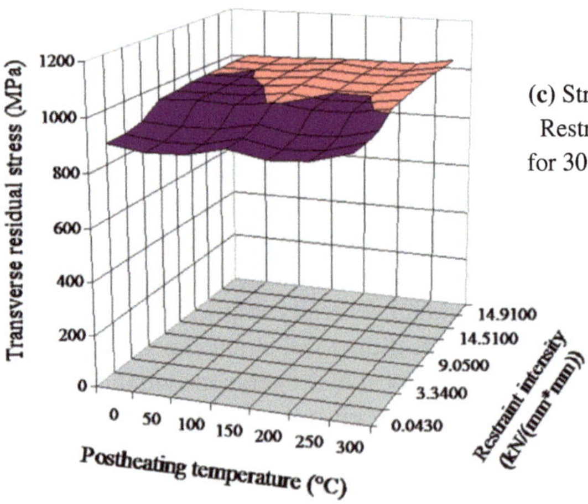


(**c**) Stress-Heating-Restraint diagram for 30 min postheating

**Fig. 28** Calculated residual stresses at high restraint intensities of $R_{Fy}$ = 14.91 kN/(mm * mm) in the 60° V root weld **a** and stress-heating-restraint diagrams for 30 min of preheating **b** and postheating **c** at different temperatures [104]

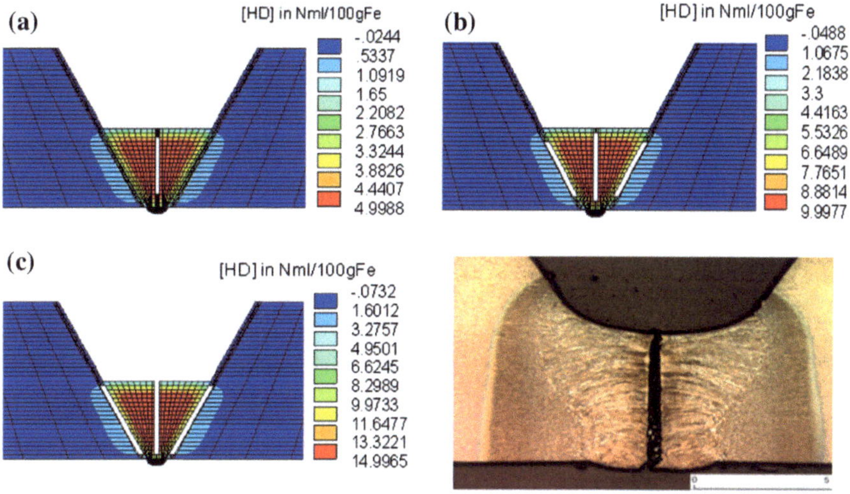


**Fig. 29** Simulated HACC in a single-pass S 1100 QL weld with various initial hydrogen concentrations of $HD_0$ = 5, 10 and 15 ml/100 g Fe picked up during welding and micrograph of the cracked IRC specimen carried out with gas metal arc welding process adding 1 % $H_2$ to the shielding gas [104]. **a** $HD_0$ = 5 ml/100 g Fe. **b** $HD_0$ = 10 ml/100 g Fe. **c** $HD_0$ = 15 ml/100 g Fe

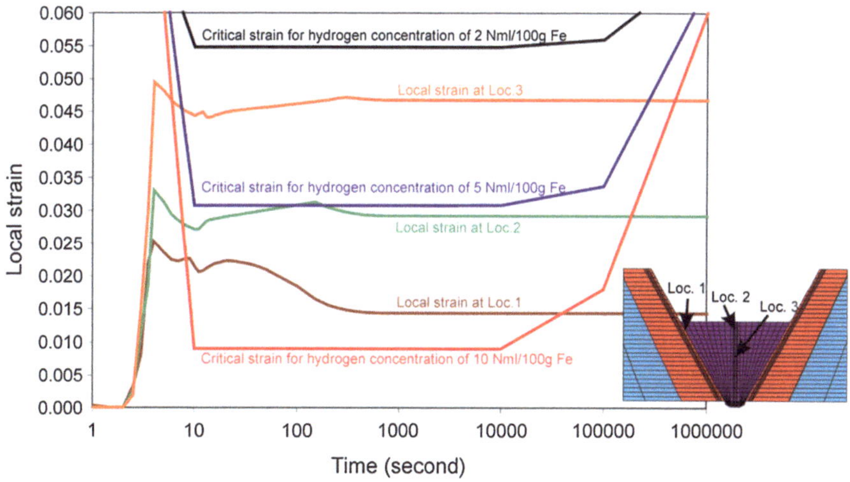


**Fig. 30** TSF diagram for numerical determination of local strain in critical regions of IRC test specimens in S 1100 QL root weld

The critical strain the weld metal might tolerate is immediately reduced by the respective concentrations traced into the weld metal. However, as can be drawn from the TSF diagram, crack safe conditions are only reached, if the critical strain increases above the level of the local strains, i.e. after a very long period of more than 500,000 s. This shows that about one day is probably conservative to evaluate the risk of delayed cracking, because the critical strain increases afterwards. But, the diagram also shows, how long the risk of HACC might persist in certain weld regions.

In order to reduce the risk for HACC, respective hydrogen removal heat treatments (HRHT) has to be applied direct after welding while reaching the respective postheat temperature during cooling. In Fig. 31 the calculated remaining hydrogen concentration in the center of the root weld after HRHT and subsequent cooling to room temperature is shown. As to be expected, with increasing temperature and time hydrogen concentration is reduced much faster from an initial hydrogen concentration of $HD_0$ = 5.3 ml/100 g Fe and the risk of HACC decreases.

The corresponding HRHT procedures are shown in Fig. 32. The diagram is normalized by relating the actual hydrogen concentration present in the welded

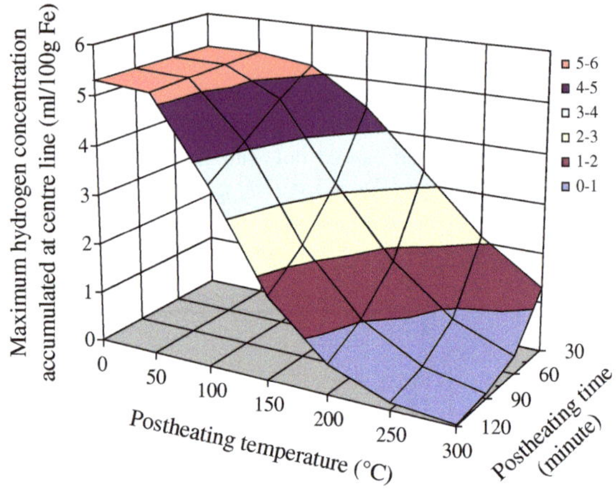


**Fig. 31** Hydrogen concentration-postheating time and temperature diagram for the root pass of a weld in S1100QL steel with an initial hydrogen concentration of $HD_0$ = 5.3 ml/100 g Fe [104, 105]

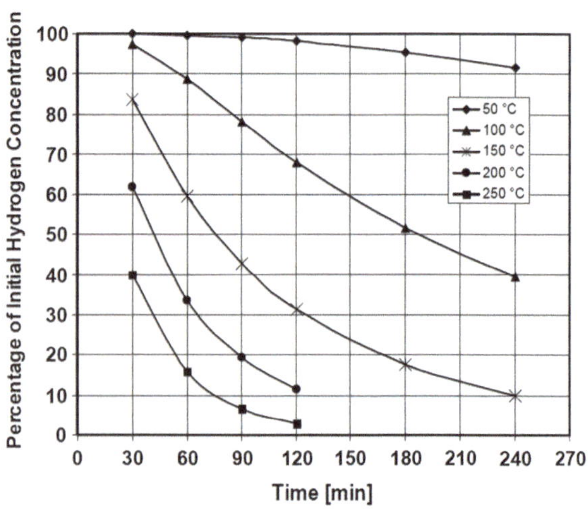


**Fig. 32** Hydrogen removal heat treatment diagram for the root weld in a 20 mm steel plate of S 1100 QL [104, 105]

joint after postweld heat treatment to the initial hydrogen concentration introduced during welding. Thus, the diagram allows determining sufficient HRHT times and temperatures for arbitrary initial concentrations, due to the fundamentals of Fick's diffusion.

By considering the critical hydrogen concentration of 1.8 ml/100 g Fe for the weld metal (WM) cracking initiation of S 1100 QL, the sufficient postheating temperature for hydrogen removal of 60° V root weld of the steel S 1100 QL at the weld centre line can be drawn from this diagram by following the 34, 18, 12 and 10 % line for an initial hydrogen concentration of 5.3, 10, 15 and 20 ml/100 g Fe, respectively.

As a few examples, the following procedures for reduction of the initial hydrogen concentration in the weld metal below a crack critical limit can be determined from the diagram:

- At $HD_0 = 5.3$ ml/100 g Fe, at least a postheating time of approximately 120 min is required for a temperature of 150 °C
- At $HD_0 = 10$ ml/100 g Fe, at least a postheating time of approximately 90 min is required for a temperature of 200 °C
- At $HD_0 = 15$ ml/100 g Fe, at least a postheating time of approximately 120 min is required for a temperature of 200 °C
- At $HD_0 = 20$ ml/100 g Fe, at least a postheating time of approximately 80 min is required for a temperature of 250 °C

The numerical simulation of multi-pass welding was carried out with a 12 mm steel plate at the same joint configuration compared to the root weld (butt weld, weld seam included angle 60°). In Fig. 33 the results of the structural analysis, hydrogen diffusion analysis and cracking analysis are shown. For a comparison of numerical results and results from IRC tests, a micrograph of the multi-pass butt weld showing a hydrogen assisted crack is visualized in Fig. 33f.

As elucidated by Fig. 33, the locations of the highest stresses do not correspond to the location of the highest strains. While the highest stresses can be found in the upper third of the model, in the HAZ right below the last layer, the highest strains occur in the root layer. Figure 33 also shows the hydrogen distribution after welding and cooling down to room temperature for about one day (approx. 90,000 s). As can be depicted from this figure, nearly one day after welding the hydrogen concentration is only reduced about approx. 18 % from the initial hydrogen concentration of $HD_0 = 5.3$ ml/100 g Fe, which corresponds to the IRC test specimen welded with 1 % $H_2$ in the shielding gas.

Thus, there is still a high risk for delayed HACC, in particular, if small additional loads are superimposed. It has to be noted that the crack appeared in the region of the weld where very high strains, but not the highest strains and also not the highest hydrogen concentrations are located. Cracking thus appears at a critical concentration-strain combination.

By applying a sufficient HRHT hydrogen can be reduced quickly to subcritical levels. The corresponding normalized HRHT diagram for the 12 mm butt-welded S 1100 QL is shown in Fig. 34. Similar to the HRHT diagram of the root weld

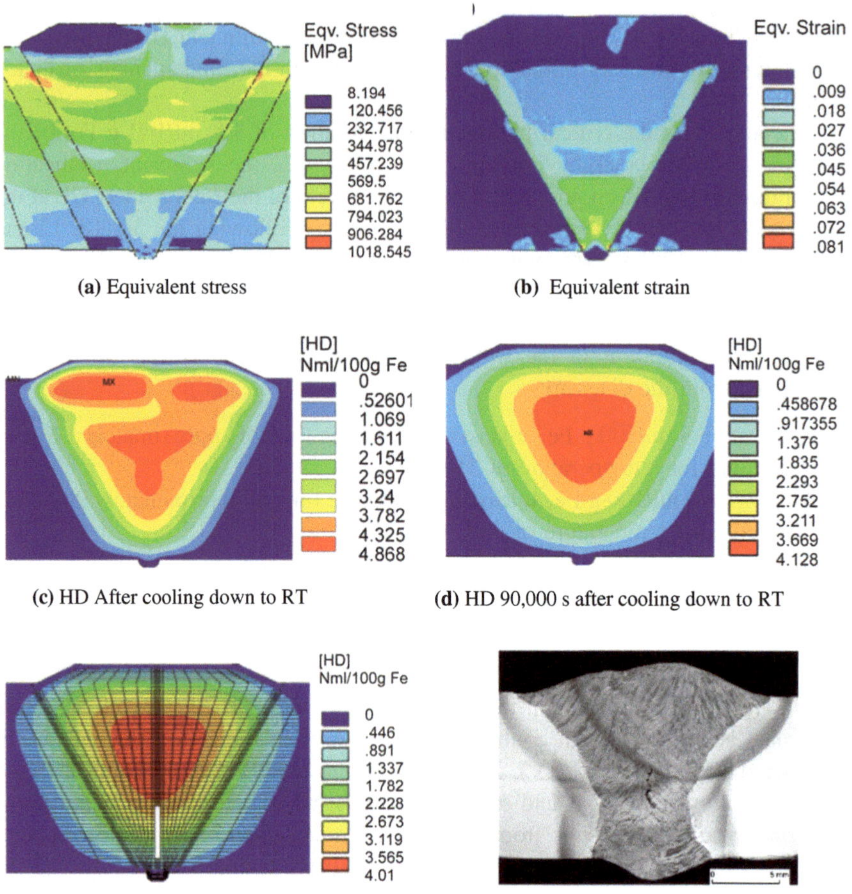


(**a**) Equivalent stress (**b**) Equivalent strain

(**c**) HD After cooling down to RT (**d**) HD 90,000 s after cooling down to RT

(**e**) HACC, controlled interpass at 120 °C (**f**) Micrograph of multi-pass butt weld

**Fig. 33** Results of the numerical simulation of HACC in multi-pass welding of S 1100 QL: **a** Equivalent stress, **b** Equivalent strain, **c** Hydrogen concentration (HD) after cooling down to room temperature (RT) ($HD_0$ = 5 ml/100 g Fe), **d** Hydrogen concentration 90,000 s (approx. 1 day) after cooling down to room temperature ($HD_0$ = 5 ml/100 g Fe), **e** simulated HACC in conjunction with IRC test at $HD_0$ = 5.3 ml/100 g Fe and application of a controlled interpass temperature of 120 °C, **f** Micrograph of HACC occurred in IRC test with 1 % $H_2$ in shielding gas and application of controlled interpass temperature of 120 °C [104–109]

(Fig. 32), the diagram can be used in order to evaluate sufficient postheating procedure for arbitrary initial hydrogen concentrations.

Numerical HACC modelling was subsequently applied to a lap joint and to a T–joint [115, 117], in order to evaluate the influence of different weld geometries on crack initiation and sufficient HRHT procedures. Figure 35 summarizes a comparison of the hydrogen distribution and occurrence of HACC in the lap, T- and butt joint at a plate thickness of 12 mm. Regarding to the influence of the plate thickness it is referred to an additional study [115, 117].

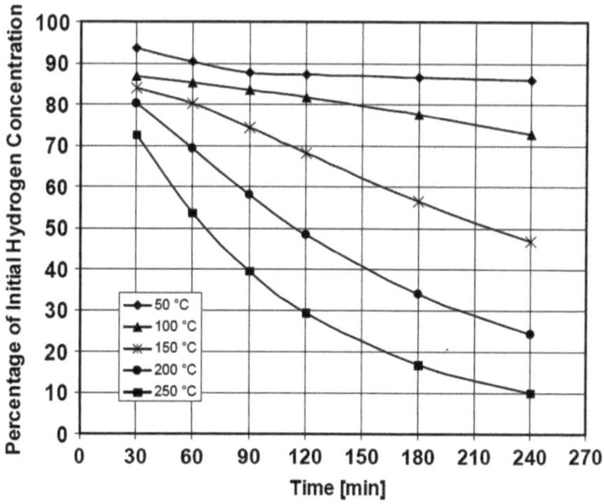


**Fig. 34** HRHT diagram showing hydrogen concentration at the end of the postheating time related to the initial hydrogen concentration dependent on the postheating temperature for the S 1100 QL multi-pass butt joint [104–109]

As can be drawn from Fig. 35, the maximum hydrogen concentration after 20 days is higher in the lap joint and, in particular, in the T-joint as compared to the butt joint. Thus, more time will be required to reduce the hydrogen concentration in the two other joints than in the butt joint to reduce the hydrogen concentration underneath a crack critical limit. Due to the larger material volume in case of the lap joint and T-joint, resulting in faster cooling compared to the butt joint, higher hydrogen levels remain in the weld after cooling down to room temperature. Regarding the butt weld, hydrogen in the root and in the intermediate layers can effuse much longer at higher temperatures and lower amounts of hydrogen in the middle of the weld have to be reduced after finishing welding.

For a comparison, the total hydrogen effusion process at room temperature for 20 days is shown in Fig. 36. The fast decay of the effusion curves at the beginning is due to the location of the maximum hydrogen concentration. At the beginning the maximum hydrogen concentration is located in the last welded pass having a short distance to the surface. Thus, the maximum hydrogen concentration is reduced very fast. With increasing effusion time the hydrogen concentration gradient decreases and the diffusion path growth longer while the maximum hydrogen concentration shifts to the middle of the weld (Fig. 33c, d).

Additionally, the results of the simulated HACC are shown in Fig. 35. Regarding the lap joint and the T-joint, it is seen that the risk for HACC shifts from the weld metal to the HAZ. This is due to the higher hydrogen level remaining in the welded zone as well as the higher residual stresses and strains resulting from the joint geometry compared to the butt joint. Even at lower initial hydrogen concentrations HACC occurs in the weld metal and HAZ of the lap joint and T-joint (Fig. 37). In order to avoid HACC in the welded joints hydrogen has to be reduced to subcritical values, so that the influence of hydrogen in possible cracking processes can be excluded. Thus, sufficient HRHT procedures were determined for both joint geometries.

In Fig. 38 the HRHT procedures for the butt joint and the lap joint are compared and in Fig. 39 the HRHT procedures for the T-joint are shown. As seen in Fig. 38, the

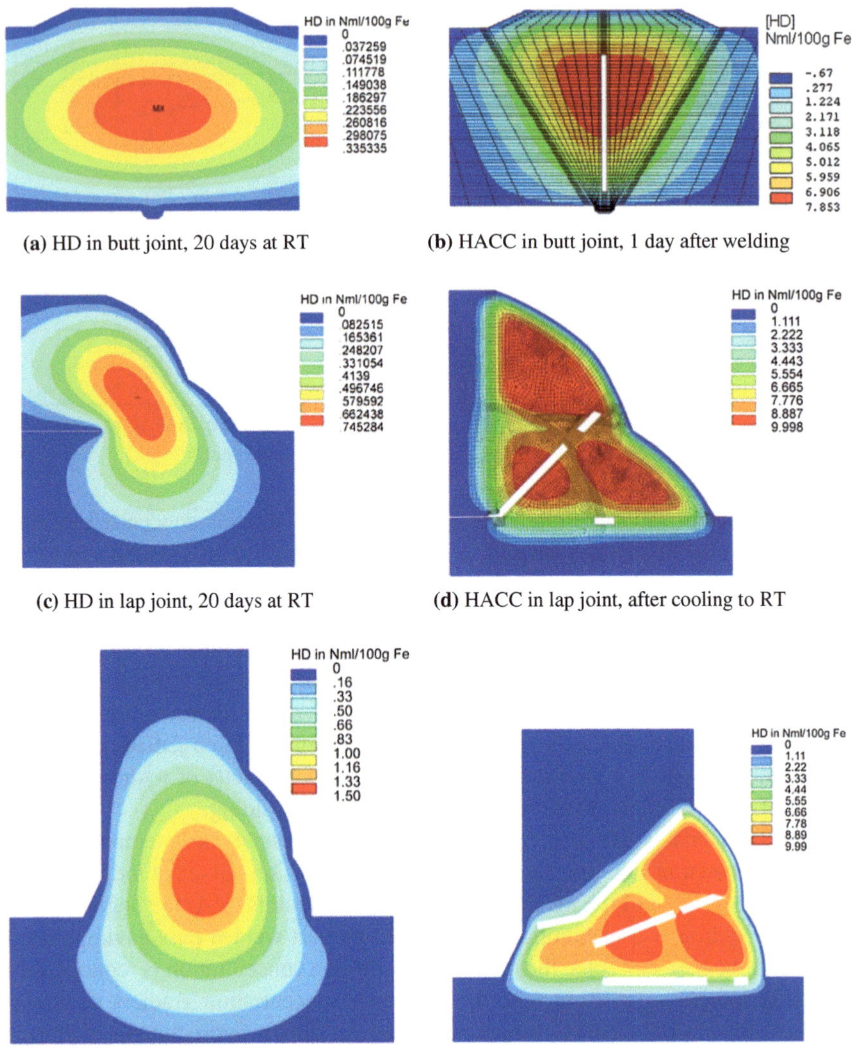


**(a)** HD in butt joint, 20 days at RT

**(b)** HACC in butt joint, 1 day after welding

**(c)** HD in lap joint, 20 days at RT

**(d)** HACC in lap joint, after cooling to RT

**(e)** HD in T-joint, 20 days at RT

**(f)** HACC in T-joint, after cooling to RT

**Fig. 35** Simulated hydrogen distribution 20 days after cooling down to room temperature for **a** the 5 pass butt joint, **c** the 3 pass lap joint and **e** the 4 pass T-joint assuming an initial hydrogen concentration of $HD_0$ = 5 ml/100 g Fe and simulated HACC **b** in the butt joint approx. 1 day after cooling down to room temperature (RT), **d** in the lap joint after cooling down to room temperature and **f** in the T-joint after cooling down to room temperature assuming an initial hydrogen concentration of $HD_0$ = 10 ml/100 g Fe [104–117]

curves overlap with increasing temperature and time. Remembering Fig. 36, the hydrogen concentration in the butt weld is reduced much faster than in the fillet weld without any heat treatment. In the case of HRHT these effects are reversed. By applying a HRHT, a larger material volume is heated up to higher temperatures. Thus,

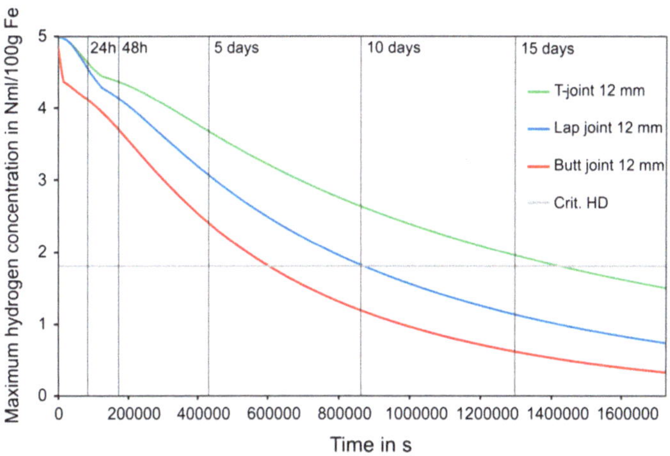


**Fig. 36** Evolution of maximum hydrogen concentration in different welded joints for 20 days effusion at room temperature without any heat treatment [115, 117]

hydrogen can diffuse into the adjacent hydrogen free material to reach the free surface. Furthermore, the ratio between weld metal surface and weld metal volume is much better in the fillet weld due to the larger included angle. Moreover, hydrogen has to be transported over a shorter diffusion distance, thus hydrogen concentration can be reduced much faster at a precipitous gradient. After a few minutes at higher HRHT-temperatures the hydrogen concentration in the fillet weld is much lower than that in the butt weld. The same behavior was observed for the T-joint.

Regarding the results shown in this section, the numerical model allows correct prediction of HACC. Thus, the model can be used determining sufficient HRHT procedures in order to avoid HACC in different weld geometries as well as to determine lifetime of welded components.

## *Simulation of HAC in Duplex Stainless Steel Microstructures at the Mesoscale*

As a further approach, the numerical model for HAC in welded components has been adapted to numerical calculations of HAC in heterogeneous microstructures, at first in Duplex stainless steels (DSS) at the mesoscale [118–122]. Due to the excellent mechanical as well as corrosion properties of DSS compared to conventional used austenitic and ferritic stainless steels, they are increasingly applied in offshore structures as well as in the chemical and petrochemical industry. However, some failure cases [116, 119] showed that the mechanisms of HAC especially in DSS have not been fully clarified which particularly concerns the frequently observed brittle HAC failure of the austenite phase. DSS consist of approx. equal portions of austenite and δ-ferrite. Both phases have different mechanical properties

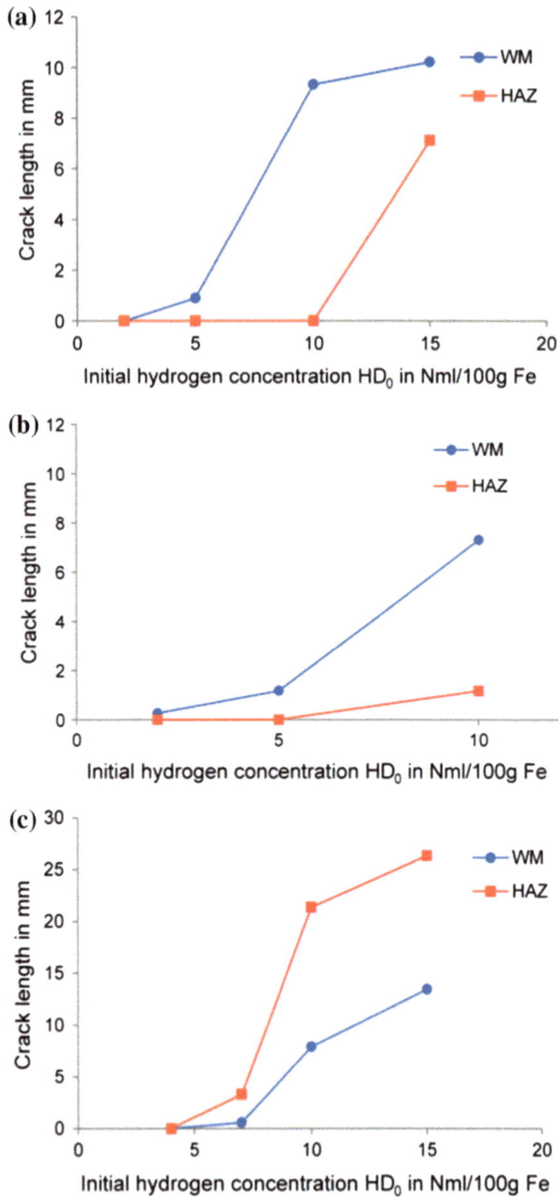


**Fig. 37** Simulated crack length in the weld metal (WM) and heat affected zone (HAZ) dependent on initial hydrogen concentration **a** in the butt joint, **b** in the lap joint and **c** in the T-joint

[123–129] as well as hydrogen transport properties [130, 131] and the key questions was, how HAC might be developed differently in the two microstructures. Thus, the sensitivity for HAC is different in both phases. In order to investigate HAC in DSS a two dimensional mesoscale model of a real DSS microstructure has been developed (Fig. 40).

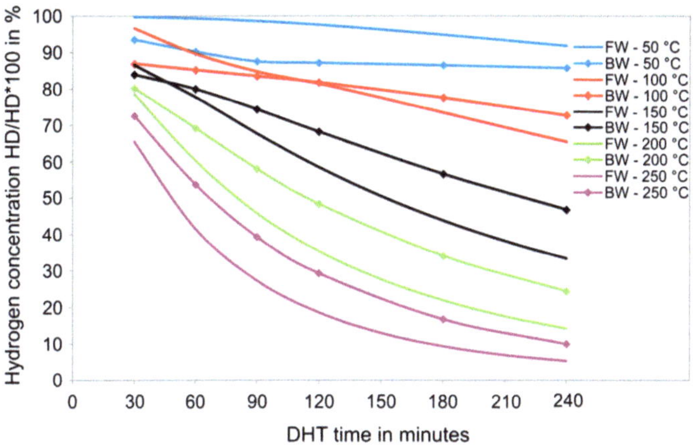


**Fig. 38** Comparison of the HRHT procedures for the 12 mm fillet weld (*FW*) and the 12 mm butt weld (*BW*) [115, 117]

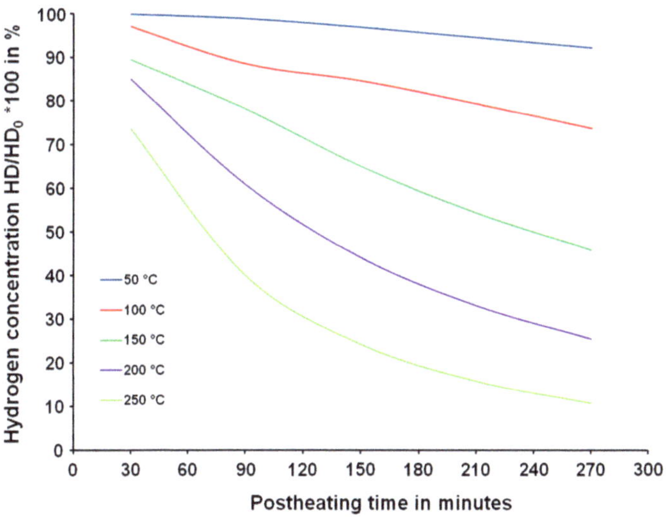


**Fig. 39** HRHT diagram for the 12 mm T-joint [115, 117]

Two different techniques have been utilized in such model to simulate HAC related crack initiation and propagation. The node release technique introducing predefined crack paths was used as first approach which entails the advantage that also sharp crack tips can be modelled. As another approach, the element elimination technique was used, similar to the earlier model [2], but without using predefined crack paths. By such procedure, cracks can expands freely in the model. The modelling in both cases is quite similar to previously described techniques, except that for the node release technique the continuous comparison of the actual to crack critical values was performed for the respective nodes alongside the crack paths

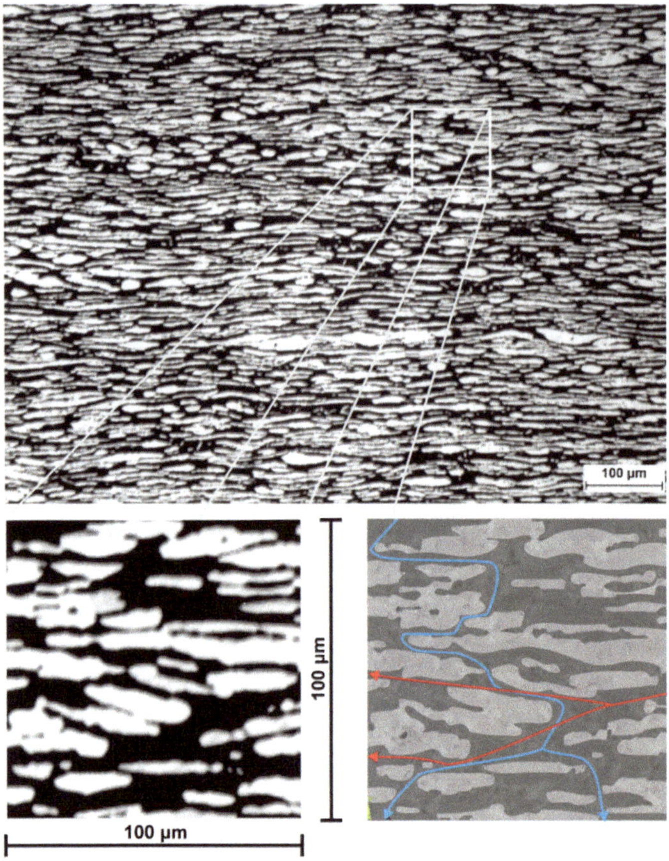


**Fig. 40** Cross section of the DSS microstructure, chosen area of 100 × 100 μm and numerical model, ferrite = dark and austenite = bright [118–122]

rather than for the elements. A detailed description of both approaches to model HAC in DSS has been reported elsewhere [115–118] and thus, needs not to be discussed in detail here.

As particular advantage, the mesoscale model could be used to investigate the hydrogen transport behavior under the influence of trapping at phase boundaries, different phase specific stress-strain behavior on stress-strain distribution as well as on HAC in the DSS microstructure and compared to experimental findings from recent literature.

Additionally to the model of the DSS base material microstructure, two simplified models of the weld metal microstructure were created [117], in order to determine hydrogen transport behavior in DSS weld microstructure.

Analysis of hydrogen transport behavior showed that hydrogen diffusion strongly depends on microstructural alignment of austenite and δ-ferrite and hydrogen transport occurs mainly by diffusion through ferrite [117, 118]. Austenite

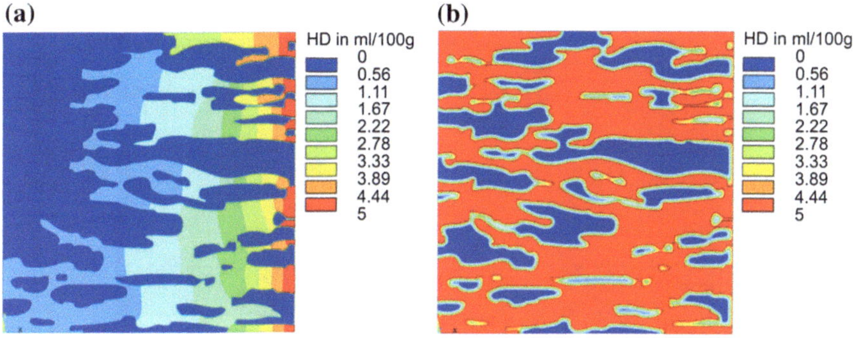


**Fig. 41** Simulated hydrogen distribution in the DSS microstructure for 100 s (**a**) and one hour (**b**) after beginning of charging with hydrogen from the right side of the model (in direction of microstructural alignment) [118–121]

traps hydrogen and acts as diffusion barrier, due to the low diffusion coefficient compared to δ-ferrite. Thus, high amounts of hydrogen will accumulate in δ-ferrite within short times. In Fig. 41, the hydrogen distribution 100 s and one hour after beginning of hydrogen charging alongside the microstructural alignment, without trapping at the phase boundaries, is shown.

Due to the high sensitivity for HAC of δ-ferrite compared to austenite, crack initiation and propagation mainly occurs in δ-ferrite and fracture of austenite links preceding cracks in δ-ferrite [118–121]. In Fig. 42, the results of HAC analysis using the node release technique are shown. As seen, cracks initiate and propagate in δ-ferrite until reaching austenitic islands (Fig. 42a, b). With increasing load fracture of austenite occur linking preceding cracks in δ-ferrite, Fig. 42c, d, similar to experimental observations and theoretical considerations [132].

In order to enable modelling without predefined crack paths, i.e. freely developing cracks, the element elimination technique was used assuming high hydrogen concentrations or low hydrogen concentrations present in the DSS microstructure (Fig. 43). Assuming high hydrogen concentrations in the numerical model, several secondary cracks are developing beside the main fracture which is a similar result than to experimental observations of Michalska et al. [133]. Moreover, a rougher fracture surface alongside both phases can be observed as compared to low hydrogen concentrations or hydrogen free DSS material [118–121]. Also, the theoretical considerations of San Marchi et al. [132] that first cleavage micro-cracks form in the ferrite inducing high stresses and strains in austenite and cracking through austenite can be visualized in the modelling. It can generally be stated that the results achieved with the numerical model are qualitatively consistent to other investigations regarding HAC in DSS.

By comparing both, the results of analyses using the node release technique with predefined crack paths with the results of analyses using element elimination technique the results differ from each other. However, the results using the element elimination technique without predefined crack paths are much closer to reality,

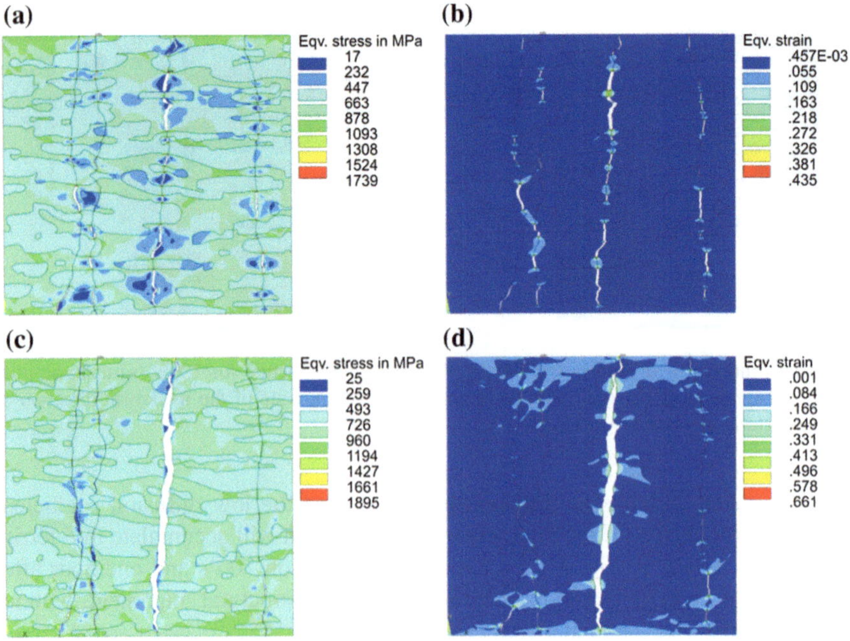


**Fig. 42** Results of numerical crack analysis assuming high hydrogen concentrations in the DSS microstructure; true equivalent stresses and strains at different stages of crack propagation [118]. **a** Equivalent stress during crack propagation. **b** Equivalent strain at crack propagation. **c** Equivalent stress of fractured model. **d** Equivalent strain of fractured model

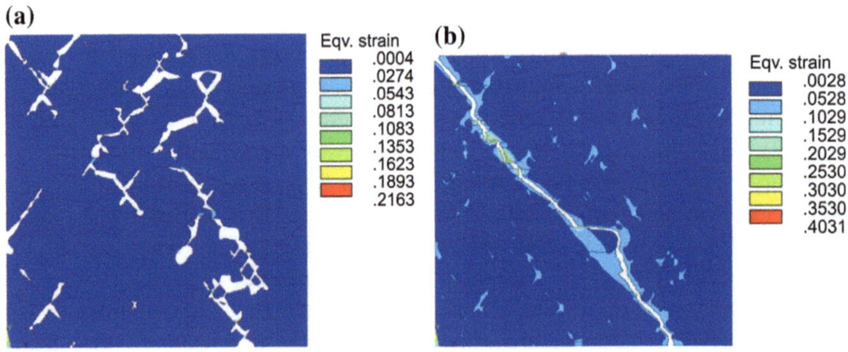


**Fig. 43** Simulated crack propagation in DSS microstructure assuming high (**a**) and low (**b**) hydrogen concentrations, true equivalent strain at final crack propagation [118–121]

i.e. experimental results. Such numerical investigations allow a deeper insight into HAC phenomenon and together with modern experiments they will continue to help clarification of the various hydrogen assisted cracking mechanisms.

## Conclusions

From the present state of the HAC modelling literature and numerical simulations, the following conclusions can be drawn:

1. Due to diverse metallurgical mechanisms very different models of hydrogen assisted cracking have been developed within the past decades. The approaches existing up to now can be categorized by their main simulated item of the three basic and local influences on hydrogen assisted cracking: mechanical load, hydrogen concentration and microstructure.
2. Hydrogen assisted cracking mechanisms are not yet fully understood. Although the theories concentrate more and more on a combination of HELP and HEDE, modelling based on the three main local influences (microstructure, load and hydrogen concentration) allows simpler and application oriented procedures. However, localized effects of HEDE and HELP might be incorporated into such modelling procedures as shown by several authors, but the various effects have to be quantified first. Moreover, detailed modelling of hydrogen uptake from different environments should be incorporated. By adoption of more recently developed analytical approaches, it should be possible to simulate the hydrogen subsurface concentration changes inside a pit or crevice entailing HASCC or even during HASCC propagation.
3. The basic numerical model for hydrogen assisted cracking has been developed on finite element calculations of the hydrogen distribution and the mechanical load. By continuous comparison of the actual hydrogen concentration and the actual mechanical load to the crack critical conditions cracking can be modelled within limited time steps and crack lengths as represented by each finite element ahead of the crack tip. In the past decade, such modelling techniques has successfully been applied to macroscopic simulation of HASCC of supermartensitic stainless steel welds in sour service, to cold cracking of high strength low alloyed steel welds and, already on a meso-scale, to duplex stainless steel base materials. Experimental and numerical results correspond fairly well and thus, the model can be regarded as satisfactorily verified.
4. Significant progress has been made in the past 10 years in establishment of numerical modelling and in using such procedures for crack avoidance measures, as for instance by defining hydrogen removal heat treatments for cold cracking avoidance.
5. Recently, and of similar importance, such modelling techniques were extended to non-predefined crack paths. This allows numerical simulations of hydrogen assisted cracking in any type of structural material and, in particular, in dual phase materials, as for instance duplex stainless steels, the microstructural effects on crack propagation can be studied.

Summarizing this contribution, the ways towards comprehensive numerical approaches for modelling of hydrogen assisted cracking have been shown. It becomes particularly obvious that reliable materials data and detailed experiments are essential for consistent validation of the various modelling approaches and their modifications. It becomes obvious that modelling of hydrogen assisted cracking in weld microstructures represents an innovative materials informatics application that requires both consistent and robust simulation techniques as well as reliable materials data [134, 135].

## References


1. Th. Boellinghaus: Modelling of Hydrogen Diffusion and Cracking in Steel Welds. In: *Mathematical Modelling of Weld Phenomena 5*, H. Cerjak (Eds.), Seggau, Austria, Maney Publishing, 1999, pp. 1019–1060
2. Th. Boellinghaus; H. Hoffmeister: Numerical Model for Hydrogen-Assisted Cracking. *Corrosion Science*, Vol. 56 (2000), No. 6, pp. 611–622
3. H.G. Nelson: Hydrogen Embrittlement. In: *Treatise on Materials Science and Technology: Embrittlement of Engineering Alloys, Volume 25*, C.L. Briant, S.K. Banerji (Eds.), Academic Press Inc, 1983, pp. 275–359
4. Th. Boellinghaus: Hydrogen Assisted Cracking of Supermartensitic Stainless Steels. In: *International Conference on Hydrogen Effects on Materials Behavior and Corrosion Deformation Interaction*, N.R. Moody, A.W. Thompson., R.E. Ricker, G.W. Was, R.H. Jones (Eds.), Jackson Lake Lodge, Moran, Wyoming, TMS-The Minerals, Metals and Materials Society, 2003, pp. 1009–1018
5. E. Wendler-Kalsch: Grundlagen und Mechanismen der Wasserstoff-induzierten Korrosion metallischer Werkstoffe. In: *Wasserstoff und Korrosion*, D. Kuron (Eds.), Bonn, Verlag Irene Kuron, 2000, pp. 7–53
6. J.P. Hirth: Effects of hydrogen on the properties of iron and steel. *Metallurgical Transactions A*, Vol. 11 (1980), No. 6, pp. 861–890
7. W. Gerberich: Modeling hydrogen induced damage mechanisms in metals. In: *Gaseous hydrogen embrittlement of materials in energy technologies - Mechanisms, modelling and future developments*, R.P. Gangloff, B.P. Somerday (Eds.), Woodhead Publishing Ltd, 2012, pp. 209–246
8. C.A. Zapffe; C.E. Sims: Hydrogen Embrittlement, Internal Stress and Defects in Steel. *Transactions of the AIME*, Vol. 145 (1941), pp. 225–259
9. N.J. Petch; P. Stables: Delayed Fracture of Metals under Static Load. *Nature*, Vol. 169 (1952), No. 4307, pp. 842–843
10. R.A. Oriani: Hydrogen Embrittlement of Steels. *Annual Review of Materials Science*, Vol. 8 (1978), No. 1, pp. 327–357
11. I.M. Robertson; P. Sofronis; A. Nagao; M.L. Martin; S. Wang; D.W. Gross; K.E. Nygren: Hydrogen Embrittlement Understood. *Metallurgical and Materials Transactions B*, Vol. 46 (2015), No. June, pp. 1085–1103
12. D. Kuron: Einfluß des Wasserstoffs auf Titan, Zirconium, Niob und Tantal. In: *Wasserstoff und Korrosion*, D. Kuron (Eds.), Bonn, Verlag Irene Kuron, 2000, pp. 212–256
13. U. Zwicker: Wasserstoffversprödung von Titan und Titanlegierungen. *Materialwissenschaft und Werkstofftechnik*, Vol. 5 (1974), No. 5, pp. 233–241
14. D. Eliezer; Th. Boellinghaus: Hydrogen effects in titanium alloys. In: *Gaseous hydrogen embrittlement of materials in energy technologies - The problem, its charackterisation and effects on particular alloy classes*, R.P. Gangloff, B.P. Somerday (Eds.), Woodhead Publishing Ltd, 2012, pp. 668–706
15. A. Troiano: The Role Of Hydrogen and other Intersticials in the mechanical Behavior of Metals. *Transactions of ASM*, Vol. 52 (1960), pp. 54–80

16. R.A. Oriani: A Mechanistic Theory of Hydrogen Embrittlement of Steels. *Berichte der Bunsengesellschaft für physikalische Chemie*, Vol. 76 (1972), No. 8, pp. 848–857
17. H.K. Birnbaum; I.M. Robertson; P. Sofronis; D. Teter: Mechanisms of Hydrogen Related Fracture - A Review. In: *Corrosion-Deformation Interactions*, M. Thierry (Eds.), Nice, France, Maney Publishing, 1996, pp. 172–193
18. G.E. Dieter: Elements of the Theory of Plasticity. In: *Mechanical Metallurgy*, New York, McGraw Hill Higher Education, 1988, pp. 69–102
19. J. Tien; A.W. Thompson; I.M. Bernstein; R.J. Richards: Hydrogen transport by dislocations. *Metallurgical Transactions A*, Vol. 7 (1976), No. 6, pp. 821–829
20. C.D. Beachem: A new model for hydrogen-assisted cracking (hydrogen "embrittlement"). *Metallurgical Transactions B*, Vol. 3 (1972), No. 2, pp. 441–455
21. E. Viyanit: Numerical Simulation of Hydrogen Assisted Cracking in Supermartensitic Stainless Steel Welds. Dissertation, Helmut-Schmidt-Universität Hamburg, 2005, 229 pages
22. H. Birnbaum; P. Sofronis: Hydrogen-Enhanced Localized Plasticity-A Mechanism for Hydrogen-Related Fracture. *Materials Science and Engineering: A*, Vol. 176 (1994), No. 1–2, pp. 191–202
23. P. Sofronis; H.K. Birnbaum: Mechanics of the hydrogen dislocation impurity interactions—I. Increasing shear modulus. *Journal of the Mechanics and Physics of Solids*, Vol. 43 (1995), No. 1, pp. 49–90
24. P. Sofronis; R. McMeeking: Numerical analysis of hydrogen transport near a blunting crack tip. *Journal of the Mechanics and Physics of Solids*, Vol. 37 (1989), No. 3, pp. 317–350
25. P. Sofronis: The influence of mobility of dissolved hydrogen on the elastic response of a metal. *Journal of the Mechanics and Physics of Solids*, Vol. 43 (1995), No. 9, pp. 1385–1407
26. S.P. Lynch: Progress Towards Understanding Mechanisms Of Hydrogen Embrittlement And Stress Corrosion Cracking. In: *Corrosion - Conference and Expo*, Nashville, Tennessee, NACE International, 2007, pp. 1694–1748
27. S. Lynch: Hydrogen embrittlement phenomena and mechanisms. *Corrosion Reviews*, Vol. 30 (2012), No. 3–4, pp. 105–123
28. P. Novak; R. Yuan; B.P. Somerday; P. Sofronis; R.O. Ritchie: A statistical, physical-based, micro-mechanical model of hydrogen-induced intergranular fracture in steel. *Journal of the Mechanics and Physics of Solids*, Vol. 58 (2010), No. 2, pp. 206–226
29. A. Barnoush; H. Vehoff: In situ electrochemical nanoindentation: A technique for local examination of hydrogen embrittlement. *Corrosion Science*, Vol. 50 (2008), No. 1, pp. 259–267
30. S. Wang; M.L. Martin; P. Sofronis; S. Ohnuki; N. Hashimoto; I.M. Robertson: Hydrogen-induced intergranular failure of iron. *Acta Materialia*, Vol. 69 (2014), pp. 275–282
31. A. Turnbull: Modelling of environment assisted cracking. *Corrosion Science*, Vol. 34 (1993), No. 6, pp. 921–960
32. M. Dadfarnia; P. Sofronis; B.P. Somerday; D.K. Balch; P. Schembri; R. Melcher: On the environmental similitude for fracture in the SENT specimen and a cracked hydrogen gas pipeline. *Engineering Fracture Mechanics*, Vol. 78 (2011), No. 12, pp. 2429–2438
33. J. Lufrano; P. Sofronis; D. Symons: Hydrogen transport and large strain elastoplasticity near a notch in alloy X-750. *Engineering Fracture Mechanics*, Vol. 59 (1998), No. 6, pp. 827–845
34. J.C. Sobotka; R.H. Dodds Jr.; P. Sofronis: Effects of hydrogen on steady, ductile crack growth: Computational studies. *International Journal of Solids and Structures*, Vol. 46 (2009), No. 22–23, pp. 4095–4106
35. Y. Liang; D. Ahn; P. Sofronis; R. Doddsjr; D. Bammann: Effect of hydrogen trapping on void growth and coalescence in metals and alloys. *Mechanics of Materials*, Vol. 40 (2008), No. 3, pp. 115–132
36. D.C. Ahn; P. Sofronis; R. Dodds: Modeling of hydrogen-assisted ductile crack propagation in metals and alloys. *International Journal of Fracture*, Vol. 145 (2007), No. 2, pp. 135–157

37. B.P. Somerday; M. Dadfarnia; D.K. Balch; K. a. Nibur; C.H. Cadden; P. Sofronis: Hydrogen-Assisted Crack Propagation in Austenitic Stainless Steel Fusion Welds. *Metallurgical and Materials Transactions A*, Vol. 40 (2009), No. 10, pp. 2350–2362
38. D. Delafosse; T. Magnin: Hydrogen induced plasticity in stress corrosion cracking of engineering systems. *Engineering Fracture Mechanics*, Vol. 68 (2001), No. 6, pp. 693–729
39. N.R. Raykar; S.K. Maiti; R.K. Singh Raman; S. Aryan: Study of hydrogen concentration dependent growth of external annular crack in round tensile specimen using cohesive zone model. *Engineering Fracture Mechanics*, Vol. 106 (2013), pp. 49–66
40. A.H.M. Krom; R.W.J. Koers; A. Bakker: Hydrogen transport near a blunting crack tip. *Journal of the Mechanics and Physics of Solids*, Vol. 47 (1999), No. 4, pp. 971–992
41. O. Takakuwa; M. Nishikawa; H. Soyama: Numerical simulation of the effects of residual stress on the concentration of hydrogen around a crack tip. *Surface and Coatings Technology*, Vol. 206 (2012), No. 11–12, pp. 2892–2898
42. W. Brocks; R. Falkenberg; I. Scheider: Coupling aspects in the simulation of hydrogen-induced stress-corrosion cracking. *Procedia IUTAM*, Vol. 3 (2012), pp. 11–24
43. A. T. Yokobori: The mechanism of hydrogen embrittlement: the stress interaction between a crack, a hydrogen cluster, and moving dislocations. *International Journal of Fracture*, Vol. 128 (2004), No. 1, pp. 121–131
44. C.V. Di Leo; L. Anand: Hydrogen in metals : A coupled theory for species diffusion and large elastic – plastic deformations. *International Journal of Plasticity*, Vol. 43 (2013), pp. 42–69
45. R. Oriani: The diffusion and trapping of hydrogen in steel. *Acta Metallurgica*, Vol. 18 (1970), No. 1, pp. 147–157
46. Y. Liang; P. Sofronis: Micromechanics and numerical modelling of the hydrogen–particle–matrix interactions in nickel-base alloys. *Modelling and Simulation in Materials Science and Engineering*, Vol. 11 (2003), No. 4, pp. 523–551
47. A. Turnbull; D.H. Ferriss; H. Anzaib: Modelling of the hydrogen distribution at a crack tip. Vol. 206 (1996), pp. 1–13
48. A. Krom; H. Maier; R. Koers; A. Bakker: The effect of strain rate on hydrogen distribution in round tensile specimens. *Materials Science and Engineering A*, Vol. 271 (1999), No. 1–2, pp. 22–30
49. S. Serebrinsky; E.A. Carter; M. Ortiz: A quantum-mechanically informed continuum model of hydrogen embrittlement. *Journal of the Mechanics and Physics of Solids*, Vol. 52 (2004), No. 10, pp. 2403–2430
50. T. Siegmund; W. Brocks: A numerical study on the correlation between the work of separation and the dissipation rate in ductile fracture. *Engineering Fracture Mechanics*, Vol. 67 (2000), No. 2, pp. 139–154
51. A. Turnbull: Hydrogen diffusion and trapping in metals. In: *Gaseous hydrogen embrittlement of materials in energy technologies - Mechanisms, modelling and future developments classes*, R.. P. Gangloff, B.P. Somerday (Eds.), Woodhead Publishing Ltd, 2012, pp. 89–128
52. I.M. Robertson; M.L. Martin; J.A. Fenske: Influence of hydrogen on the behavior of dislocations. In: *Gaseous hydrogen embrittlement of materials in energy technologies - Mechanisms, modelling and future developments classes*, R.P. Gangloff, B.P. Somerday (Eds.), Woodhead Publishing Ltd, 2012, pp. 166–206
53. D. Delafosse: Hydrogen effects on the plasticity of face centred cubic (fcc) crystals. In: *Gaseous hydrogen embrittlement of materials in energy technologies - Mechanisms, modelling and future developments*, R.P. Gangloff, B.P. Somerday (Eds.), Woodhead Publishing Ltd, 2012, pp. 247–285
54. M.R. Begley; J.A. Begley; C.M. Landis: Continuum mechanics modeling of hydrogen embrittlement. In: *Gaseous hydrogen embrittlement of materials in energy technologies - Mechanisms, modelling and future developments*, R. Gangloff, B. Somerday (Eds.), Woodhead Publishing Ltd, 2012, pp. 286–325

55. M. Dadfarnia; P. Sofronis; B.P. Somerday; D.K. Balch; P. Schembri: Degradation models for Hydrogen embrittlement. In: *Gaseous hydrogen embrittlement of materials in energy technologies - Mechanisms, modelling and future developments*, R.P. Gangloff, B. P. Somerday (Eds.), Woodhead Publishing Ltd, 2012, pp. 326–377
56. M.M. Hall Jr: Effect of inelastic strain on hydrogen-assisted fracture of metals. In: *Gaseous hydrogen embrittlement of materials in energy technologies - Mechanisms, modelling and future developments*, R.P. Gangloff, B.B. Somerday (Eds.), Woodhead Publishing Ltd, 2012, pp. 378–429
57. X. Yu; F. Gou; B. Li; N. Zhang: Numerical study of the effect of hydrogen on the crack propagation behavior of single crystal tungsten. *Fusion Engineering and Design*, Vol. 89 (2014), No. 7–8, pp. 1096–1100
58. D. Connétable; Y. Wang; D. Tanguy: Segregation of hydrogen to defects in nickel using first-principles calculations: The case of self-interstitials and cavities. *Journal of Alloys and Compounds*, Vol. 614 (2014), pp. 211–220
59. R. Matsumoto; S. Seki; S. Taketomi; N. Miyazaki: Hydrogen-related phenomena due to decreases in lattice defect energies—Molecular dynamics simulations using the embedded atom method potential with pseudo-hydrogen effects. *Computational Materials Science*, Vol. 92 (2014), pp. 362–371
60. G.S. Was; D. Farkas; I.M. Robertson: Micromechanics of dislocation channeling in intergranular stress corrosion crack nucleation. *Current Opinion in Solid State and Materials Science*, Vol. 16 (2012), No. 3, pp. 134–142
61. M. Itakura; H. Kaburaki; M. Yamaguchi; T. Okita: The effect of hydrogen atoms on the screw dislocation mobility in bcc iron: A first-principles study. *Acta Materialia*, Vol. 61 (2013), No. 18, pp. 6857–6867
62. J. Song; W. a. Curtin: Mechanisms of hydrogen-enhanced localized plasticity: An atomistic study using α-Fe as a model system. *Acta Materialia*, Vol. 68 (2014), pp. 61–69
63. S. Jothi; T.N. Croft; S.G.R. Brown; E. a. de Souza Neto: Finite element microstructural homogenization techniques and intergranular, intragranular microstructural effects on effective diffusion coefficient of heterogeneous polycrystalline composite media. *Composite Structures*, Vol. 108 (2014), pp. 555–564
64. V. Olden; A. Saai; L. Jemblie; R. Johnsen: FE simulation of hydrogen diffusion in duplex stainless steel. *International Journal of Hydrogen Energy*, Vol. 39 (2014), No. 2, pp. 1156–1163
65. R.W. Smith: Computer simulation of intergranular stress corrosion cracking via hydrogen embrittlement. *Modelling and Simulation in Materials Science and Engineering*, Vol. 8 (2000), No. 4, pp. 629–648
66. A. Traidia; M. Alfano; G. Lubineau; S. Duval: An effective finite element model for the prediction of hydrogen induced cracking in steel pipelines. *International Journal of Hydrogen Energy*, Vol. 37 (2012), No. 21, pp. 16214–16230
67. L.O. Jernkvist; a. R. Massih: Multi-field modelling of hydride forming metals. Part I: Model formulation and validation. *Computational Materials Science*, Vol. 85 (2014), pp. 363–382
68. L.O. Jernkvist: Multi-field modelling of hydride forming metals Part II: Application to fracture. *Computational Materials Science*, Vol. 85 (2014), pp. 383–401
69. a. G. Varias; J.L. Feng: Simulation of hydride-induced steady-state crack growth in metals ? Part I: growth near hydrogen chemical equilibrium. *Computational Mechanics*, Vol. 34 (2004), No. 5, pp. 339–356
70. T. Shoji; Z. Lu; H. Murakami: Formulating stress corrosion cracking growth rates by combination of crack tip mechanics and crack tip oxidation kinetics. *Corrosion Science*, Vol. 52 (2010), No. 3, pp. 769–779
71. J. Shi; J. Wang; D.D. Macdonald: Prediction of crack growth rate in Type 304 stainless steel using artificial neural networks and the coupled environment fracture model. *Corrosion Science*, Vol. 89 (2014), pp. 69–80

72. B.-H. Choi; A. Chudnovsky: Observation and Modeling of Stress Corrosion Cracking in High Pressure Gas Pipe Steel. *Metallurgical and Materials Transactions A*, Vol. 42 (2010), No. 2, pp. 383–395
73. H. Hoffmeister; O. Klein: Modeling of SCC initiation and propagation mechanisms in BWR environments. *Nuclear Engineering and Design*, Vol. 241 (2011), No. 12, pp. 4893–4902
74. H. Hoffmeister; Th. Böllinghaus: Modeling the Effects of Dissolved Hydrogen, Temperature, pH, and Global Stress on Stress Corrosion Cracking of a Cr-Ni Model Alloy in High-Temperature Water. *Corrosion*, Vol. 71 (2015), No. 5, pp. 559–573
75. H. Hoffmeister; Th. Böllinghaus: Modeling of Combined Anodic Dissolution/Hydrogen-Assisted Stress Corrosion Cracking of Low-Alloyed Power Plant Steels in High-Temperature Water Environments. *Corrosion*, Vol. 70 (2014), No. 6, pp. 563–578
76. H. Hoffmeister: Modeling the Effects of Local Anodic Acidification on Stress Corrosion Cracking of Nickel. *Corrosion*, Vol. 67 (2011), No. 7, pp. 075002–1–075002–12
77. H. Hoffmeister; J. Neuß: Assessment and Crack Tip Ductility by Tensile Slow Strain Rate Crack Start Measurements of High Alloyed Materials. In: *Stainless Steel World 2011 - Conference & Exhibition*, Maastricht, Niederlande, 2011, pp. paper 11051
78. H. Hoffmeister: Modeling of Crevice Corrosion of Pure Nickel by Coupling of Phase and Polarization Behavior at Various pH, Chloride, and Oxygen Levels. *Corrosion*, Vol. 61 (2005), No. 9, pp. 880–888
79. H. Hoffmeister: Modeling the effects of nuclear environments on crevice corrosion of pure nickel by coupling of phase and polarization behavior. *Nuclear Engineering and Design*, Vol. 239 (2009), No. 10, pp. 1795–1803
80. H. Hoffmeister: Modeling the Effect of Chloride Content on Hydrogen Sulfide Corrosion of Pure Iron by Coupling of Phase and Polarization Behavior. *Corrosion*, Vol. 64 (2008), No. 6, pp. 483–495
81. L. Mishnaevsky: Numerical Experiments in the Mesomechanics of Materials. Habilitationsschrift, Technische Universität Darmstadt, 2005, 150 pages
82. L.L. Mishnaevsky; S. Schmauder: Continuum Mesomechanical Finite Element Modeling in Materials Development: A State-of-the-Art Review. *Applied Mechanics Reviews*, Vol. 54 (2001), No. 1, pp. 49
83. N.N.: Guidelines on Materials Requirements for Carbon and Low Alloy Steels for H2S-Containing Environments in Oil and Gas Production. Maney Publishing, 2009, 56 pages, ISBN: 978–1-90654-033-3
84. Th. Boellinghaus; H. Hoffmeister: Finite element calculations of pre- and postheating procedures for sufficient hydrogen removal in butt joints. In: *Mathematical Modelling of weld Phenomena 3*, H. Cerjak (Eds.), Seggau, Austria, The Institute of Materials, London, 1997, pp. 726–756
85. Th. Boellinghaus; E. Viyanit; H. Hoffmeister: Numerical modelling of hydrogen assisted cracking. In: *Corrosion 2001*, Houston, TX, USA, NACE National Association of Corrosion Engineers, 2001, pp. Paper 01226
86. N.N.: Corrosion Resistant Alloys for Oil and Gas Production: Guidance on General Requirements and Test Methods for H2S Service. Maney Publishing, 2002, 96 pages, ISBN: 978-1-902653-55-6
87. Th. Boellinghaus; H. Hoffmeister; J. Klemme; H. Alzer: Hydrogen permeation in a low carbon martensitic stainless steel exposed to H2S containing brines at free corrosion. In: *Corrosion*, NACE International, 1999, pp. Paper 609
88. D.P.G. Lidbury: The Estimation of Crack Tip Strain Rate Parameters Characterizing Environment Assisted Crack Growth Data. In: *International Conference on Environment Assisted Cracking by the Localized Crack Environment*, R.P. Gangloff (Eds.), New York, ASME, 1983, pp. 149–172
89. Th. Boellinghaus; H. Hoffmeister; S. Dietrich: Slow Strain Rate Testing of Low Carbon Martensitic Stainless Steels. In: *Advances in Corrosion Control and Materials in Oil and Gas Production*, P.S. Jackman, L.M. Smith (Eds.), The Institute of Materials, London, 1999, pp. 274–285

90. Th. Boellinghaus; H. Hoffmeister; M. Littich: On-line Sulfide Stress Cracking Monitoring of 13 % Cr Pipe Welds at Realistic Weld Restraint Conditions in the Instrumented Restraint Cracking (IRC) Test. In: *Advances in Corrosion Control and Materials in Oil and Gas Production*, P.S. Jackman, L.M. Smith (Eds.), The Institute of Materials, London, 1999, pp. 286–303
91. H. Takabe; H. Amaya; H. Hirata, M. Ueda: Corrosion Resistance of Weldable Modified 13Cr Stainless Steel for $CO_2$ Applications. In: *Advances in Corrosion Control and Materials in Oil and Gas Production*, P. S. Jackman, L. M. Smith (Eds.), EFC Publications No. 26, IOM London, UK, pp. 219–230
92. J. J. Dufrane; E. Franceschetti; J. Heather, H. van der Winden: Weldable 13 % Chromium Steel: the Development of the Components for a Wet Gas Piping System. In: *Advances in Corrosion Control and Materials in Oil and Gas Production*, P. S. Jackman, L. M. Smith (Eds.), EFC Publications No. 26, IOM London, UK, pp. 249–258
93. H. Heuser; C. Jochum; E. Pertender; J. Toesch: Girth Welding of Supermartensitic Stainless Steels with Matching Filler Wires. In: *IIW-Doc. IX-H-515-01 / IX-2005-01*, 2001
94. Th. Boellinghaus; E. Viyanit: Numerical Simulation of Hydrogen Addisted Cracking in Girth Welds of SMSS Pipelines - Report I. In: *Mathematical Modelling of Weld Phenomena 6*, H. Cerjak (Ed.), Seggau, Austria, Maney Publishing, 2002, pp. 839–855
95. Th. Boellinghaus; E. Viyanit; H. Hoffmeister: Numerical Simulations of Hydrogen-Assisted Cracking in Girth Welds of Supermartensitic Stainless Steel Pipelines - Report II. In: *Mathematical Modelling of Weld Phenomena 7*, H. Cerjak, H.K.D.H. Bhadeshia, E. Kozeschnik (Eds.), Seggau, Austria, Verlag der Technischen Universitat Graz, 2005, pp. 847–874
96. M. Ueda; H. Amaya; H. Hirata; K. Kondo; Y. Muarata; Y. Komizo: Corrosion Resistance of 13 % Chromium Steel Welded Joints in Flow Line Applications. In: *Advances in Corrosion Control and Materials in Oil and Gas Production*, P. S. Jackman, L. M. Smith (Eds.), EFC Publications No. 26, IOM London, UK, pp. 267–273
97. Th. Boellinghaus; H. Hoffmeister; I. Stiebe-Springer; W. Florian; T. Michael: Component Testing of Welded Supermartensitic Stainless Steel Pipes. In: *International Conference EuroCorr 2000*, London, European Federation of Corrosion, 2000
98. Th. Boellinghaus; H. Hoffmeister; K. Feuerstake; H. Alzer; J. Krewinkel: Finite element calculation of hydrogen uptake and diffusion in martensitic stainless steel welds. In: *Mathematical Modelling of Weld Phenomena 4*, H.H. Cerjak (Eds.), Seggau, Österreich, The Institute of Materials, 1998, pp. 355–378
99. Th. Boellinghaus; H. Hoffmeister; C. Middel: Scatterbands for hydrogen diffusion coefficients in steel having a ferritic or martensitic microstructure and steels having an austenitic microstructure at room temperature. *Welding in the World*, Vol. 37 (1996), No. 1, pp. 16–23
100. Th. Boellinghaus; T. Kannengiesser; C. Jochum; I. Stiebe-Springer: Effect of Filler Material Selection on Stress-Strain Distribution and Hydrogen Cracking in Welded Supermartensitic Stainless Steel Pipes. In: *Corrosion - Conference and Expo*, Denver, Colorado, NACE International, 2002, pp. 20
101. Th. Bollinghaus; H. Hoffmeister; L. Reuter: Material properties of as delivered and quenched modified martensitic stainless steels dependent on hydrogen concentration. In: *Proceedings of the 1st International Conference Supermartensitic Stainless Steels*, Brüssel, Belgian Welding Institute, 1999, pp. 264–271
102. Th. Boellinghaus; H. Hoffmeister; T. Tanski: Wet H2S Performance of As Delivered and Quenched Modified Martensitic Staunless Steel Determined by SSRT. In: *Corrosion*, 1999, pp. Paper 99588
103. F. Richter: Die wichtigsten physikalischen Eigenschaften von 52 Eisenwerkstoffen. *Stahleisen - Sonderberichte*, (1973), No. 8, pp. 1–30
104. P. Wongpanya; Th. Boellinghaus; G. Lothongkum; T. Kannengiesser: Effects of preheating and interpass temperature on stresses in S1100QL multi-pass butt-welds. *IIW-Doc. IX-2240-07*, (2007), pp. 1–26

105. P. Wongpanya; T. Boellinghaus; G. Lothongkum: Effects of hydrogen removal heat treatment on residual stresses in high strength structural steel welds. *Welding in the World*, Vol. 50 (2006), No. Special Issue, pp. 96–103
106. F. Richter: Physikalische Eigenschaften von Stählen und ihre Temperaturabhängigkeit. 1983, 42 Seiten
107. P. Wongpanya; Th. Boellinghaus; G. Lothongkum: Numerical Simulation of Hydrogen Removal Heat Treatment Procedures in High Strength Steel Welds. In: *Mathematical Modelling of Weld Phenomena 8*, H. Cerjak (Eds.), Seggau, Austria, Verlag der Technischen Universität Graz, 2007, pp. 743–765
108. P. Wongpanya; Th. Boellinghaus; G. Lothongkum; H. Hoffmeister: Numerical Modelling of Cold Crack Initiation and Propagation in S 1100 QL Steel Root Welds. *Welding in the World*, Vol. 53 (2009), No. 3/4, pp. R34–R43
109. Th. Boellinghaus; P. Wongpanya: Modeling of Hydrogen Assisted Cold Cracking in High Strength Steel Welds. In: *Proceedings of the 2008 International Hydrogen Conference*, B. Somerday, P. Sofronis, R. Jones (Eds.), Jackson Lake Lodge, Wyoming, USA, ASM International, 2009, pp. 588–595
110. T. Boellinghaus; H. Hoffmeister; A. Dangeleit: A scatterband for hydrogen diffusion coefficients in micro-alloyed and low carbon structural steels. *Welding in the World*, Vol. 35 (1995), No. 2, pp. 83–96
111. D.M. Seeger: Wasserstoffaufnahme und -diffusion in Schweißnahtgefügen hochfester Stähle. Dissertation, Helmut-Schmidt-University – University of the Federal Armed Forces Hamburg, 2004, 144 pages
112. D. Brinkmann: Beitrag zur experimentellen und numerischen Verformungsanalyse von Schweißverbindungen. Dissertation, Technische Universität Braunschweig, 2000, 168 pages
113. H. Hoffmeister: Concept and procedure of the IRC test for assessing hydrogen assisted weld cracking. *Steel Research*, Vol. 57 (1986), No. 7, pp. 344–347
114. Th. Boellinghaus; H. Hoffmeister; M. Littich: Application of the IRC-Test for Assessment of Reaction Stresses in Tabular Joints with Respect to Hydrogen Assisted Weld Cracking. *Welding in the World*, Vol. 43 (1999), No. 2, pp. 27–35
115. T. Mente; Th. Boellinghaus; M. Schmitz-Niederau: Heat treatment Effects on the Reduction of Hydrogen in Multi-Layer High-Strength Weld Joints. *Welding in the World*, Vol. 56 (2013), No. 7-8, pp. 26–36
116. T.S. Taylor; T. Pendlington; R. Bird: Foinaven Super Duplex Materials Cracking Investigation. In: *Offshore Technology Conference*, Houston, Texas, Society of Petroleum Engineers, 1999, pp. 467–480
117. E. Steppan; T. Mente; Th. Böllinghaus: Numerical investigations on cold cracking avoidance in fillet welds of high-strength steels. *Welding in the World*, Vol. 57 (2013), No. 3, pp. 359–371
118. T. Mente; Th. Boellinghaus: Numerical investigations on hydrogen assisted cracking in duplex stainless steel microstructures. In: C. Cross, Th. Boellinghaus, J. Lippold (Eds.): *4th International Workshop Cracking Phenomena in Welds*, Berlin, Springer, accepted for publication, 2015
119. S. Huizinga; B. Mclaughlin; I.M. Hannah; S.J. Paterson; B.N.W. Snedden: Failure of a Subsea Super Duplex Manifold Hub by HISC and Implications for Design. In: *Corrosion*, San Diego, California, USA, NACE International, 2006, pp. 9
120. T. Mente; Th. Boellinghaus: Mesoscale modeling of hydrogen-assisted cracking in duplex stainless steels. *Welding in the World*, Vol. 58 (2013), No. 2, pp. 205–216
121. T. Mente; Th. Boellinghaus: Modeling Of Hydrogen Distributionin A Duplex Stainless Steel. *Welding in the World*, Vol. 56 (2013), No. 11-12, pp. 66–78

T. Mente; Th. Boellinghaus: Numerical Model for Hydrogen Assisted Cracking in Duplex Stainless Steel Microstructures. In: *Mathematical Modelling of Weld Phenomena 10*, C. Sommitsch, N. Enzinger (Eds.), Seggau, Austria, Verlag der Technischen Universitat Graz, 2012, pp. 337–356

123. M. Knyazeva; M. Pohl: Duplex Steels: Part I: Genesis, Formation, Structure. *Metallography, Microstructure, and Analysis*, Vol. 2 (2013), No. 2, pp. 113–121
124. J. Nilsson; G. Chai: The physical metallurgy of duplex stainless steels. In: *International Conference & Expo Duplex 2007*, Gardo, Italien, Associazione Italiana di Metallurgia, 2007, pp. 73–82
125. N. Tsuchida; T. Kawahata; E. Ishimaru; A. Takahashi; H. Suzuki; T. Shobu: Static Tensile Deformation Behavior of a Lean Duplex Stainless Steel Studied by In Situ Neutron Diffraction and Synchrotron Radiation White X-rays. *ISIJ International*, Vol. 53 (2013), No. 7, pp. 1260–1267
126. R. Dakhlaoui; A. Baczmański; C. Braham; S. Wroński; K. Wierzbanowski; E.C. Oliver: Effect of residual stresses on individual phase mechanical properties of austeno-ferritic duplex stainless steel. *Acta Materialia*, Vol. 54 (2006), No. 19, pp. 5027–5039
127. R. Dakhlaoui; C. Braham; A. Baczmanski: Mechanical properties of phases in austeno-ferritic duplex stainless steel—Surface stresses studied by X-ray diffraction. *Materials Science and Engineering: A*, Vol. 444 (2007), No. 1–2, pp. 6–17
128. J. Johansson; M. Odén; X.-H. Zeng: Evolution of the residual stress state in a duplex stainless steel during loading. *Acta Materialia*, Vol. 47 (1999), No. 9, pp. 2669–2684
129. J.J. Moverare; M. Odén: Influence of elastic and plastic anisotropy on the flow behavior in a duplex stainless steel. *Metallurgical and Materials Transactions A*, Vol. 33 (2002), No. 1, pp. 57–71
130. E. Owczarek; T. Zakroczymski: Hydrogen transport in a duplex stainless steel. *Acta Materialia*, Vol. 48 (2000), No. 12, pp. 3059–3070
131. A. Turnbull; R.B. Hutchings: Analysis of hydrogen atom transport in a two-phase alloy. *Materials Science and Engineering: A*, Vol. 177 (1994), No. 1–2, pp. 161–171
132. C. San Marchi; B.P. Somerday; J. Zelinski; X. Tang; G.H. Schiroky: Mechanical Properties of Super Duplex Stainless Steel 2507 after Gas Phase Thermal Precharging with Hydrogen. *Metallurgical and Materials Transactions A*, Vol. 38 (2007), No. 11, pp. 2763–2775
133. J. Michalska; M. Sozańska; M. Hetmańczyk: Application of quantitative fractography in the assessment of hydrogen damage of duplex stainless steel. *Materials Characterization*, Vol. 60 (2009), No. 10, pp. 1100–1106
134. World Materials Research Institute Forum (WMRIF): 10 Major trends in Materials Science and Engineering, https://www.wmrif.info
135. Materials Genom Initiative, https://www.whitehouse.gov/mgi

# Metallurgical Factors Influencing the Susceptibility of Hydrogen Assisted Cracking in Dissimilar Metal Welds for Application Under Cathodic Protection

**D. Bourgeois, B. Alexandrov, J. Lippold and J. Fenske**

**Abstract** Dissimilar metal weld (DMW) overlays of Nickel-based filler metals on low-alloy steel pipes are used in the oil and gas industry in order to eliminate the need for field post weld heat treatment (PWHT) of adjacent closure welds. Brittle failures have been reported along some of these DMW interfaces, especially in AISI 8630-IN625 weld combinations during subsea service under cathodic protection (CP). These failures have been attributed to hydrogen assisted cracking (HAC) due to local hydrogen embrittlement of susceptible microstructures that form at the fusion boundary during welding and PWHT. Testing at The Ohio State University using the delayed hydrogen cracking test (DHCT) has concluded that this type of HAC is strongly affected by the base metal/filler metal combination, and by the welding and PWHT procedures. These controlling factors determine the microstructures that form at the fusion boundary. Thermo-Calc$^{TM}$ and Dictra$^{TM}$ software was used to simulate the above-mentioned controlling factors. Based on the simulations conducted, carbon concentration near the fusion boundary was monitored closely. The results of the simulations yielded results that correlate well with DHCT experiments. Also, the controlling factors of HAC can be optimized to reduce the accumulation of carbon at the fusion boundary.

**Keywords** Welded joints · Hydrogen assisted cracking · Dissimilar metal weld · Cathodic protection · Carbon diffusion

D. Bourgeois (✉) · B. Alexandrov · J. Lippold
The Ohio State University, Columbus, USA
e-mail: desmond.j.bourgeois@exxonmobil.com

J. Fenske
ExxonMobil Development Company, Irving, USA



Th. Boellinghaus et al. (eds.), *Cracking Phenomena in Welds IV*,
DOI 10.1007/978-3-319-28434-7_19

# Introduction

## *Statement of Problem*

The oil and gas industry uses steel components that are clad with Nickel-based filler metal to eliminate the need for field post weld heat treatment (PWHT) and to provide increased corrosion resistance [1, 2]. NACE MR0175 requires PWHT to satisfy hardness requirements (<250 HV10) of components that will be used for sour service [3]. Common PWHT procedures are conducted at high temperatures for long times (for example 650 °C (1202 °F) for 5–10 h); however, they cause diffusion of carbon from the steel towards the Nickel-base filler metal [1]. The steel is depleted of carbon, which leads to a degradation of mechanical properties [4]. In addition, the carbon that diffuses towards the Nickel-base filler metal accumulates near the dissimilar metal fusion boundary. Active CP with sacrificial anodes produces atomic hydrogen that diffuses to the susceptible microstructures resulting in local embrittlement that leads to in-service failures [5]. Brittle failures have been reported in DMWs of AISI 8630 with Inconel 625 (IN625) filler metal [6]. These failures are classified as HAC and their susceptibility is strongly affected by base metal/filler metal combinations, and by the PWHT and welding procedures [1].

ASTM A182-F22 and AISI 8630 are forging grade steels that are used for subsea applications under CP [7]. AISI 8630 however was discontinued from use for subsea applications due to catastrophic failures caused by the formation of highly susceptible microstructures after PWHT. ASTM A182-F22 has not shown similar embrittlement and is widely used for subsea applications. IN625 is still being utilized as the clad weld metal for the same subsea applications.

## *Metallurgical and Microstructural Factors of HAC*

As noted earlier, the welding procedure affects the susceptibility to HAC in DMWs under CP. There are two different welding procedures, or bead sequences (BS) that have been used to deposit IN625 on the F22 substrate. BS1 is where the welding beads are deposited normal to the F22 substrate, and is called the stacked bead sequence. In contrast, the welding beads in BS3 are deposited parallel to the F22 substrate, and is called the layered bead sequence. Both of these can be seen in Fig. 1. It is important to note that based on industrial experience BS1 is more susceptible to HAC than BS3. This is due to how the welding beads are deposited on the forging between the two welding procedures. In BS1, multiple weld passes come in contact with the forging. This leads to the formation of a larger planar growth region at the weld interface in BS1 and the formation of base metal "swirls" close to or connected to the interface. In BS3, a continuous welding pass is made on the forging. After that, subsequent passes temper the previous welding bead,

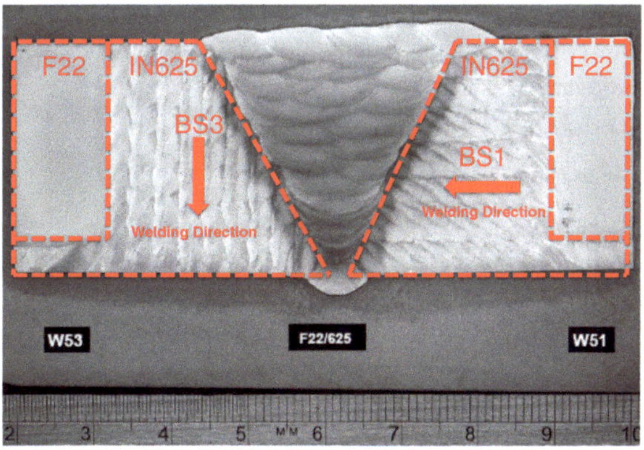


**Fig. 1** Micrograph of F22-IN625 DMW that indicates the welding directions of BS1 and BS3

allowing for less susceptible microstructures to be present once the welding is concluded.

The PWHT procedure also affects the susceptibility to HAC in DMWs under CP. A typical PWHT for components that will be subjected to subsea service and under CP is 13 h at 680 °C (1256 °F) for F22-IN625 DMWs. This accounts for one inch of thickness of the forging material per one hour of PWHT time that is close to the $A_1$ temperature of the material [3]. Alternate PWHT procedures that have been used include 2 h at 650 °C (1202 °F), 4 h at 650 °C, and 2 h at 650 °C + 6 h at 760 °C (1400 °F). The PWHT time and temperature affects the amount of carbon that diffuses towards the fusion boundary of the weld [1]. High carbon content at the fusion boundary will develop microstructures that are highly susceptible to HAC, particularly untempered martensite, high carbon austenite, and carbide precipitation [8]. The difference in carbon solubility between steel and Ni-base filler metal and the chemical concentration gradient lead to various microstructure evolutions during PWHT [9]. The effects of the PWHT are shown in Table 1, which consist of hardness maps and metallography of each corresponding PWHT. These figures show low angle micro sections (LAMS) of F22-IN625 DMWs. LAMS is a technique used to extend the fusion boundary 10 times more than normal cross section (NCS) techniques and allows the characterization of microstructural features that are difficult to observe using NCS techniques. It can be seen that for 2 h at 650 °C + 6 h at 760 °C PWHT, there are regions of high hardness near the IN625 and a carbon depleted region in the F22 steel. Fenske et al. performed a study on these regions of DMWs F22-IN625 and found that nanosized $M_7C_3$ carbides precipitated in these region causing high hardness measurements [5].

## *Delayed Hydrogen Cracking Test*

The delayed hydrogen cracking test (DHCT) is a testing method that has been developed at The Ohio State University. It allows for a test sample to

**Table 1** Correlation between PWHT and hardness

<table>
<tr><th>PWHT procedure</th><th>Metallography</th><th>Hardness map</th></tr>
<tr><td>2 h at 650 °C</td><td>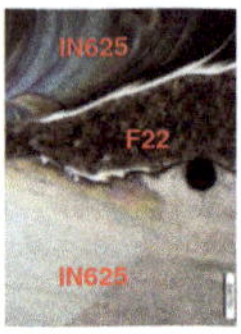
</td><td>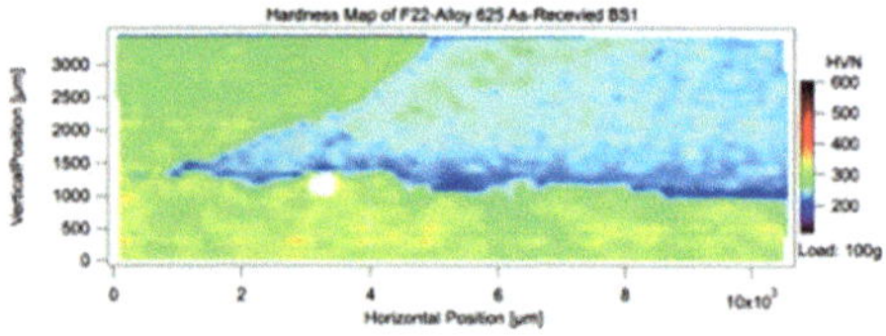
</td></tr>
<tr><td>4 h at 650 °C</td><td>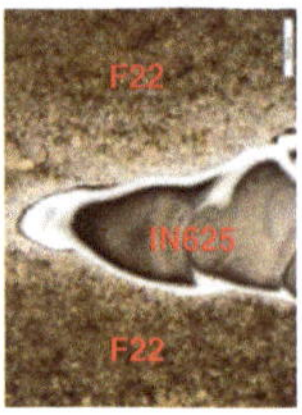
</td><td>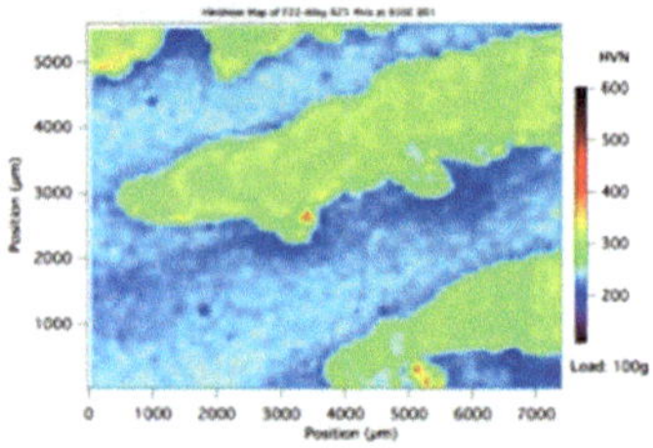
</td></tr>
<tr><td>2 h at 650 °C +<br>6 h at 760 °C</td><td>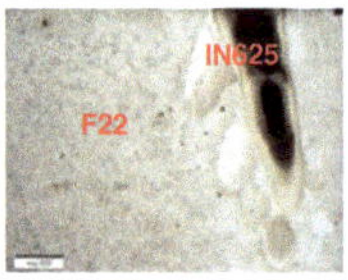
</td><td>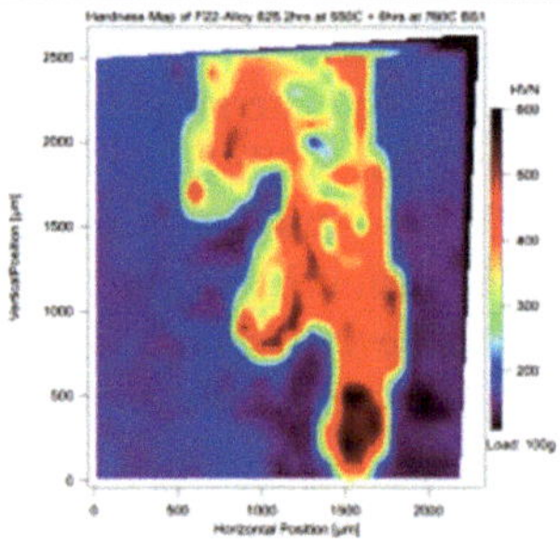
</td></tr>
</table>

simultaneously be charged with hydrogen while under a constant tensile load. For testing purposes a load of 90 % of the yield strength in F22 was applied to the test sample. This allowed for the mechanisms of HAC to cause failure and it also simulates the typical amount of residual stress that is experienced in DMWs. An electrolyte that consisted of 0.1 N $H_2SO_4$, 0.1 $Na_2S_2O_3$ (sodium thiosulfate) and 10 L of distilled water was used to charge the sample with hydrogen. Charging conditions for the DHCT setup was a current density of 10 mA/cm$^2$ to simulate active CP via an external current source. The samples that were tested contained the

**Table 2** Nominal composition of ASTM A182-F22 steel and IN625

| Material | Composition (wt%) | | | | | | | | |
|---|---|---|---|---|---|---|---|---|---|
| | C | Si | Mn | P | S | Cr | Mo | Fe | Ni |
| F22 | 0.05–0.15 | 0.50 | 0.30–0.60 | 0.040 | 0.040 | 2.00–2.50 | 0.87–1.13 | Bal | |
| IN625 | 0.10 | 0.50 | 0.50 | | | 20.0–23.0 | 8.0–10.0 | 5.0 | 58.0 |

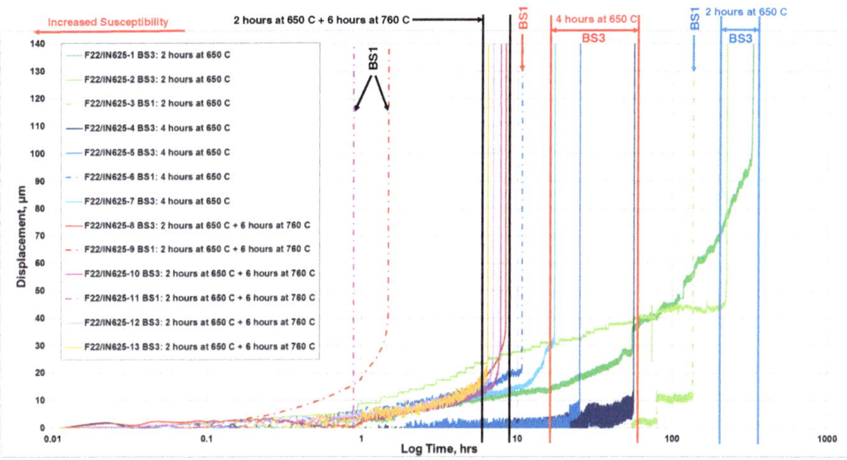


**Fig. 2** DHCT results on different welding and PWHT procedures for F22-IN625 DMWs

dissimilar weld interface of the F22-IN625 welding beads [3]. The nominal chemical compositions for each material are given in Table 2.

The DHCT has been used to successfully rank the susceptibility to HAC in DMWs under CP. This test has proven to be very sensitive to welding and PWHT procedure effects in F22-IN625 DMWs. The time-to-failure (TTF) results follow the trend that longer PWHT time and higher PWHT temperatures increase the TTF by one order of magnitude [3]. Furthermore, in terms of welding procedure, BS3 had longer TTF results in all PWHT categories relative to BS1. It has been observed that BS1 contains more planar growth region than in BS3 due to BS1 having more weld metal swirls [1, 3]. The planar growth region contains diluted region of the weld where the $A_1$ temperature is lower. When a PWHT is higher than the $A_1$ temperature of the diluted regions then untempered martensite will form on cooling from the PWHT temperature. It is because of this effect that BS1 is more susceptible than BS3 and this effect was observed during DHCT testing and metallurgical characterization. A full summary of the DHCT results are provided in Fig. 2.

## Experimental Approach

The objective of this study was to quantify the effect of the welding procedures (different levels of heat input and penetration depth), PWHT procedure (temperature and duration), and filler metal alloy composition on the distribution of carbon across the fusion boundary in F22-IN625 DMWs. The variables used in these simulations were the initial carbon composition in the IN625 filler metal, welding procedure (width of the transition zone), and PWHT procedure (varying times and temperatures). These diffusion simulations will provide a computational modeling approach for predicting the controlling factors that affect the susceptibility to HAC. Additionally, the data generated from these simulations can be used to optimize PWHT procedures to reduce and/or eliminate the microstructures that are highly susceptible to HAC.

## Results

### *Initial Modeling Conditions*

The computational modeling was carried out using Dictra™ version 24. The SSOL4 and MOB2 databases were used for prediction of thermodynamics and the diffusion of the elements, respectively. The elements that were included in the simulation were C, Cr, Fe, Ni, Mo, Si and Mn. Nb was left out of the simulations because of inconsistent and non-realistic results that occurred when it was included. Phases that were included were fcc, which corresponds to the IN625 weld metal and bcc, which correlates to the F22 steel. $M_{23}C_6$ and $M_7C_3$ were also included in the simulation to determine if the formation of stable carbides occurs the region of the fusion boundary. This effect will tie up carbon that forms microstructures highly susceptible to HAC. The Euler backwards method was used to integrate the partial differential equations associated with the simulations.

Actual DWMs contain a complex transition region as shown in Fig. 3. The width of the transition region in GTAW hot-wire DMWs is approximately 10–20 μm [1, 5]. It can be seen in the figure that there are various types of microstructures that can form during welding and after PWHT. The planar growth region is formed during welding due to a combination of low solidification growth rate and a steep temperature gradient [1]. When these conditions are not satisfied at short distance from the fusion boundary, the planar growth region breaks down to cellular and columnar dendritic solidification. Weld metal swirls form when the F22 steel is swept into the weld pool but not completely melted and mixed with the filler metal. This occurs since the F22 has a higher melting temperature than the IN625 filler

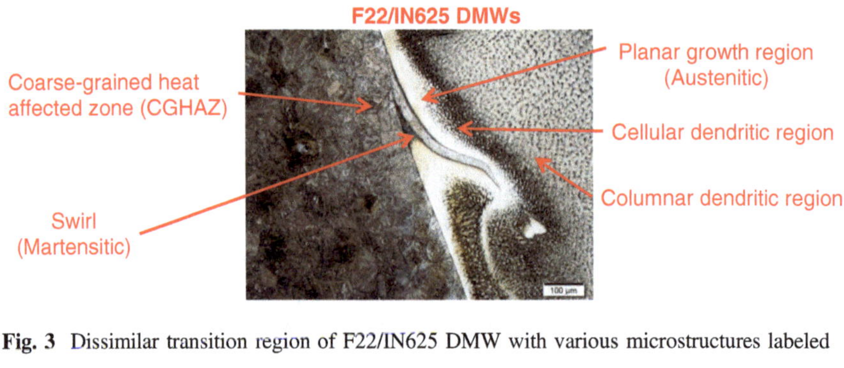


**Fig. 3** Dissimilar transition region of F22/IN625 DMW with various microstructures labeled

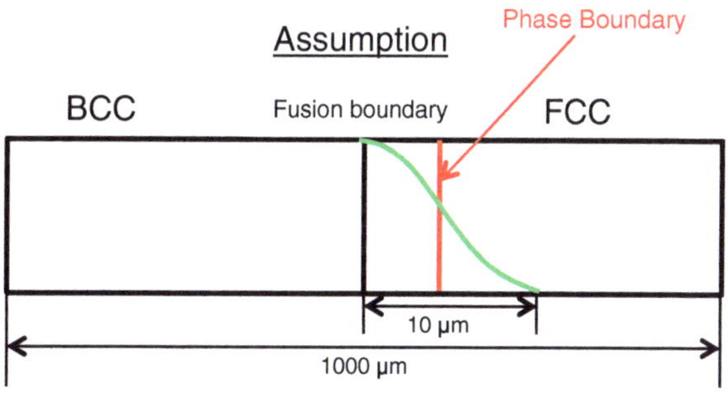


**Fig. 4** Assumed DMW transition region with composition gradient that was used for Dictra simulations

metal, 1510 °C (2750 °F) and 1350 °C (2460 °F), respectively. It is important to note that a planar growth region also forms around these swirls, similar to the fusion boundary. The coarse-grained heat affected zone (CGHAZ) forms by solid state reactions that are produced by the weld thermal cycle.

In order to simplify the simulation an assumed DMW joint was constructed in Dictra™ as shown in Fig. 4. This assumed DMW model was used because the bcc + fcc region cannot currently be modeled in Dictra™. However, the composition gradient was modeled in Dictra™ and closely resembles actual DMW joints. The difference between the actual and assumed model is that the assumed model incorporates a phase boundary that separates the bcc and fcc. The phase boundary will move based on the thermodynamic conditions. The phase boundary does not move in actual welds, but in order for a simulation to function properly in Dictra™ a phase boundary must be defined. Current work is being conducted to investigate the chemical composition of the bcc + fcc region so a three-phase model can be developed and utilized in future simulations.

## *Carbon Diffusion Parameters*

In order to understand the simulation results, carbon diffusion parameters such as diffusion coefficient, diffusion distance and chemical potential were calculated using Thermo-Calc™ and Dictra™. These parameters also give insight into how the simulation results are calculated and as shown in detail in Table 3.

Each of these parameters was calculated by varying the material, PWHT temperature and initial weight percent of carbon in the IN625 weld metal in the Thermo-Calc™ simulations. Chemical potential of carbon decreases in the F22 steel as the PWHT temperature increases. This is due to the increase of entropy of the system as the temperature increases [10]. The chemical potential of carbon for the IN625 weld metal exhibits a slight decrease as the temperature increases and there is minimal variation as the initial carbon concentration increases.

The diffusion distance of carbon in F22 increases from 3.94 to 8.79 μm as the temperature of the PWHT increases from 650 to 760 °C, respectively. This trend is

**Table 3** Carbon diffusion parameters

| Material | Temperature, °C | Weight % carbon in IN625 | Diffusion coefficient, $m^2/s$ | Diffusion distance for 10 h, μm | Chemical potential, J mol |
|---|---|---|---|---|---|
| F22 | 650 | – | $1.077 \times 10^{-10}$ | 3.94 | −7732 |
| | 760 | – | $5.307 \times 10^{-10}$ | 8.74 | −20,773 |
| IN625 | 650 | 0.055 | $4.211 \times 10^{-11}$ | 2.46 | −41,632 |
| | | 0.1 | | | −41,630 |
| | 760 | 0.055 | $3.602 \times 10^{-10}$ | 7.20 | −45,688 |
| | | 0.1 | | | −44,628 |

the same in the IN625 weld metal but slightly lower in magnitude, 2.46–7.20 μm as the temperature increases from 650 to 760 °C. Dictra™ predicts that the diffusion distance will be the same with the initial carbon concentration varying from 0.055 to 0.1 wt%. The diffusion coefficient increases with temperature in the F22 steel. However, it is slower in the IN625 weld metal but increases with temperature by an order of magnitude. Similar to the diffusion distance, Dictra™ predicts that the diffusion coefficient will not change when the carbon concentration is increased in the IN625 weld metal.

## *Effect of PWHT Hold Time*

The Dictra™ simulations that were performed replicated how the controlling factors of HAC in DMWs (PWHT procedure, welding procedure and base/filler metal combinations) interacted with each other and affected the carbon concentration at the fusion boundary. The controlling factors that were specified previously are not all of the controlling factors that cause HAC. Other factors include hydrogen concentration (governed by surrounding environment) and local stress/strain conditions. Figure 5 shows the Dictra™ simulation results for the effect of PWHT time and temperature on carbon distribution at the fusion boundary in F22-IN625 DMWs. For PWHT of 2 h at 650 °C, a maximum carbon concentration of 0.105 wt % is predicted at the phase boundary. The peak carbon concentration remains relatively constant as the time at a given PWHT temperature increases. For example, a PWHT of 4 h at 650 °C results in a carbon concentration at the phase boundary of 0.107 wt%.

## *Effect of PWHT Hold Temperature*

Temperature has a much greater effect on peak carbon concentration at the fusion boundary compared to the holding time. For the dual PWHT of 2 h at 650 °C + 6 h

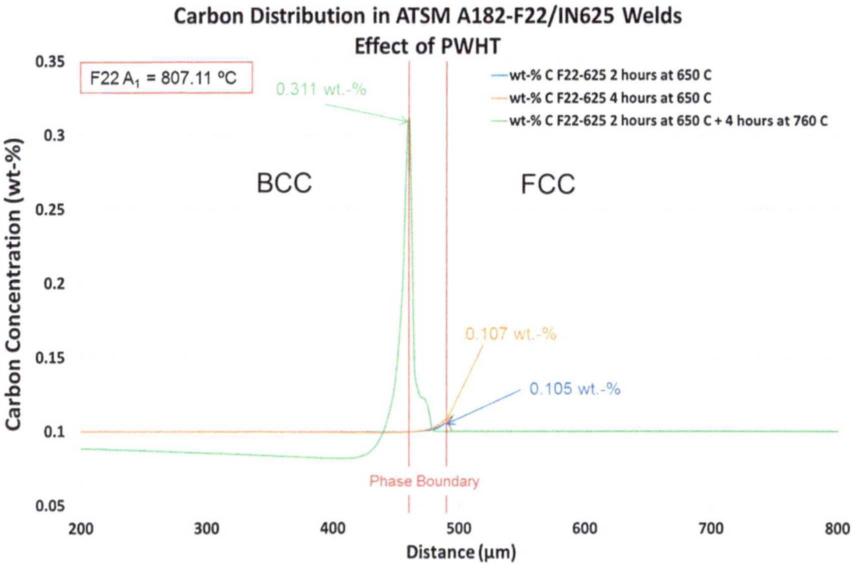


**Fig. 5** Dictra™ simulation results for how carbon distribution near the fusion boundary is affected by PWHT time and temperature

at 760 °C, the carbon concentration reaches 0.311 wt%. The magnitude of the carbon concentration at the phase boundary can be attributed to 760 °C being close to the calculated $A_1$ temperature of the F22 steel, which is 807 °C. In addition, the Timken handbook states that 0.311 wt% carbon is equivalent to 50 Rockwell C (513 HV, conversion is based on ASTM E140) [11, 12]. This prediction from Dictra™ is in good agreement with the hardness measurement values that are shown in Table 1. The 2 h at 650 °C + 6 h at 760 °C has a maximum hardness measurement of 597 HV0.1. Hardness measurements were recorded via a Leco microhardness tester LM 100AT. Testing conditions were 100 gram load, dwell time of 10 s and indent spacing of 100 μm based on ASTM E 92-82 [13].

Furthermore, these simulations show good correlation with the DHCT results. It can be seen in Fig. 2 that the DHCT samples that were PWHT for long times at high temperatures resulted in an increased susceptibility to HAC. Based on the characterization that was performed on these samples, microstructures that are highly susceptible to HAC formed at the fusion boundary due to the PWHT procedures. Dictra™ predicts a high carbon concentration at the fusion boundary and this increased level of carbon will lower the $A_1$ and $A_3$ temperatures [1]. If a PWHT temperature exceeds the $A_1$ temperature, then austenite will form at the hold temperature of the PWHT. As the DMW is cooled from the PWHT temperature untempered martensite will form. This microstructure is highly susceptible to HAC because they are sites where crack initiation occurs due to it low fracture toughness [5].

## *Width of Transition Region*

The other controlling factor that was varied in the Dictra™ simulations was the width of the dissimilar transition region. The transition region is the area that lays between the F22 steel HAZ and the IN625 weld metal. In effect, this represents a compositional transition from the F22 steel base metal to IN625 weld metal. The welding process influences the width of this region. For example, a high energy welding process such as submerged flux core arc welding (SFCAW) would create a larger transition region. However, a low heat input process such as pulse arc welding technique, would result in a small transition region. There are other variables that control the width of the transition region in a DMW such as solidification temperature range, temperature gradient in the transition region, thermophysical properties of the base and filler metals, and the magnitude of stirring in the weld pool [1]. But, due to the nature of ThermoCalc™ and Dictra™, the previously state variables were not tested in the simulations. The various transition regions that were tested were 0.1, 10, 35 and 100 μm. The 0.1 μm transition region would correspond to low heat input weld processes. Alexandrov et al. observed a range of 10–35 μm for the transition regions of gas tungsten arc welding (GTAW) hot-wire process [1]. A 100 μm transition region would correlates to a higher heat input process.

Based on the simulation results, as the width of the transition region increases the carbon concentration near the fusion boundary increases. Furthermore, when the transition region increases, a wider carbon depletion zone forms in the F22 steel. Both of these effects can be seen in Fig. 6 and are especially pronounced in the 35 μm transition simulations. It is also important to note that the 0.1 μm transition does not show any difference in peak carbon concentration at different times unlike

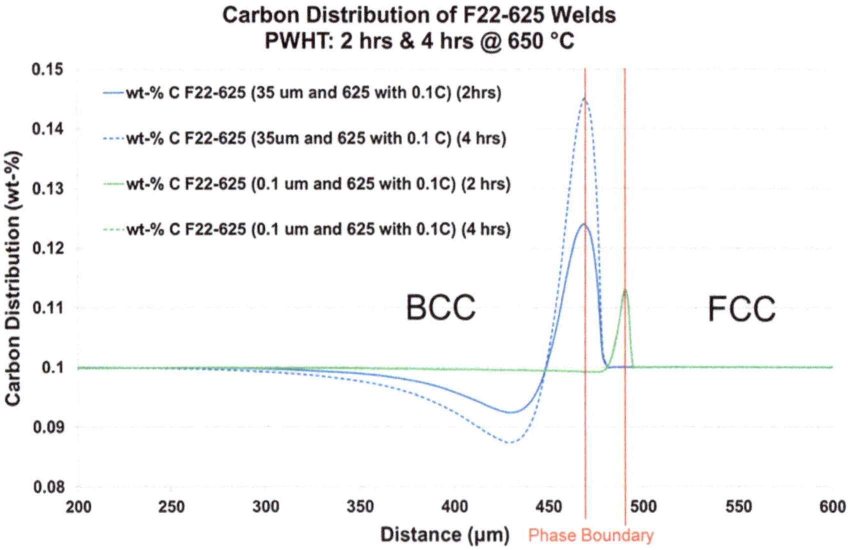


**Fig. 6** Dictra™ simulation results on effect of transition region width and PWHT time

the case for the 35 μm transition zone. It is assumed based on the simulated results that small transition regions do not have sufficient width for the carbon to diffuse into and accumulate.

### *Effect of Initial Carbon Concentration in IN625*

The initial carbon content in the IN625 weld metal was also simulated in Dictra™ at levels of 0.055 and 0.1 wt% carbon to represent the minimum and maximum carbon concentrations of IN625, respectively. Based on the simulation, as the initial carbon concentration decreases, an increase in carbon will occur at the fusion boundary. Also, as the initial carbon concentration decreases, there will be an increase in the carbon depletion in the HAZ of the F22 steel. Both of these effects can be seen in Fig. 7. This is caused by the decrease in chemical potential as the initial carbon content decreases. As the chemical potential decreases to lower values more carbon diffusion will occur from the F22 steel towards the IN625.

## Discussion

Simulated data shows that single-step PWHT with long duration and high temperatures will result in higher carbon concentration near the fusion boundary and more pronounced carbon depletion in the F22 steel HAZ. High carbon

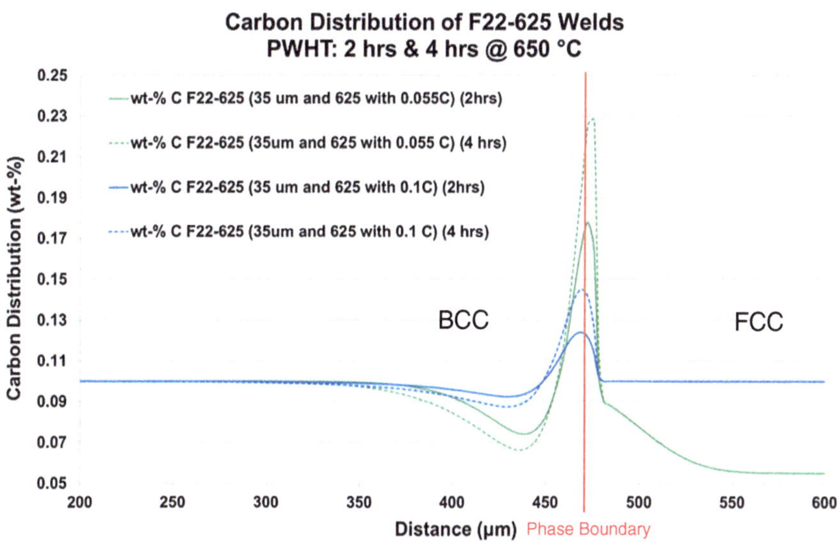


**Fig. 7** Dictra™ simulation results on effect of initial carbon concentration in IN625 weld and PWHT time

concentrations at the fusion boundary lead to microstructures that are highly susceptible to HAC such austenite that is retained at lower temperatures that later transforms to untempered martensite on cooling from the PWHT temperature. Carbon depletion in the CGHAZ of the F22 steel base metal results in a drop in hardness and local loss of strength in the steel in this region. This trend is observed in the experimental data as well; where high hardness measurement (>500 HV0.1) occurs in the transition region because of high PWHT temperatures for long durations (2 h at 650 °C + 6 h at 760 °C).

Previous tests that were performed using DHCT show that the HAC susceptibility increases as PWHT time and temperature increases. In addition, welding procedures that result in narrower or intermittent planar growth regions at the fusion boundary will have a lower HAC susceptibility. This was also confirmed by DHCT; BS3 welds contained less area of planar growth than BS1 [1]. By using Dictra$^{TM}$ simulations, carbon behavior near the fusion boundary was accurately predicted and these results can be used to model variation in PWHT and welding procedures.

Figure 8 illustrates a comprehensive summary of all of PWHT simulations of the different effects of welding procedure and initial carbon concentration in the weld metal. In simulations where the initial carbon concentration is 0.055 wt%, there is more carbon diffusion at longer PWHT times. This trend suggests that there is a larger effect of PWHT duration when the carbon concentration in IN625 is held to its minimum. As the welding procedure produces a larger transition region, the carbon concentration decreases.

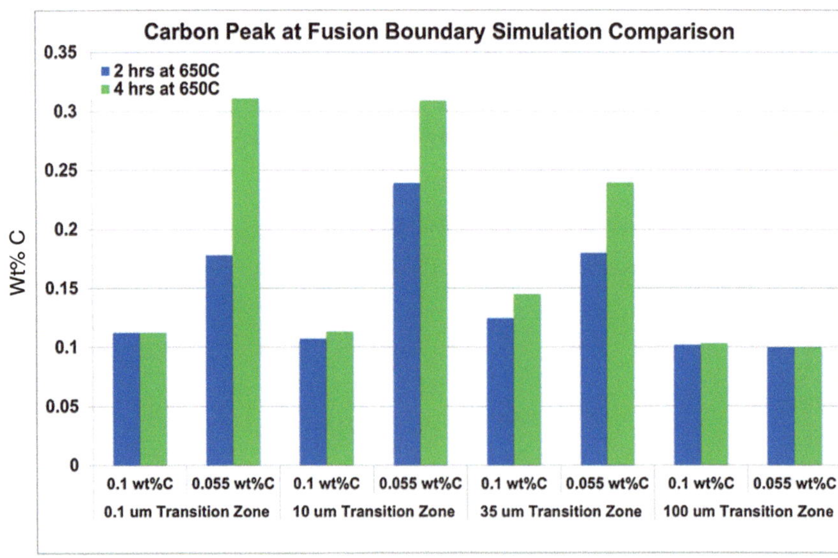


**Fig. 8** Summary of all the Dictra$^{TM}$ simulations compare with the carbon distribution near the fusion boundary

In the simulation where the initial carbon concentration of the IN625 weld metal was 0.1 wt%, there are minimal variations in carbon concentration near the fusion boundary when PWHT time is increased from 2 to 4 h. Furthermore, there are only small differences as the width of the transition region is increased. By comparing the initial carbon concentration simulation results against each other there is a large difference in carbon concentration near the fusion boundary. Simulations where the initial carbon concentration of the transition region was 0.055 wt% showed higher carbon concentrations near the fusion boundary than when the level was 0.1 wt% for transition zone widths of 0.1, 10 and 35 μm.

The carbon concentration at the fusion boundary shows no effect in wide transition regions when the PWHT time and initial carbon concentration in the IN625 weld metal are varied. However, only a few welding techniques, such as electroslag welding or submerged arc welding (SAW) with multiple torch-wire-feeder systems can achieve a 100 μm transition region. Therefore, in order to decrease the peak carbon concentration at the fusion boundary, the combination of a narrower transition region and higher carbon levels in the IN625 weld metal are warranted only for the F22-IN625 DMW combination, based on the results of these simulations. Further investigation is needed using simulations and experiments to find out how other DMW combination behave under the same simulated and experimental conditions. But by using the simulated results alone can provide a sound foundation for which metallurgical and microstructural factors should be varied in experiments.

## Conclusions

These simulations give insight to how the metallurgical and microstructure factors of HAC in DMWs affect carbon diffusion towards the fusion boundary. The results from the simulations ran in this study are in good agreement with the experimental results of the DHCT.

1. The width of the transition region can be controlled to some degree by the welding procedure. Large peak carbon concentrations near the fusion boundary and wide carbon depletion zones in the HAZ are associated with wider transition regions. Therefore, a welding process that can produce a small transition region should be chosen. This would ensure that a minimal amount of carbon could migrate towards the fusion boundary during PWHT.
2. Higher carbon concentrations in the weld metal will lead to lower peak carbon levels near the fusion boundary. Therefore, if the carbon level in IN625 is biased toward the maximum allowable value then the chemical potential between the F22 base metal and the IN625 weld metal would be high leading to a lower tendency for carbon in F22 to diffuse toward IN625.
3. If the PWHT were optimized to shorter times and lower temperatures, this will also lead to lower carbon concentrations near the fusion boundary.

4. Optimizing all of the metallurgical and microstructural factors will result in low carbon concentrations near the fusion boundary during PWHT and will lead to a decrease in the susceptibility of HAC in DMWs under CP. However, as stated before these factors are only a part of the complex number of controlling factors to HAC. Therefore, further investigation on how other controlling factors, such as hydrogen concentration, loading conditions, and local stress/strain parameters need to be considered.
5. Historically PWHTs at lower temperatures require longer times to achieve the desired microstructure and hardness. Two-step heat treatments are being studied and developed so that the first step (higher temperature) can temper the general HAZ and the second step (lower temperature) will temper the diluted region with lower $A_1$ temperatures.

**Acknowledgements** This work was supported by Exxon-Mobil through the Center for Integrative Materials Joining Science for Energy Applications (CIMJSEA) at The Ohio State University. The authors also acknowledge Thermo-Calc™ for assistance with developing the simulation models used in this study.

## References

1. Alexandrov, B.T., J.C. Lippold, J.W. Sowards, A.T. Hope, and D.R. Saltzmann, *Fusion Boundary Microstructure Evolution Associated with Embrittlement of Ni-base Alloy Overlays Applied to Carbon Steel.* 2013.
2. Fenske, J.A., M.W. Hukle, B.D. Newbury, J.R. Gordon, R. Noecker, and I.M. Robertson, *Hydorgen Induced Mechanical Property Behavior of Dissimilar Weld Metal Interfaces.* Omae2011: Proceedings of the Asme 30th International Conference on Ocean, Offshore and Arctic Engineering, Vol 3: Materials Technology2011, New York: Amer Soc Mechanical Engineers. 509–516.
3. MR0175, N.S., *Sulfide stress cracking resistant metallic materials for oilfield equipment.* Houston, TX: NACE, 2002.
4. Lundin, C., *Dissimilar metal welds-transition joints literature review.* Welding Journal, 1982. **61**(2): p. 58–63.
5. Fenske, J.A., I.M. Robertson, R. Ayer, M. Hukle, D. Lillig, and B. Newbury, *Microstructure and Hydrogen-Induced Failure Mechanisms in Fe and Ni Alloy Weldments.* Metallurgical and Materials Transactions A, 2012. **43**(9): p. 3011–3022.
6. Burk, J.D. and C.L. Ribardo. *SS: Thunder Horse and Atlantis Deepwater Frontier Developments in the Gulf of Mexico: Thunder Horse Materials Welding and Corrosion Challenges and Solutions.* in *Offshore Technology Conference.* 2010. Offshore Technology Conference.
7. ASTM A182/A182 M-14a, Standard Specification for Forged or Rolled Alloy and Stainless Steel Pipe Flanges, Forged Fittings, and Valves and Parts for High-Temperature Service, Section ASTM International, West Conshohocken, PA, 2014, http://www.astm.org, 10.1520/A0182_A0182M-14A
8. Dodge, M.F., H.B. Dong, M. Milititsky, R.P. Barnett, V.F. Marques, and M.F. Gittos, *Environment-Induced Cracking in Weld Joints in Subsea Oil and Gas Systems: Part I.* Proceedings of the ASME 31st International Conference on Ocean, Offshore and Arctic Engineering–2012, 2012. **3**: p. 305–313.

9. DuPont, J., *Microstructural evolution and high temperature failure of ferritic to austenitic dissimilar welds.* International Materials Reviews, 2012. **57**(4): p. 208–234.
10. Porter, D.A. and K.E. Easterling, *Phase Transformations in Metals and Alloys, (Revised Reprint)*1992: CRC press.
11. Timken, *Practical Data for Metallurgists.* 2011.
12. ASTM E140-12be1, Standard Hardness Conversion Tables for Metal Relationship Among Brinell Hardness, Vickers Hardness, Rockwell Hardness, Superficial Hardness, Knoop Hardness, Scleroscope Hardness, and Leeb Hardness, Section ASTM International, West Conshohocken, PA, 2012, http://www.astm.org, 10.1520/E0140
13. ASTM E92-82(2003)e2, Standard Test Method for Vickers Harndess of Metallic Materials, Section ASTM International, West Conshohocken, PA, 2010, http://www.astm.org, 10.1520/E0092-82R03E02

# Hydrogen Trapping in Supermartensitic Stainless Steel TIG Welds

**Th. Boellinghaus and D. Eliezer**

**Abstract** A number of common defects in stainless steel welding are the result of the presence of hydrogen in the weld. In addition, the service life of the stainless steel joints is significantly dependent on the presence of hydrogen in the respective environment and the susceptibility of the various weld microstructures to hydrogen degradation. Hydrogen's effects on various Tungsten Inert Gas (TIG) welded SMSS microstructures are investigated by means of X-ray diffraction (XRD) and optic (OM) and electron microscopy (SEM). A number of methods for estimating the amount of absorbed hydrogen have been employed. Hydrogen interaction with structural defects and the characteristics of hydrogen desorption have been studied by means of thermal desorption spectroscopy (TDS). The effects of the respective microstructure on hydrogen absorption and desorption behavior is discussed in detail. The powerful abilities of TDS for studying the absorption/desorption behavior and trapping effects in supermartensitic stainless steel TIG welds is also examined.

## Introduction

Supermartensitic stainless steels are essentially low carbon iron based alloys which contain chromium, nickel and molybdenum. They contain about the same amount of chromium as conventional martensitic stainless steels of the AISI 410 and 420 series, but their carbon content has been reduced. The addition of up to 6.5 wt% nickel substantially decreases the martensite transformation towards lower temperatures and thus, the weld microstructures might contain considerable amounts of retained and annealing austenite. With respect to fitness for purpose selection three

Th. Boellinghaus
Federal Institute for Materials Research and Testing (BAM), 12205 Berlin, Germany

D. Eliezer (✉)
Ben-Gurion University of the Negev, 84105 Beer-Sheva, Israel
e-mail: dan.eliezer@gmail.com



Th. Boellinghaus et al. (eds.), *Cracking Phenomena in Welds IV*,
DOI 10.1007/978-3-319-28434-7_20

different levels of nickel and molybdenum are commercially available. Lean alloys are intended to be used in sweet service and contain nickel below 2 wt% and molybdenum below 1 wt%. Fabricated as low carbon (LC) steels with carbon content below 0.02 wt% and with nickel and molybdenum amounts between 2 and 4 wt% and about 1 wt%, respectively, medium alloys have been developed

for intermediate sour service applications. At increasing $H_2S$ levels and decreasing pH extra low carbon (ELC) fat grades are applied. These materials contain even less carbon, i.e. under 0.01 wt%, but more nickel up to 6.5 wt% and also more molybdenum up to 2.5 wt% [1–4]. The base materials are usually heat treated in several stages and can consist of considerable portions of annealing and retained austenite dependent on the nickel content [2].

Supermartensitic alloys have been commercially available since the early 90s. They have increasingly been applied to many industries. In the oil and gas industry, for example, it is exclusively used. For economic reasons selection of materials which fit exactly to the intended applications becomes increasingly important for the oil and gas industry. For example, supermartensitic stainless steels represent an attractive cost reducing alternative to higher alloyed and more expensive materials, and are increasingly applied as flowline materials at several North Sea oil and gas fields. At these locations, welded components are subjected to sour service conditions, providing a potential risk for hydrogen uptake, as well as degradation and cracking of supermartensitic stainless steels.

Another potential failure risk for welded SMSS is Hydrogen Assisted Cold Cracking (HACC), and it must be avoided during welding. This failure type also often occurs after welding has been completed and, although extensive, may be difficult to detect [5]. Even small hydrogen concentrations trapped inside the material might lead to failure. Consequently, a comprehensive knowledge of hydrogen-trapping interactions is necessary to make any decision and/or judgment whether a trap site, or a particular trapped hydrogen content, is detrimental for safe service operation of welded joints of such materials.

In this paper it will be shown how the thermal desorption spectroscopy (TDS) applications in quantitative studies of hydrogen trapping and release behavior can help us better understand hydrogen trapping in supermartensitic stainless steel TIG welds. The physical nature of hydrogen trapping and the interactions between a hydrogen atom and a trap site are briefly reviewed. The theoretical overview is followed by experimental examples of TDS applications on supermartensitic stainless steels. The paper also addresses recent TDS results obtained from the study of hydrogen evolution from supermartensitic stainless steels.

Eliezer [18, 19] uses a model accounting for diffusion and trapping based on McNabb and Foster's work [20]. Trapping was accounted for at the end of each time-step in solving the diffusion equation by calculating the equilibrium between lattice and trap sites. Eliezer and Bollinghaus have applied TDS technology in many structural systems and the advantages were described in detail in a number of papers [23–28].

## *Various Trap–Hydrogen Interactions*

At long range relative to the atomic spacing, a hydrogen atom interacts with the defects in crystals through its elastic strain field. At short range, it interacts chemically through localized bonding with matrix atoms [21]. Due to the difficulty in making a distinction among the various chemical interactions, the local interactions are described phenomenologically in terms of a binding energy to the defect, and usually only one trap energy level is assigned to the defect.

If the binding energy, $E_B$, is small the corresponding traps are referred to as reversible traps and can act as either hydrogen sinks (which capture the hydrogen atoms from weaker traps), or as hydrogen sources (which deposit hydrogen atoms to stronger traps). On the other hand, large $E_B$ give rise to irreversible traps which normally will not release hydrogen. The binding energy of 60 kJ/mol H for an edge dislocation or a grain boundary is generally regarded as the upper limiting value for a reversible trap [22].

The objective of the present study was to investigate the hydrogen interactions of GTA welded supermartensitic stainless steel microstructures in detail, with a substantial focus on the adsorption and desorption characteristics as well as the hydrogen trapping mechanism.

## Experimental Approach

The chemical composition of the tested supermartensitic stainless steel is listed in Table 1. In order to investigate the hydrogen effects on the microstructure as well as on the absorption and desorption behavior of hydrogen in supermartensitic stainless steel, the alloy was exposed to hydrogen in three different modes: electrochemical hydrogenation, gaseous-phase hydrogenation and hydrogen introduced by TIG welding via a mixed Ar + $H_2$ shielding gas.

Hydrogen charging was carried out electrochemically (cathodic charging) at room temperature. The cathodic charging technique generates H+ on the sample's surface. A power supply (galvanostat) impresses a constant current on the electrolytic cell, such that the specimen is cathodic relative to an inert electrode such as platinum (Pt). The anode, a platinum wire, is located symmetrically to the cathodic polarized specimen, thus distributing the potential uniformly around the specimen. The specimens were cathodically hydrogenated at identical charging times, using a current density of 0.5 $mA/mm^2$. The charging was carried out in 1 N $D_2SO_4$ (deuterosulphuric acid) electrolyte + 0.25 g $NaAsO_2$ (sodium arsenide).

**Table 1** Measured chemical composition of the SMSS (wt%)

| C | S | P | Mn | Si | Ni | Cr | Mo | N |
|---|---|---|---|---|---|---|---|---|
| 0.06 | 0.0009 | 0.021 | 1.870 | 0.294 | 6.498 | 11.65 | 2.330 | 0.009 |

Gaseous phase charging was carried out in a $H_2$ UHP 99.99 atmosphere using a pressure of 1 atm during 6.5 h at room temperature and 650 °C.

Charging the weld metal with hydrogen was performed by Tungsten Inert Gas (TIG) using the parameters as summarized in Table 2.

Prior to electrochemical and gas phase hydrogen exposure, samples were cut from the SMSS plates and were mechanically polished up to 0.05 μ. The total content of hydrogen/deuterium absorbed in the alloy was measured by a LECO RH-404 hydrogen determinator system, using the test protocol described elsewhere [6]. Hydrogen/Deuterium evolution and trapping characteristics were studied by TDS, a technique that involves measurement of the desorption rate of gas atoms, soluted or trapped in the material, while heating the sample at a known rate [7, 8]. After ultrasonic cleaning in ethanol, the deuterium charged specimens were placed in the specimen holder; the system was sealed and pumped down to 10 μPa. The heating rates (ramps) was 3 K $min^{-1}$, and the temperature range was 25–600 °C, with both parameters being programmed into a temperature controller. While heating, the mass spectrometer was placed in a continuous mode for scanning atomic masses in the range of 3.5–4.5 amu.

Specimens for optical microscopy were prepared by mechanical polishing through 0.05-μm alumina. The samples were finally etched using one of the etchings:

1. 5 g Cupric chloride in 100 cc Hydrochloric acid + 100 cc Ethyl alcohol + 100 cc Water.
2. 10 g Ferric chloride in 20 cc Hydrochloric acid + 4 cc Nitric acid.

The use of different etchants was necessary to highlight specific features and phases and thus obtain complementary information on microstructures. Digital photographs were taken using a Leica DMR digital camera. The specimen's surface was observed using a JEOL JSM 5600 scanning electron microscope (SEM), operated at an accelerating voltage of 15 kV.

**Table 2** Welding parameters for SMSS

| Shielding gas type | Shielding gas (l/min) | Current (A) | Voltage (V) | Weld speed (mm/s) |
|---|---|---|---|---|
| Argon + 0 % hydrogen | 10 | 145 | 10 | 5 |
| Argon + 2 % hydrogen | 10 | 145 | 11.5 | 5.3 |
| Argon + 5 % hydrogen | 10 | 145 | 11.5 | 5.3 |
| Argon + 7.5 % hydrogen | 10 | 145 | 11.5 | 6.5 |

The microstructure and phase characterization of the above specimens were studied by means of X-Ray diffraction (XRD) analyses, using a Rigaku Type 2000 X-ray diffractometer with a CuKα radiation ($\lambda$ = 1.54059 Å) and graphite monochromator on diffracted beam. The data was collected in an angular range of $41° < 2\theta < 54°$, $61° < 2\theta < 94°$ by steps of 0.05° ($2\theta$) with a constant counting time of 4 s by step.

# Results and Discussion

## Hydrogen Absorption

The amount of hydrogen absorbed in the alloy in the as-received state and after exposure to hydrogen was measured by means of LECO analysis. The results are summarized in Table 3. The hydrogen contents in different welding areas of the TIG welded specimens welded with gas shield atmospheres containing different hydrogen percentages are presented in Fig. 1.

From Table 3, it can be seen that the different ways by which the hydrogen was introduced to the alloy significantly influences the amount of absorbed hydrogen.

It has to be emphasized that already the base material contains about 4 ppm in the as- delivered condition. It can generally be stated that the higher the fugacity of hydrogen in the atmosphere, the larger hydrogen concentrations are absorbed in the alloy. The hydrogen content absorbed in electrochemical charged specimen is the highest and differs about two orders of magnitude from the as-received specimen. This result can be attributed to high hydrogen concentrations produced on the sample surface suppressing most of the recombination during electrochemical charging, as reported in former studies [2]. The distribution of hydrogen in the volume is dependent on the hydrogen charging technique, i.e. if hydrogen is introduced from the surface as during electrolytic or gaseous-phase charging or is introduced by a welding arc and homogenously distributed, e.g., weld metal. The

**Table 3** Hydrogen content in SMSS alloy

| Sample | Absorbed hydrogen (ppm) |
|---|---|
| As received | 4.06 ± 0.28 |
| Charged for 24 h with deuterium | 165 ± 11.55 |
| [a]Gas phase hydrogenation | 2.16 ± 0.11 |
| [b]Gas phase hydrogenation | 1.87 ± 0.15 |
| TIG welded Ar 0 % $H_2$ | 6.7 ± 0.34 |
| TIG welded Ar 2 % $H_2$ | 6.2 ± 0.42 |
| TIG welded Ar 5 % $H_2$ | 5.83 ± 0.25 |
| TIG welded Ar 7.5 % $H_2$ | 8.64 ± 0.60 |

[a](R.T., 6.5 h, 1 atm)
[b](650 °C, 6.5 h, 1 atm)

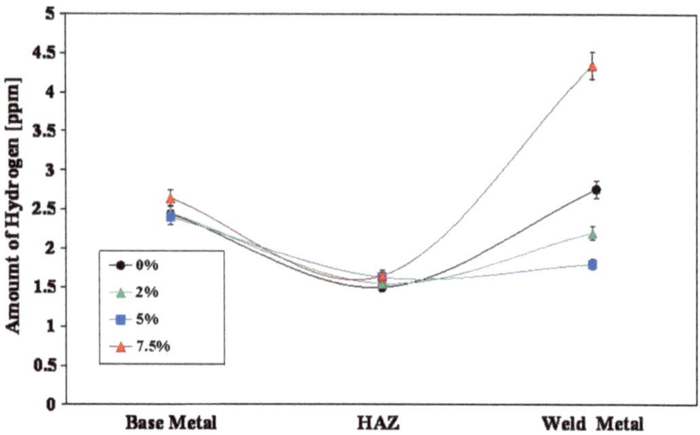


**Fig. 1** Hydrogen contents in different regions of SMSS TIG weld in gas shield atmospheres containing different hydrogen percentages

internal hydrogen distribution is also significantly dependent on the various weld microstructures due to their different diffusivities and solubilities [2, 9].

It has to be emphasized that the hydrogen amount absorbed during gas-phase hydrogenation is considerably low and even lower than in the as delivered base material. This is predominantly attributed to the prior degassing in the process. However, it can tentatively be concluded from this result which shows some consistency to a similar experimental study with carbon mild steels [10, 11] that at low pressures hydrogen will not enter the material from a gaseous phase even at elevated temperatures of 650 °C, as long as the material is not considerably plastically deformed under a static or dynamic mechanical load.

From Fig. 2, it can be seen that the different weld microstructures significantly influence the amount of hydrogen absorbed in the specimen due to their different solubility. The lowest concentrations have sometimes been measured in the heat affected zone (HAZ) which has to be attributed to the heat introduced into the HAZ during welding and allowing a considerable amount of hydrogen to desorption. However, this result stands in contrast to a recent investigation of the HAZ of multilayer TIG welded SMSS of similar chemical composition where the highest or at least similarly high hydrogen concentrations have been found in the HAZ as compared to the weld metal, as shown in Fig. 2. Such high hydrogen levels in the HAZ have probably to be attributed to a high fraction of retained or annealing austenite in the HAZ of about 35 %.

Such annealing and austenite forming effects are not present in the mono-layer weld investigated in this study and thus, less hydrogen is accumulated in the HAZ. Also, in contrast to the previous study with multilayer welds [2], somewhat higher hydrogen concentrations than in the HAZ have been found in the weld metal of the mono-layer weld investigated here. The hydrogen content in the weld metal exceeds that of the base material only if very high hydrogen percentages are added in the shielding gas, i.e., at a fraction of 7.5 % in argon.

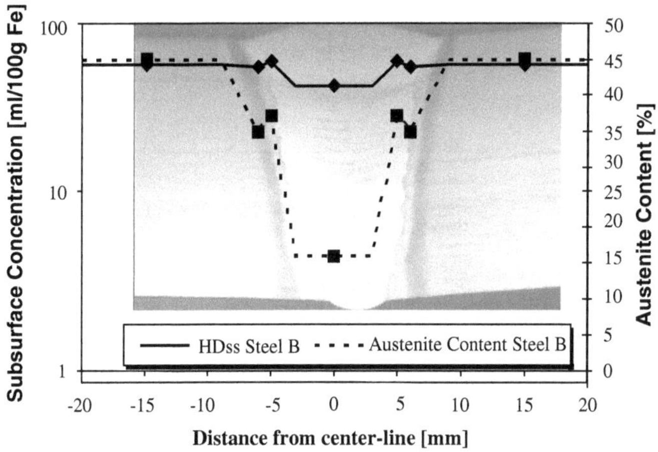


**Fig. 2** Solubility evaluated by the sub-surface concentration of hydrogen dependent on the austenite content in the weld microstructures of a similarly nickel and molybdenum alloyed SMSS multilayer steel weld [2]

## *Hydrogen's Effects on Microstructure*

The exposure parameters extensively influence the amount, quality and depth of hydrogen absorption. This means, as a consequence of the hydrogenation process, different microstructural morphologies might be obtained. Also, the respective phase transformations are dependent on the way and the amount of hydrogen introduced into the material. For instance, hydrogen represents an interstitial and might shift the martensite formation towards lower temperatures during cooling after welding. In this regard, the manner in which hydrogen ingresses into the material might also influence respective cracking at a later time [5].

The initial microstructure and XRD pattern of the studied material are shown in Figs. 3 and 4. The phase analysis of SMSS XRD patterns reveals the existence of two phases BCC-martensite and FCC-austenite. Cathodic charging of deuterium into SMSS induced microstructural changes, as it becomes evident by the results shown in Fig. 4. The $\gamma$ phase reflections exhibit significant broadening, but the fcc lattice parameter remains unchanged.

The $\alpha$ phase reflections, on the other hand, exhibit negligible broadening and the bcc lattice parameter contracts by approximately 0.8 %. This result can be attributed to the fact that hydrogen has very different diffusion rates in austenitic and in martensitic microstructures. The rate is higher in body centered cubic lattices, and remains high even at low temperatures. Therefore, the retained and continuous $\gamma$ phase of the SMSS alloy offers a blocking site for hydrogen diffusion [12].

Despite the fact that hydrogen caused significant strain in the material, optical and electronical microscopy did not find any evidence for micro cracking regardless of the hydrogenation mode. This result stands in contrast to Pressouyre [12], who

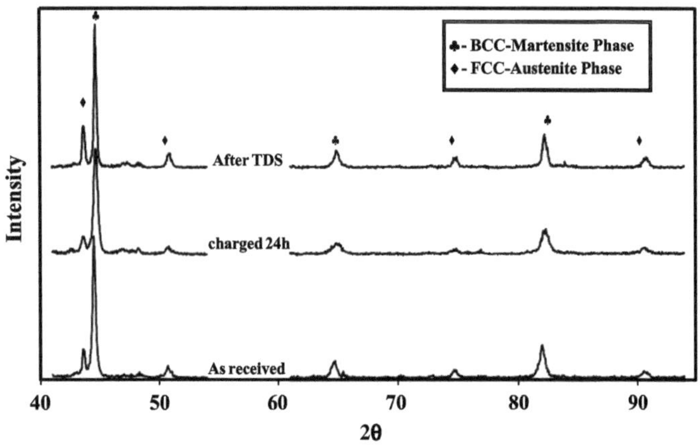


**Fig. 3** X-ray diffraction patterns of electrochemical charged SMSS alloy with deuterium for 24 h

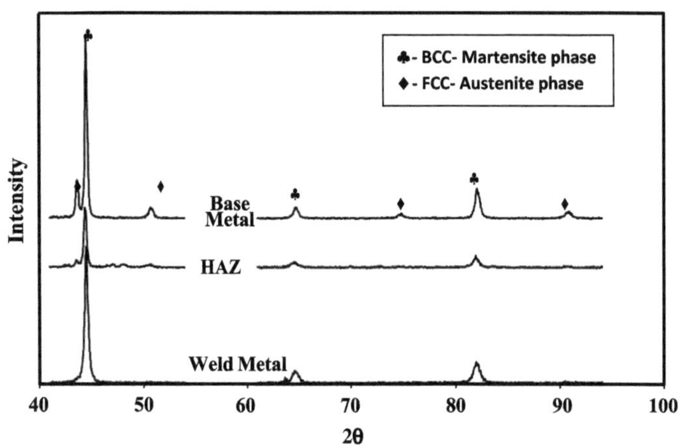


**Fig. 4** X-ray diffraction patterns of TIG welds in gas shield atmospheres containing Ar + 7.5 % $H_2$, from different welding areas

indicated that cracking will appear in the metal if the hydrogen concentration exceeds a certain critical limit. This means that in SMSS weld microstructures hydrogen has assisting influence on cracking, but might not originate cracking.

The microstructure and XRD pattern of the different weld microstructures are shown in the Figs. 4 and 5. The phase analysis of SMSS XRD patterns reveals the existence of two phases BCC-martensite and FCC-austenite according to the WRC 1992 Diagram [13]. As to be expected from such a mono-layer weld, the base material contains a larger amount of retained or annealing austenite in comparison to the HAZ area, while the weld metal completely transformed into martensite.

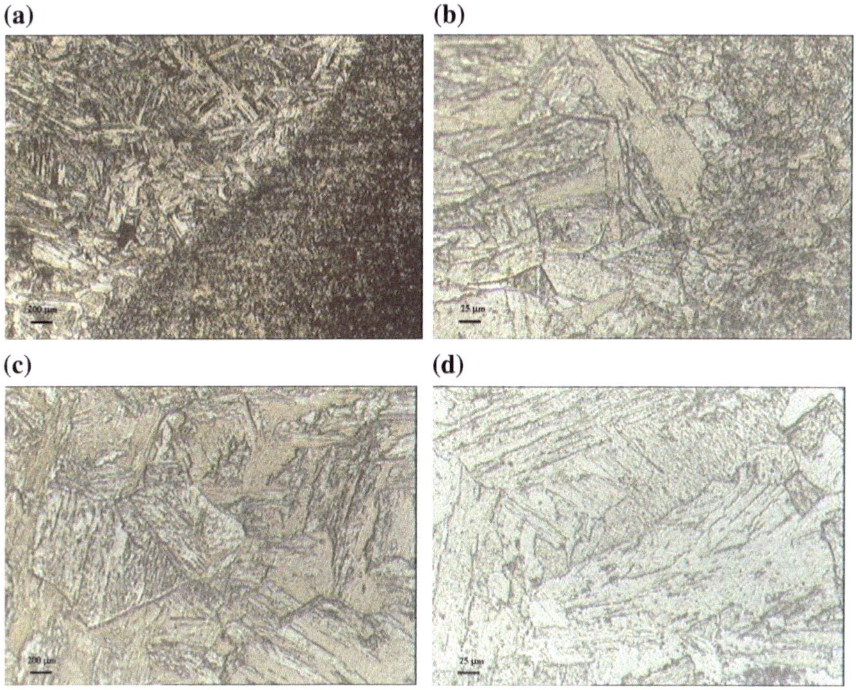

**Fig. 5** Optical micrographs of the SMSS TIG weld microstructures shielded with Ar + 7.5 % $H_2$. **a** Weld metal and HAZ at the fusionline, **b** HAZ, **c** Weld metal, **d** Weld metal

Figure 6 exhibits X-ray the diffraction patterns of TIG welded SMSS alloy, from the weld metal, at different hydrogen percentages in the atmosphere. As the hydrogen percent in the atmosphere increases, more hydrogen is introduced into the completely martensitic lattice and consequently, the α-phase reflections exhibits significant broadening.

Some results of XRD patterns after gas-phase hydrogenation are shown in Fig. 7. Obviously, the soluted hydrogen concentrations at such charging conditions are too small to cause any broadening of γ and α phase reflections, i.e. no lattice parameter expansion occurs at such conditions.

## *Desorption Characteristics*

In contrast to conventional hydrogen measurement and extraction technologies used for welding applications, like the mercury procedure or carrier gas hot extraction, the TDS technique represents a far more suitable, reliable and expressive way to characterize the hydrogen evolution process and to assess hydrogen trapping characteristics in a microstructure. It has thus been utilized within this study.

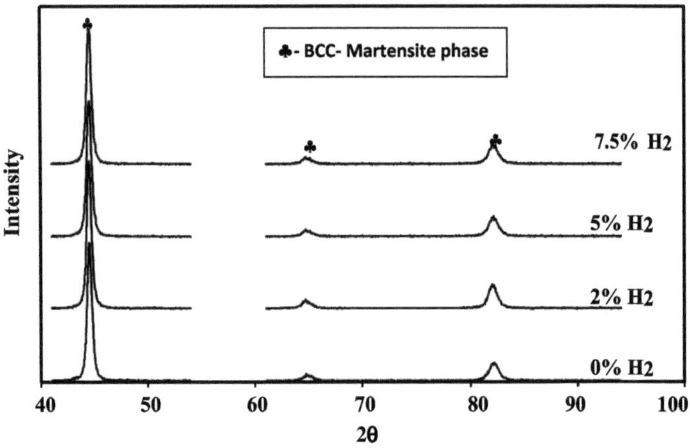


**Fig. 6** X-ray diffraction patterns of the TIG welded SMSS at different hydrogen percentages in the atmosphere

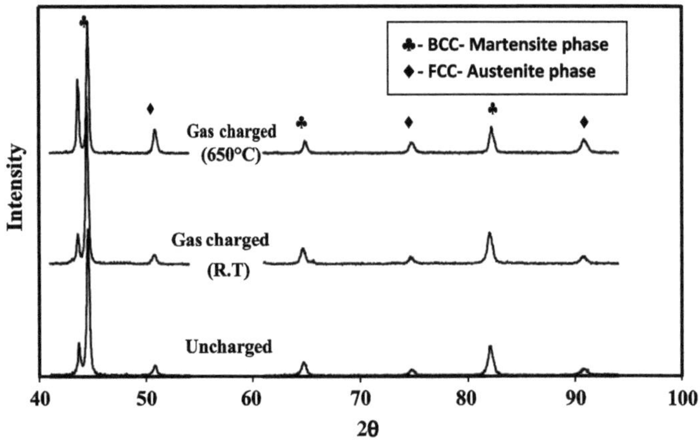


**Fig. 7** X-ray diffraction patterns (1 atm, 6.5 h) of SMSS after of gas-phase hydrogenation, charged at room temperature and 650 °C

The spectra analyses were supported by data obtained from a variety of other experimental techniques, such as LECO hydrogen quantity analyses, XRD and microstructure investigation by means of optic and electron microscopy [6, 7].

The trapping phenomena in steels are not well understood and the connection with hydrogen assisted cracking is not clear. For instance, up until this point, it has not been clarified as to whether hydrogen deeply trapped at a Titanium Carbide might not be (re-)activated during straining the microstructure and thus, contributes to cracking. Furthermore, it has been shown that the hydrogen solubility increases and its diffusivity decreases with increased chromium carbide precipitation in SMSS [14], and it can be anticipated, but it is not evident, that such hydrogen accumulations contribute to intergranular cracking [1]. However, it is difficult to obtain unambiguous experimental information on the meaning of traps for cracking, partly because of the small hydrogen concentration involved. However, there is general agreement in the literature that the lattice diffusivity of hydrogen in steel is strongly retarded by the presence of microstructural inhomogeneities including grain boundaries, dislocations, carbides, and nonmetallic particles, all acting as potential H trapping sites [15–17].

Characteristics of hydrogen desorption from the SMSS specimens are presented in the Figs. 8 and 9. The calculated results from these TDS graphs are summarized in the Tables 4 and 5. For all of the SMSS specimens subjected to hydrogen in the as-received, TIG welded and gas phase hydrogenated condition, the TDS spectra exhibit a common feature that initial degassing occurs until the vacuum reaches $10^{-7}$ to $10^{-8}$ mbar. In addition, all the spectra exhibited a first peak at relatively low

**Table 4** Parameters of thermal desorption analysis from TIG welded SMSS at a heating rate of 3 °C/min

| Specimen | Peak # | Temp. at desorption peak (°C) | Maximal desorption rate ($\times 10^{14}$ H-atoms/g s) | Half height peak width (°C) | Calculated activation energy (kJ) |
|---|---|---|---|---|---|
| As received | | | | | |
| As received | I | 30 | 11.3 | 50.78 | 4.2 |
| | II | 92.34 | 3.39 | 37.50 | 5.1 |
| | III | 130 | 2.80 | 25.78 | 6.5 |
| Ar + 0 % $H_2$ | | | | | |
| Base metal | I | 32.3 | 7.90 | 51.23 | 4.56 |
| | II | 88.11 | 3.19 | 30.20 | 5.72 |
| | III | 121 | 2.81 | 31.50 | 6.65 |
| | IV | 143.85 | 2.63 | 28.43 | 6.68 |
| Weld metal | I | 30.94 | 8.74 | 50.31 | 4.54 |
| | II | 82.34 | 3.40 | 28.59 | 5.70 |
| | III | 110.94 | 2.94 | 27.97 | 6.62 |
| | IV | 162.19 | 2.57 | 202.50 | 6.70 |
| Ar + 7.5 % $H_2$ | | | | | |
| Base metal | I | 32.5 | 7.91 | 42.34 | 4.56 |
| | II | 89.38 | 3.19 | 30.00 | 5.72 |
| | III | 120 | 2.72 | 37.50 | 6.66 |
| | IV | 143.75 | 2.57 | 28.59 | 6.68 |
| Weld metal | I | 34.84 | 9.33 | 47.03 | 4.57 |
| | II | 92.5 | 3.56 | 36.56 | 5.73 |
| | III | 117.5 | 3.24 | 27.19 | 6.63 |
| | IV | 172.34 | 3.42 | 108.91 | 6.70 |
| | V | 302.34 | 3.65 | 137.19 | 9.10 |

**Table 5** Parameters of thermal desorption analysis from SMSS electrochemical charged with deuterium for 24 h at a heating rate of 3 °C/min

| Peak # | Temp. at desorption peak (°C) | Maximal desorption rate ($\times 10^{14}$ D-atoms/g s) | Half height peak width (°C) |
|---|---|---|---|
| As-received charged with deuterium for 24 h | | | |
| I | 77.5 | 112 | 38.44 |
| II | 155 | 167 | 172.50 |
| Weld metal of TIG welded in Ar + 7.5 % $H_2$ charged with deuterium for 24 h | | | |
| I | 80 | 128 | 40.63 |
| II | 152.34 | 195 | 148.13 |

temperatures, i.e. around room temperature, indicating a larger loss of hydrogen by the height. These peaks are probably associated with hydrogen escaping from the specimen's surface at a relatively low activation energy of approximately 4.5 kJ.

A second evolution peak occurred in all spectra over a wide temperature range from about 80 °C up to 200 °C with activation energy of approximately 5.7–6.7 kJ.

The spectra for the as- received steel (Fig. 8a), 1 atm gaseous-charged material (Fig. 8b), weld metal and the base material produced without hydrogen in the shielding gas (Fig. 9a) and base material of the weld shielded with 7.5 vol.% hydrogen in the argon atmosphere (Fig. 9b), have nearly an identical shape. As to

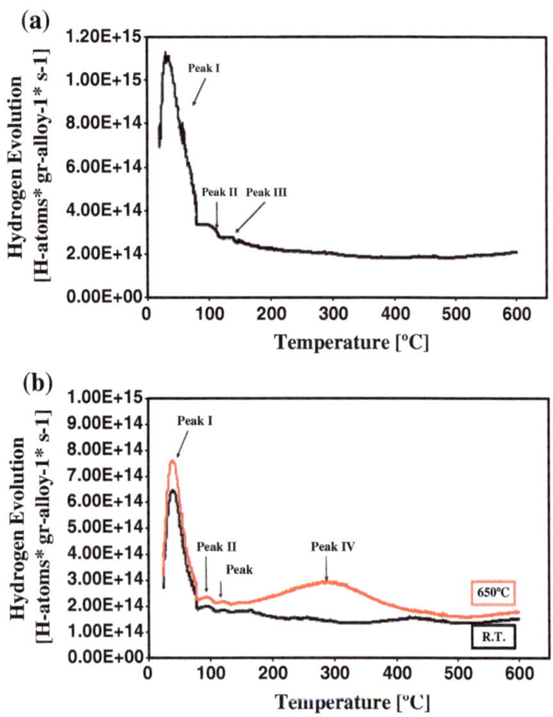


**Fig. 8** TDS spectrum of SMSS at different charging temperatures and at a heating rate of 3 °C/min

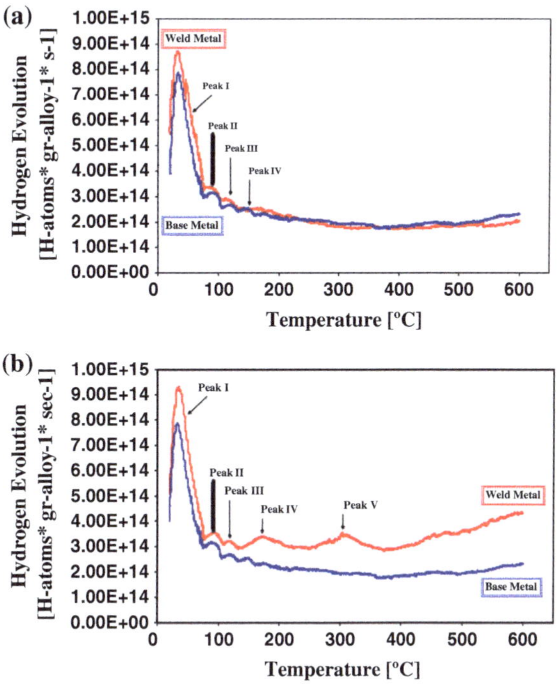


**Fig. 9** TDS spectrum for different SMSS weld microstructures of alloy, TIG welded in different shielding gas atmospheres at a heating rate of 3 °C/min. **a** Ar + 0 % $H_2$, **b** Ar + 7.5 % $H_2$

be expected, no additional hydrogen was taken up in the material during welding without hydrogen. Also, no hydrogen was traced in the base material during welding with 7.5 vol.% in the shielding gas. More important, nearly no additional hydrogen was taken up during charging at room-temperature in a gaseous hydrogen environment.

However, for the 7.5 vol.% hydrogen shielded weld metal and after gas-phase hydrogenation at high temperature, it was observed that the hydrogen evolution reaches a characteristic peak at approximately 285–300 °C and to cover a wide temperature range between ΔT = 140 and 360 °C with an activation energy of approximately 8.5–9.1 kJ. In comparison to the literature, the activation energy values calculated close to migration energy; in pure annealed Fe, $E_m$ is 7 kJ/mol (0.07 eV/atom) and $D_L$ is $1.3 \times 10^{-3}$ $mm^2/s$ at 25 °C [14].

In addition, a large peak was observed at approximately 150 °C that covers a wide range of temperature between ΔT = 25 and 400 °C. It has to be emphasized that such a peak occurs after hydrogenation of the as-delivered material as well of the weld metal previously welded in a 7.5 vol.% hydrogen-argon atmosphere. This evidently shows that this hydrogen evolution peak has to be attributed to electrochemically introduced hydrogen. Unfortunately, the activation energy corresponding to this peak could not be determined, since for each measurement taken the results obtained were different.

But, it can be assumed that the hydrogen evolution associated to this peak is the outcome of several defects that act as trapping sites in particular for electrochemically introduced hydrogen.

The preliminary metallographic investigations reveal existence of different inhomogeneities in the martensitic microstructure of these welds and it has to be expected that these defects act as various trapping sites. For example, the largest activation energy value obtained from the respective desorption peak calculations is approximately 9 kJ/mole, indicating that the respective trap belongs to a H-dislocation elastic stress field.

With respect to the various potential trapping sites in supermartensitic stainless steel weld microstructures, the following considerations should be undertaken: Desorption energy is defined as the sum between the trap binding energy and the activation energy for lattice diffusion ($E_d = E_b + E_m$). When the activation energy for lattice diffusion ($E_m$) is very large relative to the activation or binding energy of a trap ($E_b$), a diffusion controlled hydrogen evolution will be measured.

As previously mentioned, the theoretical values and the measured values of the activation energy are very close. This means that the bulk diffusion is not negligible and thus, the theory of Lee and Lee does not apply in this case. The approach of Lee and Lee, based only on the hydrogen evolution from trap sites, is attractive because of its theoretical simplicity. However, diffusion is not accounted for and, correspondingly, there is likewise no accounting for the possibility of hydrogen released from one trap becoming available to other traps. This simply means that the Lee and Lee model is not suitable to explain hydrogen evolution or to calculate the activation energy for the investigated supermartensitic stainless steel weld microstructures.

In addition, it was reported in former studies that in fcc materials, the binding energy of hydrogen to defects is small compared to its interstitial activation energy; in bcc material, the defect binding energy is as high as an order of magnitude above this fundamental migration energy. This implies that defects will dominate the hydrogen transport and location in bcc material, but will be much of less importance in fcc material. To summarize, the present results show in the relatively small activation energies measured, that the retained austenite is very much affecting hydrogen transport in the investigated SMSS weld microstructures. However, it also has to be anticipated that the hydrogen evolution from these microstructures is diffusion controlled [8]. This means that the hydrogen is not trapped irreversibly in all the supermartensitic weld microstructures, regardless of the charging process.

## Conclusions

Investigating the effect of different hydrogen exposure of a welded supermartensitic stainless steel to obtain a better understanding of microstructural influences on the absorption/desorption behavior and the interactions between hydrogen atoms and the possible trap sites, the following conclusions can be drawn:

1. The amount of absorbed hydrogen is significantly dependent on the different modes of hydrogenation. During electrochemical charging, the absorbed hydrogen concentrations can reach much higher levels than during charging in a gaseous hydrogen atmosphere or during welding SMSS in a hydrogenated shielding gas with up to 7.5 vol.% $H_2$.
2. The absorbed hydrogen concentration significantly depends on the weld microstructure. In the investigated mono-layer weld, lower concentrations are soluted in the HAZ as compared to the weld metal and to the base material. This has to attributed to respectively higher content of retained and annealing austenite produced during production in the base material and also to a higher amount of retained austenite in the weld metal. However, due to reheating and annealing effects producing respectively more austenite in the HAZ, much higher hydrogen concentrations can be taken up in the HAZ multilayer joints.
3. The influence of hydrogen percent in the atmosphere on the amount absorbed appeared only in the atmosphere containing 7.5 % $H_2$.
4. Further investigations using other approaches also accounting for diffusion controlled hydrogen transport are needed to show which trapping sites are actually occurring in the weld microstructures.

## References


1. Th. Boellinghaus: Hydrogen Assisted Cracking in Supermartensitic Stainless Steels, Hydrogen Effects on Material Behavior and Corrosion Defformation Interactions, Wyoming, TM, pp. 1009–1018
2. D. Seeger, Th. Boellinghaus: Hydrogen Permeation in Supermartensitic Stainless Steel Weld Microstructures, Conf. Corrosion 2004, NACE, Houston, paper 04142
3. E. Folkhard: Metallurgie der Schweißung hochlegierter Stähle, Springer, 1984
4. Th. Boellinghaus, H. Hoffmeister: Hydrogen Permeation in Supermartensitic Stainless Steels Dependent on Heat Treatment and Chemical Composition, Conf. Corrosion 2000, NACE, Houston, paper 00141
5. R. Pargeter: Evaluation of Necessary Delay before Inspection for Hydrogen Cracks, Welding Journal 2003, No. 11, pp. 321–329
6. Tal-Gutelmacher, E.; Eliezer, D.; Eylon, D. (2004). *Materials Science and Engineering A*, **381**, 230
7. E. Tal-Gutelmacher, N. Eliaz, D. Eliezer, D. Zander, L. Jastrow, U. Koester, *Mat. Sci. Eng. A.*, **358(1–2)**, (2003), 219
8. S.-M. Lee, J.-Y. Lee: (1986). *Metallurgical Transactions A*, **17A**: 181
9. Rozenak, P., Eliezer, D. (1983). *Materials Science and Engineering*, **61**: 31.
10. M. Kesten, H. Gräfen: Druckwasserstoffangriff auf unlegierte und niedriglegierte Stähle im Temperaturbereich unterhalb 200 °C, Wasserstoff und Korrosion, 2nd Edition, D. Kuron (ed.), Bonner Studien Reihe, Bonn, 2000, pp. 101–126
11. H. Spähn: Druckwasserstoffangriff auf unlegierte und niedriglegierte Stähle im Temperaturbereich oberhalb 200 °C, Wasserstoff und Korrosion, 2nd Edition, D. Kuron (ed.), Bonner Studien Reihe, Bonn, 2000, pp. 127–183
12. G. M. Pressouyre, F. M. Faure. (1988). Hydrogen Embrittlement: Prevention and Control. 353.

13. D. J. Kotecki, T. A. Siewert: WRC-1992 Constitution Diagram for Stainless Steel Weld Metals: A Modification of the WRC-1998 Diagram, Welding Journal, Vol. 71 (1992), No. 5, pp. 171-S-178-S
14. Th. Boellinghaus, H. Hoffmeister: Hydrogen Permeation in Supermartensitic Stainless Steels Dependent on Heat Treatment and Chemical Composition, Conf. Corrosion, NACE, Houston, 2000, paper 00141
15. Th. Boellinghaus, H. Hoffmeister, A. Dangeleit: A Scatterband for Hydrogen Diffusion Coefficients in Micro-Alloyed Low Carbon Structural Steels, Welding in the World, Vol. 35, 1995, No. 2, pp. 83-96
16. Shewmon: Diffusion in Solids, McGraw-Hill, USA, 1960
17. Th. Boellinghaus, H. Hoffmeister, C. Middel: Scatterbands for Hydrogen Diffusion Coefficients in Steels Having a Ferritic or Martensitic Microstructure and Steels Having an Austenitic Microstructure at Room Temperature, Welding in the World, Vol. 37, 1996, No. 1, pp. 16-23
18. E. Abramov, D. Eliezer, Metall. Mater. Trans. A25 (1994) 949.
19. E. Tal-Gutelmacher, D. Eliezer and E. Abramov, "Thermal Desorption Spectroscopy (TDS)—Application in Quantitative Study of Hydrogen Evolution and Trapping in Crystalline and Non-Crystalline Materials", Materials Science and Engineering A, 445, 625-31 (2007).
20. A. McNabb, P.K. Foster, Trans. TMS-AIME 227 (1963) 618.
21. J.P. Hirth, in: R.A. Oriani, J.P. Hirth, M. Smialowski (Eds.), Hydrogen Degradation of Ferrous Alloys, Noyes Publications, Park Ridge, NJ, 1985, p. 131.
22. G.M. Pressouyre, I.M. Bernstein, Metall. Trans. 12A (1981) 835.
23. D. Eliezer, E. Tal-Gutelmacher, C.E. Cross and Th. Boellinghaus, " Hydrogen Absorption and Desorption in a Duplex-Annealed Ti-6Al-4 V Alloy during Exposure to Different Hydrogen-Containing Environments", *Materials Science and Engineering A*, 433, 298-304 (2006).
24. E. Tal-Gutelmacher, D. Eliezer and Th. Boellinghaus, "Investigation of Hydrogen-Deformation Interactions in Beta-21S Titanium Alloy using Thermal Desorption Spectroscopy", *J. Alloys & Compounds*, 440, 1-2, 204-9 (2007).
25. E. Tal-Gutelmacher, D. Eliezer and Th. Boellinghaus, "Hydrogen's Absorption/Desorption Behavior in Gaseous-Phase Charged Duplex-Annealed Ti-6Al-4 V Alloy", *Materials Science Forum*, 546-9, 1367-72, (2007).
26. E. Tal-Gutelmacher, D. Eliezer and Th. Boellinghaus, "Hydrogen Behavior in GTA Welded Ti-6Al-4 V and Beta-21S Aerospace Applicative Titanium Alloys", *Materials Science Forum*, 546-9, 1413-20, (2007).
27. G. Ben-Hamu, D. Eliezer, C.E. Cross and Th. Bollinghaus, "The Relation between Microstructure and Corrosion Behavior of GTA Welded AZ31B Magnesium Sheet", *Materials Science and Engineering A*, 452-453, 210-8 (2007).
28. D. Eliezer, Y. Nissim and Th. Kannengießer, "Effects of Shielding with Various Hydrogen-Argon Mixtures on Supermartensitic Stainless Steel TIG Welds", *Materials Testing*, 306-315, 52 (2010).

# Part VII
# Elevated Temperature Solid-State Cracking

# Stress-Relief Cracking in Simulated-Coarse-Grained Heat Affected Zone of a Creep-Resistant Steel

**Katherine Strader, Boian T. Alexandrov and John C. Lippold**

**Abstract** Cracking has been reported in newly constructed water wall panels of fossil power plants during startup testing. Both high hardness (exceeding 350 HV) and high level of welding residual stress have been reported in welds of waterwall panels made of T23 and T24 steels. Stress-relief cracking (SRC) is being considered as a possible failure mechanism during high temperature exposure such as PWHT. High temperature exposure of non PWHT-ed welds of Grade T23 and T24 steels leads to hardening in the weld and coarse-grained heat-affected zone (CGHAZ). It has been suggested that such a hardening mechanism can lead to stress-relief cracking (SRC). The objective of this study is to evaluate the susceptibility to SRC in the coarse grained heat affected zone (CGHAZ) of Grade T24 steel utilizing a Gleeble-based SRC test developed at The Ohio State University. The strain-age cracking test developed at The Ohio State University was modified in order to better replicate the conditions of PWHT in highly restrained welds and quantify the stress-relief cracking susceptibility in creep resistant steels. In addition to reduction in area and time to failure, the modified test allows quantification of the stress and strain that cause failure during SRC testing. This test utilizes the Gleeble® 3800 thermo-mechanical simulator. SRC testing of simulated-CGHAZ in Grade T24 Steel has revealed ductile failure for the sample tested at 600 °C, predominantly intergranular with ductile features for the sample tested at 650 °C, and brittle intergranular failures for the samples tested at temperatures of 675 °C and above. For PWHT above 600 °C at residual stress levels close to the yield stress, the CGHAZ in Grade T24 steel welds may be susceptible to SRC.

K. Strader (✉) · B.T. Alexandrov · J.C. Lippold
The Ohio State University, Columbus, USA
e-mail: strader.30@buckeyemail.osu.edu



Th. Boellinghaus et al. (eds.), *Cracking Phenomena in Welds IV*,
DOI 10.1007/978-3-319-28434-7_21

## Introduction

Steam power plants use heat generated from primary fossil fuels to heat and produce high temperature, high pressure steam [1]. In order to increase efficiency and decrease greenhouse gas emissions in modern boilers, the operating parameters such as the pressure and temperature of steam must be increased. Ultra Supercritical (USC) Boilers operate at pressures at or above 24.8 MPa and temperatures >593 °C [2]. Conventional ferritic-bainitic steels such as Grades 12 and 22 do not have high enough creep rupture strength and require larger wall thickness for use in waterwall tubing of USC boilers. Grades 23 and 24 steels were developed to be welded without preheat or post-weld heat treatment (PWHT) in thin-wall components while having superior creep strengths so that higher allowable stresses and higher operating parameters could be experienced without failure [3]. Presently, the service temperature of water wall panels like those made of Grade T23 and T24 steels reaches about 560 °C [4].

Grades 23 and 24 steels were developed based on the classic Grade 22 with microalloying additions that provide higher creep rupture strengths and allow higher operating temperatures [5]. Both Grade 23 and 24 have reduced carbon content in order to help improve weldability. In addition to the lowered carbon content, Grade 23 has additions of tungsten, vanadium, niobium, nitrogen and boron, and reduced molybdenum content. Grade 24 has additions of titanium, vanadium and boron. Vanadium, niobium and titanium are used mainly as precipitation strengthening elements to form MC-type carbides. Boron is used for enhanced creep strength and tungsten may be substituted for molybdenum in order to provide solid-solution strengthening.

Cracking was reported in Grades T23 and T24 welds in water wall panels of newly constructed power plants. Currently investigated failure mechanisms for the cause of cracking include stress-corrosion cracking (SCC) and hydrogen assisted cracking (HAC) [6–9]. Both failure mechanisms occur when a critical combination of susceptible microstructure, high residual stresses and corrosive environment (exposure to oxygen or hydrogen) is present [8]. Both high hardness (exceeding 350 HV) and high level of welding residual stress have been reported in welds of waterwall panels made of T23 and T24 steels [5, 6, 10]. SCC in such welds may be caused by high oxygen concentrations in the boiler water [7, 9, 11]. A possible HAC mechanism may be related to $H_2S$ evolution during acid cleaning of the waterwalls, or to hydrogen evolution as a product of the Schikorr reaction (magnetite formation on tube ID) at the first stage of service [6, 9].

PWHT that would relieve residual stresses and reduce hardness in the weld zone is currently considered for resolving the cracking problem. However, it has been shown that high temperature exposure of non PWHT-ed welds in T23 and T24 steels leads to hardening in the weld metal and coarse-grained heat-affected zone (CGHAZ) [12, 13]. It has been suggested that such a hardening mechanism can lead to stress-relief cracking (SRC).

In several creep-resistant, precipitation-strengthened steels like those studied in this research, SRC is a known cause of many weld failures [14]. SRC is generally intergranular cracking that occurs in the heat-affected zone (HAZ) or weld metal of welded assemblies during exposure to high temperature service or PWHT. During PWHT, residual stresses may be relieved via plastic deformation of the material [15]. The CGHAZ is known to be the most susceptible region of a steel weldment to SRC. If a material's microstructure has strong grain interiors that are resistant to plastic deformation along with weak grain boundaries, strain may be localized at the grain boundaries.

During the arc welding process the base material closest to the fusion zone reaches temperatures close to the melting point which takes it into the austenite phase field [15]. While in the austenite phase field, pre-existing carbides, carbonitrides, nitrides and some inclusions dissolve into the matrix and the magnitude of dissolution is dependent on the welding parameters. If dissolution occurs to a great extent, this allows austenite grains to grow to large sizes. During fast cooling, carbon and other dissolved alloying elements may remain trapped in solution while the austenite transforms to bainite or martensite. Upon elevated temperature exposure to PWHT or elevated service temperatures, carbides like $M_3C$, $M_{23}C_6$, $M_6C$, and MC may precipitate out and may nucleate on dislocations within grain interiors which causes precipitation strengthening and secondary hardening. These precipitates are typically incoherent with the matrix, are stable at higher temperatures, retard dislocation movement and restrict relaxation of residual stresses. Carbides may also form on the prior austenite grain boundaries. The matrix adjacent to these boundaries may become depleted of alloying elements creating a denuded or precipitate-free zone which is softer and more ductile.

Some investigations have shown that one SRC mechanism is likely to be due to impurities, especially phosphorus, segregating to grain boundary/carbide interfaces or carbide-free grain boundary areas, especially under high thermal tensile stresses (developed on cooling after PWHT) [13]. The phosphorous concentration is highest at the grain boundary/carbide interfaces, so it is here where intergranular cracking initiates. Without proper PWHT, the strength of the grain boundary/carbide interfaces decreases. This decrease in strength along with the segregation leads to decohesion along these boundaries. Some main factors for why SRC occurs includes higher material hardness (high thermal stress), slow cooling rates after PWHT (promotes impurity segregation) and larger grain sizes in the weld metal.

SRC has been said to occur in creep-resistant steels due to a precipitation-strengthened matrix along with a softer Cr- or C-depleted zone that forms along the prior austenite grain boundaries [13, 15]. This is likely caused by coarse, incoherent precipitates; a soft denuded zone and/or elemental segregation. This leads to stress relief not by plastic deformation of the grains, but by cracking along the prior austenite grain boundaries.

Most of the available tests for evaluation of the susceptibility to SRC in creep resistant alloys utilize the Gleeble® thermo-mechanical simulator. The Belgian Welding Institute (BWI) developed a SRC testing procedure that involves straining of a test sample to failure at a particular temperature in the PWHT temperature range. The reduction in area of the tested sample is used as a criterion for ranking SRC susceptibility [16].

The SRC test developed at Lehigh University is a constant displacement test [17]. A test sample is heated to a selected test temperature and loaded in tension to a predetermined displacement value that corresponds to the 0.2 % offset yield strength at the test temperature. The time to failure at a constant displacement at a particular test temperature is used as a SRC susceptibility criterion in this test.

A test for evaluation of susceptibility to strain-age cracking was developed at The Ohio State University and applied for Ni-base superalloys [18]. This is a constant displacement test that involves application of yield strength level stress on cooling during HAZ thermal simulation. This is followed by simulation of PWHT under constant displacement and hot ductility testing. The susceptibility to SRC is evaluated based on the reduction in area and on temperature/reduction in area curves.

The objective of this study was to develop a SRC testing procedure that closely replicates the conditions of PWHT in highly restrained welds of creep resistant steels, to evaluate the susceptibility to SRC in the CGHAZ of Grade T24 steel, and to determine safe PWHT conditions for welds in this steel.

## Materials and Experimental Procedures

### *Materials*

Grade T24 steel was delivered in the normalized and tempered condition. The ASTM specification compositional range is shown in Table 1. Grade T24 tubing with an outside diameter (OD) of 44.5 mm and a wall thickness of 7 mm was evaluated in this study.

**Table 1** Specified ASTM chemical composition range for T24 (wt%)

| | T24 |
|---|---|
| C | 0.05–0.1 |
| Mn | 0.3–0.7 |
| P | Max 0.02 |
| S | Max 0.01 |
| Si | 0.15–0.45 |
| Cr | 2.2–2.6 |
| Mo | 0.9–1.1 |
| Ti | 0.05–0.1 |
| V | 0.2–0.3 |
| B | 0.0015–0.007 |
| N | Max 0.012 |
| Al | Max 0.02 |

## *OSU Stress-Relief Cracking Test*

The strain-age cracking test developed at The Ohio State University was modified in order to better replicate the conditions of a localized PWHT on the weld zone region in highly restrained welds and quantify the stress-relief cracking susceptibility in creep resistant steels. In addition to reduction in area and time to failure, the modified test allows quantification of the stress and strain that cause failure during SRC testing. This test utilizes the Gleeble® 3800 thermo-mechanical simulator. A schematic of the modified SRC testing procedure is shown in Fig. 2.

In this study, CGHAZ simulation on samples of Grade T24 tubing was performed using the Gleeble® 3800 thermo-mechanical simulator. Dogbone-shaped samples were resistively heated at a rate of 100 °C/s to a peak temperature of 1350 °C and then free-cooled with a $t_{8/5} \approx 12$–13 s. To prevent oxidation, the test chamber was evacuated to a pressure of $5 \times 10^{-3}$ torr. Low-force jaws were utilized to ensure that no load is applied to the sample during CGHAZ simulation. Low-force jaws allow free thermal expansion and contraction of the sample on heating and on cooling, therefore limiting induced stresses during the thermal cycle.

The rest of the OSU SRC testing procedure utilizes mechanical jaws with serrated grips. The distance between the sets of serrated stainless steel grips, which determines the restraint distance, was set to 42.5 mm as shown in Fig. 1.

Using force control mode at room temperature, the test sample is loaded with 90 % of the CGHAZ yield strength (0.9YS) in order to simulate high level welding residual stresses. The room temperature tensile strength in the simulated-CGHAZ of Grade T24 steel was determined using the Gleeble® at an extension rate of 0.5 mm/min.

Once the test sample is loaded at 0.9YS, the test switches from a force control mode to a displacement control mode and the stroke is fixed in order to simulate a high level of weld restraint. Thus, The OSU SRC test reproduces the worst case scenario of high level residual stresses in highly restrained welds of water wall panels.

**Fig. 1** *Top view* of The OSU SRC test setup: dog-bone sample in serrated stainless steel hot grips with thermocouple and ceramic rods on an extensometer

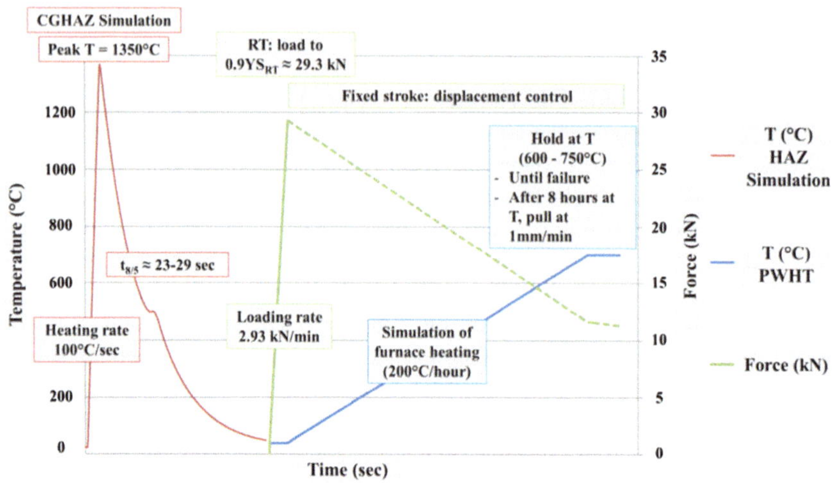


**Fig. 2** The CGHAZ-simulation thermal history and OSU SRC Test

PWHT is simulated locally in the region of the weld zone with a gauge section of about 6 mm under constant displacement mode by heating the test sample at a rate of 200 °C/h to the test temperature and holding for 8 h. If no failure occurs in the 8 h period, the sample is strained to failure at a rate of 1 mm/min. The test outputs are time to failure, stress at failure, strain at failure and reduction in area. The strain over a uniformly heated gauge section is monitored using a strain gauge and recorded throughout the whole test duration (Fig. 2).

## Sample Preparation

Four inch long dogbone samples for CGHAZ-simulation and SRC testing were extracted along the tube axis of Grade T24 tubes. The sample geometry is shown in Fig. 3.

A Type K (chromel-alumel) thermocouple was spot-welded onto the center of each dogbone sample in a plane perpendicular to the longitudinal axis as shown in Fig. 4. The thermocouples were used to monitor the thermal history as well as provide program feedback. Each thermocouple wire is covered with fiberglass insulation in order to prevent short circuiting.

An extensometer was used to measure the strain being applied to the sample during testing. Two wires were spot welded onto each sample, each at a position of 3 mm from the centerline of the dog-bone sample. These two wires are cut to a height of about 1 mm tall (or less) so they will hold rigid and will not deform when the ceramic extensometer rods are pressed up against the sample while applying

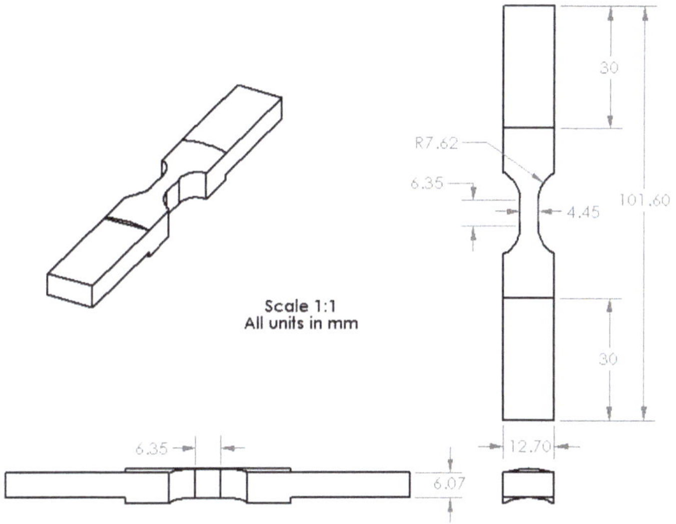


Fig. 3 Schematic illustration of samples for SRC testing

Fig. 4 Thermocouple and extensometer placement on dogbone samples, gauge length and restraint distance shown

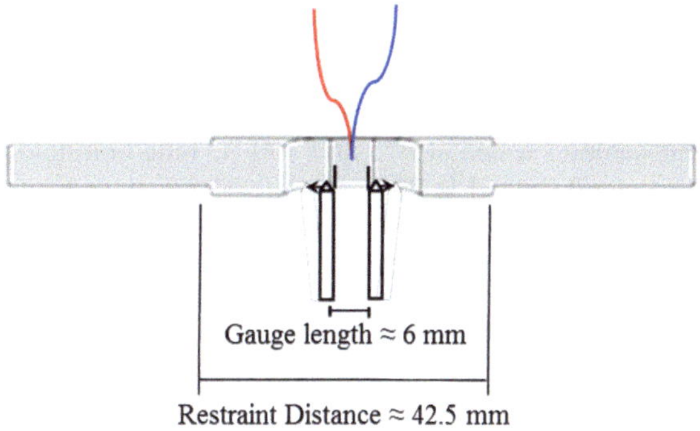


pressure to the wires as shown in Fig. 4. The distance between these wires, which is the gauge length, is 6 mm. The gauge length was measured using vernier calipers for each sample tested and was taken into account for strain calculations.

The 6 mm gauge length was determined by performing temperature distribution tests where one thermocouple was placed at the centerline of the sample and other thermocouples were placed at certain distances away from the centerline. It was found that within 3 mm to each side of the centerline the temperature varied no more than 6 °C when held at temperatures between 550 and 750 °C. Based on this study, it was concluded that a 6 mm gauge section is within an acceptable range for elevated temperature tensile testing and SRC testing.

### *Reduction in Area*

After each dog-bone test sample failed, the percent reduction in area was calculated. The smaller half of the dog-bone specimen was analyzed with a binocular microscope and photos were taken. These images were analyzed using imaging software. The cross-sectional area was found by using a wand tool that allows it to count the area of pixels in the image and this area is compared to the number of pixels representing the scale bar. The percent reduction in area is then found as the ratio of the final cross-sectional area and the initial machined cross-sectional area.

## Results and Discussion

In order to quantify the stress-relief cracking susceptibility, many variables were measured and/or calculated including the applied stress at room temperature, stress and strain at the holding temperature, time to failure, stress and strain at failure, total strain for the duration of the test and reduction in area. The total strain is the strain experienced for the whole duration of the test while the strain at failure only accounts for the strain experienced at the testing temperature.

A total of six samples of simulated CGHAZ in Grade T24 steel were tested and an overview is given for the samples tested at 600, 650, 700, and 750 °C. Samples tested at 600 and 650 °C survived the full 8 h hold and had to be pulled to failure, the samples tested at 675 and 700 °C failed on holding at the test temperature, and the sample tested at 725 and 750 °C failed on heating.

### *SRC Test at 600 °C*

The CGHAZ sample tested at 600 °C failed in a completely ductile manner with no signs of intergranular failure which is representative for SRC, Figs. 5 and 6. There is apparent reduction in the cross-sectional area visible in Fig. 5. The decarburized layer formed during tube production is made visible by etching with 5 % Nital. Figure 6 clearly shows elongation of the grains near the fracture surface from being pulled to failure after the 8 h holding at 600 °C.

Figure 7 shows the stress and strain versus time to failure at 600 °C. On heating, the applied stress decreases from 987 MPa at room temperature to 446 MPa at 600 °C. This is due to generation of thermal expansion stress in the gauge section during heating under constant displacement. Notice that the strain is very low for the full duration of the 8 h test and there is essentially no stress reduction. This shows that there was little or no stress relief occurring during holding at a temperature of 600 °C. The stress at failure, ($\sigma_{failure}$) was 815 MPa, the strain at failure ($\varepsilon_{failure}$) was 32.5 %, and the reduction in area was 44 %.

**Fig. 5** Longitudinal section, ductile failure, test temperature 600 °C, $\sigma_{failure}$ = 815 MPa, $\varepsilon_{failure}$ = 32.5 % (pulled to failure)

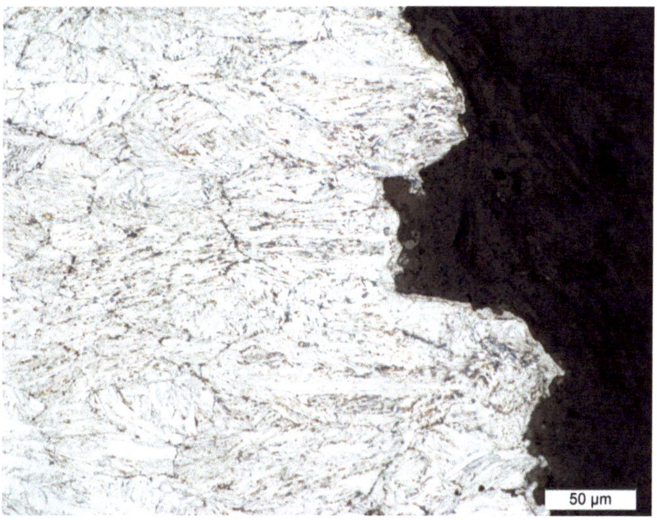


**Fig. 6** Longitudinal section, elongated grains visible near fracture surface, test temperature 600 °C, $\sigma_{failure}$ = 815 MPa, $\varepsilon_{failure}$ = 32.5 % (pulled to failure)

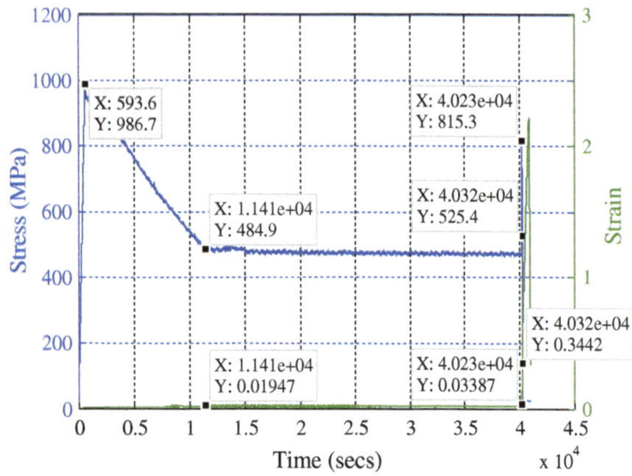


**Fig. 7** Stress and strain versus time, test temperature 600 °C, $\sigma_{failure}$ = 815 MPa, $\varepsilon_{failure}$ = 32.5 % (pulled to failure)

## *SRC Test at 650 °C*

The simulated-CGHAZ sample tested at 650 °C failed in a mostly brittle intergranular manner as shown in Figs. 8, 9 and 10 even though it was pulled to failure after lasting the full 8 h at temperature. Some intergranular cracking along with slight elongation of a few grains closest to the fracture surface is visible in Fig. 9. Figures 10 and 11 are SEM images. Figure 10 shows mainly intergranular failure with some ductility present and a few unidentified blocky particles which could be titanium carbide precipitates. Figure 11 shows ductile dimples more clearly along with a blocky particle.

**Fig. 8** Longitudinal section, mostly brittle failure, test temperature 650 °C, $\sigma_{failure}$ = 669 MPa, $\varepsilon_{failure}$ = 1.1 % (pulled to failure)

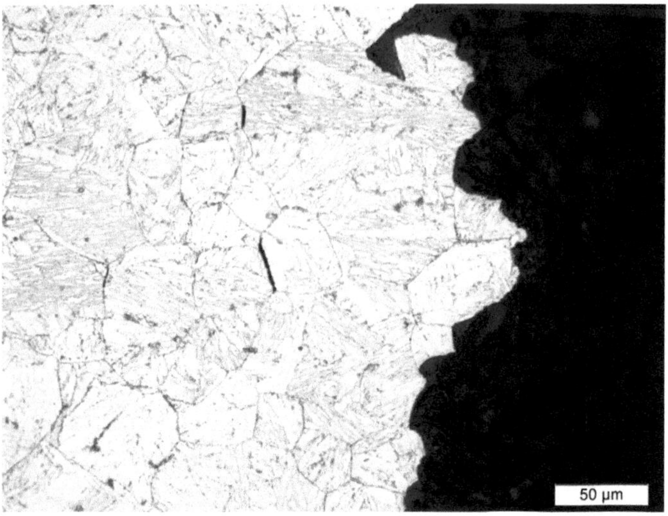


**Fig. 9** Longitudinal section, some intergranular cracking visible, slight elongation of some grains, test temperature 650 °C, $\sigma_{failure}$ = 669 MPa, $\varepsilon_{failure}$ = 1.1 % (pulled to failure)

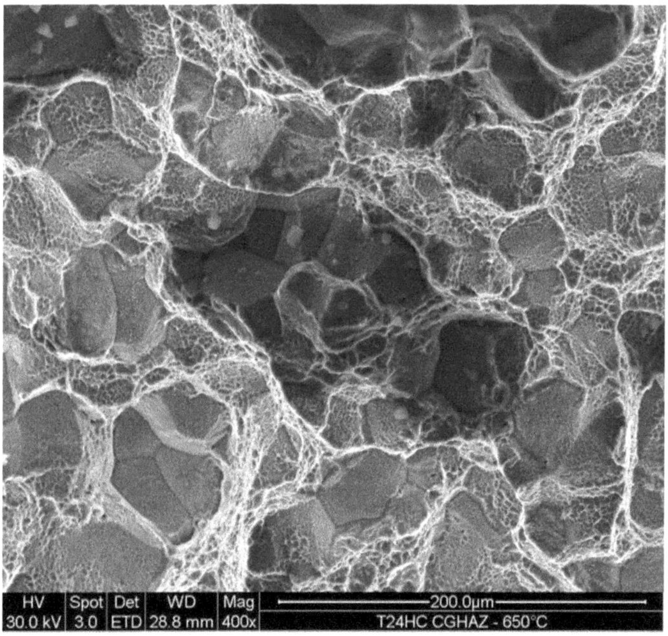


**Fig. 10** SEM—Mainly intergranular failure, some ductility present, unidentified blocky particles, test temperature 650 °C, $\sigma_{failure}$ = 669 MPa, $\varepsilon_{failure}$ = 1.1 % (pulled to failure)

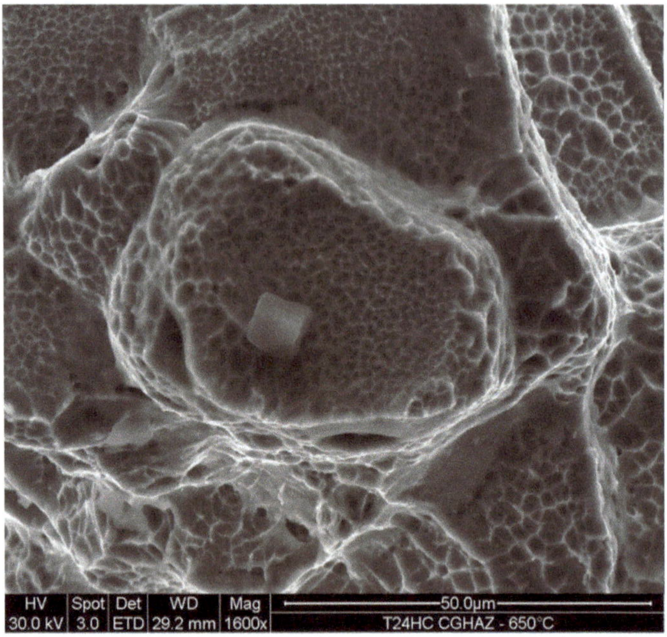


**Fig. 11** SEM—Ductile dimples, unidentified blocky particles, test temperature 650 °C, $\sigma_{failure}$ = 669 MPa, $\varepsilon_{failure}$ = 1.1 % (pulled to failure)

Figure 12 shows the stress and strain versus time to failure at 650 °C. Notice that the strain was very low for the full duration of the 8 h test and there is a small reduction in the stress. It can be assumed that embrittlement had occurred and some of the cracks were formed during holding, which led to a slight decrease in the stress during holding. The assumption is supported by the low strain at failure (only 1.1 %) and low reduction in area (8 %). The stress at failure was 669 MPa.

**Fig. 12** Stress and strain versus time, test temperature 650 °C, $\sigma_{failure}$ = 669 MPa, $\varepsilon_{failure}$ = 1.1 % (pulled to failure)

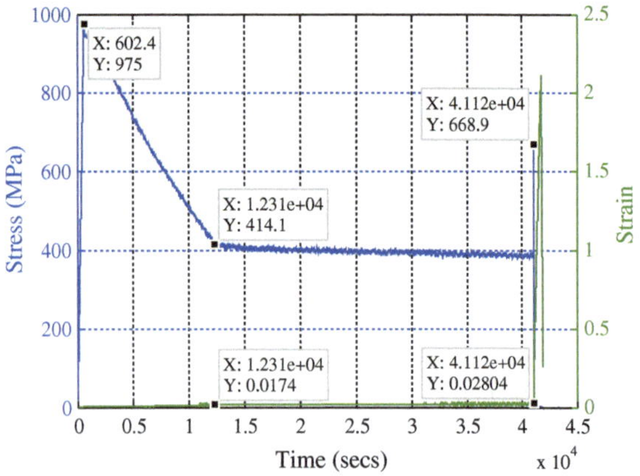

## *SRC Test at 700 °C*

The CGHAZ sample tested at 700 °C failed during testing after being held for 11 min at temperature in a brittle intergranular manner as shown in Figs. 13, 14, and 15. The SEM image of the fracture surface in this sample in Fig. 15 does not show any ductile features as those found in the sample tested at 650 °C. Unidentified particles, which are likely titanium carbide, are seen in the sample tested at 700 °C much like in the sample tested at 650 °C. There are no signs of ductile elongation in the test sample (Fig. 13). Figure 14 shows evidence of brittle intergranular cracking in the HAZ.

Figure 16 shows the stress and strain versus time at 700 °C. There was a large decrease in stress and a significant increase in strain during the 11 min hold at 700 °C until the sample failed. The short time to failure, extremely low strain at failure (0.07 %), reduction in area (7 %), and the fully intergranular fracture mode provide evidences that the failure mechanism was SRC. The stress at failure was 406 MPa.

## *SRC Test at 750 °C*

The CGHAZ sample intended for testing at 750 °C failed at 729 °C on heating in a brittle intergranular manner as shown in Fig. 17.

Figure 18 shows the stress and strain versus time curves in this sample. The sample underwent embrittlement and failed by SRC during heating. The stress at failure was 361 MPa, the reduction in area was 7 %, and the total strain was 3.5 %.

**Fig. 13** Longitudinal section, brittle failure, test temperature 700 °C, $\sigma_{failure}$ = 406 MPa, $\varepsilon_{failure}$ = 0.07 % (failed on holding after 11 min)

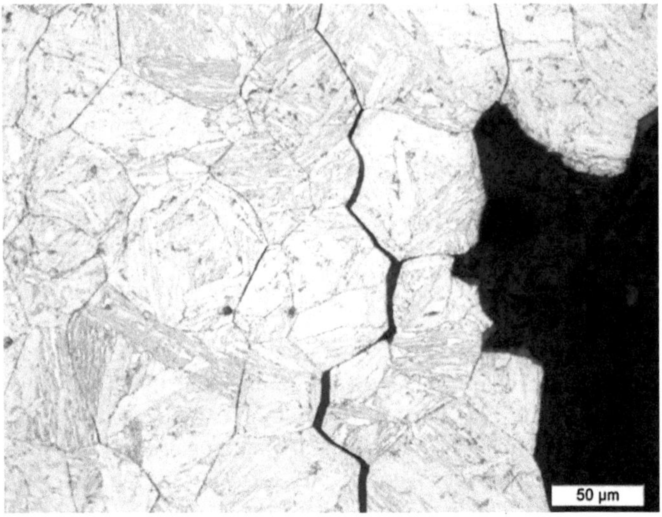


**Fig. 14** Longitudinal section, intergranular cracking, test temperature 700 °C, $\sigma_{failure}$ = 406 MPa, $\varepsilon_{failure}$ = 0.07 % (failed on holding after 11 min)

**Fig. 15** Intergranular cracking, unidentified particles present, test temperature 700 °C, $\sigma_{failure}$ = 406 MPa, $\varepsilon_{failure}$ = 0.07 % (failed on holding after 11 min)

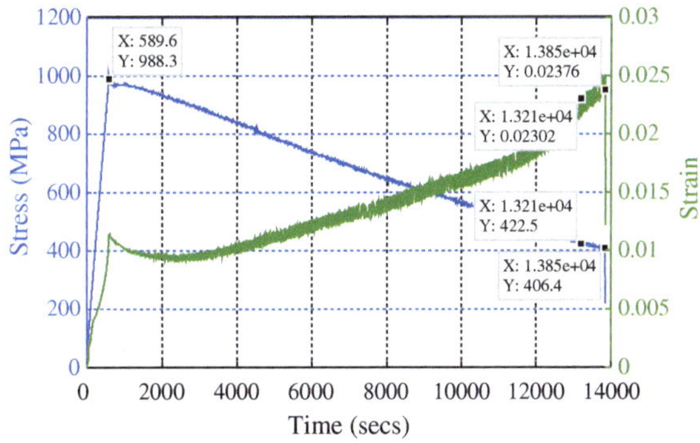


**Fig. 16** Stress and strain versus time, test temperature 700 °C, $\sigma_{failure}$ = 406 MPa, $\varepsilon_{failure}$ = 0.07 % (failed on holding after 11 min)

**Fig. 17** Longitudinal section, brittle failure, test temperature 729 °C, $\sigma_{failure}$ = 361 MPa, $\varepsilon_{failure}$ = 0 % (failed on heating)

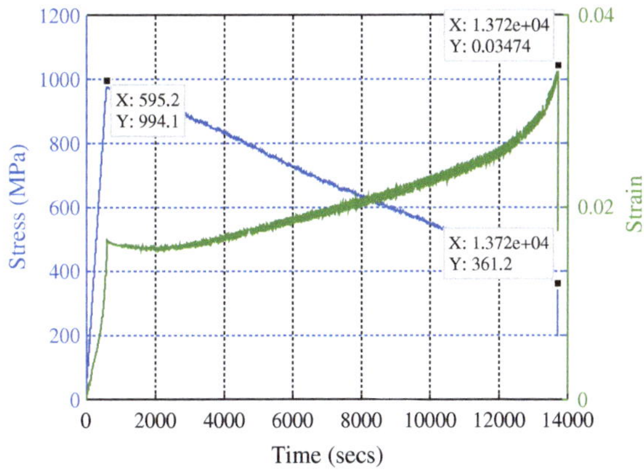


**Fig. 18** Stress and strain versus time, test temperature 729 °C, $\sigma_{failure}$ = 361 MPa, $\varepsilon_{failure}$ = 0 % (failed on heating)

## *Summary of the SRC Test Results*

The results for all SRC tests performed on simulated-CGHAZ samples of Grade T24 steel are summarized in Table 2 and in Figs. 19 and 20. The applied tensile stress at room temperature is relatively consistent. The initial stress at holding temperature is significantly lower than the stress originally applied at room temperature due to thermal expansion in the gauge section during heating under displacement control.

The samples tested at 600 and 650 °C did not fail during holding at test temperature and were pulled to failure, Fig. 19. The samples tested at 675 and 700 °C failed during holding at test temperature. The samples with intended testing temperatures of 725 and 750 °C failed on heating correspondingly at 725 and 729 °C.

The time to failure, stress and strain at failure, total strain and reduction in area decrease significantly at testing temperatures above 600 °C, Table 2 and Figs. 19 and 20. The failure mode changes from ductile at 600 °C to predominantly intergranular with ductile features at 650 °C and to purely intergranular at higher testing temperatures.

**Table 2** Summary of results—The OSU SRC Test for T24 simulated-CGHAZ (D—ductile; IG—intergranular)

| Test T (°C) | Applied stress at RT (MPa) | Stress at holding T(MPa) | Time to failure (h) | Strain at failure (%) | Total strain (%) | RA (%) | Stress at failure (MPa) | Failure mode |
|---|---|---|---|---|---|---|---|---|
| 600 | 987 | 485 | 8, pulled | 32.5 | 34.4 | 44 | 815 | D |
| 650 | 975 | 414 | 8, pulled | 1.1 | 2.8 | 8 | 669 | IG + D |
| 675 | 988 | 416 | 3.54 | 0.7 | 2.8 | 7 | 390 | IG |
| 700 | 988 | 423 | 0.18 | 0.07 | 2.4 | 7 | 406 | IG |
| 725 | 989 | N.A. | On heating at 725 °C | N.A. | 2.3 | 7 | 385 | IG |
| 750 | 994 | N.A. | On heating at 729 °C | N.A. | 3.5 | 7 | 361 | IG |

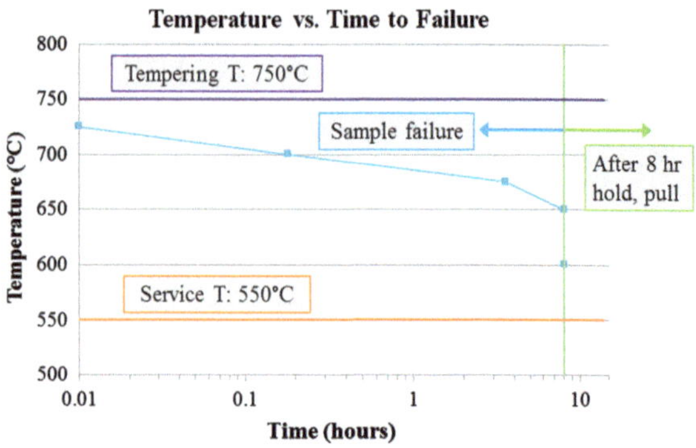


**Fig. 19** Temperature versus time to failure at holding temperature for simulated-CGHAZ in Grade T24 steel

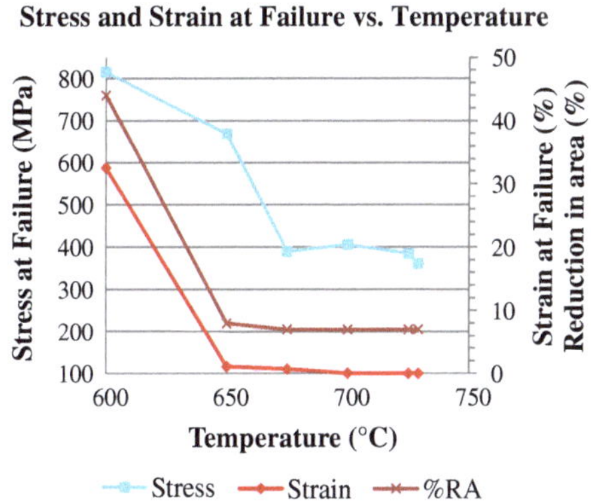


**Fig. 20** Stress and strain at failure versus temperature for simulated-CGHAZ in Grade T24 steel

These results show that the failure mechanism in simulated-CGHAZ of Grade T24 steel for these particular testing conditions is stress relief cracking. The sample tested at 650 °C did not fail for eight hours at a tensile stress of 414 MPa and had to be strained to failure. However, the low strain at failure and low reduction in area provide evidence that stress relaxation embrittlement had already occurred during the holding stage at 650 °C. The on-heating failures at 725 and 729 °C show that stress relief embrittlement may also occur during slow heating to high temperatures under high tensile stress and high restraint.

There is a trend of gradual reduction in the time to failure and strain at failure with increasing the testing temperature to 725 °C. The total strain and reduction in area above 600 °C and the stress at failure above 650 °C remain almost constant. Consequently, the time to failure, and the strain at failure (defined in this study as the strain experienced at testing temperature) are more sensitive indicators of susceptibility to SRC.

The stress relief cracking in the tested material is potentially related to intragranular strengthening caused by re-precipitation of carbides that have been dissolved during the CGHAZ simulation and to simultaneous embrittlement along the prior austenite grain boundaries. Further high level metallurgical characterization work is needed to clarify the embrittlement mechanism in the tested material.

## Summary and Conclusions

1. A new SRC test procedure has been developed at OSU that replicates PWHT in welds of highly restrained components loaded with high residual stresses. The time to failure, stress and strain at failure, total elongation and reduction in area are quantified in this procedure and can be utilized as indicators for ranking susceptibility to SRC. The time to failure and strain at failure appear to be the most sensitive indicators of SRC susceptibility in the tested material.

2. Simulated CGHAZ in Grade T24 steel is not susceptible to SRC during PWHT at 600 °C under high restraint, even when preloaded with yield level tensile stress at room temperature.
3. Stress relief embrittlement occurs in the simulated CGHAZ of this steel during PWHT at 650 °C and higher temperatures. The kinetics of embrittlement significantly accelerates with increasing the PWHT temperature. This is evidenced by the decreasing time to failure and strain at failure during PWHT at 650, 675 and 700 °C, and by complete sample failures during heating to 725 and 729 °C.
4. The results of this study show that highly restrained welds in Grade T24 steel that are loaded with high level residual stresses, such as in water wall panels of ultra-super critical fossil power plants, may be potentially susceptible to stress relief cracking during PWHT above 600 °C.

**Acknowledgments** The authors would like to acknowledge the sponsors of this research: Alstom, American Electric Power, Babcock & Wilcox, ESAB, Vallourec, and voestalpine Böhler Welding. Thanks to Emeric Suma for help with metallographic preparation.

## References

1. J. B. Kitto and S. C. Stultz, Steam/its generation and use. 41st edition., Barberton, Ohio: The Babcock & Wilcox Company, 2005.
2. S. A. David, J. A. Siefert and Z. Feng, "Welding and Weldability of Candidate Ferritic Alloys for Future Advanced Ultrasupercritical Fossil Power Plants," *Science and Technology of Welding and Joining,* vol. 18, no. 8, pp. 631-651, 2013.
3. P. Nevasmaa, J. Salonen and et. al., "Reheat Cracking Susceptibility of P23 (7CrWVMoNb9-6) Steel Welds Made Using Matching and Mis-Matching Filler Metals," in *9th Liege Conference on Materials for Advanced Power Engineering*, 2010.
4. Q. Zhao, "High Temperature Corrosion of Water Wall Materials T23 and T24 in Simulated Furnace Atmospheres," *Chinese Journal of Chemical Engineering,* vol. 20, no. 4, pp. 814-822, 2012.
5. J. Arndt, The T23/T24 Book: New Grades for Waterwalls and Superheaters, Vallourec & Mannesmann Tubes, 2nd Edition, 2000.
6. L. Mráz and et. al., "Application of Creep Resistant Steel T24 for welding membrane walls," IIW. Doc. IX-C-1029-13, 2013.
7. "T24 Experience: an Hitachi Power Europe perspective," *Modern Power Systems,* October 2012.
8. C. Ullrich, S. Heckmann, W. Tillmann, T. Bodmer and G. Gierschner, "Stress Corrosion Cracking of T24 - Service Experience and Investigation Program," in *ETD Seminar*, London, May 2013.
9. S. Huysmans, F. Vanderlinden and E. De Bruycker, "Weldability Aspects of T24 Boiler Tubing in the context of Stress Corrosion Cracking," in *ETD Seminar*, London, May 2013.
10. K. Park, S. Kim, J. Chang and C. Lee, "Post-weld Heat Treatment Cracking Susceptibility of T23 Weld Metals for Fossil Fuel Applications," *Materials and Design,* vol. 34, pp. 699–706, 2012.
11. C. Ullrich, S. Heckmann, W. Tillmann, T. Bodmer and G. Gierschner, "Stress Corrosion Cracking of T24 - Project Achievements and Ongoing Investigations," in *ETD Seminar*, London, May 2013.

12. P. Mohyla and et. al., "Contribution to Research of Weldability of Modern Low-Alloy Creep Resistant Steels," *Acta Metallurgica Slovaca,* vol. 9, no. 3, pp. 210-216, 2003.
13. N. H. Heo, J. C. Chang and S.-J. Kim, "Elevated temperature intergranular cracking in heat-resistant steels," *Materials Science and Engineering A.,* vol. 559, pp. 665-677, 2013.
14. A. Dhooge and A. Vinckier, "Reheat Cracking - A Review of Recent Studies," *International Journal of Pressure Vessels and Piping,* vol. 27, pp. 239-269, 1987.
15. J. G. Nawrocki, "The Mechanism of Stress-Relief Cracking in a Ferritic Alloy Steel," *Welding Journal,* vol. 82, no. 2, pp. 25s-35s, Feb 2003.
16. A. Dhooge and J. Vekeman, "New Generation 2.25Cr Steels T/P 23 and T/P 24 Weldability and High Temperature Properties," *Welding in the World,* vol. 49, pp. 31-49, 2005.
17. J. G. Nawrocki, J. N. DuPont, C. V. Robino and A. R. Marder, "The Stress-Relief Cracking Susceptibility of a New Ferritic Steel - Part 1: Single-Pass Heat-Affected Zone Simulations," *Welding Research Supplement,* pp. 355s-362s, December 2000.
18. S. Norton, "Development of a Gleeble Based Test for Postweld Heat Treatment Cracking in Nickel Alloys," in *MS, Thesis*, The Ohio State University, 2002.

# Testing Approaches for Stress Relaxation Cracking in Gamma-Prime Strengthened Ni-Base Alloys

**David C. Tung and John C. Lippold**

**Abstract** Increasing the efficiency and reducing the emissions of coal-fired power plants is most economically accomplished by increasing maximum steam temperatures and pressures. Designs have progressed beyond the temperatures where typical power plant steels are useful and Ni-based superalloys are being investigated for use in the highest temperature areas. Construction with superalloys is common in the aerospace industry where solution annealing heat treatments following welding may be done in a furnace. In power plant construction, solution annealing is difficult with standard postweld heat treatment techniques and directly aging weldments without solution annealing has been studied. The practice of directly aging weldments does not allow for residual stress relaxation to occur before gamma prime precipitation. Simultaneous stress relaxation and precipitation can lead to stress relaxation cracking via a creep mechanism. There is very limited experience with superalloys regarding power plant construction and there is some evidence that stress relaxation cracking could be a problem. Stress relaxation cracking testing approaches, which are not standardized, can be categorized as self-restrained or externally-loaded. Tests typical of each approach are currently underway. Externally-loaded tests typically achieve failure in a short duration by applying uniaxial loads (or strains) and this practice allows some level of quantification. Self-restrained tests require much longer times but can be used to study triaxial stress states. Ongoing testing of a number of Ni-base alloys including 740H, 282, 617, 718 and Waspaloy is reviewed.

## Introduction

Increasing steam temperature and pressure in coal-fired power plants is a perpetual goal driven by the pursuit of increasing efficiency and reducing fuel consumption and emissions [1, 2]. Temperature and pressure regimes are qualitatively described

D.C. Tung (✉) · J.C. Lippold
The Ohio State University, Columbus, OH, USA
e-mail: david.tung35@gmail.com



Th. Boellinghaus et al. (eds.), *Cracking Phenomena in Welds IV*,
DOI 10.1007/978-3-319-28434-7_22

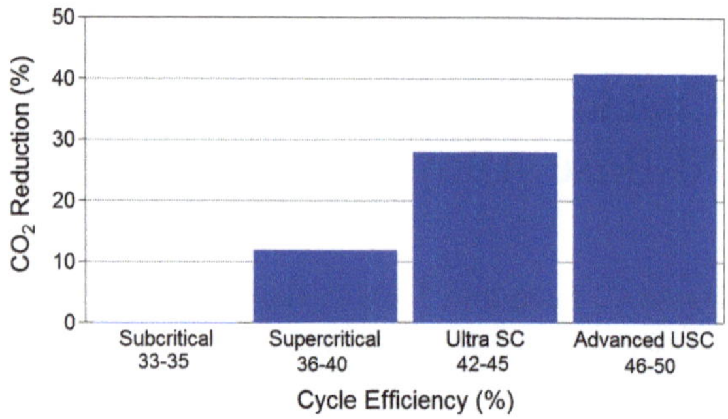


**Fig. 1** Effect of cycle efficiency on $CO_2$ emissions. Adapted from Zachary [3]

based on their relationship with the critical point of water. The current practice is to indicate higher temperatures and pressures by simply preceding the name of the previous level with a more emphatic adjective. In this way, the next generation of high efficiency coal-fired power plants is referred to as *advanced-ultra-supercritical* (AUSC). The corresponding steam temperature and pressure is 760 °C (1400 °F) and 35 MPa (5000 psi) [1, 2]. Figure 1 indicates that AUSC plants are expected to reduce $CO_2$ emissions by 13 % relative to modern ultra-supercritical (USC) plants because of an increase in efficiency [3]. It is important to note that in order to reduce $CO_2$ emissions with a carbon capture and sequestration system (CCS), a significant efficiency penalty would be taken, ultimately resulting in decreased power output.

Transitions to higher temperatures and pressures are often limited by material properties. The creep strength enhanced ferritic (CSEF) steels that are common in the construction of USC plants will not be suitable for the temperatures associated with AUSC plants [4]. One obstacle to AUSC plant construction is the fabrication of a suitable main header pipe; materials challenges include extrusion, welding, and postweld heat treatment (PWHT). These materials challenges, as well as creep rupture strength and corrosion resistance, must be considered during alloy selection [5].

Materials for AUSC plants must demonstrate 100,000 h creep rupture strength of 100 MPa (14,500 psi) or greater at 750 °C [2, 4, 6–8]. Based on the engineering materials available for power plant construction, plants that will operate at AUSC levels must employ precipitation strengthened Ni-base alloys such as Inconel 740 or Haynes 282, as shown in Fig. 2 [1]. Alloy 740H[1] is the leading candidate material. The compositions of these alloys and other alloys in this study are given in Table 1.

Precipitation strengthened alloys are typically thoroughly homogenized before aging and this is most readily done in a furnace [16–18]. Unfortunately, large-scale power plant construction makes furnace heat treatments impossible. Construction heat treatments are typically done with ceramic heating blankets that are wrapped around large components. These blankets typically do not have the same heating

[1]Alloy 740 was modified to improve its weldability and microstructural stability [9]. The modified composition is designated with the letter H, however H here is an arbitrary designator that reflects a broad composition modification and does not reflect carbon level as it does with stainless steels [10].

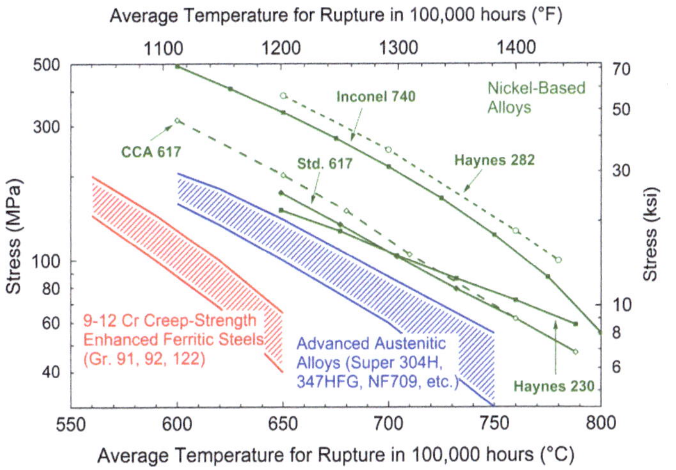


**Fig. 2** Creep rupture data for common and candidate power plant materials [1]

**Table 1** Nominal compositions of alloys in this study

| Element | 617B [11] | 740 [11] | 740H [12] | 282 [13] | 230 [14] | 800 [15] |
|---|---|---|---|---|---|---|
| Ni | Balance | Balance | Balance | Balance | Balance | 33 |
| Cr | 22 | 25 | 25 | 20 | 22 | 21 |
| Co | 12 | 20 | 20 | 10 | 5 | – |
| Mo | 9 | 0.5 | 0.5 | 8.5 | 2 | – |
| Fe | 1.5 | 2 | 0.7 | 1.5 | 3 | Balance |
| Mn | 0.3 | 1 | 0.3 | 0.3 | 0.5 | – |
| Al | 1 | 0.9 | 1.3 | 1.5 | 0.3 | 0.4 |
| Ti | 0.4 | 1.8 | 1.5 | 2.1 | – | 0.4 |
| Nb | – | 2 | 1.5 | – | W = 14 | – |
| Si | 0.3 | 0.7 | 0.15 | 0.15 | 0.4 | – |
| C | 0.06 | 0.04 | 0.03 | 0.06 | 0.1 | 0.1 |
| B | 0.004 | 0.004 | 0.001 | 0.005 | 0.015 | – |

capabilities as furnaces and maximum temperatures are limited. It is very desirable to reduce heat treatment duration as much as possible to expedite construction or repair and reduce associated costs. Temperature limitations and the desire to minimize PWHT duration led to the study of a single, sub-solvus PWHT for nickel-base superalloys considered for AUSC power plant construction. This PWHT, 800 °C/4 h, was meant to directly age the weld metal and heat affected zone (HAZ) after welding without a solution annealing procedure. Without the stress relief that would occur during a solution annealing treatment, residual stress driven cracking mechanisms, such as reheat or strain-age cracking during PWHT or stress relaxation cracking during service, become a concern.

Stress relaxation cracking is a cracking mechanism associated with structures which are not stress relieved before they begin service. Residual stresses relax over time and accelerate creep, leading to premature failure. Microstructural features of relaxation cracking are similar to reheat and strain-age cracking, phenomena which occur in much shorter times, typically during heating to a PWHT temperature. All three of these cracking mechanisms are characterized by intergranular cracking in either the coarse grain HAZ near the fusion boundary or in the weld metal. There is a considerable amount of published information regarding these mechanisms, including several thorough reviews [19–25], the most recent of which is by Lippold [26]. In the HAZ near the fusion boundary, the thermal excursion solutionizes the microstructure. During cooling some re-precipitation occurs and residual stresses develop. As the structure is heated again, either to the service temperature (stress relaxation cracking, SRC), tempering temperature (reheat cracking, RHC), or solution annealing temperature (strain-age cracking, SAC), re-precipitation occurs and may develop a microstructure susceptible to one of these mechanisms. Susceptible microstructures typically exhibit intense intragranular precipitation and a weaker precipitate configuration on the grain boundaries. Grain boundaries are typically the weakest link in a microstructure and an unfavorable precipitate distribution in the HAZ, such as that which occurs in relaxation/reheat/strain-age cracking, exacerbates this problem. This weaker configuration on the grain boundary could be a lack of precipitates due to generally higher solid solubility, a continuous precipitate layer, or a precipitate-free zone just adjacent to the boundary caused by solute element absorption by large intergranular precipitates. Impurity segregation to the boundary generally increases cracking susceptibility.

The distinguishing characteristic of relaxation cracking is the time required for failure. Strain-age and reheat cracking occur in short times during the PWHT cycle (on the order of 10 h) and this is somewhat convenient from an inspection and testing perspective since cracking can be readily detected and defective components can be repaired. Relaxation cracking occurs after very long times in service (on the order of 10,000 h or longer) which makes detection much more complicated and repair much more expensive. The schematic thermal cycles in Fig. 3 illustrate this difference. Strain-age cracking occurs during heating to an annealing temperature as shown in Fig. 3a. In Fig. 3b, relaxation cracking would occur in the absence of the short PWHT step. Since this PWHT is at a relatively low temperature, it is possible that insufficient residual stress relief will occur and relaxation cracking will still be a danger. An important point raised by Siefert and Tanzosh in their review of relaxation cracking is that the time of cracking is often impossible to know; the time of detection is what is ultimately reported [21].

Stress relief heat treatments are an effective solution to this problem (provided that strain-age or reheat cracking do not occur) since it is the additional creep strain due to stress relaxation drives relaxation cracking [23]. Superalloy weldments are typically solution annealed before they are aged which simultaneously eliminates microsegregation and residual stress. Since it is standard practice to eliminate residual stress, SRC is normally not a problem in superalloys. In AUSC plant construction, which uses superalloys in a non-traditional role, achieving stress relief

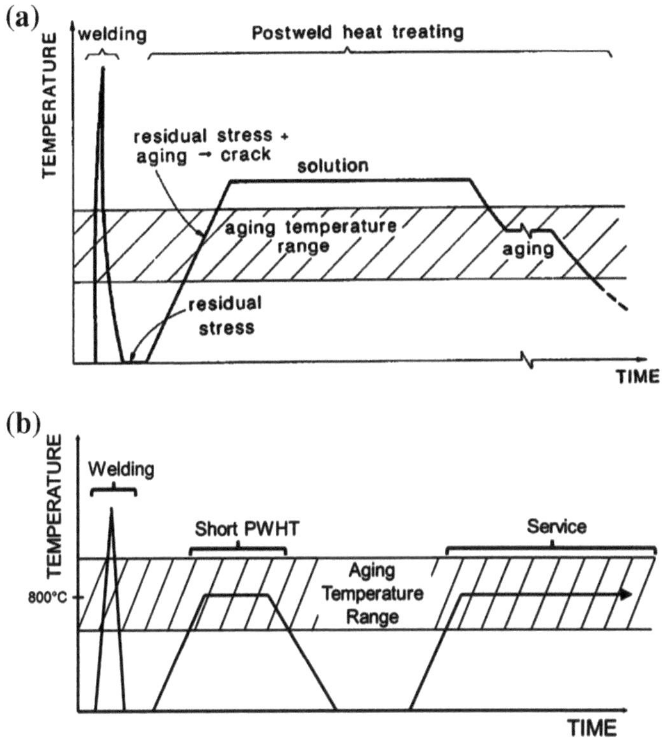


**Fig. 3** Comparison of strain age and stress relaxation cracking thermal cycles adapted from Kou [27]. **a** Strain age cracking. **b** Relaxation cracking concern in AUSC plant construction

is very difficult and therefore concerns regarding SRC have arisen. These concerns are largely due to experience with Alloy 617, the details of which have not been reported until recently [11].

Relaxation cracking in thick walled Alloy 617 was first detected 15,000 h after a 700 °C component test facility began operation in Europe (COMTES700) and additional cracking in thick section Alloy 617 was found in following years [11]. Following initial detection, the average detection time for relaxation cracks in Alloy 617 was less than 1 year and was occasionally discovered due to leaking steam. Cracking was generally associated with high levels of additional operating stresses and occasion-ally originated interdendritically in weld metal. In some cases, a high concentration of Cr and Mo rich carbides was reported although no mention of precipitate free zones was made. Besides the mention of intergranular/interdendritic, no detailed description of the crack path was made. Some cracking was associated with areas of cold work. The most severe cracking was found in circumferential repair welds in thick material only a few months after the repair was done. A PWHT was not applied following either the original weld or the repair welds. It is reported that a 980 °C/3 h PWHT following repair welding eliminated the formation of large cracks but small cracks were still detected. A laboratory study determined that an 1160 °C/1 h heat treatment of service exposed material was possible with construction techniques and this heat treatment solutionized $\gamma'$ and coarsened very fine $M_{23}C_6$ carbides. Bechetti and DuPont found that a similar heat treatment will homogenize Alloy 740H weld metal [28]. Before heat treatment, service exposed material showed many 70 nm $\gamma'$ and significantly smaller $M_{23}C_6$ carbides. The overall effect of this heat treatment would significantly increase ductility and relieve residual stress, presumably avoiding relaxation cracking. Thin-walled Alloy 617 components were not affected and this attests to the importance of stress triaxiality regarding relaxation cracking. Based on these findings, suitable repair procedures were developed and it was ultimately mandated that thick-section welds must be postweld heat treated. Fortunately, a PWHT is already planned for Alloy 740H and it has been successfully performed on 76 mm (3 in.) thick material [5] but its efficacy regarding SRC has not yet been determined.

Relaxation cracking has been prevented with stress relief heat treatments as low as 900 °C in Alloy 617 [21] but heat treatment temperature is alloy dependent. A PWHT temperature of 930 °C was investigated with respect to Alloy 740H weldments and it was found not to be appropriate for several reasons. The first is that stress relief is most effective above the solvus of strengthening precipitates which is over 1000 °C for AUSC superalloys. As an example, a calculated TTT diagram for 740H is shown in Fig. 4. The second is that microsegregation in 740H weld metal decreases the alloying concentration in the dendrite cores to such a degree that the solvus for this location is below 930 °C. This PWHT will produce a high $\gamma'$ density along interdendritic boundaries and leave $\gamma'$ entirely absent from dendrite cores due to the depressed solvus, as shown in Fig. 5. Because of this problem, it is not advisable to use this heat treatment to prevent SRC. The direct aging PWHT planned for 740H is not likely to relieve any significant level of residual stress because the temperature is too low and $\gamma'$ will be formed which will further prevent stress relaxation.

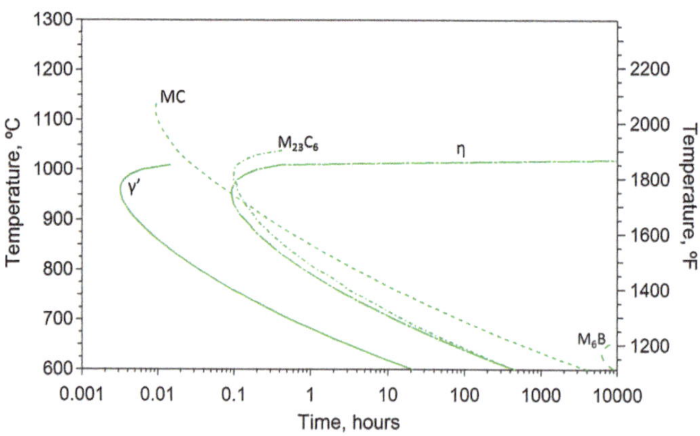


**Fig. 4** Alloy 740H nominal composition TTT diagram calculated with JMatPro v5.1. Even though η phase is predicted to form in relatively short times, this is not observed experimentally [29]

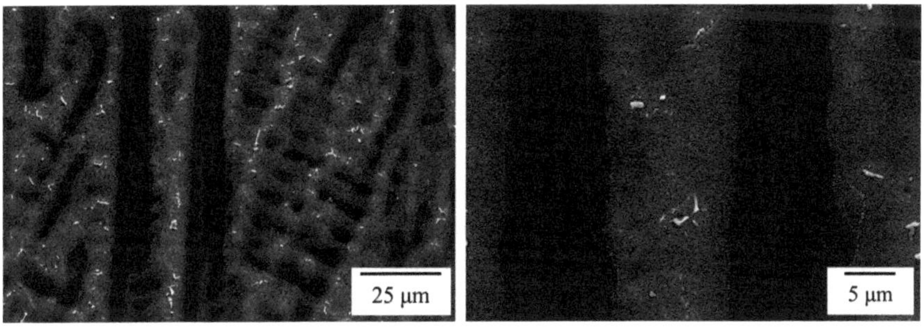


**Fig. 5** Absence of $\gamma'$ in dendrite cores of Alloy 740H weld metal which was aged at 930 °C for 3 h

Since it has already been demonstrated that Alloy 740H can withstand the intended direct age PWHT in full section thickness (3 in., 76 mm) without consequence [5], the need for SAC testing is obviated and concern is shifted to longer times where SRC may affect creep life. Appropriate testing approaches should consider this shift in time scale.

## Procedures

There are many different testing approaches for SRC but none of them were designed specifically with superalloys in mind because relaxation cracking is mainly a problem with low alloy steels and stainless steels. Stress relaxation cracking generally does not affect superalloys because they are always solution annealed before aging. If the solution annealing temperature can be reached without cracking by the SAC mechanism, residual stress driven cracking will no longer affect superalloy weldments. As explained earlier, superalloys for AUSC construction will not be solution annealed and relaxation cracking is now a relevant concern.

Testing approaches to SRC can be separated based on loading strategy. There are many tests which use externally applied loads such as those studied by Craver (cantilever beam) [30], Norton (Gleeble) [31], Christoffel (thermal expansion loaded three point bend) [32], and VanWortel (screw loaded three point bend) [23], among many others. Externally applied loads can be very large in order to cause failure in a shorter time but this makes it very likely to introduce an inappropriate creep or voiding mechanism. Since stress relaxation cracking is a creep phenomenon, a representative test will reproduce the appropriate creep and voiding mechanism. For a thorough discussion of creep void formation, the reader is referred to the review by Svensson and Dunlop [33]. Another disadvantage of many externally loaded tests is a simple, uniaxial stress state. This is a problem because weld residual stresses are typically triaxial and triaxiality affects creep performance.

In order for the results of externally loaded tests to be meaningful, they must be benchmarked against materials that have known strengths or weaknesses. The advantage of these tests, however, is the possibility for more 'analog' data; i.e. a curve can be developed from the resulting time, load, or ductility, and the simple stress state which simplifies analysis.

The alternative option to externally-loaded tests is self-restrained tests. Borland (circular weld) [34] and Turski (pre-strained compact tension) [35] both developed self-restrained tests which are applicable to SRC. The main advantage of these tests is that the stresses are actually complex, triaxial residual stresses which are more representative of the actual stress state than applied uniaxial stresses. The importance of this was seen in the report detailing Alloy 617 cracking; no cracks were observed in thin section material [11]. Another important point regarding these tests is that both have been very accurately modeled with finite element software using the ductility exhaustion approach found in the R5 assessment procedure [35–37]. This is important because accurate modeling allows interpretation of self-restrained tests beyond a simple pass/fail conclusion. The ductility exhaustion approach in R5 is interesting and unique because it considers both cavity nucleation and growth and also accounts for different levels of stress triaxiality by considering hydrostatic, maximum principal, and von Mises stresses [36]. The method investigated by Turski has an additional advantage due to ease of inspection. Not only is the compact tension sample geometry well understood, it is easy to monitor in-situ with a potential drop method and generally easy to inspect otherwise.

Strain-age cracking testing procedures are only indirectly applicable to relaxation cracking since SAC is a short term phenomenon. Nevertheless, they serve as a good starting point and the procedure used by Norton [31] was chosen for further investigation. The Gleeble-based procedure used by Norton was selected because it uses a specific test duration and also allows for the observation of residual stress changes. Norton was able to use residual stress (load) data to make conclusions about precipitate nucleation and growth rates. By using a Gleeble, the thermo-mechanical history can be well controlled and the microstructures that evolve can be very useful in understanding the mechanism of relaxation. Important microstructural information includes precipitation rate and grain boundary precipitate nature.

The procedure used in this work was modeled very closely after Norton's procedure when he studied SAC in Waspaloy and Alloy 718 [31]. This approach simulates the entire thermal and mechanical cycle of the HAZ from welding through the aging stage of a PWHT. An elevated-temperature tensile test is also executed after a certain aging time while the sample is still at the PWHT temperature. This experimental method is illustrated in Fig. 6. The similarities between the SAC susceptible thermal cycle shown in Fig. 3 and the temperature curve in Fig. 6 should be noted.

At $t_1$ in Fig. 6, the temperature represents the approximate maximum temperature that the HAZ reaches during welding. In the current study, a heating rate of 100 °C/s was used to bring the samples to a peak temperature of 1200 °C where they were held for 30 s. Between $t_1$ and $t_2$, the sample cools naturally and a force is

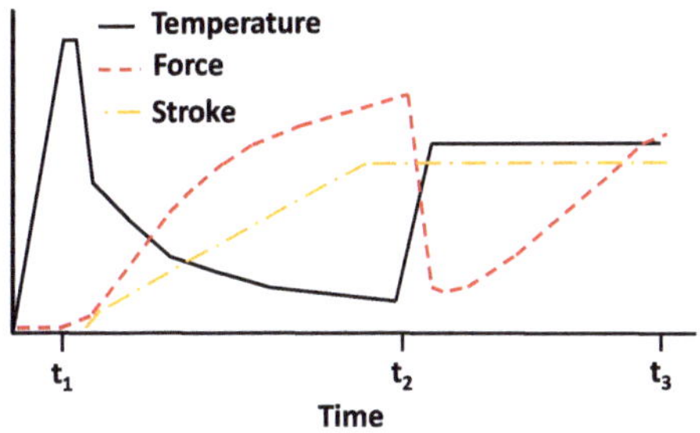


**Fig. 6** Thermal and mechanical profiles developed by Norton for investigation of SAC in Alloy 718 and Waspaloy [31]

applied. This simulates the development of residual stress. It is important to note that Norton applied this force by specifying a displacement and during preliminary testing it was found that this generated a stress greater than the yield stress of all alloys in the study. The force generated through stroke control was also very inconsistent. In the present work, the average load generated through stroke control was used as the load set point. This force was 1000 kgf (2200 lbf) which produces a stress of 35.6 MPa (78.6 ksi) over the sample's 19 mm (3/4 in.) long gauge section which has a diameter of 4.7 mm (3/16 in.). Note that after the load has been applied, the stroke will ideally remain constant for the remainder of the test. At $t_2$, the sample is raised to the test temperature and stress decreases due to both relaxation and sample expansion. After the initial relaxation, Norton observed stress increases during the PWHT due to the formation of precipitates with a negative lattice misfit, as shown in Fig. 7. Nucleation is delayed in Alloy 718 because it is primarily strengthened by Nb-bearing $\gamma''$ which is known for its sluggish precipitation kinetics. Once $t_3$ is reached, the sample is pulled to failure in an elevated-temperature tensile test. This is done in order to end the test after a consistent, reasonable amount of time and collect a data point regarding ductility.

The self-restrained testing approach used in the present study was developed following the procedure detailed by Turski et al. [35] where the authors compressively

**Fig. 7** Stress response observed by Norton during SAC testing [26]

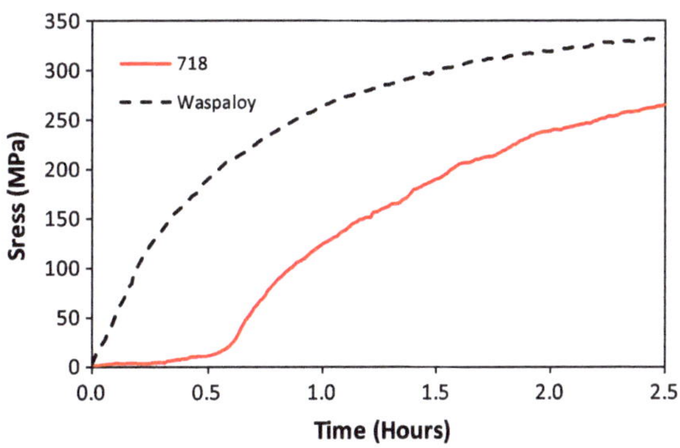

pre-strained a compact tension sample and subjected the unloaded sample to a 4,500 h heat treatment. Compressively yielding a compact tension sample will leave the yielded notch root in tension when the sample is unloaded. The original work by Turski characterized the residual strain profile following compression with high energy X-ray and neutron diffraction and used this information to establish a finite element model of the residual stress and strain produced by compression. A similar finite element model was used to find the thinnest sample geometry that would have the same residual stress, strain, and triaxialtiy as the original geometry. The details of this work have been reported elsewhere [38]. It was found that the thickness could be reduced from 25 mm (1 in.) to 10 mm (0.4 in.) without significantly sacrificing residual stress, strain, or triaxiality. Figure 8 compares the residual strain profiles of the original and modified geometries along a path immediately behind the notch root on the mid-thickness and quarter-thickness planes.

The test matrix shown in Table 2 was planned using the modified sample geometry [38] and crack growth is monitored using the potential drop method [39]. This matrix is based on known incidences of relaxation cracking in Alloy 617 [11, 23], interest in alloys 740H and 282 for AUSC applications [5], and the anticipated relaxation cracking resistance of Alloy 230 as a solely solid solution strengthened alloy. The material which was pre-aged for 10,000 h was base metal left over from different work and only alloys 740H and 230 were available.

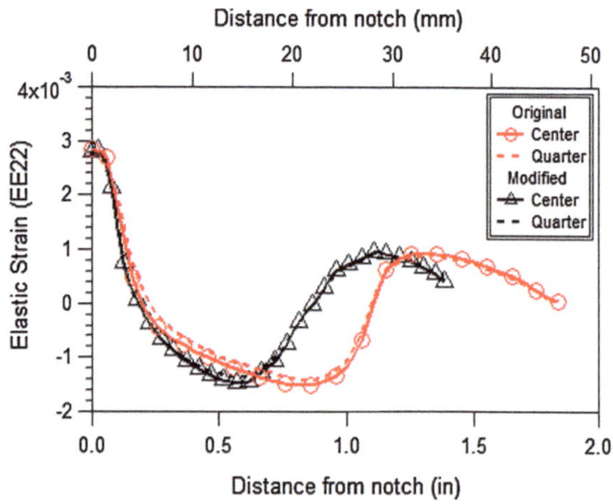


**Fig. 8** Calculated residual strain profiles immediately behind the notch of a compressively pre-strained compact tension sample [38]. Note that the modified geometry is not only thinner, it is also shorter, hence the horizontal shift in the profile

**Table 2** Testing matrix for self-restrained samples

| | Simulated HAZ | Base metal | 800 °C/10,000 h |
|---|---|---|---|
| 740H | X | X | X |
| 617 | X | X | |
| 282 | X | | |
| 230 | X | | X |

## Results

Reproducing the Gleeble-based testing following Norton was challenging due to machine control issues. The Gleeble jaws are capable of moving very quickly and applying large loads and in order to do this, some sacrifices in precision must be made. For most testing, this is irrelevant but over the course of a multi-hour hold, some drift can be observed. This drift occurred in the tensile direction and was on the order of 1 % strain. A small stroke drift is acceptable; the real problems occur during load measurement. The drift was not constant but oscillated as it drifted, further obscuring the actual load data. It is likely that Norton did not experience this behavior since he used an older Gleeble model. One advantage of Norton's testing approach was observing stress relaxation and subsequent increase due to precipitation (Fig. 7) but this has not been possible because of the effect of stroke oscillation and drift on the load data. Even though precise load data showing the effect of precipitation couldn't be collected, testing was continued because a procedure was developed that produced a somewhat consistent drift.

Ductility measurements from successful tests are reported in Fig. 9. It is clear that many more samples are required for meaningful conclusions to be drawn, however; the microstructures generated in this testing are especially pertinent to this discussion since they were generated at the proposed AUSC service temperature. The ductility ranking that was observed in this testing was expected based on the typical inverse relationship between strength and ductility. This is further reflected in the observed precipitate density near the fracture: similar precipitate densities were observed in Waspaloy, Alloy 718, and Alloy 740H, while considerably less precipitation was observed in alloys 617 and 800. One inconsistency found here is a higher precipitate density in Alloy 617 even though it also shows a higher ductility than Alloy 800. The ductility measurements made in Alloy 740H are consistent with similar testing conducted by Ramirez [40] for 0, 1, and 2 h at the same temperatures, also in Alloy 740H.

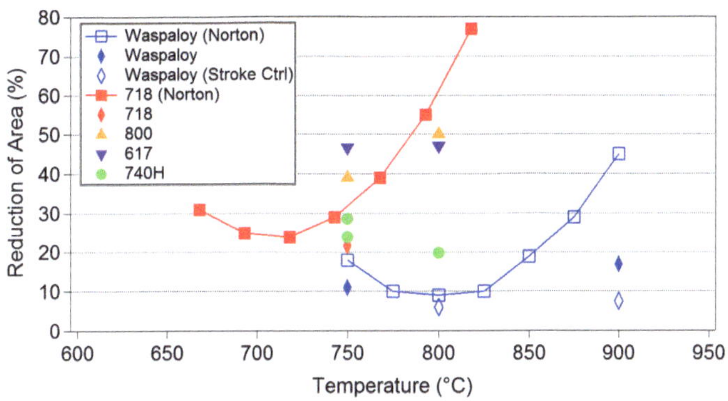


**Fig. 9** Ductility measurements from the testing procedure developed following the work of Norton [31]

As increasing ductility was observed, increasing voiding was also observed. It is important to mention that without interrupted testing, it is not possible to know if this voiding occurred during the hold at temperature or the tensile test at temperature but some conjectures can be made on the basis of void shape. Figure 10 shows the precipitation and voiding in these alloys near the fracture. The least voiding can be observed in Alloy 718 and Waspaloy. Except for the large edge crack in Alloy 740H, it shows very similar voiding to Alloy 800. Alloy 617 clearly shows the most voiding.

Based on the Gleeble testing, three interesting microstructural features were observed: (1) evidence of shear banding in Alloy 617, (2) precipitate free zone

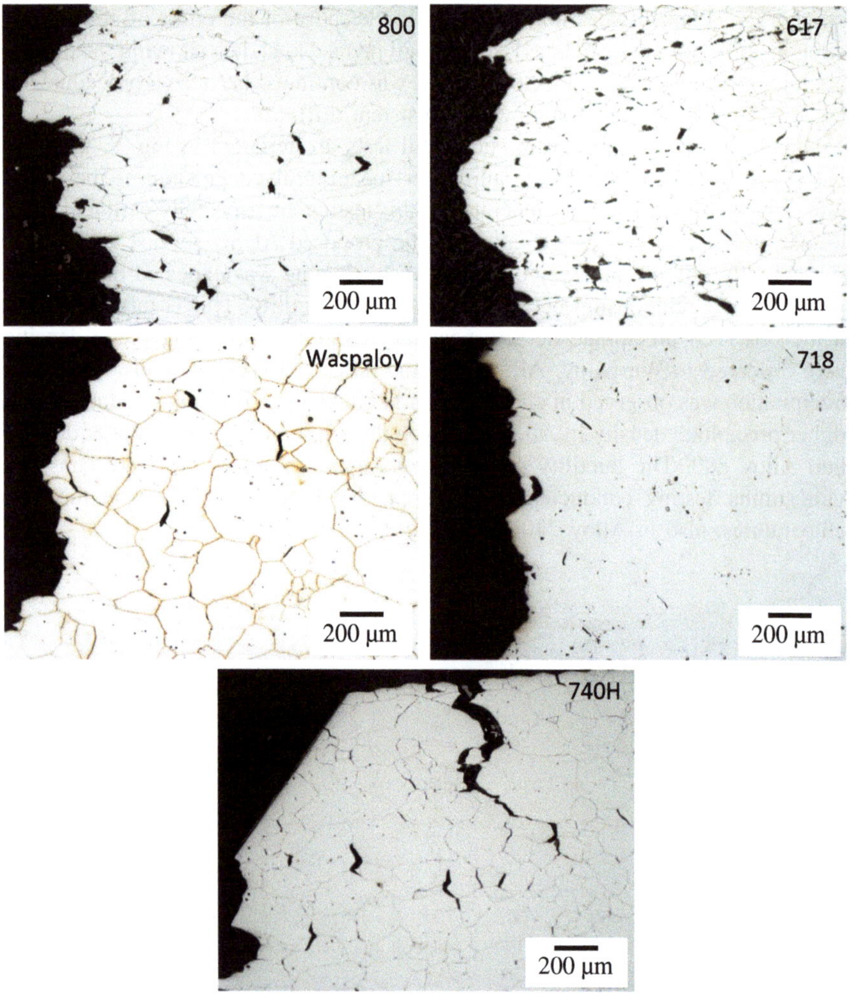


**Fig. 10** Voiding behind Gleeble SRC samples tested at 750 °C for 3 h

formation in Alloy 617, and (3) extensive discontinuous precipitation in Alloy 740H. Electron micrographs of the features observed in Alloy 617 are shown in Fig. 11. The unique precipitation pattern in Alloy 617 could be evidence that precipitation rates were accelerated beyond normal intragranular precipitation rates because some microstructural features, which are hypothesized to be shear bands, acted as preferential nucleation sites. Increased precipitation rate would limit how much residual stress could relax before the alloy fully hardens, thus increasing susceptibility to SRC. Precipitate free zone formation in this alloy would simply increase SRC susceptibility by creating a weak, narrow region adjacent to grain boundaries where strain can accumulate.

Discontinuous (or cellular) precipitation has been reported previously in Alloy 740H [41]. This microstructure forms when precipitates on a mobile grain boundary, such as $\gamma'$ in the case of Alloy 740H, coarsen in a direction perpendicular to the boundary as it migrates. Solute can easily diffuse down grain boundaries and this assists precipitate growth. Coarsening of these precipitates will lead to very wide precipitate free zones between them. This has been identified as the mechanism for precipitate free zone formation in this alloy [41]. Figure 12 shows an

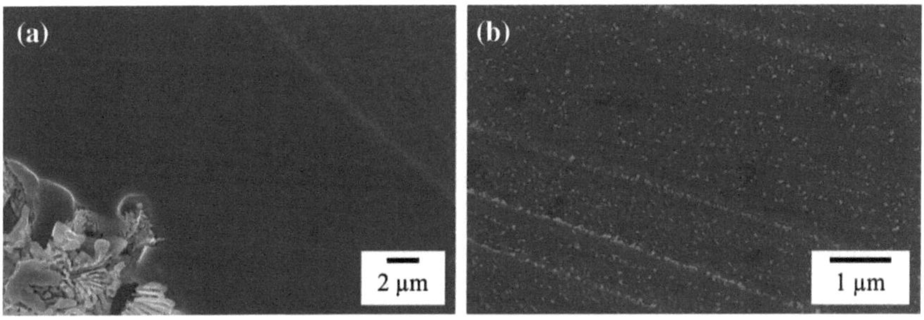


**Fig. 11** Alloy 617 micrographs showing a precipitate free zone next to a large precipitate (**a**) and precipitate ordering that could be due to shear banding (**b**)

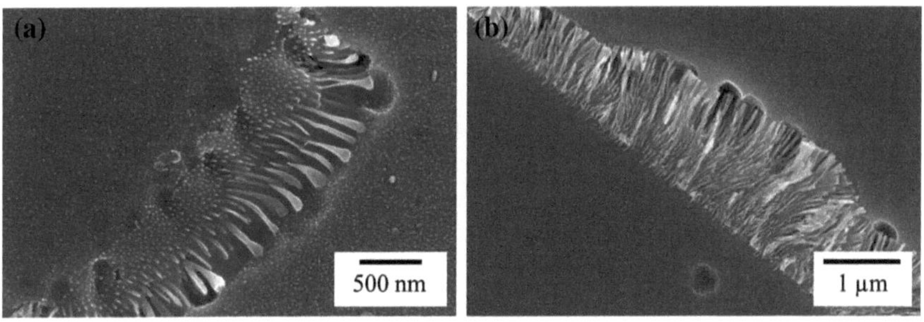


**Fig. 12** Examples of discontinuous precipitation in Alloy 740H. Weld metal aged 800 °C/ 4 h (**a**) and a Gleeble sample tested at 750 °C/ 3 h (**b**). The discontinuous reaction is much coarser in the weld metal example

example of this reaction in directly aged Alloy 740H weld metal and in an Alloy 740H Gleeble sample. The precipitate free zones associated with the coarsening of these cells are expected to increase SRC susceptibility by providing a microstructural “soft zone” where strain can accumulate. For more details on discontinuous precipitation reactions, the reader is referred to two exhaustive reviews; one by Williams and Butler [42] and another by Manna et al. [43].

Regarding the self-restrained relaxation cracking test, a significant challenge was encountered during simulation of the HAZ microstructure. The HAZ microstructure was to be simulated using the Gleeble because the modified thickness (10 mm, 0.4 in.) allowed the sample to fit in the jaws and the nature of resistive heating would ensure that the highest temperatures would occur immediately behind the notch. This obviates the problem of locating a notch in a particular microstructure. Interestingly, cracks formed in these samples during HAZ simulation at a nominal temperature below 1250 °C. By varying thermocouple location, it was found that the area immediately behind the notch was being significantly overheated relative to the control thermocouple location. The control thermocouple was placed 3 mm behind the notch in order to avoid affecting the behavior of the notch in the subsequent creep test. This overheating is attributed to the skin effect, which is the concentration of high-frequency alternating current near the surface of conductors due to opposing eddy currents. This effect only occurred during heating and the temperature immediately behind the notch diminished as the bulk of the sample reached steady-state. This temperature profile is associated with the large current applied to the sample during heating and reduced current required to maintain the steady-state temperature. Temperature profiles collected during several tests are shown in Fig. 13. Based on the temperature profiles, it can be concluded that the notch was cooling and contracting while the bulk sample was heating and expanding, creating a tensile stress on the notch.

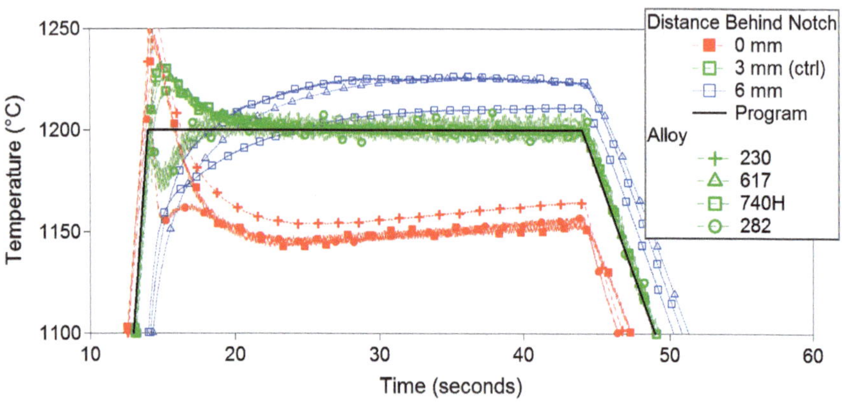


**Fig. 13** Temperature profiles collected during HAZ simulations. Note the very high degree of overshoot immediately behind the notch

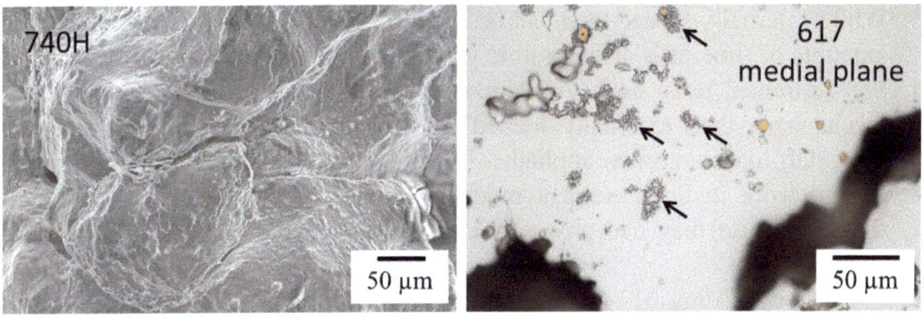


**Fig. 14** Representative fractrographic and metallographic evidence of liquation observed during investigation of HAZ simulation samples

All of the HAZ simulation samples were sectioned on the mid-thickness plane and prepared metallographically. The other halves of the samples were used for fractography after the notch root was bent open. This investigation found evidence of liquation. The peak temperature due to the skin effect was accurately recorded to be at least 1250 °C which is only 30 °C below the reported nil-strength temperature of 1280 °C in Alloy 740 [40]. The actual temperature most likely exceeded 1250 °C, thus liquation is a very reasonable conclusion. The extent of cracking correlated with composition as one would expect: the precipitation strengthened alloys (617, 740H, 282) showed similar crack depths (2–2.5 mm) and the solid solution strengthened alloy 230 showed very little cracking (0.4 mm). Figure 14 shows the fracture surface of an Alloy 740H sample and the cross section of an Alloy 617 sample, both of which show evidence of liquation.

Approaches to prevent this problem in the future include the following: reducing the heating rate so heating is more controlled and more uniform, compressing the sample during heating so liquated boundaries cannot be opened by tensile stresses, and consideration of different heating methods. Testing was begun with only base metal samples and this test is running. Periodic visual examination has not identified any cracking at this time. A thorough investigation will be concluded after 1500 h at temperature. Samples with a simulated HAZ will be included in the next round of testing.

## Conclusions

1. The use of superalloys in power plants creates the possibility of stress relaxation cracking, a phenomenon typically avoided in this alloy system by solution annealing.
2. Postweld heat treatments in superalloys must be low enough to fully harden the microstructure or high enough for homogenization; otherwise microsegregation will create very unfavorable weld metal microstructures.

3. While externally-loaded tests can be used to generate analog data in reasonable amounts of time, it is also possible to produce the incorrect creep mechanism.
4. Self-restrained tests are better suited to offer accurate stress states and creep mechanisms than externally loaded tests where unrepresentative (large and uni-axial) loads are often applied.
5. Recent work in finite element modeling has made self-restrained tests more useful because they can now provide information beyond a simple pass/fail conclusion.
6. Even though Alloy 617 showed a very high ductility, it also showed the most voiding and its microstructure appears to be very susceptible to relaxation cracking.
7. It is well known that Alloy 718 and Waspaloy have very different responses to strain age cracking but they both appeared very similar and very resistant in Gleeble-based testing.
8. Overheating due to the skin effect caused liquation cracking in Gleeble HAZ simulations.

**Acknowledgements** The authors would like to acknowledge Jim Tanzosh and Joe Dierksheide of the Babcock and Wilcox Power Generation Group for financial support of this project. Special thanks to Brian Baker, Ronnie Gollihue, and Jack DeBarbadillo of Special Metals for providing material. This project is part of the National Science Foundation's Industry/University Collaborative Research Center known as the Center for Integrative Materials Joining Science for Energy Applications (CIMJSEA).

## References

1. R. Viswanathan, et al., "US Program on Materials Technology for Ultra-Supercritical Coal Power Plants," *Journal of Materials Engineering and Performance,* 14 (3) 2005. 281-292.
2. R. Viswanathan, J.P. Shingeldecker, and R. Purgert, "Evaluating Materials Technology for Advanced Ultrasupercritical Coal-Fired Plants," *Power,* (8) 2010.
3. J. Zachary, "Options for Reducing a Coal-Fired Plant's Carbon Footprint, Part II," *Power,* July 2008.
4. J.J. DeBarbadillo, B.A. Baker, and L. Klingensmith, "Nickel-Base Superalloys for AUSC Power Plants" BaoSteel Conference 2010, 2010.
5. B.A. Baker, R.D. Gollihue, and J.J. deBarbadillo, "Fabrication and Heat Treatment of Weld Joints in Inconel Alloy 740H Superalloy Steam Header Pipe and Superheater Tubing" Welding and Repair Technology for Power Plants: Tenth International EPRI Conference, Marco Island, FL. 2012.
6. G.D. Smith, and H.W. Sizek, "Introduction of an Advanced Superheater Alloy for Coal-Fired Boilers" Corrosion, 2000.
7. R.J. Smith, B.A. Baker, and G.D. Smith, "Nickel Alloys for the Next Generation of Advanced Steam Boilers" VGB Materials and Quality Assurance Conference, Dortmund, Germany. 2004.
8. R. Blum, "Materials Development for Power Plants With Advanced Steam Parameters-Utility Point of View," *Materials for Advanced Power Engg., D. Coursouradis et al. Ed,* 1991. 15-30.
9. S.Q. Zhao, et al., "Research and Improvement on Structure Stability and Corrosion Resistance of Nickel-Base Superalloy INCONEL Alloy 740," *Materials & Design,* 27 (10) 2006. 1120-1127.

10. J.J. DeBarbadillo, private communication with author, Lehigh University, 11 April 2012.
11. G. Schmidt, et al., "Component Test Facility for a 700C Power Plant (Comtes700)" European Comission: Directorate - General for Research and Innovation, Directorage G - Industrial Technologies, Unit G.5 - Research Fund for Coal and Steel. 2013.
12. C.J. Cowen, P.E. Danielson, and P.D. Jablonski, "The Microstructural Evolution of Inconel Alloy 740 During Solution Treatment, Aging, and Exposure at 760 °C," *Journal of Materials Engineering and Performance,* 20 (6) 2011. 1078-1083.
13. *Haynes Alloy 282 Technical Bulletin*. 2008, Haynes International.
14. *Haynes Alloy 230 Technical Bulletin*. 2009, Haynes International.
15. *Inconel Alloy 800 Technical Bulletin*. 2004, Special Metals.
16. M.J. Donachie, and S.J. Donachie, *Chapter 8: Heat Treating*, in *Superalloys: A Technical Guide*. ASM International, 2002.
17. J.N. DuPont, J.C. Lippold, and S.D. Kiser, *Welding Metallurgy and Weldability of Nickel-Base Alloys*. John Wiley & Sons, Hoboken, NJ. 2009.
18. M. Prager, and C.S. Shira, "Welding of Precipitation-Hardening Nickel-Base Alloys," *Welding Research Council Bulletin,* 128 (February) 1968.
19. A. Dhooge, and A. Vinckier, "Reheat Cracking - Review of Recent Studies (1984-1990)," *Welding in the World,* 30 (3/4) 1992. 44-71.
20. A. Dhooge, and A. Vinckier, "Reheat Cracking - Review of Recent Studies," *Welding in the World,* 24 (5/6) 1986. 2-18.
21. J.A. Siefert, and J.M. Tanzosh, "Stress Relaxation Cracking Literature Review" EPRI Conference, 2010.
22. L.E. Shoemaker, et al., "Fabricating Nickel Alloys to Avoid Stress Relaxation Cracking" NACE Corrosion Conference and Expo, 2007.
23. H. Van Wortel, "Control of Relaxation Cracking in Austenitic High Temperature Components" NACE 2007, 2007.
24. J.C. Van Wortel, "Prevention of Relaxation Cracking by Material Selection and/or Heat Treatment Final Report" TNO. 2000.
25. R. Thamburaj, W. Wallace, and J.A. Goldak, "Post-Weld Heat-Treatment Cracking in Superalloys," *International Metals Reviews,* 28 (1) 1983. 1-22.
26. J.C. Lippold, "Elevated Temperature, Solid-State Cracking in Welds" Cracking Phenomena in Welds IV, 2014. p.
27. S. Kou, *Welding Metallurgy.* John Wiley & Sons, Hoboken, NJ. 2003.
28. D.H. Bechetti, et al., "Homogenization and Dissolution Kinetics of Fusion Welds in Inconel Alloy 740H," *Metallurgical and Materials Transactions A,* 45A (June) 2014. 3051-3063.
29. J.P. Shingledecker, N.D. Evans, and G.M. Pharr, "Influences of Composition and Grain Size on Creep-Rupture Behavior of Inconel Alloy 740," *Journal of Materials Engineering and Performance,* 2012.
30. K.A. Craver, *Evaluation of a Stress Relaxation Cracking Susceptibilty Testing Technique, Master's Thesis*, in *Welding Engineering*. 2012, The Ohio State University.
31. S.J. Norton, *Development of a Gleeble Based Test for Post Weld Heat Treatment Cracking in Nickel Alloys, Master's Thesis*, in *Welding Engineering*. 2002, The Ohio State University: Columbus, Ohio.
32. R.J. Christoffel, "Cracking in Type 347 Heat-Affected Zone During Stress Relaxation," *Welding Journal,* June 1962. 251s-256s.
33. L.-E. Svensson, and G.L. Dunlop, "Growth of Intergranular Creep Cavities," *International Metals Reviews,* 77 (2) 1981. 109-131.
34. J.C. Borland, "Cracking Tests for Assessing Weldability," *British Welding Journal,* 7 1960. 508-512.
35. M. Turski, et al., "Residual Stress Driven Creep Cracking in AISI Type 316H Stainless Steel," *Acta Materialia,* 56 2008. 3598-3612.
36. M.W. Spindler, "The Multiaxial Creep Ductility of Austenitic Stainless Steels," *Fatigue and Fracture of Engineering Materials and Structures,* 27 2004. 273-281.

37. M.W. Spindler, "The Use of Borland Specimens to Reproduce Reheat Cracking in Type 316H" Int. Conf. on High Temperature Plant Integrity and Life Extension, Robinson College, Cambridge University. 2004.
38. D.C. Tung, and J.C. Lippold, "Modeling a Stress Relaxation Cracking Test for Advanced Ultra Supercritical Alloys" Advances in Materials for Fossil Power Plants, 7th International Conference, 2013.
39. ASTM, *E 647 - 05 Standard Test Method for Measurement of Fatigue Crack Growth Rates*, in *A6. Guidelines for Electric Potential Difference Determination of Crack Size*. 2005.
40. J.E. Ramirez, "Evaluation of Susceptibility of Alloy IN740 to HAZ Stress-Relaxation Cracking," *Welding Journal,* 92 (4) 2013. 89s-100s.
41. D.H. Bechetti, *Microstructural Evolution and Creep Rupture Behavior of Inconel Alloy 740H Fusion Welds*, in *Materials Science and Engineering*. 2013, Lehigh University.
42. D.B. Williams, and E.P. Butler, "Grain Boundary Discontinuous Precipitation Reactions," *International Metals Reviews,* 80 (3) 1981. 153-183.
43. I. Manna, S.K. Pabi, and W. Gust, "Discontinuous Reactions in Solids," *International Materials Reviews,* 350 (2) 2001. 53-91.

MIX
Papier aus verantwortungsvollen Quellen
Paper from responsible sources
FSC® C105338

If you have any concerns about our products,
you can contact us on
**ProductSafety@springernature.com**

In case Publisher is established outside the EU,
the EU authorized representative is:
**Springer Nature Customer Service Center GmbH**
**Europaplatz 3, 69115 Heidelberg, Germany**

Printed by Libri Plureos GmbH
in Hamburg, Germany